T0406125

Handbook of the Anthropocene

Nathanaël Wallenhorst • Christoph Wulf
Editors

Handbook of the Anthropocene

Humans between Heritage and Future

Volume II

Editors
Nathanaël Wallenhorst
Université Catholique de l'Ouest (UCO)
Angers, France

Christoph Wulf
FB Erziehungswissenschaft
Freie Universität Berlin
Berlin, Germany

ISBN 978-3-031-25909-8 ISBN 978-3-031-25910-4 (eBook)
https://doi.org/10.1007/978-3-031-25910-4

© The Editor(s) (if applicable) and The Author(s), under exclusive license to Springer Nature Switzerland AG 2023, corrected publication 2024
This work is subject to copyright. All rights are solely and exclusively licensed by the Publisher, whether the whole or part of the material is concerned, specifically the rights of translation, reprinting, reuse of illustrations, recitation, broadcasting, reproduction on microfilms or in any other physical way, and transmission or information storage and retrieval, electronic adaptation, computer software, or by similar or dissimilar methodology now known or hereafter developed.
The use of general descriptive names, registered names, trademarks, service marks, etc. in this publication does not imply, even in the absence of a specific statement, that such names are exempt from the relevant protective laws and regulations and therefore free for general use.
The publisher, the authors, and the editors are safe to assume that the advice and information in this book are believed to be true and accurate at the date of publication. Neither the publisher nor the authors or the editors give a warranty, expressed or implied, with respect to the material contained herein or for any errors or omissions that may have been made. The publisher remains neutral with regard to jurisdictional claims in published maps and institutional affiliations.

This Springer imprint is published by the registered company Springer Nature Switzerland AG
The registered company address is: Gewerbestrasse 11, 6330 Cham, Switzerland

Acknowledgements

We would like to thank all the authors for their contributions. Special thanks go to our editor Christi Lue, who believed in this project from the start and honoured us with her trust, and to Dr. Möller, Deputy Secretary General of the German Commission for UNESCO. We also thank the Université catholique de l'Ouest (Angers, France) for the grant to supervise the translations. Finally, we would like to thank Liz Hamilton for her extremely valuable and thorough help with the translations.

Anthropocene, the Concept of the 21st Century – A General Introduction

What Happened to the Future?

For millennia, the prospect of the future has offered hope. Men and women have gazed, in fascinated anticipation, towards the horizon of the future. 'What new inventions, creations and exploration will humanity achieve in years to come?', 'What wonders await?' and other such questions have engaged our ancestors for centuries.

Human achievements over time have been staggering. We have discovered that the earth is not the centre of the universe. The invention of the steam engine sparked the Industrial Revolution, which changed the face of the world. Quantum physics has totally reshaped the way in which we think. The advent of antibiotics has meant that diseases which would once have been a death sentence are now curable. In the space of only a few short decades since setting foot on the moon, we have built communications networks that keep us connected wherever we are on the planet. Today, thanks to internet search engines, we have practically all human knowledge just a few clicks away. More astonishing still is that we are now able to actually edit the human genome. Dolly the sheep was cloned less than 25 years ago; yet today CRISPR-Cas9 allows us to slice and splice parts of the human genome.

Time has also brought many nasty shocks. The Industrial Revolution, encouraged by Descartes' exhortation that we become 'masters and possessors of nature', was in full swing in the early 20th century, when war broke out on a scale never seen before. Economic globalisation saw that war spread across the face of the planet like a plague. Barely two decades later came a second global conflict, but where WWI had left 18 million dead, manmade technology meant 60 million lost their lives in WWII! However, despite the twists and turns and bumps along the way, the modernist era was characterised by the marriage of technical and social progress. The commitment to build a brighter future elevated vast swathes of humanity out of grinding poverty.

However, three prosperous decades later, when 'grand narratives' (Lyotard, 1979) collapsed and ushered in the era of postmodernism, the timelines began to unravel. Technical and social progress no longer go hand in hand. The idea of linear progress – today is better than yesterday, and tomorrow will be better still – no longer holds true. In addition, the new geological era, the Anthropocene – meaning human activity has now inexorably altered the planet's habitability – threatens the very survival of humanity (Bonneuil & Fressoz, 2013; Bourg & Papaux, 2015; Federau, 2017; Magny, 2019; Wallenhorst, 2019, 2020, 2021; Wallenhorst & Wulf, 2022; Kamper & Wulf, 1989, 1994; Gil & Wulf, 2015; Wulf & Zirfas, 2020; Wulf, 2022a). Our future – whether it holds shock or awe – is fading from view. Greta Thunberg's words to her young peers put the point across in resonant fashion: 'Why should I be studying for a future that soon may be no more?' This is a sobering hypothesis: we may soon have no future at all.

Voices – mainly those of Californian technology giants – occasionally proclaim the opposite: we *do* have a future, and a bright one at that. The future will be *great* or *bigger than ever*, we are told by a chorus of such figures, including: Mark Zuckerberg, Facebook co-founder and CEO; Larry Page and Sergey Brin, founders of Google; Jeff Bezos, Amazon founder and CEO; and Elon Musk, whose ground-breaking businesses include SpaceX, Tesla and Neuralink (the 2016 startup aiming to interface the human brain with digital technology). The future is undoubtedly digital. Ultimately, Musk tells us, this is nothing to fear. His ambition is to establish a human colony on Mars within the foreseeable future, using Artificial Intelligence (AI). Neoliberal capitalism has destroyed our 'shared home', so Musk wants to build us a new one, on Mars, harnessing the power that the coming digital and technoscientific revolution promises. There are no limits – planetary, corporeal or cognitive – that cannot be overcome by human engineering. We shall soon be able to merge with machines, and so massively increase our capabilities. The future will be in the hands of digital technology. Here we have a second hypothesis: our future will be a Promethean digital endeavour. The *hubris* of *Homo oeconomicus*, maximising individual interests, will prevail. The stratospheric rise of the so-called Big 5 (or GAFAM, standing for Google, Amazon, Facebook, Apple and Microsoft) and their Chinese counterparts (BATX – Baidu, Alibaba, Tencent and Xiaomi) is indicative of this trend. These companies are shaping the future for their billions of consumers across the globe.

Clearly, we need a future and must obviously work diligently to bring it about. However, surely that future is not the preserve of those Promethean few who have access to a spacecraft capable of reaching Mars. Might we not have a different future – one which is post-Promethean, post*hubris*, postcapitalist or simply convivial? We may, even now, be seeing such a world begin to emerge. Here is a third hypothesis: the future could be marked by an ethos of coexistence. Of course, in reality, the future is unlikely to be at either extreme of a scale, but rather, somewhere in between. By examining the world today, we can extract clues as to how it will look tomorrow.

The future is not a given – far from it. Perhaps the most pressing questions we face relate to the fate of humanity. *Who will we become? How can we continue to be human in the current context, marked by climate change, ecosystem collapse, but also the augmented humans, profound social transformations and rampant radicalisation?* Our living environment has been inexorably altered, transhumanist theories abound, and researchers are seeking to surpass human limitations through technology. In this world, it is increasingly complex – even, sometimes, impossible – to 'form a society together'. We can no longer take our humanity for granted, but it is crucial to retain it. How do we define ourselves, and what do we want to make of ourselves? This anthropological question is addressed in this handbook, based on three conceptual and epistemic options.

Firstly, it is by becoming more earthly – more connected to the earth – that *Homo sapiens* will become progressively more human. Indeed, the anthropological viewpoint discussed herein is that humanity should immerse itself fully in the natural world. This stands in opposition to the theories, which have been dominant in Western thinking since the Age of Enlightenment (Papaux, 2015), whereby humanity is hierarchically superior to the natural world, and should therefore maintain distance from it.

Secondly, breaking with the essentialism and substantialism currently popular in Western thinking, *Homo sapiens* becomes human as a result of, through and in relationships (with others, with future generations and with non-humans). Humanity is defined not just intrinsically and in static terms, but also on the basis of what it could become and what it makes of itself within its environment. In this context, the modern boundaries between nature and culture break down, as demonstrated by Philippe Descola (2005).

Thirdly, the human race will have a great deal to gain – not least, its own assured survival – from post-Promethean practices.

One thing appears clear: the hypotheses concerning the future and how to prepare for it are shaped by this new context which we must take ownership of and to which we keep returning – the Anthropocene. It concerns what is happening to the climate, changes in the biosphere and the dynamics of our societies. It is the concept of the 21st century, the concept of which it is necessary to have an in-depth understanding in order to understand the transformations that are taking place in the Earth system, to understand the world and allow ourselves to continue the process of becoming human.

Anthropocene: Improvisation of the Term at a Conference in Mexico in 2000

In February 2000 in Mexico, during exchanges at the conference of the IGBP – *International Geosphere-Biosphere Programm*e – an interdisciplinary group studying the Earth system, there was discussion of the Holocene, the preceding epoch (lasting from over 11,700 years ago more or less to the current day). Someone stood up and said, 'Stop using the word Holocene. We're not in the Holocene anymore. We're in the…the…the… [searching for the right word] …the Anthropocene!' Later he would say, 'I suddenly thought this was wrong. (…) I just made up the word on the spur of the moment. Everyone was shocked. But it seems to have stuck' (Keats, 2011, p. 19). We can say that the proposal of the term 'Anthropocene' by Crutzen in 2000 was pure improvisation (Zalasiewicz et al., 2017a, p. 56).

Having used the term at the IGBP conference, Paul Crutzen contacted the American biologist, Eugene Stroemer, to suggest collaborating with him on a paper for the IGBP review, because Stroemer had been using this term informally since the 1980s. Here we read how Jacques Grinevald (2007, p.243), the French philosopher who was working in Geneva, reports Stroemer's proposal: 'I started to use the term "Anthropocene" in the 1980s but I never formalised it before Paul [Crutzen] contacted me.' It is clear in this one-page article that one of the characteristics of the invention of the term 'Anthropocene' was that it was put forward before it was accorded a scientific profile and before a precise, exhaustive definition was suggested. Later, in 2002, Crutzen picked up on these elements in a short article in the review *Nature*, where he summarises all the environmental changes brought about by humans since the industrial revolution as evidence that we are entering the 'Anthropocene': a tenfold increase in the world population between 1700 and 2000 along with a similar increase in cattle; depletion of fossil resources and the release of CO_2 into the atmosphere; a large increase in species extinction, by a factor of 1000.

A second feature of the concept of the Anthropocene, besides the fact that it was an improvisation, is that it refers to a geological epoch and is thus part of those time periods that are customarily defined by stratigraphers on the basis of their observations of soils. But one of the first uses of this term was proposed by Paul Crutzen, a Dutch geochemist working in Germany, independently of stratigraphic observations. Originally, the term Anthropocene was not suggested by geologists based on stratigraphic evidence but with reference to changes in processes of the Earth system brought about by human activity. Therefore, in its original form, the concept of the Anthropocene is a systemic concept which has its roots in Earth system science rather than in geology.

The third interesting point is that Crutzen was awarded the Nobel Prize for Chemistry in 1995, which gave a large amount of publicity to this new term in the media and in the scientific world. The fact that a word was coined without being precisely defined (even in 2011 Steffen, Grinevald, Crutzen and McNeil recognised that the term Anthropocene still had an informal flavour) and without the usual respect for academic rules is seen by many as a reason for the success of this term, also in the way it was taken up by a very heterogeneous mixture of disciplines before it became widely used by the stratigraphers. Thus, as Ellis (2017) recognised, we may say that the concept of the Anthropocene was created by Eugene Stroemer, then publicised by Paul Crutzen. Its establishment as a scientific concept, based on stratigraphic observations, has been the fruit of work by many different Earth system scientists, the most notable of whom are Steffen and Zalasiewicz.

To begin with, it was non-geologists (chemists, physicists, biologists) who suggested creating a new geological epoch. This audacity initially took the geologists by surprise, but in 2008 the Geological Society of London instituted an official working group on the Anthropocene with the aim of getting recognition from the International Union of Geological Sciences (IUGS). This working group, composed of brilliant research scientists, published dozens of articles, in the field of Earth system science and in geology, from the establishing of irrefutable truth to stratigraphic foundations, from the end of the Holocene Epoch and from the beginning of the Anthropocene (Zalasiewicz et al., 2011, 2014, 2017a). Thus, the Anthropocene is a new geological epoch, characterised by a change in the habitability conditions of the Earth, impacting in particular human life in society as we know it.

The scientific debates are exciting and are structured within three disciplines which examine three concepts of the Anthropocene: Earth System science, geological stratigraphy and finally the whole field of humanities and social sciences.

The Concept of the Anthropocene Within Earth System Science

The first point is the extent of the systemic upset we have caused to Earth's systems. All publications in this field examine when we passed, or will pass, the boundaries of what the planet can cope with, and the room for manoeuvre that we humans have in which to safely take action. The functioning of the Earth is a complex system, characterised by sudden, irreversible changes when certain bio-geological processes are forced upon it. There are certain boundaries that cannot be passed if we do not wish the Earth system to transgress certain systemic thresholds. There are nine boundaries to have been identified (Rockström et al., 2009; Steffen et al., 2015; Persson, 2022; Erlandsson, 2022): climate change, rates of loss of terrestrial and marine biodiversity, the alteration of biogeochemical cycles such as nitrogen and phosphorous, the impoverishment of the stratospheric ozone layer, acidification of the oceans, the global use of fresh water, land usage, chemical pollution and the discharge of aerosols into the atmosphere. The first two have a special status. Crossing just one of these boundaries is enough to tip the entire Earth system into a way of functioning that is considerably less propitious for life. *Example:* today we find that our principal allies in our economic development over the last two centuries, soil and the plants that grow in it, which act as carbon reservoirs absorbing some of the CO_2 emissions that have been released by human activities, not only capture less CO_2 but may also be releasing more than they actually capture. The Amazonian jungle, for a long time considered to be the 'lungs of the planet', even shows a positive carbon balance (releasing more carbon than it absorbs) in years that are particularly dry (Gatti et al., 2014). Other articles followed, confirming these results (Brienen, 2015; Doughty, 2015; Esquivel-Muelbert, 2019, 2020). We cannot consider ourselves to be outside nature. There are innumerable studies which show how deeply embedded we are in our environment. Barnosky et al. (2011) write about the sixth mass extinction; Rockström et al. (2009) and Steffen et al. (2015) stress the importance of staying within the planet's boundaries. Barnosky et al. (2012) and Steffen et al. (2018) warn that our biosphere is approaching a systemic tipping point, and the planet is in danger of becoming a 'hothouse Earth', incapable of supporting human society. Im et al. (2015, 2017), Mora et al. (2017) and Bador et al. (2017) forecast heatwaves of increasing frequency and intensity in years to come, etc. This branch of the debate also includes publications on what corrective action we can take, by geoengineering, or 'planetary stewardship' (Crutzen, 2006; Steffen et al., 2011b). It is directly connected to normative and prospective political discussion. Often, the question 'What can we do?' is asked. The irreversible corruption of our planet's ability to support life is a matter of concern. Thus, we expect this vein of Earth system science will continue to produce studies grounded in the humanities and social sciences, on how society can be reorganised to exist in the altered biosphere (Eckersley, 2017; Arnsperger & Bourg, 2017; Curnier, 2017; Federau 2017; Lesourt, 2018; Wallenhorst, 2019, 2021, 2022; Wallenhorst & Pierron, 2019; Hétier, 2021; Hétier et Wallenhorst, 2021, 2022a, b, c; Prouteau et al., 2022; Wulf 2022a).

The Concept of the Anthropocene Within the Sciences of Geology and Stratigraphy

The second main branch of debate concerns the date of entry into the Anthropocene. Innumerable articles have been published in this field, although, as they are less 'spectacular' and more technical, they are less well known to the general public. This debate among experts in Earth system science usually draws on the multidisciplinary practice of stratigraphy. While they may indeed be less widely known, these articles are of singular importance. It is these studies (rather than those which focus on humans' indelible and systemic impact on the biosphere) that will perhaps determine the 'official' start of the new geological epoch. That official date, when it is known (and if the International Union of Geological Sciences officially ratifies this new geological epoch) will be included in school curricula and will not be edited to suit a particular political agenda. As pupils study the geological timescale early in secondary school (in different countries), the Anthropocene will become part of the curriculum once officially included in that timeline.

Geologists think of Earth's history in terms of different slices of time, based on shifts in the planet's overall condition, markers of which can be seen in sediment cores. They identify recognisable segments on the basis of climate, sea level and biodiversity. Stratigraphic markers are used to define chronostratigraphic units, giving geologists a common language in which to advance their knowledge of the planet's history. In the debate on how to date the onset of the Anthropocene, Earth system researchers are leading the way, identifying features visible today which we know (or can assume with a high degree of probability) will still be visible in hundreds of thousands – or even millions – of years. We do not yet know precisely what the Anthropocene has in store for the next few hundred or few thousand years; also, the most remarkable anthropic changes made to Earth's system have yet to manifest themselves. However, it is important for Earth system scientists to put a date on the dawn of the new geological epoch, based on sufficiently solid stratigraphic evidence. Most boundaries are defined by identifying a specific point on Earth in a stratotype known to date from a specific geological era. The markers are known as GSSPs: Global Boundary Stratotype Sections and Points. They may be rocks, sediments or glaciers which developed during a given geological period. Stratigraphic commissions aim to locate such points, which are also known as 'golden spikes', because once agreed upon, the stratigraphic section is marked with a plaque, sometimes on the head of a golden spike driven into the rock face.

One of the difficulties in defining a GSSP for the Anthropocene is that the full effects of human activities on the Earth system are not immediately apparent but are spread out across decades or even centuries (Lewis & Maslin, 2015, p. 173). A GSSP places a date on a physical marker that correlates to other, secondary markers. It must be at a particular site on the planet but have proven correlation with a new global context; it must also have 'complete continuous sedimentation with adequate thickness above and below the marker' (Lewis & Maslin, 2015, p. 172).

Indeed, a series of indicators can help date the onset of the Anthropocene, such as deposits from human activities, shifts in the biorecord, geochemical shifts, oceanic shifts (such as changes in ocean geochemistry, ocean biodiversity and sea level) and catastrophic events (be they 'natural' or caused by humans). Isotopes[4] can also contribute to the debate. Changes in isotope levels can yield data about the climate or the chemical composition of the atmosphere. For example, lead (Pb) isotopes clearly indicate the Ancient Greek civilisation, which was marked by innovation and the use of heavy metals. Human activities have altered not only the atmospheric concentrations of different gases but also isotope levels. Whilst the isotope record evidences anthropogenic changes over the past few millennia, the isotopes alone do not clearly demarcate the Anthropocene from the Holocene (Dean et al., 2014, p. 284).

Waters et al. (2014) provide evidence of a stratigraphic basis for the Anthropocene, by identifying a collection of anthropogenic sediments. In their view, there is a clear stratigraphic boundary between the Holocene and the Anthropocene: 'Humans are altering the planet, including long-term global geological processes, at an increasing rate' (p. 137). Stratigraphic indicators of a shift from the Holocene to the Anthropocene have recently been found in sediments from a lake in west Greenland, containing plastics, radionuclides, fly ash, pesticides, reactive nitrogen and metals (Waters et al., 2016, p. 137). They view the anthropic signatures of the new-dawned Anthropocene, distinguishing it from the Holocene, as the combination of 'accelerated technological development, rapid human population growth and increased resource consumption' (Waters et al., 2016, p. 139).

The stratigraphic principles involved in geological methodology are particularly rigorous and based on reason. What is astonishing is how much of human activities the Earth retains. In the scientific literature of the last 20 years, various possible dates have been suggested for the beginning of the Anthropocene: the Stone Age with the controlled use of fire (Doughty, 2013); the emergence of agriculture with the control over ecosystems and the progressive change in the chemical composition of the atmosphere (Ruddiman, 2013; Ruddiman et al., 2014); the encounter between the ancient and the new world which transformed the ecosystems (importation of different species) and reduced the world population (genocide and pandemics) (Lewis & Maslin, 2015); the steam engine and associated capitalism, becoming increasingly hegemonic (Crutzen, 2002; Steffen et al., 2011); the explosion of atomic bombs with radionuclides spreading to all four corners of the earth (Lewis & Maslin, 2015); or when we entered an age of increased consumption with its preprogrammed obsolescence (Steffen et al., 2015). There is no shortage of hypotheses when it comes to dating our entry into this new geological epoch – and the debates surrounding this have been particularly lively (Autin & Holbrook, 2012; Finney & Edwards, 2016; Zalasiewicz et al. 2012, 2017b; Edwards, 2018). Furthermore, each one of these debates contains valuable anthropological insights: human beings are characterised by their control over ecosystems, their imperialism, their unreserved exploitation of resources to the extent of determinedly sawing off the branches which give them life, they are true economic animals seeking to maximise individual interests, they are Promethean, considering all boundaries as obstacles to be overcome, they are capable of associating techno-scientific capabilities with military power, etc.

The Concept of the Anthropocene Stemming from the Humanities and Social Sciences

Since around 2010, a number of criticisms of the concept of the Anthropocene have emerged from the humanities and social sciences. They begin by questioning the *anthropos* of the Anthropocene. Crutzen, in his 2002 article, stipulates that the Anthropocene is the work of 25% and not all of humanity; however, in most of the articles by Earth system researchers, the *anthropos* responsible for our entry into this new geological epoch appears somewhat undifferentiated, far removed from the critical contributions of history and the sciences. The narrative of the Anthropocene produced by Earth system science is constructed within the framework of species and nature. At the Earth Summit in Rio in 1992, the political declaration of the various governments refers to 'common but differentiated responsibilities' (CBDR). It is a question of a weighting that is not completely missing in the narrative of the Anthropocene, but which does not give sufficient consideration to this differentiation.

In the following quotation from the main authors of the concept of the Anthropocene, Zalasiewicz et al., (2010), we see their analysis of the ultimate cause of our entry into to the Anthropocene and the link between human population growth and the use of fossil fuels in the Industrial Revolution: 'First, how have the actions of humans altered the course of Earth's deep history? The answers boil down to the unprecedented rise in human numbers since the early nineteenth century – from under a billion then to over six billion now, set to be nine billion or more by midcentury. This population growth is intimately linked with massive expansion in the use of fossil fuels, which powered the Industrial Revolution, and allowed the mechanization of agriculture that enabled those additional billions to be fed.' This analysis may seem quite obvious at first glance, but it is worth examining the reasons for the progressive hegemony of fossil fuels that were at the heart of the Industrial Revolution. In an article from 2013, the Swedish historian Andreas Malm shows the way in which the choice of the steam engine rather than hydraulic energy in Great Britain during the 19th century was due to the power of capital over labour. The advantage of the steam engine is that it can be deployed close to towns and workers, and the fact that it does not matter what environment it is in and whether it is close to hydraulic plants. Although in the first few decades, the steam engine's use of fossil fuel resulted in considerably lower energy yields, the capitalist industrialists preferred to use the steam engine because it enabled the energy of the workers to be harnessed to make better use of the workforce. Thus, fossil energy was not created by an undifferentiated *anthropos*.

The humanities and social sciences insist, therefore, on the importance of social and societal forces for the Earth system. Malm and Hornborg remind us that between 1850 and 2000, the "capitalist" (or "northern") countries accounted for 18.8% of the world population and were responsible for 72.7% of CO_2 emissions. Just after 2000, 45% of the poorest people on the planet were responsible for 7% of CO_2 emissions; at the very same time, 7% of the richest people emitted 50%. Furthermore, the British environmentalist David Satterthwaite (who was a member of GIEC) shows in a precise study (2009) that people's responsibility for the emission of greenhouse gases varies by a ratio of 1:1000 according to where they live across the world and the life choices they make. Hence the amount of emissions in tons of carbon per person is 10.94 in Qatar; 4.71 in the USA; 1.57 in France; 0.01 in Chad and Mali. It is this that tells Malm and Hornborg that 'As long as there are human societies on Earth, there will be lifeboats for the rich and privileged' (Malm & Hornborg, 2014, p. 66). They therefore ask this simple question: 'Are these basic facts reconcilable with a view of humankind as the new geological agent?'

Following all these early critical studies concerning the concept of an undifferentiated *anthropos* being the origin of our entering the Anthropocene, Steffen *et al.* gradually introduced elements of differentiation into their analyses, but this remains fairly sketchy and does not go as far as proposing a different concept to the Anthropocene. We can see this in figure 4 (2011b). However, it is possible to see that if the curves of the graph gradually show a differentiated *anthropos*, the accompanying narrative is characterised by a relative lack of differentiation (compared to the contributions of the social sciences).

It is, therefore, not possible to consider the current changes in the Earth system, and the changes that are to come, simply as the product of human nature or of the activity of an undifferentiated or abstract humanity. It is not a question of natural inevitability but a question of social, political, economic or historical causes that can be clearly identified. Earth system science alone is not able to explain the deep-rooted causes for our entry into the Anthropocene. *Anthropos* is not necessarily at the heart of the way the Earth system functions – or in any case would need further in depth analysis.

The changes in the biosphere are evidence of how fragile the political and social world as we know it, and for which we are educating our children, has become. There is little time left to us to ensure that the Earth system can remain within the ranges of climatic variation and the functioning of the biosphere that were those of the Holocene and thus to avoid leaving its trajectory with no possible turning back for millions of years to come. The timescale we are given shows the urgency of the situation. Moreover, the risks currently incurred by the anthropic violation of our ecosystems, our use of the soil, the climate system, the chemical composition of the atmosphere and the oceans or biogeochemical fluxes, all of these are transcendental dangers to the whole of human life in society. The Anthropocene will continue to make us very fearful (something of which we became aware during the Covid 19 pandemic). Finally, lurking in the shadows of the strictly geological threat of the Anthropocene, we can see a threat that is directly political. In fact, in circles that are aware of the real risks of climate change and the collapse of our ecosystems, it is impossible to deny the seductive nature of 'strong' powers, hoping that they will succeed where neo-liberal democracies have failed.

To conclude, it is now up to each disciplinary field to develop its acceptance of the Anthropocene, in other words to take ownership of all the knowledge at their disposal in order to identify the paradigmatic breaks to be constructed within each of the disciplines.

The Future of Humanity Based on the Materiality of the New Geological Context

With human beings having such a central role to play on our planet in the Anthropocene, anthropological research continues to gain in importance. In such research, the human being is rarely seen an entity that is separate from the world, a *homo clausus*. It is far more common to see the human being as entangled with nature, with other creatures and other human beings. There is an increasing awareness of the *dual historicity and culturality* of anthropology. In anthropological research, themes and objects are examined at a specific time in history within a specific cultural frame of reference. There is a focus on both the past and the present, and frequently assumptions and projections of future developments are made. Historical, cultural and philosophical anthropology has offered some important contributions to our understanding of humanity in recent decades (Kamper & Wulf, 1994; Wulf & Kamper, 2002; Wulf, 2013, 2022a; Wallenhorst & Wulf, 2022) that show the importance of anthropological knowledge for our understanding of the Anthropocene. This is also true of the research into the negative developments of the Anthropocene and the global efforts to lessen their impact. Particularly important are the attempts to combine local and regional views with global perspectives. Many problems of the Anthropocene can only be understood and dealt with by taking into consideration the interaction of local and global dimensions. Agenda 2030, with its Sustainable Development Goals, is an example of this.

In view of the demands of the Anthropocene, anthropological research projects use *plural forms of knowledge* that are often interconnected and yet also follow their own developmental logic. They acquire their knowledge from many scientific disciplines for which they use various research methods. Anthropological knowledge is often multi-, inter- and trans-disciplinary. Because of the situation in which our planet currently finds itself, this knowledge is also often inter- or transcultural. Often different, ever-changing points of reference are at the basis of global knowledge which make these research projects more complex. It is a challenge for contemporary anthropological research to make connections between several scientific disciplines and paradigms. For example, investigations by the earth sciences are important now – these include geological, stratigraphic, Earth system science and multi-paradigmatic investigations. New knowledge that is important for our understanding of the Anthropocene is also emerging in almost all sciences.

Anthropology is an *open science*. It has no pre-determined subject area. Generally, its research projects have had to do with the actions and behaviour of human beings in the Anthropocene and contribute to a better understanding of human beings. In principle, this includes all areas of human action and behaviour in relation to the world and other people. Anthropological knowledge is multi-paradigmatic; it includes the natural sciences, and in the humanities and social sciences, it embraces both qualitative and quantitative knowledge. New forms of knowledge arise in anthropology that may also be paradoxical and self-contradictory. If we presuppose a broad understanding of science this can result in new forms of anthropogenic knowledge that change the accepted way of seeing nature, the world and human beings. Examples of this are to be found in climate research, genetics and Artificial Intelligence.

New forms of knowledge are also arising in the cultural and social sciences which are important for the idiographic understanding and the nomothetic explanation of the effects of humans on the planet. Some of these come about in relation to the Sustainable Development Goals and others because of new historical or cultural perspectives. There is also the formation of hybrid forms of knowledge in which dynamics that differ from one culture to another affect each other. This happens to the same extent both in the arts and sciences, and these forms of knowledge create complex processes or "grand narratives" that are contingent on them, and new interpretations, as mentioned above (Lyotard, 1979).

By showing how knowledge relates to human beings and how they act with regard to nature and the world, the anthropological perspective can help to make it easier to understand knowledge in the Anthropocene that is often hard to access and make sense of (Council of Science and Humanities, 2006). Anthropological research projects attempt to dissolve traditional boundaries between the sciences and the humanities through new forms of connecting and combining knowledge. They can also clearly show the importance of anthropological processing of scientific knowledge for bringing the Sustainable Development Goals to fruition.

When faced with the complexity of the Anthropocene, there needs to be *critical reflection* on the limitations of anthropological research. The question is to what extent it is possible, in discourses on the trans-human for example, to see beyond the anthropological perspective (More & More, 2013). This gives rise to doubt. Although the anthropological perspective is often expanded in an appropriate way, the anthropological viewpoint is not, as a matter of course, transcended. Even when these discourses criticise the central role of human beings as the reference point of research, they do refer to them and can transcend neither anthropology nor human beings. For many years now, the goal of anthropological research is no longer abstract human beings, detached from their historical and cultural context. The goal is far more historical-cultural research that has as its focus the specific conditions of human life in the Anthropocene (Kamper & Wulf, 1994; Wulf, 2013, 2022a, b; Wulf et al., 2010, 2011; Wallenhorst & Wulf, 2022).

With this cultural-historical anchorage, the ambition of this *Handbook of the Anthropocene* is to look ahead and to try to identify the main lines of a prospective anthropology. This dynamic joins the structuring elements of the 'new humanism', as defined in 2010 by UNESCO, in a tradition marked by utopian pragmatism, which emphasises humanity's potential creativity, in terms both of devising new projects and goals and of redefining ourselves. These articles attempt to marry new humanism with ecology-based thinking and biogeochemical materiality. Such thinking (which is not seen in new humanism) represents a breakaway from anthropocentrism. This is singularly important when thinking about humanisation – the focus is not on the essence of humanity, but on our relationship with nature (one of dependency), and on accommodation of the human race (or the 'human adventure') as an integral part of the Earth system.

The Anthropocene: A Critical but Potentially Constructive Challenge to Shape the World

This collective work is a step toward prospective anthropological reflection, but it is by no means intended to be exhaustive. Finally, rather than being restricted to a single scientific discipline, this compendium opens doors and encourages debate.

When we refer to the Anthropocene, this often has negative connotations. We think of the great problems of the modern age that have arisen as a consequence of industrialisation, the acceleration of life and the increasing abstraction of modern ways of life. These include, for example, climate change, the acidification of the oceans, the destruction of biodiversity, the development of nuclear energy, the depletion of non-renewable energy and destruction of the environment. Most of this destruction is the result of an increasing number of people wanting a 'comfortable' life, based on exponential consumption, coupled with population growth. Now, however we realise that the ways of life involved in this have led to a high degree of violence towards nature, other people and also individual existence. This has created a situation in which many people see the future of humankind and the planet as being under threat. Given this situation, there is an increase in endeavours that rest on people's creative ability and require intensive efforts to prevent such a negative outcome. If, in the Anthropocene, humans have such a profound influence on everything that concerns the planet, they can surely also use this influence to improve the situation – and, as we have mentioned, the question of temporalities is decisive here. With the work on the 17 Sustainable Development Goals (SDGs), this would seem possible. Indeed, in these articles, education is presented as a tool to help make choices in relation to the medium- and long-term changes we face (Wallenhorst et al., 2018; Wallenhorst & Pierron, 2019; Wulf, 2022a). However, it is by no means certain that these efforts will be successful. And yet it seems reasonable and promising for the nations of the global community to want to work towards achieving these goals in as far-reaching a way as possible. Here it is clear that hopes and utopian dreams have a role to play. These SDGs require fundamental changes in politics, the economy, public awareness and education. It is clear that the education and socialisation of the future generations is important. And for this reason, one of the goals is expressly aimed at education and personal development.

The global community needs to be transformed on the basis of the SDGs. Such a development extends across the following five areas that need to be improved by means of the SDGs. These are *people* (poverty and hunger, dignity, equality, a healthy environment), *planet* (protection of the ecosystems), *peace* (inclusion, peace, justice), *prosperity* (bringing prosperity to all people by means of economic and technical development) and *partnership* (cooperation). These tasks should be fulfilled through the principles of universality, inseparability, inclusion, accountability and partnership (Wallenhorst, 2019; Wulf, 2022a; Wallenhorst & Wulf, 2022, 2023). In every area a programme is being developed, the success of which hinges on correcting the undesirable developments caused by humans. Here educational processes play an important part. They oscillate between perfection and incorrigibility (Kamper & Wulf 1994). Violence, which may be explicit or implicit, plays a central role in these processes. Considerable energy and strength is required to achieve the necessary social transformations and violence is therefore sometimes unavoidable. Paradoxical conflicts that are very difficult to resolve often arise in these processes, and education creates important methods of dealing with these conflicts in a constructive way. There is a need for all-embracing processes of education and learning and the anthropological research that will make them possible. There is still uncertainty about the extent to which these processes are possible in the long term in non-democratic societies. We need to create new images of nature, the world and humanity that are not simply based on utopian ideas but also include elements of human resistance and incorrigibility. The goal is not only to change the individual and collective imaginary but also to change actions and behaviour. There needs to be a critical but constructive change in people's dominant habits, which have at their centre, for example, consumption and utility. There also need to be new approaches to nature and the world (Wulf, 2013, 2022a; Paragrana, 2014). Mimetic processes, i.e. processes of creative imitation of examples of successful practice, are important here. Creative mimetic processes that embrace the sensuality of the body can help to form new habits (Gebauer & Wulf, 1995, 1998). On the one hand, these are the product of embodied practical knowledge, and on the other hand, they also lead to new forms of practice and structures. Rituals and ritualisations play an important role in this (Wulf, 2010; Wulf & Zirfas, 2014). For the SDGs to produce the social transformations for which they strive, they must be performative, i.e. they must lead to the staging and performance of social behaviour, that is geared towards the critical but constructive possibilities inherent in human actions and behaviour in the Anthropocene. These processes require us to embody within us the anthropogenic actions and behaviour to which we aspire (Kraus & Wulf, 2022) and also to realise the importance of what is implicit and tacit (Kraus et al., 2021).

It should be pointed out that the concept of the Anthropocene has been the subject of so many empirical studies, research and scientific publications in almost all disciplinary fields that it has now acquired an autonomy with respect to official geological institutions. The Anthropocene, regardless of its formalisation, can now be understood as a total social fact with a proven systemic scientific basis (Mauss, 1923–24), which is why it is important to explore it in this *Handbook of the Anthropocene*.

This *Handbook* is divided into six sections which constitute six facets of the Anthropocene that it is necessary to understand if we are to consider and prepare a long-term future.

1. The planet: caught between biogeophysical knowledge and the uncertainty of our adventure
2. Moving towards new epistemological paradigms linking the sciences and humanities
3. Human beings: bridging nature and culture
4. Societies: Prometheanism and post-Prometheanism
5. Profound long-term changes: education, apprenticeships and socialisation
6. Peace and violence

The almost 280 articles that make up this *Handbook of the Anthropocene* do not by any means cover the whole spectrum of the Anthropocene and its impact on human societies in a comprehensive way. We would have liked to add many more entries. But the multidisciplinary project had to end somewhere. This end is relative and temporary: as of now this task is continuing in the preparation of an *Encyclopaedia of the Anthropocene* for Springer-Nature. This work will mobilise the authors of this handbook and many others (from all continents and all disciplinary fields) to continue to circumscribe all the novelties and ruptures of the new geological era that is now ours.

References

Arnsperger, C., & Bourg, D. (2017). *Ecologie intégrale, Pour une société permacirculaire*. PUF.

Autin, W. J., & Holbrook, J. M. (2012). Is the Anthropocene an issue of stratigraphy or pop culture? *GSA Today, 22(7)*, 60–61.

Bador, M., et al. (2017). Future summer mega-heatwave and record-breaking temperatures in a warmer France climate. *Environmental Research Letters, 12*, 1–2.

Barnosky, A. D., et al. (2011). Has the Earth's sixth mass extinction already arrived? *Nature, 471*, 51–57.

Barnosky, A. D., et al. (2012). Approaching a state shift in Earth's biosphere. *Nature, 486*, 52–58.

Bonneuil, C., & Fressoz, J.-B. (2013). *L'événement anthropocène*. Seuil.

Bourg, D., & Papaux, A. (Eds.). (2015). *Dictionnaire de la pensée écologique*. PUF.

Brienen, R. J. W., et al. (2015). Long-term decline of the Amazon carbon sink. *Nature, 519*, 344–348.

Council of Science and Humanities. (2006). *Recommendations for the development and promotion of the humanities in Germany*. Wissenschaftsrat.

Crutzen, P. J. (2002). Geology of mankind: 'The Anthropocene'. *Nature, 415*, 23.

Crutzen, P. J. (2006). Albedo enhancement by stratospheric sulfur injections: a contribution to resolve a policy dilemma? *Climate change, 77*, 211–219.

Curnier, D. (2017). *Quel rôle pour l'école dans la transition écologique? Esquisse d'une sociologie politique, environnementale et prospective du curriculum prescrit*. Doctoral thesis in Environmental Sciences, University of Lausanne.

Dean, J. R., Leng, M. J., & Mackay, A. W. (2014). Is there an isotopic signature of the Anthropocene? *The Anthropocene Review, 1*, 276–287.

Descola, P. (2005). *Par-delà nature et culture*. Gallimard.

Doughty, C. E. (2013). Preindustrial human impacts on global and regional environment. *Annual Review of Environment and Resources, 38*, 503–527.

Doughty, C. E., et al. (2015). Drought impact on forest carbon dynamics and fluxes in Amazonia. *Nature, 519*, 78–82.

Eckersley, R. (2017). La démocratie à l'ère de l'Anthropocène. *lapenseeecologique.com, 1(1)*, 1–19.

Edwards, L. E. (2018). What is the Anthropocene? *Eos*, https://eos.org/opinions/what-is-the-anthropocene

Erlandsson, L. W. et al. (2022). A planetary boundary for green water. *Nature reviews Earth and Environment*, 26 April.

Esquivel-Muelbert, A., et al. (2019). Compositional response of Amazon forests to climate change. *Global Change Biology, 25*, 39–56.

Esquivel-Muelbert, A., et al. (2020). Tree mode of death and mortality risk factors across Amazon forests. *Nature communications, 12*, 1–20.

Federau, A. (2017). *Pour une philosophie de l'Anthropocène*. PUF.

Finney, S., & Edwards, L. E. (2016). The 'Anthropocene' epoch: scientific decision or political statement? *GSA Today, 26*(3–4), 4–10.

Gatti, L. V., et al. (2014). Drought sensitivity of Amazonian carbon balance revealed by atmospheric measurements. *Nature, 506*, 76–80.

Gebauer, G., & Wulf, C. (1995). *Mimesis. Culture, art, society*. University of California Press.

Gebauer, Gunter and Wulf, Christoph (1998). *Spiel, Ritual, Geste. Mimetisches Handeln in der sozialen Welt.* : Rowohlt.

Gil, I. C., & Wulf, C. (Eds.). (2015). *Hazardous future: Disaster, representation and the assessment of risk*. De Gruyter.

Hétier, R., & Wallenhorst, N. (Eds.). (2021). L'éducation au politique en Anthropocène, *Le Télémaque, 58*.

Hétier, R., & Wallenhorst, N. (2022a). The COVID-19 pandemic: a reflection of the human adventure in the Anthropocene. *Paragrana, 30*, 41–52.

Hétier, R., & Wallenhorst, N. (2022b). *Enseigner à l'époque de l'Anthropocène*. Le Bord de l'eau.
Hétier, R., & Wallenhorst, N. (2022c). Promoting embodiment through education in the Anthropocene. In A. Kraus & C. Wulf (dir.) *Learning bodies – Tact, emotion, performance. A european handbook*. Palgrave MacMillan.
Im, E.-S., Pal, J. S., & Eltahir, E. A. B. (2017). Deadly heat waves projected in the densely populated agricultural regions of South Asia. *Science Advances, 3(8)*, 1–7.
Kamper, D., & Wulf, C. (1989). *Looking back at the end of the world*. Massachusetts Institute of Technology.
Kamper, D., & Wulf, C. (Eds.). (1994). *Anthropologie nach dem Tode des Menschen. Vervollkommnung und Unverbesserlichkeit*. Suhrkamp.
Keats, J. (2011). Anthropocene. In J. Keats (Ed.), *Virtual words* (pp. 18–22). University Press of Oxford.
Kraus, A., & Wulf, C. (Eds.). (2022). *Palgrave handbook of embodiment and learning*. London Palgrave Macmillan.
Kraus, A., Budde, J., Hietzge, M., & Wulf, C. (Eds.). (2021). *Handbuch Schweigendes Wissen. Erziehung, Bildung, Sozialisation und Lernen* (2nd ed.). Beltz Juventa.
Lesourt, E. (2018). *Survivre à l'Anthropocène*. PUF.
Lewis, S. L., & Maslin, M. A. (2015). Defining the Anthropocene. *Nature, 519*, 171–180.
Lyotard, J.-F. (1979). *La condition postmoderne*. Les éditions de minuit.
Magny, M. (2019). *Aux racines de l'Anthropocène*. Le Bord de l'eau.
Malm, A. (2013). The origins of fossil capital: From water to steam in the British cotton industry. *Historical Materialism, 21*(1), 15–68.
Malm, A., & Hornborg, A. (2014). The geology of mankind? A critique of the Anthropocene narrative. *The Anthropocene Review, 1*, 62–69.
Mauss, A. (2012) (or. ed. 1923–1924). *Essai sur le don. Forme et raison de l'échange dans les sociétés archaïques*. PUF.
Mora, C., et al. (2017). Global risk of deadly heat. *Nature climate change, 7*, 501–506.
Nature. (2011). Editorial. The human epoch. Official recognition for the Anthropocene would focus minds on the challenges to come. *Nature, 473*, 254.
Papaux, A. (2015). Homo faber. In D. Bourg & A. Papaux (Eds.), *Dictionnaire de la pensée écologique* (pp. 536–540). PUF.
Paragrana. (2014). Internationale Zeitschrift für Historische Anthropologie. *Art and Gesture, 23*(1).
Persson, L., et al. (2022). Outside the safe operating space of the planetary boundary for novel entities. *Environmental Science & Technology, 56*(3), 1510–1521.
Prouteau, F., Hétier, R., & Wallenhorst, N. (2022). *Critique, utopia and resistance: three functions of pedagogy of resonance in the Anthropocene. Vierteljahrsschrift für Wissenschaftliche Pädagogik.*
Renaud, H. (2021). *L'humanité contre l'Anthropocène. Résister aux effondrements*. PUF.
Rockström, J. (2015). Bounding the planetary future: Why we need a great transition. *Great Transition Initiative, 9*, 1–14.
Rockström, J. W., et al. (2009). A safe operating space for humanity. *Nature, 461*, 472–475.
Ruddiman, W. F. (2013). The Anthropocene. *The Annual Review of Earth and Planetary Sciences, 41*, 45–68.
Ruddiman, W. F., et al. (2014). Does pre-industrial warming double the anthropogenic total? *The Anthropocene Review, 1*, 1–7.
Steffen, W., et al. (2011a). The Anthropocene: Conceptual and historical perspectives. *Philosophical Transactions of the Royal Society, 369*, 842–867.
Steffen, W., et al. (2011b). The Anthropocene: From global change to planetary stewardship. *Ambio, 40*(7), 739–761.
Steffen, W., et al. (2015). The trajectory of the Anthropocene: The great Acceleration. *The Anthropocene Review, 2(1)*, 81–98.
Steffen, W., et al. (2016). Stratigraphic and Earth system approaches to defining the Anthropocene. *Earth's Future, 4*, 1–22.

Steffen, W., et al. (2018). Trajectories of the Earth system in the Anthropocene. *Proceedings of the National Academy of Sciences, 115(33)*, 8252–8259.
Visconti, G. (2014). Anthropocene: another academic invention? *Rendiconti Lincei: Science Fisiche e Naturali, 25(3)*, 381–392.
Wallenhorst, N. (2019). *L'Anthropocène décodé pour les humains*. Le Pommier.
Wallenhorst, N. (2020). *La vérité sur l'Anthropocène*. Le Pommier.
Wallenhorst, N. (2021). *Mutation. L'aventure humaine ne fait que commencer*. Le Pommier.
Wallenhorst, N. (2022). *Qui sauvera la planète ?* Actes Sud.
Wallenhorst, N., & Pierron, J.-P. (Eds.). (2019). *Éduquer en Anthropocène*. Le Bord de l'eau.
Wallenhorst, N., & Wulf, C. (2022). *Humains. Un dictionnaire d'anthropologie prospective*. Vrin.
Wallenhorst, N., Prouteau, F., & Coatanéa, D. (Eds.). (2018). *Éduquer l'homme augmenté*. Le Bord de l'eau.
Wallenhorst, N., Robin, J.-Y., & Boutinet, J.-P. (2019). L'émergence de l'Anthropocène, une révélation étonnante de la condition humaine? In N. Wallenhorst & J.-P. Pierron (Eds.), *Éduquer en Anthropocène* (pp. 23–36). Lormont.
Waters, C. N., et al. (2014). Evidence for a stratigraphic basis for the Anthropocene. In R. Rocha, J. Pais, J. Kullberg, & S. Finney (Eds.), *STRATI 2013* (pp. 989–993). Springer Geology, Springer.
Waters, C. N., et al. (2016). The Anthropocene is functionally and stratigraphically distinct from the Holocene. *Science, 351*, 137–147.
Wulf, C. (2010). *Der Mensch und seine Kultur* (2nd ed.). Anaconda.
Wulf, C. (2013). *Anthropology. A continental perspective*. University of Chicago Press.
Wulf, C. (2022a). *Education as human knowledge in the Anthropocene. An anthropological perspective*. Routledge.
Wulf, C. (2022b). *Human beings and their images. Imagination, mimesis and performativity*. Bloomsbury.
Wulf, C., & Kamper, D. (Eds.). (2002). *Logik und Leidenschaft. Erträge Historischer Anthropologie*. Reimer.
Wulf, C., & Zirfas, J. (Eds.). (2014). *Handbuch Pädagogische Anthropologie*. Springer VS.
Wulf, C., & Zirfas, J. (Eds.). (2020). Den Menschen neu denken. *Paragrana: Internationale Zeitschrift für Historische Anthropologie*, 29/2020/1.
Wulf, C., et al. (2010). *Ritual and identity: The staging and performing of rituals in the lives of young people*. Tufnell Press.
Wulf, C., Suzuki, S., et al. (2011). *Das Glück der Familie: Ethnografische Studien in Deutschland und Japan*. Springer VS.
Zalasiewicz, J., Williams, M., Steffen, W., & Crutzen, P. (2010). The new world of the Anthropocene. *Environmental Science & Technology, 44*, 2228–2231.
Zalasiewicz, J., et al. (2011). The Anthropocene: A new epoch of geological time? *Philosophical Transactions of the Royal Society, 369*, 835–841.
Zalasiewicz, J., et al. (2012). Response to Autin and Holbrook on 'Is the Anthropocene an issue of stratigraphy or pop culture?'. *GSA Today, 22*(7), e21–e22.
Zalasiewicz, J., et al. (2014). When did the Anthropocene begin? A mid-twentieth century boundary level is stratigraphically optimal. *Quaternary international, 30*, 1–8.
Zalasiewicz, J., et al. (2017a). Anthropocene: Its stratigraphic basis. *Nature, 541*, 289–289.
Zalasiewicz, J., et al. (2017b). Making the case for a formal Anthropocene Epoch: an analysis of ongoing critiques. *Newsletters on Stratigraphy, 50*, 205–226.

Contents of Volume II

Section V The Profound Changes in the Long Term: Education, Learning, and Socialization

Nathanaël Wallenhorst and Christoph Wulf

Part XVIII Developing Subjectification

Part XIX Developing Socialization

Contents of Volume I

Part XII Humanity as Symbolisation

Part XIII Humanity as Creation

About the Editors

Nathanaël Wallenhorst is Professor at the Catholic University of the West (UCO). He is Doctor of Educational Sciences and Doktor der Philosophie (first international co-supervision PhD) and Doctor of Environmental Sciences and Doctor in Political Science (second international co-supervision PhD). He is the author of twenty books on politics, education and anthropology in the Anthropocene. Books (selection): *The Anthropocene Decoded for Humans* (Le Pommier, 2019, *in French*); *Education in the Anthropocene* (ed. with Pierron, Le Bord de l'eau 2019, *in French*); *The Truth About the Anthropocene* (Le Pommier, 2020, *in French*); *Mutation: The Human Adventure Is just Beginning* (Le Pommier, 2021, *in French*); *Who Will Save the Planet?* (Actes Sud, 2022, *in French*); *Vortex: Facing the Anthropocene* (with Testot, Payot, 2023, *in French*); *Political Education in the Anthropocene* (ed. with Hétier, Pierron and Wulf, Springer, 2023, *in English*); and *A Critical Theory for the Anthropocene* (Springer, 2023, *in English*) e-mail:nathanael.wallenhorst@uco.fr.

Christoph Wulf is Professor of Anthropology and Education and a member of the Interdisciplinary Centre for Historical Anthropology, the Collaborative Research Centre (SFB, 1999–2012) 'Cultures of Performance', the Cluster of Excellence (2007–2012), 'Languages of Emotion' and the Graduate School 'InterArts' (2006–2015) at the Freie Universität Berlin. His books have been translated into 20 languages. For his research in anthropology and anthropology of education, he received the title *professor honoris causa* from the University of Bucharest. He is Vice-President of the German Commission for UNESCO. *Major research areas:* historical and cultural anthropology, educational anthropology, imagination, intercultural communication, mimesis, aesthetics, epistemology and Anthropocene. Research stays and invited professorships have included the following locations, among others: Stanford, Tokyo, Kyoto, Beijing, Shanghai, Mysore, Delhi, Paris, Lille, Modena, Amsterdam, Stockholm, Copenhagen, London, Vienna, Rome, Lisbon, Basel, Saint Petersburg, Moscow, Kazan and Sao Paulo e-mail: christoph.wulf@fu-berlin.de.

Contents (Alphabetical Order)

Section IV
Societies Caught between Prometheanism and Post-prometheanism

Nathanaël Wallenhorst and Christoph Wulf

The concept of the Anthropocene that comes from Earth system science and has become consolidated in the field of geology through the use of stratigraphic methodology, has gradually shifted into the humanities and social sciences where it has become the subject of many studies: the critical reception of geo-scientific works, the philosophical and political conceptualisation of the Anthropocene and also the vulgurisation of the scientific and philosophical apsects of this concept. These atlases, dictionaries, collections and other works have an important function, in that they gradually permit disciplinary boundaries to be crossed in the collaboration between different disciplines, and explore a variety of consequences in the areas of anthropology, society, economy, psychology and education.

The term "Anthropocene" is a neologism which spread like wildfire and became a global success. The reason this term is finding such notoriety in scientific circles – as much in Earth sciences as in the humanities and social sciences, amongst militant ecologists as well as in the media, is because it is inextricably bound up with the existence of each and every one of us. The Anthropocene is a new geological epoch resulting from the impact of human activity on the whole Earth system. This neologism means that we have changed the habitability conditions of the planet in a permanent way. Basically, the Anthropocene asks what we are doing to ourselves, to others and other things, and to all that connects us together. The Anthropocene does this with a power that is unprecedented, all of a sudden revealing what our civilisations are founded on, with the building up of agricultural surplus, the changes in the

N. Wallenhorst (✉)
Catholic University of the West, Angers, France
e-mail: nathanael.wallenhorst@uco.fr

C. Wulf
Freie Universität Berlin, Berlin, Germany
e-mail: christoph.wulf@fu-berlin.de

modern industrial, capitalist world, the acceleration in human population growth coinciding with the growth of consumerism with its built in obsolescence, the prodigious progress of technoscience – starting with the invention and expolosion of nuclear bombs. The Anthropcene is telling us that we have the capacity to destroy ourselves. The Anthropocene illuminates the dark face of our adventure, in tne very place where we have sometimes been tempted to see advances for humanity.

We can currently identify four great controversies concerning the Anthropocene. The first concerns the notion of responsibility. Who is responsible for the Anthropocene? Is it humanity as a whole or just part of it? The fact that today's debates concerning dating the beginning of the Anthropocene are so lively is because they introduce this question of responsibility. The concept of the Capitalocene (Bonneuil & Fressoz, 2013; Malm & Hornborg, 2014; Haraway, 2015; Moore, 2016) was constructed, for political purposes, to put the spotlight on this question of responsibility that the concept of the Anthropocene masks. The second controversy, which is less common in scientific debate, takes place in the public arena between the catastrophe theorists (a huge disaster or *collapse* is about to come and we must change everything with immediate effect) and the conservative attitudes which deny that anything is changing (*climate deniers*), or who minimise the role of human beings or the consequences of climate change (*climate delayers*). The third controversy is more recent and also takes place in the public arena. The climate strikes of younger people, initiated by the young Swede Greta Thunberg, brought about a polarity never seen before. Here the issue of the climate and the Anthropocene was presented as a conflict between generations, the young being the accusers. They take on board directly the notion of intergenerational justice that is practically non-existent in democracy. The *Manifesto for climate justice* (Notre affaire à tous, 2019) or the *Manifeste étudiant pour un réveil écologique (Student manifesto for ecologogical awakening)* (2018) are illustrations of this.

If we study what has been written about the Anthropocene in Earth system science as well as in the humanities and social sciences, it is clear that there a fourth controversy that we can see as a polarisation between Promethean and post-Promethean ideas, which Bruno Latour calls humans versus terrestrials (2015). This controversy is the departure point for the articles in this section. Political Promethean ideas are a result of the ideas of geoengineering developed by certain research scientists, which are heavily technical and rather pessimistic as to the possibility of humanity being able to survive the Anthropocene other than by taking control of the Earth system (which is known as "Earth system management"). Political post-Promethean ideas question the effects of the human adventure on Earth in order to change its direction. We find here ideas about changing the way we live and a kind of anthropological change (Meadows, 2013; Malm & Hornborg, 2014; Lövbran et al., 2014; Eckersley, 2017; Wallenhorst, 2019, 2020, 2021, 2022; Hétier & Wallenhorst, 2022; Wallenhorst & Wulf, 2022; Testot & Wallenhorst, 2023; Kamper & Wulf, 1989; Kamper & Wulf, 1994; Wulf & Zirfas, 2014, 2020; Wulf, 2013, 2022).

Promethean Ideas

The Swedish political scientist highlights the the fact that radical measures must be taken to survive the upheavals of the Anthropocene in the way human soocieties are organised. The first of these two options is to research other planets that are hospitable to human life and to develop the cutting edge technology required to get there and render these spaces favourable to life. It is, therefore, a case of crossing planetary boundaries. The second option, on the other hand, amounts to respecting these boundaries in a very clear way. He uses two metaphors to describe these two strategies of sustainability - StarTrek and Ecotopia. We would put forward the adjectives Promethean and post-Promethean as the best way to encapsulate these directions our societies can take in the Anthropocene. Promethean refers to the myths of Prometheus and is to be understood as the search for power founded on transgressions. In contrast, post-Promethean, refers to changes in the way we live that rest on finitude, founded on not transgressing boundaries. We should say here that the current use of the term Promethean may suggest the Faustian figure of twentieth century technological man and the way technology is used in the West. Building on the analyses of the French intellectual Jacques Ellul, the adjective Promethean may be associated with the techno-scientific adventure. Using the adjective Promethean adds here an additional dimension to the technical use: the search for power in the context of transgression.

Whereas most of the thinking in the field of political ecology emphasises the need for us to change the way we live, in the last few we have seen the emergence of modernist political thinking, promoting constructivist ecology, acknowledging that we have entered the Anthropocene and proposing a resolutely Promethean stance. Worthy of mention here is the *Accelerate Manifesto* by Srnicek and Williams (2013). Then the *Ecomodernist manifesto* (2015) appeared, in the same techno-scientific, Promethean vein, but from a liberal, capitalist angle (and not post-capitalist like the *Accelerate Manifesto*), inspired by the ecomodernism of the Breakthrough Institute, which advocates a form of ecological Prometheanism and is critical of political post-Promethean ecology. The Breakthrough Institute puts forward a vision of the Anthropocene in which the way we live is not put into question. We are dealing with a form of hypermodernity and a re-evauation of technical progress which can be directed towards permitting humanity to continue to live on planet Earth. This perspective suggests that it is up to Earth system scientists to use their ability to be particularly creative in experimenting with techniques on a global scale for the planet.

One of the first authors to have used the expression "good Anthropocene" is the American biologist Erle C. Ellis in an article that appeared in 2011 in the *Breakthrough Journal* (Ellis, 2011). This terminology does not mean that the environmental conditions of the Anthropocene are "good" or favourable to human life or any other form of life. It refers to a positive, enthusiastic view of the *anthropos* having become a geological force through mastery of his techniques. The "good Anthropocene" involves a narrative that "celebrates the death of nature as something external to us" and finds reconciliation in the concept of green growth, which was developed by those who Christophe Bonnet identifies as post-environmentalist technophiles. In this narrative, nature, and also the human species, are conceived of as a social -technical-economic construct opening the door to transhumanism. The Anthropocene, which follows on from the acquisition of power by humanity, is "good" because this power will allow us to improve the world and the human condition. The tone is positive and technophilic, along the lines of the *Ecomodernist manifesto*. This group of authors underlines that our entry into the Anthropocene can be a catalytic event which will mobilise humanity: "A good Anthropocene demands that humans use their growing social, economic, and technological powers to make life better for people, stabilize the climate, and protect the natural world" (Ecomodernist manifesto, 2015, p. 6). The aim of the 18 intellectuals in this group is to describe their vision "for putting humankind's extraordinary powers in the service of creating a good Anthropocene" (Ecomodernist manifesto, 2015, p. 7). In their view, the slowing down of climate warming is first and foremost a question of technology. Their supporters go as far as envisaging "terraformation", or, in other words, the transformation of the atmosphere of Mars in order to create a planet that is hospitable to human life. A characteristic of ecomodernism is that it espouses modernist logic as well as depoliticising the modern age and naturalising the process of technological modernisation. Technological evolution is therefore seen as being inevitable. Once progress is possible, it is invested as something to be explored without human beings having had the chance to ask whether it is desirable.

Within the Promethean conception of politics in the Anthropocene there is a a great deal of research into technological progress in the field of geoengineering, which can be defined as "the deliberate manipulation or 'engineering' of an Earth System process" (Steffen et al., 2011, p. 752). The term geoengineering appeared for the first time in the first number of the review *Climatic Change* in a study by the Italian physicst Cesare Marchetti, on the modalities of regulating CO_2 in the atmosphere. In the same year, the Russian climatologist Mikhaïl Budyko further developed this theme, considering that it is our human responsibility "to develop a plan for climate modification that will maintain existing climatic conditions, in spite of the tendency toward a temperature increase due to man's economic activity" (Budyko, 1977, p. 244). Several contributions from Earth system science on the subject of the Anthropocene envisage the possibility of using geoengineering on a world-wide scale (Crutzen, 2002, p. 23; Steffen et al., 2007, 2011; Ellis, 2011).

Having coined the term Anthropocene in 2002, Paul J. Crutzen addresses the field of environmental rehabilitation. In fact he mentions that the Anthropocene "will require appropriate human behaviour at all scales, and may well involve internationally accepted, large-scale geo-engineering projects, for instance to 'optimize' climate" (Crutzen, 2002, p. 23). At a conference at the University of Yale, reported in an online article of 2011, he goes still further, saying "It is no longer us against 'Nature,' it's we who decide what Nature will be" (Crutzen & Schwägerl, 2011).

Will Steffen, Paul J. Crutzen and John R. McNeil (Steffen et al., 2011) make a plea for humanity to control the Earth system as "stewards". These three authors believe that the coming decades will be crucial for the consequences of the Anthropocene. Over the last few years we have seen the emergence of the concept of planetary stewardship, which refers to us taking control of the Earth system and its main regulatory mechanisms (Steffen et al., 2011; Crutzen & Schwägerl, 2011). This stewardship involves technical intervention in the biogeochemical cycles, or geoengineering. For many authors, deploying climate engineering techniques is a solution that should only be a last resort. Moreover, the question of feasibility, the problem at the heart of using geoengineering as a cure, is aptly summarised by Schneider who fears the "cure could be worse than the disease" (Schneider, 2008, p. 14).

There are two major types of mechanisms of geoengineering that are envisaged. The first involves controlling greenhouse gases through intervention in the carbon cycle – CDR (or Carbon Dioxide Removal). The second is solar radiation management (SRM) through controlling the amount of heat entering the atmosphere. One of the big questions geoengineering poses is that of who is in charge. Who can make the decision to bring about such effects on life within the whole of the biosphere? Before we can begin to imagine such geoengineering projects is it not more worthwhile to study in depth the social, economic and political systems set up by human beings and to see the Anthropocene as a saving grace to make us change the way we live?

Post-Promethean Views

The second type of approach for our societies in the Anthropocene can be categorised as "post-Promethean". *The Convivialist Manifesto* which has now been signed by more than 3700 intellectuals, the first version of which was the work the French sociologist Alain Caillé and which is in the spirit of what is known as the MAUSS movement (*Mouvement anti-utilitariste en sciences sociales*). A similar approach is that of the British economist Kate Raworth, who proposes a model of human survival that she calls the "doughnut" model. In this visual image she has a space of safety and justice for human beings, seen as being at the centre of all the planetary boundaries that must not be crossed (an extension of Rockström's nine planetary boundaries, 2009), and, in between, the "social foundation" – the standards for humans and society, without which human life is no longer possible – access to water, to work, to energy, to an income, to social equity, to education, to health, to democratic participation, to social resilience (Raworth & Economics, 2017).

The term "post-Promethean" covers numerous articles about the Anthropocene. In recent years we have seen the appearance of a number of manifestos, most of which have been written by academics. They have been edited by one or two or signed by the whole group as "think tanks". Their emphasis is on demanding, denouncing, opposing and on the dream of seeing the emergence of alternatives to the current economic, hegemonic ways of life that are founded on capitalism. These manifestos tell us that it is possible for the world to be different. Or in any case we should hope for a different world. These contributions, each one being a new voice in the public debate and presenting itself as a new political path, are all part of a renewal in political thinking, because all areas of social life (health, industry, education, the military, leisure, transport, food, etc.) are suffering from the consequences of the major changes brought about by the Anthropocene. In addition to the *Convivialist Manifesto* we must also mention (and this list is by no means exhaustive) the *Manifest für das Anthropozän* (*Manifesto for the Anthropocene*, 2015), the *Animalist Manifesto* (2017), the *Manifesto for Climate Justice* (2019) and *Ecologie intégrale: Le manifeste* (2019). They explicitly mobilise the concept of the Anthropocene and what it involves, the changing of the Earth system in an anthropic way. Each has its own point of view, each tries to unravel the concept of possible global change: the implementation of politics and poetics for life (*Manifesto for the Anthropocene*), recognising animals as being political subjects *(Animalist Manifesto),* the struggle against productivism (*Manifesto for Climate Justice)*, the possibility of revitalising politics (*Ecologie intégrale: Le manifeste*) or taking control of our *hubris* through recognising the way we live together (*Convivialist Manifesto*).

The political responsibility for preparing for the furure is certainly one of the major challenges posed by the Anthropocene. In fact, the Anthropocene "prompts us to contemplate the possibility and meaning of the unthinkable: an Earth without us" (Eckersley, 2017, p. 15). Our entering the Anthropocene makes us realise how eminently problematic is the trajectory of our civilisation (Bourg, 2013, p. 70). We now have how to think of our future in the consciousness that our power is finite (Villalba, 2015, p. 59). The responsibility that has now devolved to us is not by any means individual but concerns the whole of humanity, who have become responsible for the Earth and who now have the power to irreversibly change nature (Bourg, 2013, p. 69). For the American philosopher Roy Scranton the choice is a clear one: "We can continue acting as if tomorrow will be just like yesterday, growing less and less prepared for each new disaster as it comes, and more and more desperately invested in a life we can't sustain. Or we can learn to see each day as the death of what came before, (…) If we want to learn to live in the Anthropocene, we must first learn how to die" (Scranton, 2015, p. 3).

Faced with the boundaries of the biosphere and the prospect of a collapse of the ecosystems causing a decrease in global food production, researchers have put forward the concept of *Earth system governance*. This is difference from stewarship of the planet in that the repertoire of actions which may allow the human adventure to continue in the Anthropocene is different from that of geoengineering. In political science the idea of Earth sytem governance has emerged as a reaction to the concept of the Anthropocene, accepting it both on an analytical and a normative level. It was put forward in 2007 by the Dutch political scientist Frank Biermann and is a resolutely political reaction to the Anthropocene with no technically oriented solution. It involves working on five challenges concerning the changes brought about anthropically in the Earth system: the "persistent uncertainty" about the Earth system, the new intergenerational dependencies; the interdependencies of the subsystem and the system; the spatial interdependencies with their global environmental and social consequences (a deterioration in one part of the world can have social consequences right across the world) and a heightened possibility of the whole of humanity being affected. The concept of "Earth system governance" does not signify the governance of the Earth, nor does it signify the governance of biogeochemical processes of the Earth system. It refers to the human impact on the Earth system and the governance of human societies, giving particular weight to the long-term effects of these on the Earth system (Biermann, 2014, p. 59).

The concept of Earth system governance has similarities to Dominique Bourg's proposal of "long-term governance". This is an alternative to the concept of stewardship by geoengineering. He proposes examining current scientific data on changes in the Earth system and looking at their implications for human life and our societies, so that we can prepare ourselves to face the dangers that are awaiting us. His approach is political and does not advocate the use of technology or geoengineering. On the contrary, it presupposes the capacity for anthropological evolution. The concept of long-term governance re-establishes politics rather than replacing it by the technoscience of Earth system management. "Long-term" here refers to the great physical and biological parameters of the Earth system, those that are referred to by the idea of the planetary boundary (Bourg, 2017, p. 5). Several elements that are currently undergoing change will have an effect for the next millenia: this is the case for the mean temperature that will be reached over the course of the next century, the amount of CO_2 in the atmosphere, the disruption of biogeochemical cycles, the level of the seas, the acidification of the oceans and the extinction of species, the use of land, etc. Our long-term future is determined by biogeophysical elements which will coninue to be present for millions of years to come, of that there is no doubt. In this way the "long-term" defined by Bourg differs from the forecasts of scenarios of societal structures for the coming decades. From the information that we have at our disposal reagrding the long-term, it is important to establish our political priorities.

Rethinking the Structures of Politics – Finding a Middle Way Between Democracy and Authoritarianism

One of the big questions posed by the concept of the Anthropocene to our societies is how to know if it will be a source of democratic renewal or if it will be a danger to democracy (Eckersley, 2017, p. 4). The controversies surrounding the Anthropocene have been popularised to a certain extent by the media which has led to the concept becoming more widely known in the public and political arenas. This concept allows us to re-examine political ecology and forms a basis for social criticism. The Anthropocene "has clear implications, not only in the way we grasp the complexity of the world in theory, but also in the modalities according to which we must think its political transformation" (Deléage, 2010, p. 23).

In view of the long-term challenges presented above, on the political scene the present seems preferable to the future. Politics often give the impression of reacting to events rather than preparing for a future that is getting ever harder to predict (Kemp, 2013). The relationship between politics and time-scales leads us to the question of responsibility. Responsibility twoards the future is a central theme in the work of the German philopsopher Hans Jonas. Jonas views it as an "ontological imperative". His "ethics of responsibility" take into consideration the long-term effects of how we live and act with regard to future generations, i.e. to persons who do not yet exist but who we know will exist. In a way, human beings must feel responsible for what is to become of humanity. Hans Jonas does not have the slightest hesitation in turning democratic values upside down by proposing solutions that are in part authoritarian and which restrict individual freedoms when these freedoms threaten the survival of the human adventure. In his view, the democratic system would not be sufficiently capable of taking those decisions early enough to manage the urgency of the ecological situation. It is only an authoritarian power that can address this challenge and respond to this imperative in a limited amount of time, and that is capable of putting science and experts at the centre and acting in our long-term interests. (Nevertheless, of course we cannot say of Hans Jonas that he does not distance himself critically from authoritarianism, since he has not inspired authoritarian factions and his thinking is not part of the authoritarian school; Bourg & Whiteside, 2017, p. 7).

The responsibility of politics to prepare for the future is certainly one of the fundamental challenges posed by the Anthropocene. In contrast to authoritarian controls, the Anthropocene can also be accompanied by a reform of institutions in order to achieve democratic renewal. Environmental questions generate acts of violent repression, in democratic countries as well, as the journalist Élisabeth Schneiter shows in her reports of the increase in assassinations of militant ecologists across the world (22 per year, or 4 per week) (Schneiter, 2018).

Our entry into the Anthropocene, the fact that it is taken into consideration in the structure of political life in the light of the harmful long-term effects on the way we live, may make us think differently about democracy and give it a new dynamic. A number of authors foresee the creation of a third chamber that would be characterised by deliberation methods and would have as its mission the protection of future generations. In 1991, in *We Have Never Been Modern* (Latour, 1991) Bruno Latour sketched out an idea of democratic reorganisation around a higher and a lower chamber. He developed this idea further in 1999 in *Politics of Nature* (*Politiques de la nature)* (Latour, 1999). American political scientist John Dryzek and Australian political scientist Simon Niemeyer propose the establishment of a "chamber of discourse" which would allow citizens to be better represented and to devote themselves to environmental challenges (Drysek & Niemeyer, 2008). The British philosopher Rupert Read also proposes the idea of a chamber devoted to long-term challenges and to the protection of future UK citizens (Read, 2012). Dominique Bourg and Kerry Whiteside introduce and develop the idea of a "chamber of the future" on several occasions (Bourg & Whiteside, 2011, 2017; Bourg et al., 2017).

Recently, in *Inventer la démocratie du 21e siècle (Inventing democracy of the twenty-first century,* this third chamber was presented by Domminique Bourg and his co-authors as a "Citizens Assembly of the Future", consisting of a place for reflecting on all the citizens' initatives, evaluating the contribution of citizens to ecological transition and for experimenting in a democratic way with paths leading to the desired future (Bourg et al., 2017, p. 43). This Citizens Assembly of the Furture aims to let everyone have their say and to produce specific sociological profiles of regular participants "super enlightened citizens" (Dufour, 2012, p. 33) or individuals who have a personal interest in taking part in the debate by randomly drawing lots. The idea is to shape, educate and document citizens. In any case, the advice of the Citizens Assembly of the Future is purely consultative, which seems problematical – we are all too aware of the frustration felt by citizens which arises from the consultative process, where their views are often misconstrued or paid little heed in the flow of the debate. The citizens participate and think that the time they are devoting to these debates will in fact give them the opportunity to have a powerful impact on the decision. When they realise that that is not the case, that their opinion is purely consultative, they distance themselves from the formal situation and may end up contesting views in more violent and perhaps radical ways (Mazeaud & Talpin, 2010). Is there not a risk that citizen participation will be seen as an empty shell, without any real impact on decisions? In the many citizen participation schemes, citizens deliver an opinion to which the elected representatives pay little attention without taking the trouble to try to understand it, which increases even more the frustration of those who have taken the trouble to take part. It is necessary for there to be more transparency about how decisions are taken. This Citizens Assembly of the Future, even though it appears utopian in parts, gives back to citizens their key place in the debates on the Anthropocene.

Societies, with their communities and institutions, now form an integral part of the Earth system. It is up to us to determine whether the way in which we function and structure our lives impedes or promotes the future of human life in society within the biosphere. It is this that unites the different articles in this section.

References

Biermann, F. (2014). The Anthropocene: A governance perspective. *The Anthropocene Review, 1*, 57–61.

Bonneuil, C., & Fressoz, J.-B. (2013). *L'événement Anthropocène*. Seuil.

Bourg, D. (2013). Position. La guerre écologique. *Esprit*, juillet.

Bourg, D. (2017). Gouverner le long terme. *Points de vue, 1*(1), 5. PUF. *lapenseeecologique.com*

Bourg, D., & Whiteside, K. (2011). Ecologie, démocratie et représentatio. *Le Débat, 164*, 145–153.

Bourg, D., & Whiteside, K. (2017). Écologies politiques: essai de typologie. *La Pensée Écologique, 1*(1). *lapenseeecologique.com*

Bourg, D., Augagneur, F., Blondiaux, L., Cohendet, M. A., Fourniau, J. M., François, B., & Prieur, M. (2017). *Inventer la démocratie du XXIe siècle*. Les liens qui libèrent.

Budyko, M. (1977 (ed. 1974)). *Climatic changes*. American Geophysical Union.

Crutzen, P. J. (2002). Geology of mankind: 'The Anthropocene'. *Nature, 415*, 23.
Crutzen, P. J., & Schwägerl, C. (2011). Living in the Anthropocene: Toward a new global ethos. *Yale Environment, 360*. http://e360.yale.edu/features/living_in_the_anthropocene_toward_a_new_global_ethos, consulté le 25 octobre 2017.
Deléage, J.-P. (2010). En quoi consiste l'écologie politique? *Écologie & Politique, 40*, 21.
Drysek, J., & Niemeyer, S. (2008). Discursive representation. *American Political Science Review, 102*(4), 481–493.
Dufour, P. (2012). Politique de la rue contre politique des urnes? Le mouvement étudiant québécois du printemps 2012 et la question de la représentation politique. *Savoir/Agir, 22*(4), 33–41.
Eckersley, R. (2017). La démocratie à l'ère de l'Anthropocène. *La Pensée Écologique, 1*(1), 1–19. *lapenseeecologique.com*
Ecomodernism. (2015). *An ecomodernist manifesto*. www.ecomodernism.org
Ellis, E. C. (2011). The planet of no return: Human resilience on an artificial earth. *The Breakthrough Journal, 2*, 37–44.
Haraway, D. (2015). Anthropocene, capitalocene, plantationocene, chthulucene: Making kin. *Environmental Humanities, 6*, 159–165.
Hétier, R., & Wallenhorst, N. (2022). *Enseigner à l'époque de l'Anthropocène*. Le Bord de l'eau.
Kamper, D., & Wulf, C. (1989). *Looking back at the end of the world*. Massachusetts Institute of Technology.
Kamper, D., & Wulf, C. (Eds.). (1994). *Anthropologie nach dem Tode des Menschen. Vervollkommnung und Unverbesserlichkeit*. Suhrkamp.
Kemp, P. (2013). *Le Prince*. Seuil.
Latour, B. (1993 [1991]). We have never been modern. Harvard University Press, .
Latour, B. (1999). *Politiques de la nature. Comment faire entrer les sciences en démocratie*. La Découverte.
Lövbran, E. et al. (2014). Taking the human (sciences) seriously: Realizing the critical potential of the Anthropocene. In *ECPR general conference*, Glasgow, 6th September 2014.
Malm, A., & Hornborg, A. (2014). The geology of mankind? A critique of the Anthropocene narrative. *The Anthropocene Review, 1*, 62–69.
Mazeaud, A., & Talpin, J. (2010). Participer pour quoi faire? *Sociologie, 3*(1), 357–374.
Meadows, D. (2013). Il est trop tard pour le développement durable. In Agnès Sinaï (dir.), *Penser la décroissance* (pp. 195–210). Presses de Sciences Po.
Moore, J. W. (2016). *Anthropocene or Capitalocene?* History and The Crisis of Capitalism, Oakland, PM Press.
Raworth, K., & Economics, D. (2017). *Seven ways to think like a 21st century economist*. Chelsea Green Publishing.
Read, R. J. (2012). *Guardians of the future: A constitutional case for representing and protecting future people*, Green House. https://www.greenhousethinktank.org/uploads/4/8/3/2/48324387/guardians_inside_final.pdf
Schneider, S. H. (2008). Geoengineering: Could we or should we make it work? *Philosophical Transaction of the Royal Society, 366*, 14.
Schneiter, É. (2018). *Les héros de l'environnement*. Seuil.
Scranton, R. (2015). *Learning to die in the Anthropocene. Reflections on the end of a civilization*. City Light Publisher.
Srnicek, N., & Williams, A. (2013). Accelerate. Manifesto for an accelerationist politics. *Critical Legal Thinking*, 14th May 2013 (http://criticallegalthinking.com/2013/05/14/accelerate-manifesto-for-an-accelerationist-politics).
Steffen, W., Crutzen, P. J., & McNeil, J. R. (2007). The Anthropocene: Are humans now overwhelming the great forces of nature? *Ambio, Royal Swedish Academy of Sciences, 36*(8), 614–621.
Steffen, W., et al. (2011). The Anthropocene: From global change to planetary stewardship. *Ambio, 40*(7), 739–761.
Testot, L., & Wallenhorst, N. (2023). *Vortex. Face à l'Anthropocène*. Payot.

Villalba, B. (2015). Au fondement matériel de la démocratie. *Revue Projet, 344*, 56.
Wallenhorst, N. (2019). *L'Anthropocène décodé pour les humains*. Le Pommier.
Wallenhorst, N. (2020). *La vérité sur l'anthropocène*. Le Pommier.
Wallenhorst, N. (2021). *Mutation. L'aventure humaine ne fait que commencer*. Le Pommier.
Wallenhorst, N. (2022). *Qui sauvera la planète ?* Arles.
Wallenhorst, N., & Wulf, C. (Eds.). (2022). *Humains. Un dictionnaire d'anthropologie prospective*. Vrin.
Wulf, C. (2013). *Anthropology. A continental perspective*. University of Chicago Press.
Wulf, C. (2022). *Education as human knowledge in the Anthropocene. An Anthropological perspective*. Routledge.
Wulf, C., & Zirfas, J. (Eds.). (2014). *Handbuch Pädagogische Anthropologie*. Wiesbaden.
Wulf, C., & Zirfas, J. (Eds.). (2020). Den Menschen neu denken. *Paragrana: Internationale Zeitschrift für Historische Anthropologie*, 29/2020/1.

Nathanaël Wallenhorst is Professor at the Catholic University of the West (UCO). He is Doctor of Educational Sciences and Doktor der Philosophie (first international co-supervision PhD), and Doctor of Environmental Sciences and Doctor in Political Science (second international co-supervision PhD). He is the author of twenty books on politics, education, and anthropology in the Anthropocene. Books (selection): *The Anthropocene decoded for humans* (Le Pommier, 2019, *in French). Educate in Anthropocene (*ed. with Pierron, Le Bord de l'eau 2019, *in French*). *The Truth about the Anthropocene* (Le Pommier, 2020, *in French*). *Mutation. The human adventure is just beginning* (Le Pommier, 2021, *in French*). *Who will save the planet?* (Actes Sud, 2022, *in French*). *Vortex. Facing the Anthropocene* (with Testot, Payot, 2023, *in French*). *Political education in the Anthropocene* (ed. with Hétier, Pierron and Wulf, Springer, 2023, *in English*). *A critical theory for the Anthropocene* (Springer, 2023, *in English*).

Christoph Wulf is Professor of Anthropology and Education and a member of the Interdisciplinary Centre for Historical Anthropology, the Collaborative Research Centre (SFB, 1999–2012) "Cultures of Performance," the Cluster of Excellence (2007–2012) "Languages of Emotion," and the Graduate School "InterArts" (2006–2015) at the Freie Universität Berlin. His books have been translated into 20 languages. For his research in anthropology and anthropology of education, he received the title "*professor honoris causa*" from the University of Bucharest. He is Vice-President of the German Commission for UNESCO. *Major research areas:* historical and cultural anthropology, educational anthropology, imagination, intercultural communication, mimesis, aesthetics, epistemology, Anthropocene. Research stays and invited professorships have included the following locations, among others: Stanford, Tokyo, Kyoto, Beijing, Shanghai, Mysore, Delhi, Paris, Lille, Strasbourg, Modena, Amsterdam, Stockholm, Copenhagen, London, Vienna, Rome, Lisbon, Basel, Saint Petersburg, Moscow, Kazan, Sao Paulo.

Part XV
Societies and Life

Agriculture

Bertrand Valiorgue

Abstract This article investigates the impact of the Anthropocene on agriculture. If agriculture is usually accused of emitting greenhouse gas and reducing biodiversity we show that agriculture might be seen as the first victim of the Anthropocene. This article shows the different negative impacts of the Anthropocene on agriculture and questions the future of this fundamental activity at the basis of the human civilization.

Agriculture is the basis of all human societies. Since its invention over 10,000 years ago, this essential activity has undergone continuous transformations that accelerated in the second half of the twentieth century. Agriculture is the basis of three essential activities for the maintenance of human physiology and civilization: food, drug (through herbal medicinal products) and clothes (through animal and plant textile fibres). Many authors have noted the coincidence between the entry into the geological era of the Holocene (12,000 years ago), the development of agriculture and the blossoming of human civilization (Hamilton et al., 2015). The stable and temperate climate that established with the Holocene allowed the development of an important biodiversity that humans gradually domesticated in order to meet their basic needs. It was the geological era of the Holocene that allowed human civilization to develop because it made agriculture possible. The entry into the Anthropocene fundamentally shakes up this favourable situation for agriculture as it is massively and negatively affected by the ongoing transformations of the Earth system. These impacts occur at several complementary levels.

Transformation of Hygrometry Animals and plants have vital water requirements in order to be able to develop. The presence of abundant water is a characteristic of certain regions of the terrestrial globe which is brought to evolve with the changeover in the Anthropocene. These regions, which are accustomed to rainfall and in

B. Valiorgue (✉)
Emlyon Business School, Ecully, France
e-mail: bvaliorgue@em-lyon.com

© The Author(s), under exclusive license to Springer Nature Switzerland AG 2023

N. Wallenhorst, C. Wulf (eds.), *Handbook of the Anthropocene*,
https://doi.org/10.1007/978-3-031-25910-4_177

which we logically find a large part of the agricultural areas, will have to deal with water scarcity. Rarefaction which will considerably complicate agricultural activity when it is not simply going to compromise it. This water scarcity will lead to conflicts of use around a resource that has become increasingly scarce and that some will want to appropriate in order to continue its activities (Hoekstra, 2019).

Heat and Biological Cycles Plants need access to light and certain temperatures in order to develop and grow. The absence of light and too low temperatures are prohibitive. Conversely, overexposure to light and excessively high temperatures are also particularly harmful. The plants mobilized by agricultural activity, and in particular the four main food crops (wheat, corn, rice, soybeans), need a temperate climate and a certain threshold of exposure to light (Lobell et al., 2011). Outside of this spectrum, farming is simply not feasible. The increased heat is redistributing the cards in agricultural production around the world. Underprivileged regions will have to stop farming, hitherto privileged regions will experience drop in yields and new production areas will emerge (Dar & Laxmipathi Gowda, 2013). Finally, heat peaks are particularly difficult to manage for farm animals which are parked inside buildings which are transformed into ovens. Without an adaptation of these buildings and/or a change in practices, we must expect significant production drops and peaks in mortality during heatwave periods (Lengnick, 2015).

Extreme Weather Phenomena and Agricultural Disasters For Anthropocene specialists, the climate will be marked by an increase in the frequency of occurrence of extreme events such as storms, droughts, heat waves, floods (Bonneuil & Fressoz, 2016; McGuire, 2013). When they arise, these events destroy the plantations and wipe out a whole year of work. The regions that will be hit by these events will simply be deprived of crops and will have to meet their food needs by drawing on other production areas. Tensions and volatility on prices are expected given the more frequent occurrence of these extreme events. They will also considerably weaken the economic models of farms because they involve loss of income which is very difficult to compensate for and to ensure (Raffray, 2019). The volatility of the Anthropocene climate will directly reflect on the financial markets, which will be nervous, fluctuating and deeply destabilizing. An abnormal drought in Russia can generate a rise in the price of cereals which will result in famine in Africa.

Atmospheric Change and Nutritional Value In addition to disrupting the biological cycle of plants, the Anthropocene also affects the nutritional value of agricultural products such as cereals and potatoes. Indeed, an increase in the CO2 content in the atmosphere results in a decrease in the protein content of cereals which lose part of their nutritional value. In the absence of new plant varieties, it will therefore be necessary to produce more cereals in order to generate the same protein intake (Lobell et al., 2011). This transformation of the intrinsic quality of plants will also impact industrial chains upstream of agricultural production. Companies in the agri-food sector have indeed calibrated processing tools according to certain nutritional values of agricultural raw materials. This drop in nutritional values will require adjustments to these tools and a potential increase in the cost of food items.

Loss of Biodiversity Agricultural activity is often accused of being the cause of biodiversity loss. In fact, it is the transformations of the Earth system that induce living conditions that are now inappropriate for certain plant and animal species (Pimm et al., 2014). This loss of biodiversity means a considerable depletion of the genetic resources that are the basis of agricultural activity. 75% of the world's food production is generated from twelve plant species and five animal species. The loss of biodiversity means that we are losing animal and plant alternatives that could beneficially replace the species that we have overexploited so far and that are adapted to the geological epoch of the Holocene. Legumes (peas, beans, lentils, peanuts) are, for example, underexploited plant species compared to cereals, although they have the capacity to store nitrogen in their roots and do not need additional inputs. Tubers (potatoes, yams, Jerusalem artichoke, tapioca, cassava) and roots (carrots, radish, rutabaga, turnip, celery) also have under-exploited nutritional potential.

On a more general level, the setting in motion of the Earth system shifts agriculture from a risky activity to an uncertain one. Agriculture is a risky activity in the sense that when carrying out a biological cycle (animal or plant), it is normal to see a certain number of hazards emerge that will negatively impact the activity. The occurrence of illnesses, heat waves or even extreme events are among the risks of the profession. Farmers have learned to face and manage these risks. For example, they have developed in-depth knowledge regarding of plant and animal diseases. They have developed and learned to use care methods (pesticides, fungicides, herbicides, antibiotics, etc.). They have also developed a whole range of institutions which make it possible to smooth the effects of these risks (subsidies, market regulation, storage capacities) in order to limit the phenomena of food shortage. They also have developed insurance mechanisms that guarantee income even if crops or animals have been destroyed by an extreme climatic event or an accident.

Everything changes with the arrival of the Anthropocene because agriculture (re) becomes a fundamentally uncertain activity.

The plants and animals that humans have taken thousands of years to domesticate are less and less adapted to the characteristics of the emerging Earth system. It means two things. On the one hand, the knowledge acquired to master the biological cycles of current plant and animal species is becoming increasingly obsolete because this knowledge is only valid under stable and known pedoclimatic conditions which are those of the Holocene. A new climate in fact means a new behaviour of plants and animals, new pathologies and therefore new practices which must be invented, tested and learned. On the other hand, this new Earth system logically implies to be interested in new varieties of plants or animals more adapted to the new climate and which have not been or very little mobilized. It is therefore necessary to embark on the exploitation of plant and animal cycles, the characteristics of which farmers have little knowledge of. Knowledge of animal and plant species currently used is thus becoming less and less relevant and knowledge of animal and plant species more suited to the Anthropocene era is yet to be built. The Anthropocene creates a new context for agriculture that involves reviewing knowledge and forging new institutions (Valiorgue, 2020).

References

Bonneuil, C., & Fressoz, J.-B. (2016). *L'événement Anthropocène: la Terre, l'histoire et nous*. Seuil.

Dar, W., & Laxmipathi Gowda, C. (2013). Declining agricultural productivity and global food security. *Journal of Crop Improvement, 27*(2), 242–254.

Hamilton, C., Bonneuil, C., & Gemenne, F. (2015). *The anthropocene and the global environmental crisis*. Routledge.

Hoekstra, A. (2019). *The water footprint of modern consumer society*. Routledge.

Lengnick, L. (2015). *Resilient agriculture: Cultivating food systems for a changing climate*. New Society Publishers.

Lobell, D., Schlenker, W., & Costa-Roberts, J. (2011). Climate trends and global crop production since 1980. *Science, 333*(6042), 616–620.

McGuire, B. (2013). *Waking the giant: How a changing climate triggers earthquakes, tsunamis, and volcanoes*. Oxford University Press.

Pimm, S., Jenkins, C., Abell, R., Brooks, T., Gittleman, J., Joppa, L., et al. (2014). The biodiversity of species and their rates of extinction, distribution, and protection. *Science, 344*(6187), 1246752.

Raffray, M. (2019). Gestion des risques climatiques : luxe ou nécessité. *Paysans & Société, 378*, 26–31.

Valiorgue, B. (2020). *Refonder l'agriculture à l'heure de l'Anthropocène*. Le Bord de l'Eau.

Bertrand Valiorgue is Professor at emlyon business school. He has a PhD in management and work more specifically on sustainable issues in the agricultural sector. He is the author of several articles and books and notably *Refonder l'agriculture à l'heure de l'Anthropocène* (Le Bord de l'Eau, 2020).

Bioeconomics

Sylvie Ferrari

Abstract This article examines how the Anthropocene era leads us to reconsider the place of economic activities within the Biosphere. For this purpose, we focus on Nicholas Goergescu-Roegen's approach called Bioeconomics that highlights a new link of the economic system to both nature and time. Such a perspective brings useful insights for humanity that has to irrevocably live within planetary boundaries.

Bioeconomics is a new approach to the relationship between economic activity and the environment that was developed by Nicholas Georgescu-Roegen (1906–1994), a famous economist of the twentieth century who was also a mathematician, philosopher and science historian. It was in 1975 that the concept of bioeconomics first appeared in his work with the article entitled 'Energy and economic myths' (Georgescu-Roegen, 1975). It argued that the economic sphere has to be considered as a subset of the biosphere, thereby providing a key to understanding the historical dimension of human development given the limited access to a stock of resources that can be drawn from nature. This global perspective nourished a new paradigm that was at odds with both the neoclassical view of economics that arose in the middle of the nineteenth century and the mechanistic epistemology of Newtonian physics (Grinevald, 1992; Georgescu-Roegen, 1995; Mayumi, 2001; Missemer, 2013).

Specifically, bioeconomics derived insights from thermodynamics to account for the way in which economic activities transform energy and matter taken from nature. This introduction and generalization of 'the entropy law' in the study of economic phenomena led Nicholas Georgescu-Roegen to build a new evolutionary paradigm in which energy, matter, irreversibility, qualitative changes and historical time play a major role. Numerous writings have subsequently given an account of the biophysical foundations of economics and, in particular, two famous

The original version of the chapter has been revised. A correction to this chapter can be found at https://doi.org/10.1007/978-3-031-25910-4_280

S. Ferrari (✉)
BSE, UMR CNRS and University of Bordeaux, Pessac, France
e-mail: sylvie.ferrari@u-bordeaux.fr

© The Author(s), under exclusive license to Springer Nature Switzerland AG 2023, corrected publication 2024

N. Wallenhorst, C. Wulf (eds.), *Handbook of the Anthropocene*,
https://doi.org/10.1007/978-3-031-25910-4_178

publications: *The entropy law and the economic process* (1971) and *Energy and economic myths* (1976).

The first dimension of bioeconomics is of a physical nature and concerns the introduction of the law of entropy into the economic sphere. According to this law, also called the second principle of thermodynamics, the entropy of an isolated system is continuously increasing. The reference to a closed system implies the emergence of an equilibrium characterized by a state where all energy is dissipated. Dissipation is inevitable and it is this qualitative change in energy that allows us to understand the transformations within economic systems over a long period of time. From a physical point of view, economic activities create nothing and only transform available energy and matter through irreversible dissipation. From an economic point of view, there is a difference between the elements that enter the economic process and those that get out of it: valuable elements characterized by low entropy (resources taken from nature) enter the economic process while worthless elements characterized by high entropy (waste, pollutants) leave it. Entropy is therefore a means of measuring the qualitative difference between useful resources and useless waste. This in turn guides reflection about the very raison d'être of economic processes. In reality, there is an immaterial flow called "the enjoyment of life" which is inseparable from the removal of low entropy from the environment.

To sum up, no economic activity of production or consumption can be dissociated from the laws that apply to the biosphere.

Designating the economic system using the concept of process is not neutral and refers to the second dimension of bioeconomics, the biological dimension. Bioeconomics carries within it the biological origin of the economic process as humans take part in it. The economic process is seen as a continuation or an extension of biological evolution, which is made possible by other means that are no longer endosomatic but exosomatic. Based on the notions introduced by the mathematician and biologist Alfred Lotka, endosomatic evolution refers to the biological evolution of living species while exosomatic evolution refers to the use of detachable organs, external to the person and produced from energy and material resources (tools, technologies...). From the bioeconomic perspective, it is exosomatic instruments that have enabled humanity to undertake increasing withdrawals of low entropy resources from nature. The economic struggle is therefore centred on the incessant search for these scarce resources for the satisfaction of human needs.

It is interesting to note that the evolutionary conception of bioeconomics has its origins in the work of Joseph A. Schumpeter (1954) on development, and in particular on the role played by technical progress in the structural transformation of the economy. At the heart of these transformations, which punctuate changes in the state of the economic process, are technologies. The exosomatic evolution of humanity has thus been accompanied by the increasing production of technologies that rely upon quantities of energy and matter drawn from the earth's stocks of resources. However, the quantities of energy and matter accessible are necessarily finite given the implications of thermodynamics. Not only is the efficiency of energy-mobilizing transformations not unlimited, but the amount of low entropy available in the environment can only be used once by humans. Thus, the power of technical progress is a major factor in the irreversible transformations of nature by societies.

Another implication of exosomatic evolution is that it is accompanied by permanent social conflicts and inequalities between rich and poor countries. According to Nicholas Georgescu-Roegen, the origin of economic inequalities lies in the exosomatic dimension of lifestyles (Georgescu-Roegen, 1977b, 1978): social division generates social conflicts through the appropriation of technologies. At the global level, inequalities arise from the control of the exploitation of natural resources by a few countries to the detriment of other less developed countries.

Bioeconomics sheds new light on the spatial and temporal organization of economic activities that are intrinsically linked to the elements of the biosphere and the laws that order them. Economics, as a human activity motivated by the satisfaction of needs, constantly draws upon and depletes the resources of the biosphere whose quantities (stocks and flows) and qualities are irreversibly altered.

But climate change, as well as the depletion and rarefaction of natural resources, the accumulation of pollution and environmental degradation are all major challenges for societies today (Angus, 2018; Magny, 2019; Steffen et al., 2011). Entering into the Anthropocene era invites us to reconsider more particularly the place of economic activities and, through them, the sustainability of the economic processes at stake in order to redefine trajectories to keep within planetary boundaries. These limits refer to critical thresholds for the main biophysical variables (carbon dioxide, nitrogen, phosphorus, among others) that guide the climate and the biosphere and underpin social well-being (Barnosky et al., 2012; O'Neill et al., 2018; Steffen et al., 2015). Such a context inexorably leads us to rethinking the links between the biosphere and economic activities. Bioeconomics is more relevant than ever.

Within this context, bioeconomics can generate a new perspective upon the foundations of the ecological crisis that is now endangering contemporary societies. Enriched by the dual heritage of evolutionary biology and thermodynamic physics, the economic process is changing in nature and meaning: the energetic and material side is related to the law of entropy, while the immaterial side translates into the flow of the enjoyment of life! By considering the dialectical nature of the economic process, Nicholas Georgescu-Roegen steers the analysis of the development of societies from a historical, ecological, cultural and technological perspective (Georgescu-Roegen, 1971). In a related way, dynamic time with its reversible trajectories is abandoned in favour of historical time, which carries within it irreversibility. This conception of time, which definitively links successive generations together, must lead to the imposition of ecological and ethical constraints before any economic choice is made. If not, it will not be possible to stop the acceleration of the process of dissipation of energy and matter that has increased since our entry into the Anthropocene era. From the perspective of historical time, degrowth, like the renunciation of any possibility of infinite growth in a finite world, is inevitable (Georgescu-Roegen, 1976): a 'declining state' is a necessity.

Bioeconomics invites us to reconsider economics as a new relationship to nature and time. If the purpose of economic activity is the satisfaction of human needs and therefore ultimately the conservation of the human species over a long period of time, then some principles of ethics and justice must be introduced into economics (Georgescu-Roegen, 1970). Nicholas Georgescu-Roegen proposes a detailed bio-economic program to save resource stocks in order to increase the life span of humanity (1975, 1977a). To achieve this, the minimization of flows to maintain the stocks needed to meet the needs of successive generations and the redistribution of nature's resources on an intragenerational scale are both necessary.

Thus, in the face of planetary boundaries, the philosophical scope of bioeconomics reveals the need for a new ethics, an ethics of limits, to rethink the links between the transformation of nature and the evolution of societies in the Anthropocene. First of all, it is necessary to adopt a holistic ecological approach that questions the habitability of the biosphere. There is in Nicholas Georgescu-Roegen's work an ecocentric environmental ethics that is open to the biosphere and where moral actions must aim at its protection over time. The idea of the permanence of the biosphere is reminiscent of the Gaia hypothesis developed in the work of James Lovelock (1979). One can also note the importance given to interdependence as an expression of solidarity between man and nature, a solidarity that is also present in the environmental ethics of the German philosopher Hans Jonas (1993/1979). Moreover, the ethics of limits also implies moving towards an economy of sufficiency for access to a good life for all. This is necessarily a condition for the prior redistribution of wealth on the scale of the biosphere in order to be able to distribute goods and ills more fairly among the inhabitants of the planet. The implementation of resource conservation strategies to preserve the quality of life of future generations is a path that implies limiting the needs, but not the well-being, of the better-off present generations. This perspective is a possibility within the framework of bioeconomics that is fully compatible with various approaches to a degrowth society (Van den Bergh & Kallis, 2012) or with "doughnut economics" following K. Raworth's work (2017). These different tracks can contribute to more justice and allow a good life for all while respecting planetary boundaries.

Basically, the real anthropological challenge for economics in the twenty-first century is to redefine its ultimate purpose – the conservation of the human species over a long period of time – while remaining in solidarity with the biosphere, the living and inanimate elements that inhabit it, in accordance and compliance with the biophysical principles that govern nature.

References

Angus, I. (2018). *Face à l'Anthropocène. Le capitalisme fossile et la crise du système terrestre*. Ecosociété.

Barnosky, A. D., Hadly, E. A., Barnosky, J., Berlow, E. L., Brown, J. H., Fortelius, M., Getz, W. M., Harte, J., Hastings, A., & Marquet, P. A. (2012). Approaching a state shift in Earth's biosphere. *Nature, 486*(7401), 52–58.

Van den Bergh, J., & Kallis, G. (2012). Growth, A-growth or degrowth to stay within planetary boundaries? *Journal of Economic Issues, 46*(4), 909–919.

Georgescu-Roegen, N. (1970). *La science économique. Ses problèmes et ses difficultés*. Dunod.
Georgescu-Roegen, N. (1971). *The entropy law and the economic process*. Harvard University Press.
Georgescu-Roegen, N. (1975). Energy and economic myths. *Southern Economic Journal, 41*(3), 347–381.
Georgescu-Roegen, N. (1976). *Energy and economic myths*. Pergamon Press.
Georgescu-Roegen, N. (1977a). What thermodynamics and biology can teach economists. *Atlantic Economic Journal, 5*(1), 13–21.
Georgescu-Roegen, N. (1977b). Inequality, limits and growth from a bioeconomic viewpoint. *Review of Social Economy, 25*, 361–375.
Georgescu-Roegen, N. (1978). De la science économique à la bioéconomie. *Revue d'Economie Politique, 88*(3), 37–382.
Georgescu-Roegen, N. (1995). In J. Grinevald & I. Rens (Eds.), *La Décroissance. Entropie, Écologie, Économie*. Éditions Sang de la terre.
Grinevald, J. (1992). La révolution bioéconomique de Nicholas Georgescu-Roegen. *Stratégies énergétiques, Biosphère et Société*, 23–34.
Jonas, H. (1993). *Le principe responsabilité, une éthique pour la civilisation technologique* (Translation of *Das Prinzip Verantwortung* (1979), 3rd edition). Editions du Cerf.
Lovelock, J. (1979). *Gaia: A new look at life on earth*. Oxford University Press.
Magny, M. (2019). *Aux racines de l'anthropocène. Une crise écologique reflet d'une crise de l'homme*. Edition Le Bord de l'eau.
Mayumi, K. (2001). *The origin of ecological economics. The bioeconomics of Georgescu-Roegen*. Routledge.
Missemer, A. (2013). *Nicholas Georgescu-Roegen: Pour une révolution bioéconomique*. ENS Éditions.
O'Neill, D. W., Fanning, A. L., Lamb, W. F., & Steinberger, J. K. (2018). A good life for all within planetary boundaries. *Nature Sustainability, 88*(1), 88–95.
Raworth, K. (2017). *Doughnut economics: Seven ways to think like a 21st-century economist*. Random House Business.
Schumpeter, J. A. (1954). *History of economic analysis*. Allen & Unwin.
Steffen, W., Grinevald, J., Paul, C., & McNeill, J. (2011). The Anthropocene: Conceptual and historical perspectives. *Philosophical Transactions of the Royal Society A: Mathematical, Physical and Engineering Sciences, 369*(1938), 842–867.
Steffen, W., Richardson, K., Rockström, J., Cornell, S. E., Fetzer, I., Bennett, E. M., Biggs, R., Carpenter, S. R., de Vries, W., de Wit, C. A., Folke, C., Gerten, D., Heinke, J., Mace, G. M., Persson, L. M., Ramanathan, V., Reyers, B., & Sörlin, S. (2015). Planetary boundaries; Guiding human development on a changing planet. *Science, 347*(6223), 1259855-1-12.

Sylvie Ferrari is Full Professor in Economics at the University of Bordeaux in France. She is a member of the Research Unit Bordeaux School of Economics (BSE - UMR CNRS affiliated research unit). She holds a PhD in Economics (University of Toulouse, France, 1992). Her research falls within the field of ecological economics. Her most recent publications include: (2021), 'N. Georgescu-Roegen's Bioeconomic Ethics, *Cahiers d'Economie Politique*, 2021/1 (n° 79), pp. 213–242 (*in French*); (2017), 'A functional approach of ecosystems for a sustainable management of coastal wetlands in the Gironde Estuary region (France)', in *European Union and Sustainable Development, Challenges and Prospects*, Diemer A. et al. ed., Oeconomia, p. 95–119. (with S. Lavaud S. and J.-C. Pereau).

Bioeconomy

Christine Rösch

Abstract This article reflects the bioeconomy in the Anthropocene, based on the definition given by Georgescu-Roegen in his book 'The Entropy Law and the Economic Process' (1971). He criticised neoliberal economists for ignoring the fact that an ever-increasing use of resources will lead to the exhaustion of the Earth's capacities. To change the economic practices, he developed a bioeconomies approach linked to degrowth (La Décroissance 1979). Today, it is turned into its opposite as the bioeconomy promises efficient and consistent growth by changing the resource basis from finite and fossil resources to biological ones. The bioeconomy is said to be the economic concept of the future, which can contribute to a 'good life for all within the planetary boundaries' (Georgescu-Roegen Nicholas. (1971). *The entropy law and the economic process*. Harvard University Press and De Gruyter. ISBN 9780674281653) However, with its current orientation towards growth, the bioeconomy is exacerbating ecological and social crises. This article proposes to take the 'sustainable bioeconomy' concept seriously, which was first developed by Hans Carl von Carlowitz in his book 'Sylvicultura oeconomica' (1713) to prevent this abuse.

The Earth has entered the Anthropocene – a potentially new geological epoch in which humanity is the dominant creative force on the planet. Proof of this is that the so-called 'anthropogenic mass' of man-made materials is growing exponentially. The mass of everything man-made – from concrete footpaths and glass and metal skyscrapers to plastic bottles, clothes and computers – is now roughly equal to the mass of all living things on Earth (Elhacham et al., 2020). A growing population and increasing demand for food, energy and materials continue to put a strain on the planet's finite resources. The need for solutions to improve resource and energy efficiency, reduce waste and provide alternatives for scarce resources has never been greater. These trends are evolving against a backdrop of continued growth in global

C. Rösch (✉)
Karlsruhe Institute of Technology, Karlsruhe, Germany
e-mail: christine.roesch@kit.edu

© The Author(s), under exclusive license to Springer Nature Switzerland AG 2023

N. Wallenhorst, C. Wulf (eds.), *Handbook of the Anthropocene*,
https://doi.org/10.1007/978-3-031-25910-4_179

emissions, which has led to heightened debates around climate change and possible solutions. Niko Paech proposes a post-growth theory as an alternative economic system to avoid the ecological damage caused by a growth-based economy (Paech, 2015).

The recognition for a more efficient, consistent and sufficient use of resources and the need to replace existing economic concepts and activities based on fossil and finite resources with renewable resources is not new. The 'optimistic' thinking about the infinite substitutability of exhaustible resources to satisfy human needs through 'technological progress', which implies and legitimise the mania for growth, was already criticised by the US economist Nicholas Georgescu-Roegen (1906–1994). Given the limited access to a stock of the biosphere's resources that can be drawn from nature and its ongoing depletion and qualitative change, Georgescu-Roegen developed the first theory about bioeconomies as the relationship between human economic activity and the environment. In his book 'The Entropy Law and the Economic Process', published in 1971, he draws parallels between biological and economic processes. Both processes are open systems and must fulfil the second law of thermodynamics. According to this law, the entropy of a closed system is continuously increasing. Valuable elements characterised by low entropy (resources taken from nature) enter the economic process while worthless elements of high entropy (waste, pollutants) leave it. Thus, energy can never be returned to its original state after transformation processes have taken place.

The bioeconomy concept from Georgescu-Roegen sheds new light on the spatial and temporal organisation of economic activities that are intrinsically linked to the elements of the biosphere and the laws that order them. The bioeconomy concept replaces the linear, perpetual-growth models of mainstream economics with a (primarily, but not exclusively) circular model, which comprehensively considers the ecological conditions of all economic processes. From the bioeconomic perspective, exosomatic instruments have enabled humanity to undertake increasing withdrawals of low entropy resources from nature. According to the mathematician and biologist Alfred Lotka (1880–1949) exosomatic evolution refers to detachable organs external to the person and produced from energy and material resources (e.g. tools, technologies). This development is linked to the biological evolution of living species. For thousands of years, the growing human race secured its food supply primarily by expanding arable and pasture land. That changed with the development of exosomatic scientific and technical tools, which enabled the Industrial Revolution: synthetic fertilisers, machines and breeding successes led to a highly mechanised, industrial agriculture with enormous yields. The exosomatic evolution of humanity has thus been accompanied by the increasing production of technologies that rely upon quantities of energy and matter drawn from the Earth's stocks of resources. Through digitalisation and artificial intelligence, the (bio)economy, the ongoing search for exosomatic enabling the satisfaction of human needs is accelerated. The power of technical progress is considered a significant factor in the irreversible transformations of nature by societies.

In contrast to the bioeconomy concept of Georgescu-Roegen, linked to the degrowth concept, the present interpretation of the bioeconomy concept holds the promise of sustainable growth and consumption despite the continuing growth of the world's population and scarcity of resources. This understanding is based on the belief in technological progress and relies on the development and convergence, i.e. integration, of biological, biotechnological and digital innovations. These peculiar dynamics of human economic activity are contrasted with the control principles of nature and the concept of sustainable development and planetary boundaries. Long before the great studies of the Club of Rome and the Brundtland Report of 1987, the demand for a 'sustainable' bioeconomy as a strategy for sustainable development was first recognised by Hans Carl von Carlowitz and published 1713 in his book 'Sylvicultura Oeconomica'. His concept of sustainable forest management implies that not more wood is harvested than will grow back again. His findings are based on his experiences as chief mining captain responsible for maintaining the mining industry as an essential motor for the prosperity of the Saxon state. However, the mining industry consumed vast quantities of wood as mine timber and firewood, and the rapidly increasing population demanded vast amounts of wood. The forests in the region were subsequently exploited for decades without any restraint to satisfy this demand. The procurement of wood became the bottleneck in the supply of raw materials for the critical mining industry. To secure the future supply of raw wood for the coal and steel region, Carlowitz recommended a 'continuous, steady and sustained use' of the exploited forests and, due to the decades-long growth of forests, embedded this in an intergenerational approach. In this view, this bioeconomic principle can serve the community's welfare and obligates the economies to treat nature with care and respect future generations' needs and intergenerational responsibility. More than 300 years later, in 1987, the Brundtland Report 'Our Common Future' further developed this concept to a general definition that 'sustainable development is the development that meets the needs of the present without compromising the ability of future generations to meet their own needs.' (Hauff, 1987).

The modern 'sustainable bioeconomy' concept refers to a knowledge-based, sustainable, efficient and cascading use of biological resources such as plants, animals, and microorganisms. It is considered a catalyst for systemic change with new ways of producing and consuming resources while respecting the planetary boundaries and moving away from a linear economy based on extensive use of fossil and mineral resources to a circular bioeconomy. This new bioeconomy narrative paints the picture of a future economy based on renewable energies and biological resources that will deliver 'green' economic growth and enable modern societies to phase out fossil fuels and build a sustainable future (Fukase & Martin, 2020). However, there are doubts that growth-based bioeconomies can be made sustainable by decoupling environmental throughput and gross domestic product (Krätzer & Franz-Theo, 2014). Likewise, there is a critical debate over whether the transformation of modern societies toward bio-based economic activities involves overcoming unlimited economic growth. Bio-based economies rely on materials whose availability is subject to biophysical limits and cyclical regenerative processes that cannot be expanded or accelerated at will. It is uncertain whether the accumulation of ever-increasing amounts of energy and goods can continue in an economy based on renewable biological resources.

The break with the logic of accumulation, extraction and expansion and the shift from fossil to renewable biological resources may be a starting point for a fundamental transformation of modern societies. The social organisation of work and care activities, consumption patterns and people's mindsets might change, or it might become clear that they need to change towards sustainable development. This shift in attitude has already begun and was triggered by the depletion of natural resources and environmental degradation, particularly climate change and biodiversity loss, which are major challenges for societies. Entering into the Anthropocene, the bioeconomy concept requires responsible, resilient, and sustainable use of resources for economic processes to stay within the planetary boundaries. Johan Rockström et al. (2009) identified nine processes that regulate the stability and resilience of the Earth system and proposed quantitative planetary boundaries for these processes within which humanity can continue developing and thriving for generations to come. These limits refer to critical thresholds for the main biophysical variables (carbon dioxide, nitrogen, phosphorus, among others) that guide the climate and the biosphere and underpin social well-being. Crossing these boundaries increases the risk of generating large-scale abrupt or irreversible environmental changes. With a view to the planetary boundaries, it has to be noted that the bioeconomy is not per se better for the environment because it uses biological rather than fossil resources. Therefore, the concept of bioeconomy must be aligned with sustainable development criteria concerning land use and consumption patterns. In addition, rapid, concerted and sustained changes in lifestyle and resource use that cut across all levels of society and the economy are required. At the same time, the concrete actions of key players in politics, science and industry seem to assume that the bioeconomy will allow them to continue with business as usual. Still, it is unclear if the transition to bioeconomy strengthens, transforms, or undermines the growth paradigm and if overcoming it is necessary for a socially viable 'bio-based society'. The concept of bioeconomy can shift mentalities, social structures and society-nature relations. This transformation will create new social conflicts and 'winners' and 'losers', altering existing social inequalities at the local, national, and global level and leading to new dependencies between the global South and North.

A long-term governance framework for the bioeconomy is needed limiting, regulating and prioritising the sustainable and resilient use of biological resources and avoiding economic valorisation of all living things and the economisation of even the last remnants of untouched nature (Wagner, 2015). Since land is the quarry and the factory for most bioeconomy processes, the growing demand for non-food biomass is increasingly competing with the land requirements for food security and biodiversity conservation. In the Anthropocene, humankind has fundamentally transformed land use and the terrestrial biosphere to increase the area of agricultural land and pasture. Around the year 1700, most of the world's land was still primarily in its natural state; today, only about 23% of the global land surface can be designated as wilderness (WBGU, 2020). As convincingly shown by global assessments and progress reports in recent years, humanity is increasingly destroying this natural life-support system, partly due to the growing global demand for land and terrestrial ecosystem services. The pressure on terrestrial ecosystems from overuse and competition for use has never been more significant than it is today. The way land use is managed by agriculture to provide feedstock for the bioeconomy has massive and irreversible impacts on the environment, particularly on climate change and extinction of biological diversity. Yet, the land and its biologically productive ecosystems are under more pressure than ever before. Around a quarter of the Earth's ice-free land surface is affected by human-caused degradation. There is a risk that the shift to sustainable and resilient bioeconomies could accelerate these degradation processes. The response to this is again the intensified search for new exosomatic instruments, such as the cultivation of algae, bacteria and other microorganisms in technical systems on non-arable land, contributing to highly efficient, resilient circular production and the valorising of wastes (Rösch et al., 2018; Roßmann & Rösch, 2020). Doubts are justified as to whether the massive and partly irreversible impacts of man-made economic activities in the Anthropocene can be sufficiently reduced by innovations and new technologies to implement the sustainable concept of bioeconomy on a global scale while the world population will increase from 7.4 billion in 2017 to 9.7 billion in 2050 (United Nations, 2019). This growth, mainly based on developing countries and changes in income distribution will raise world food demand by 2050 by about a third and have potentially significant implications for world agriculture and the bioeconomy (Cirera & Masset, 2010; Fukase & Martin, 2016). The demand for agricultural resources will grow as diets shift from starchy staples towards animal-based products and fruits and vegetables (Godfray, 2011).

The dietary upgrading of developing countries imposes more significant burdens on agricultural resources since the production of more diversified, particularly animal-based, diets requires much more agricultural output than plant-based diets. Beef production requires 15 times more land than cereals production per kilogram per year (Gerbens-Leenes & Nonhebel, 2002).The task of feeding a growing world population has historically been accompanied by a continuous increase in land-intensive dietary habits. As a result, there are warnings against growing global competition for land use. Since land-use and agricultural activities are the core elements of the bioeconomy, maximising food production for feeding a rapidly growing human population while minimising critical resource use and soil quality degradation is a major challenge for global sustainability. Low-input sustainable agriculture and digitalisation of agriculture are proposed to reduce environmental trade-offs of agricultural production and enhance agricultural systems' resilience. Once again, it is not the change in consumption patterns but technical progress that is expected to solve the problem of resource scarcity and the decoupling of resource use from environmental degradation. This brings us back to Georgescu-Roegen's criticism of human economic activities in 1971. The need to change consumption patterns is considered to be unattainable from a socio-economic perspective. Instead, the confidence that technological progress can solve the problem of resource scarcity and the decoupling of resource use from environmental degradation continues to be expressed by politicians and industry.

References

Cirera, X., & Masset, E. (2010). Income distribution trends and future food demand. *Philosophical Transactions of the Royal Society, B, 365*, 2821–2834. https://doi.org/10.1098/rstb.2010.0164

Elhacham, E., Ben-Uri, L., Grozovski, J., et al. (2020). Global human-made mass exceeds all living biomass. *Nature, 588*, 442–444. https://doi.org/10.1038/s41586-020-3010-5

Fukase, E., & Martin, W. (2016). Who will feed China in the 21st century? Income growth and food demand and supply in China. *Journal of Agricultural Economics, 67*(1), 3–23.

Fukase, E., & Martin, W. (2020, August). Economic growth, convergence, and world food demand and supply. *World Development, 132*, 104954. https://doi.org/10.1016/j.worlddev.2020.104954

Georgescu-Roegen Nicholas. (1971). *The entropy law and the economic process*. Harvard University Press and De Gruyter. ISBN 9780674281653.

Gerbens-Leenes, P. W., & Nonhebel, S. (2002). Consumption patterns and their effects on land required for food. *Ecological Economics, 42*(1–2), 185–199.

Godfray, H. C. (2011). Food for thought. *Proceedings of the National Academy of Sciences, 108*(50), 19845–19846.

Hauff, V. (Hrsg.). (1987). Unsere gemeinsame Zukunft. Der Brundtland-Bericht der Weltkommission für Umwelt und Entwicklung, Eggenkamp, Greven, ISBN 9783923166169.

Krätzer, A., & Franz-Theo, G. (2014). Irrweg Bioökonomie. edition unseld 51, Suhrkamp Verlag Berlin.

Paech, N. (2015). Befreiung vom Überfluss. Auf dem Weg in die Postwachstumsökonomie, 11. Auflage, München, oekom Verlag.

Rockström, J., Steffen, W., Noone, K., Persson, Å., et al. (2009). A safe operating space for humanity. *Nature, 461*, 472–475. https://doi.org/10.1038/461472a

Rösch, C., Roßmann, M., & Weickert, S. (2018). Microalgae for integrated food and fuel production. *Global Change Biology / Bioenergy, 11*(SI, 1), 326–334. https://doi.org/10.1111/gcbb.12579

Roßmann, M., & Rösch, C. (2020). Key-narratives of microalgae nutrition: Exploring futures through a public Delphi survey in Germany. *Science and Public Policy, 47*(1), 137–147. https://doi.org/10.1093/scipol/scz053

United Nations, Department of Economic and Social Affairs, Population Division. (2019). World Population Prospects 2019: Highlights (ST/ESA/SER.A/423). https://population.un.org/wpp/publications/files/wpp2019_highlights.pdf

von Carlowitz, H. C. (1713). Sylvicultura oeconomica. Hausswirthliche Nachricht und Naturmäßige Anweisung zur Wilden Baum-Zucht. Faksimile der Erstauflage. Leipzig.

Wagner, H.-G. (2015). Bioökonomie – Über die Pervertierung eines grünen Paradigmas. ZfSÖ Zeitschrift für Sozialökonomie, 52. Jahrgang, Folge 186/187, Oktober 2015 ISSN 0721-0752, S.57–69.

WBGU – German Advisory Council on Global Change. (2020). Rethinking land in the Anthropocene: From separation to integration. BerlinWestlund, Hans, Nilsson Pia. Agriculture's transformation and land-use change in a post-urban world: A case study of the Stockholm region. *Journal of Rural Studies*. Available online 27 July 2019. https://doi.org/10.1016/j.jrurstud.2019.07.002.

Dr. Christine Rösch is Head of the Research Group Sustainable Bioeconomy at the Institute for Technology Assessment and Systems Analysis (ITAS) of the Karlsruhe Institute of Technology (KIT). She is trained in Agriculture Biology and has a PhD in Agricultural Policy. She led many European and national inter- and transdisciplinary research projects. She is a well-known expert on the technology and sustainability assessment of scientific and technological innovations and developments in the field of bioeconomy and a member of bioeconomy advisory boards. She is the author of many peer-reviewed bioeconomy articles with high impact factors and the books "Technikfolgenabschätzung – Handbuch für Wissenschaft und Praxis" (2021, in German) and "Bioeconomy for Beginners" (2020) and "Bioökonomie im Selbststudium: Nachhaltigkeit und ökologische Bewertung" (2020, in German).

Capital

Pierre-Louis Choquet

Abstract In this article, we emphasize the relevance of the concept of capital in order to describe the mode of organizing socio-economic relations that has become globally dominant since the industrial revolution. Capital, we suggest, is neither a physical thing nor a mere idea – but a concrete process, in which a variety of agents (human and nonhuman alike) are dynamically enrolled, legally reframed as assets, and subjected to a pecuniary valuation. Insofar as it relates the actual value of these assets to their expected income-yielding capacity, capital thus instantiates a temporal strategy that gives consistency to the future in order to preempt and sell its potentialities in the present. Relations of production and exchange are thus entirely organized in order to deliver these future promises, and their spatialization appears mostly to be a by-product, entirely constrained by the temporal demands of investment. Propelled by the consumption of fossil fuels, the unfolding of capital has become a phenomenon of planetary proportions – thus pushing the entire Earth system to the brink.

The concept of capital (in Latin, *capitale*) emerged in the urban centers of the Latin West, in a cultural universe still marked by the prohibition of usury. As Italian merchants increasingly engaged in overseas trade throughout the western Mediterranean world, capital came to designate the dynamic form of money that was invested in their limited partnerships (*commenda*). Since the undertaking of such risky commercial enterprises was deemed beneficial to the wider community, the financial interests paid on the initial deposit were judged licit (Piron, 2020). In Europe, the habit of examining possible futures and rating the levels of risks associated with them thus accompanied the slow emergence of a monetarized sphere of exchanges, which induced a profound transformation of mental universes.

The successive waves of Iberian and Dutch colonization significantly contributed to expanding the scale of the process – which remained highly fragile, and

P.-L. Choquet (✉)
Centre de Sociologie des Organisations, SciencesPo, Paris, France
e-mail: pierrelouis.choquet@sciencespo.fr

© The Author(s), under exclusive license to Springer Nature Switzerland AG 2023

N. Wallenhorst, C. Wulf (eds.), *Handbook of the Anthropocene*,
https://doi.org/10.1007/978-3-031-25910-4_180

fraught with contingencies. As they initiated the transatlantic slave trade to develop sugarcane plantations in the Americas, the Spanish and the Portuguese were among the first to systematically isolate, abstract, and replicate the income-yielding capacities of homogenized landscapes – hence reframing their generative potentialities (significantly enhanced by coerced labor) as subservient moments in cycles of investment. From very early on, capital did not, therefore, designate a mere economic process, but also a way of assembling the social fabric and of inserting it dynamically into the wider natural environment (Hornborg, 2017). While drawing on this model, the Dutch, in turn, innovated by aggregating the variegated branches of their colonial ventures under a unified legal entity, the Dutch East India Company, which had its equity locked-in and its shares traded on the open market: these did not represent the expected gains from such or such discrete project, but rather the expected capacity of an organization to secure durable streams of income. Behind this additional layer of institutional sophistication (which enabled to scale up equity by pooling money from a multitude of savers), the underlying logic of capitation remained the same: at the end of value chains, generative potentialities were seized and turned into legal assets.

As European colonial superpowers extended their domination to the four corners of the earth, it is in Great Britain that the spark of the industrial revolution triggered a new metamorphosis of capital. At the end of the eighteenth century, the domestication of the energy potential of fossil fuels (i.e., through the use of internal combustion engines) suddenly allowed to relativize the immemorial constraints that eco-productive spaces had placed on human activities, by rendering available relics of solar energy coming from outside space and time. The materiality of coal allowed to concentrate factories in urban centers: its massive use reconfigured the spatiality of relations of production and exchange, in a sense that simultaneously eroded the bargaining power of organized labour and accrued the leverage of capital owners (Malm, 2016). At the same time, British corporate law went through a series of much debated reforms which gradually codified entity shielding, asset lock-in, and limited liability as the baseline prerogatives of the modern firm (Ciepley, 2013). After years of trial and error, it became increasingly clear that the incorporation of businesses (which brought about an unprecedented experience in risk-shifting, from the private to the public) unleashed economic growth. This legal innovation was copied by most industrializing economies as soon as they stabilized in the second half of the nineteenth century, allowing the capital process to scale up: its unfolding was not anymore the doing of individual entrepreneurs – it now occurred at the level of large business organizations developing tight relationships with state administrations and benefiting from the services of an increasingly structured finance sector (commercial banks, insurance) (L'Italien, 2016). In a few decades, the concomitant rise of material and immaterial technologies (steam machines and corporate law, to put it short) contributed to making the future more predictable, thus leading to the industrializing of its colonization. Around the middle of the twentieth century, the growing importance of oil (in addition to coal) in the global energy system anchored further this dynamic: as its steady supply short-circuited further social conflicts and made exchanges more fluid and predictable, the generalization of calculative practices gradually effectuated "the national economy" as a specific entity to be managed (Mitchell, 2014). In the wake of World War II, nearly all countries embarked one after another on the road to fossil-fuelled powered industrial development: the removal of trade barriers catalyzed a restructuration of production networks across continents, thus creating interdependencies that were further reinforced by the construction of integrated financial markets in the 1980s. With global volumes of liquidities skyrocketing, a wider range of free generative potentialities started to be routinely scrutated, and legally reframed as appropriable income-yielding entities. In the last decades, the deployment of the capital process has kept reaching new terrains, subjecting entire segments of the natural world to its logic of investment and return: in particular, the spatial appropriation and commidification of the land reframed its existence as conditional to the temporal promise of pecuniary valuation. When it takes control of the present in the name of the future (e.g., by organizing the siphoning off of natural resources in exchange of planned cashflows), capital obliquely manifests the depth of Hegel's claim that "the truth of space is time". In this respect, it operates as a secularized *eschaton* – one that actualizes a particular version of the commandment to "be in the world without being of it", thus disvaluing space as the realm of vulgar immanence (Viveiros de Castro & Danowski, 2020).

The journey from the urban centers of thirteenth-century Italy to the variegated marketplaces of today's globalized economy was, of course, not written in advance. But even though the capital process emerged in a highly-specific historical context, it is quite clear that its fundamental logic – that of getting a hold on a radically uncertain future – did strike an archaic chord in the human condition, thus showing a remarkable capacity to adapt contingent socio-cultural contexts. The past eight hundred years have therefore been marked by countless series of mutations, ramifications, recompositions, which have gradually turned capital into a polymorphic, mind-defeating phenomenon, expanding at multiple scales at the same time – permeating the smallest pores of our lives while shaping the trajectories of entire societies.

For these reasons, some have considered capital a confusing *non sequitur*, an idealized abstraction imbued with metaphysical overtones, a scapegoat devised by its critics only to be demonized. From a critical realist standpoint, however, we suggest capital denotes a deeper level of reality: neither a first principle lying behind its constitutive agencies, nor a teological attractor situated beyond them, it rather consists of a relational process that supervenes upon them. Just like a Van Gogh painting is something more – yet nothing less – than the colour pigments that are laid on the canvas, capital is something more – yet nothing less – than its constitutive agencies (on supervenience, see List & Pettit, 2013): as such, it therefore animates both material and ideational components. This, we suggest, is certainly the middle-ground view that Marx had in mind when he claimed that "capital is not a thing, but a social relation between persons, established by the instrumentality of things" (1992: 932). Its transient substance is therefore "unsubstantial", entirely dispensed through relations – thus fully disclosed and non-mysterious. Yet these relations cannot be made intelligible outside of their processual deployment, which unfolds in time: as Marx underlines, it is indeed fundamentally in history that capital "preserves itself [...] by constantly multiplying itself", which it does by "subordinating all elements of society to itself, or in creating out of it the organs which it still lacks." (1993: 270; 278). This salience of this temporal motif authorizes to construe capital as proceeding from a logic of investment (rather than of production or of exchange): its propulsive dynamism does not come from the replication of the past, but from the anticipation of the future.

This is, at least, what the enormous growth of credit suggests. When a line of credit is created, it is indeed nothing but a form without a content, in want of being filled. Repaying the debt (principal and interests) thus requires putting to work a variety of generative potentialies and taking control of them in order to secure a stable stream of income – often with only secondary concern for the social and environmental costs incurred. With the unfolding of the capital process, it is therefore "not so much the past, but the future that weighs on the brains of the living" (Levy, 2017), as these are forced to hold the diffracted promises that make up the economy in which they are all entangled.

To conclude, capital instantiates a temporal strategy that gives consistency to the future in order to preempt and sell its potentialities in the present. Guaranteeing the possibility of a fluid, continuous back and forth movement between the "here and now" and the "yet to come" appears as a limiting condition to the realization of the promise of valuation. A necessary correlate of the unfolding of the capital process (that is, one that proceeds organically from it) is therefore the endless expansion of a blind logic of control, which authorizes to construe capital not as a mere mode of investing, producing, or consuming – but more generally as a mode of exerting power (Nitzan & Bichler, 2009).

References

Ciepley, D. (2013). Beyond public and private: Toward a political theory of the corporation. *American Political Science Review, 107*(01), 139–158. https://doi.org/10.1017/S0003055412000536

Hornborg, A. (2017). *Global magic: Technologies of Appropriation from ancient Rome to wall street*. Palgrave Macmillan.

L'Italien, F. (2016). *Behemoth capital*. Nota Bene.

Levy, J. (2017). Capital as process and the history of capitalism. *Business History Review, 91*(3), 483–510. https://doi.org/10.1017/S0007680517001064

List, C., & Pettit, P. (2013). *Group agency: The possibility, design, and status of corporate agents*. Oxford University Press.

Malm, A. (2016). *Fossil capital: The rise of steam power and the roots of global warming*. Verso.

Marx, K. (1992). *Capital: Volume 1: A critique of political economy*. Penguin Classics.

Marx, K. (1993). *Grundrisse: Foundations of the critique of political economy*. Penguin Classics.

Mitchell, T. (2014). Economentality: How the future entered government. *Critical Inquiry, 40*(4), 479–507. https://doi.org/10.1086/676417

Nitzan, J., & Bichler, S. (2009). *Capital as power: A study of order and creorder*. Routledge.

Piron, S. (2020). *L'Occupation du monde : Tome 2, Généalogie de la morale économique*. Zones Sensibles.
Viveiros de Castro, E., & Danowski, D. (2020). The past is yet to come. *e-flux* Journal, no #114 (décembre). https://www.e-flux.com/journal/114/364412/the-past-is-yet-to-come/

Pierre-Louis Choquet is a post-doctoral researcher at the Centre de Sociologie des Organisations (Sciences Po), after having earned its PhD at the School of Geography & the Environment (University of Oxford).

Degrowth

Timothée Parrique

Abstract This article examines the idea of degrowth, a concept in political ecology used to envision a democratically planned downscaling of production and consumption in affluent regions of the world as a means to avoid ecological breakdown, decrease inequality, and improve quality of life. Since its inception at the beginning of the 2000s in France, the idea of degrowth has sparked a worldwide social movement which has revamped critiques of capitalism, globalization, and modernity. To better grasp its contours, I synthesize the paradigm of degrowth in four essential features: a resource-saving macroeconomic diet, the abolition of extreme wealth, a redirection of economic activities towards concrete needs, and a decentralization of power.

If critiques of economic growth date back as far as the 1960s and 1970s with books like *The Costs of Economic Growth* (1967), *Small is beautiful* (1973), and most famously, *Limits to growth* (1972), the concept of degrowth as understood today was born in France at the beginning of the 2000s. In 2002, a group of anti-consumerist activists coined the term "sustainable degrowth" to problematize the association between sustainable development and economic growth. The provocative slogan caught on and spread among critical activists and scholars in France, Belgium, Québec, Switzerland, Catalonia, and Italy, leading to a number of initiatives, from symposiums and publications to diverse social and political formations. In 2008, "décroissance" became "degrowth" as the first degrowth international conference took place in Paris. Seven international conferences later, degrowth has become a proper field of academic studies as well as an international social movement with organizations such as Research & Degrowth in Spain, Polémos in Québec, the Movimento per la Decrescita Felice in Italy, or la Maison Commune de la Décroissance in France.

T. Parrique (✉)
Stockholm Resilience Centre, Stockholm University, Stockholm, Sweden

School of Economics and Management, Lund University, Lund, Sweden
e-mail: timothee.parrique@fek.lu.se

© The Author(s), under exclusive license to Springer Nature Switzerland AG 2023

N. Wallenhorst, C. Wulf (eds.), *Handbook of the Anthropocene*,
https://doi.org/10.1007/978-3-031-25910-4_181

There are now around 600 academic articles in English on the topic and a few dozen books,[1] containing a diversity of definitions of the term (for a literature review, see Kallis et al., 2018). Even though disagreements abound on details, the essence of the idea could be captured as such: a democratically planned downscaling of production and consumption in affluent countries aiming to reduce environmental pressures and inequality while improving well-being.

A Macroeconomic Diet As the name should make it clear, it is first and foremost a macroeconomic diet, a degrowth of the material size of the economy. Historically, this is the oldest denotation of the term, coming back to the work of ecological economics pioneer Nicholas Georgescu-Roegen, especially *The Entropy Law and the Economic Process* (1971). If the economy abides to the laws of physics, including those of thermodynamics, then the pursuit of infinite growth on a finite planet is a delusional quest. Past a certain size, the biophysical metabolism of an economy starts to run out of resources or overwhelm the carrying capacities of ecosystems. Because there are limits to how much economic growth can be decoupled from environmental degradation (Parrique et al., 2019), the situation of macroeconomic obesity that characterizes most high-income nations requires a serious downsize of the volume of economic activities. This is why the symbol of the movement is the snail, because it always keeps its shell at just the right size to be large enough without becoming too heavy to carry.

The Abolition of Extreme Wealth If patterns of resource use are unequal, so should be their decrease. The shrinking that degrowth promotes targets in priority affluent classes and rich countries. "Living simply so that others may simply live" is a good way to summarize the second feature of the concept: social justice as a requirement for sustainability. Degrowth criticizes the concept of Anthropocene for assuming that the altering of Earth's processes is caused by a homogenous humanity. They rather speak of a "capitalocene" or a "growthocene" to stress that the bulk of the biocrisis is the collateral damage of the "imperial mode of living" of a few (Brand & Wissen, 2021). As already pointed out by Maria Mies's *Patriarchy and Accumulation on a World Scale* (1986), economic growth is based on the unsustainable exploitation of "reproductive labor" such as unpaid housework, subsistence farming, informal work, and unpriced environmental amenities. The eradication of poverty and the avoidance of climate breakdown demands to put an end to acquisitive patterns of wealth accumulation that are built on an unfair appropriation of resources and labor. A macroeconomic diet in rich regions of the world is a way to redistribute ecological resources as to leave the global South with a safe operating space to prosper.

[1] To only mention a few: *Degrowth: A vocabulary for a new era* (2014), *The case for degrowth* (2020), *Less is more: How degrowth will save the world* (2020), *Degrowth* (2018), *Prosperity without growth: Foundations for the economy of tomorrow* (2017), *Post Growth: Life after Capitalism* (2021), *Farewell to growth* (2010), *Exploring degrowth: A critical guide* (2020), *Post-Growth Living: For an Alternative Hedonism* (2020), *Liberation from Excess: The road to a post-growth economy* (2012), *Managing without growth: Slower by Design, not Disaster* (2019).

A Focus on Needs and Well-Being Renouncing on fast and furious growth in the global North does not only make sense from a sustainability and global justice perspective, it does also from the point of view of well-being. Income correlates with measures of quality of life only up to a point, after which the correlation disappears, leaving it for other factors like health, education, and conviviality to affect well-being. Past these thresholds, growth can become "uneconomic" (Daly, 1997), creating more problems than it solves, for example in the form of long working hours in "bullshit jobs" (Graeber, 2018) or anxiety linked to a positional competition for status goods. Time is limited, and hours spent in the sphere of market production to increase purchasing power are hours not spent in individual and community activities to maintain living power. Instead of obsessing about money, old capitalist economies should be repurposed to serve new functions such as the satisfaction of needs, the preservation of social-ecological health, and the pursuit of happiness. In that spirit, degrowth aims to decouple human well-being from economic growth or, in others words, to enable different forms of "prosperity without growth" (Jackson, 2017).

A Decentralization of Power How to fairly distribute a carbon budget? What constitutes acceptable levels of inequality? Where should the line be drawn between needs and wants? Degrowth argues that these questions should be up to democratic deliberation. Instead of leaving these decisions to the market or to governing elites, degrowth calls for the establishment of a governance system composed of a fractal of democratic forums, in the style of Murray Bookchin "libertarian municipalism." If the unsustainability of the current situation is the result of power asymmetries in decision-making, bringing more stakeholders to the table is a way of preventing situations of exploitation while ensuring that economic activities benefit to the common good. One of the degrowth canon is that direct democracy at the micro level of the village and the firm together with representative democracy at the level of the city, the region, and the state is a precondition for sustainability, justice, and well-being. Degrowth can thus be understood as a call for collective self-limitation (Kallis, 2019), a permanent, democratic deliberation about questions of political economy informed by social morals and ecological realities.

Resonating through these four features is a critique of economic growth, not only as a phenomenon but also as an ideology. The name "degrowth" is controversial by design: a missile word to decolonize the social imaginary from the idiom of economism, the mentality that reduces human development to the narrow pursuit of monetary accumulation, whether measured by income, profits, or GDP. This puts this idea of degrowth in opposition to, not only capitalism and neoliberalism, but also against productivism in all the forms it takes. It is a counter-ideology, a device of estrangement to show that what is often though to be natural and universal (the pursuit of monetary growth) is in fact a recent social construction that is both unfair and unsustainable.

Degrowthers have spent decades imagining societies of "frugal abundance" (Latouche, 2010). Smaller, slower, more local economies whose biophysical metabolism remains within sustainable limits and where production is democratically organized around the satisfaction of needs. This post-capitalist thought experiment mobilizes a variety of already existing ideas: commons for governance, agroecology, cooperatives, reciprocal networks of care, alternative indicators of prosperity, convivial low-techs, practices of voluntary simplicity, procedures for participatory democracy, special-purpose currencies, among an array of other institutions who, if weaved together, could form an alternative economic system (Parrique, 2019).

The society degrowth envisions is meant to educate our desire for alternative futures. But more than a utopia, it is also an active social movement pushing for change in the present. Over the years, advocates of degrowth have mobilized a number of policy proposals: work time reduction to liberate time and share jobs; cap-and-share schemes to phase out fossil fuels; extraction moratoria and laws against ecocide to protect nature; bans on advertising and planned obsolescence to lower the pressure to consume; ceilings on income and wealth to limit inequality; regulations against speculative finance; local job guarantees to fight unemployment; universal basic services schemes and guaranteed incomes to prevent destitution, among an array of other policy instruments. These together form an agenda for change and a transition pathway to a post-growth society.

No concept is an island and degrowth exists among a tapestry of other critical concepts (Burkhart et al., 2020). Transition Towns in the UK, the Social and Solidary Economy in France, Buen Vivir in South America, the economy of permanence in India, the idea of a Steady-state economy in the United States, voluntary simplicity in Québec, or the recent "lying flat" movement in China. All these paradigms converge on a critique of the primacy of monetary concerns over the well-being of people and the health of ecosystems. Together, they constitute a precious toolbox of alternative economic imaginaries.

References

Brand, U., & Wissen, M. (2021). *The Imperial mode of living: Everyday life and the ecological crisis of capitalism*. Verso Books.

Burkhart, C., et al. (2020). *Degrowth in movement(s): Exploring pathways for transformation*. Zero Books.

Daly, H. (1997). *Beyond growth: The economics of sustainable development*. Beacon Press.

Georgescu-Roegen, N. (1971). *The entropy law and the economic process*. Cambridge, MA: Harvard University Press.

Graeber, D. (2018). *Bullshit jobs: A theory*. Simon & Schuster.

Jackson, T. (2017). *Prosperity without growth: Foundations for the economy of tomorrow*. Routledge.

Kallis G., 2019. Limits: Why Malthus was wrong and why environmentalists should care. Standford Briefs.

Kallis, G., et al. (2018). Research on degrowth. *Annual Review of Environment and Resources, 43*, 291–316.

Latouche, S. (2010). *Farewell to growth*. Polity Press.

Parrique, T. (2019). *The political economy of degrowth.* PhD, economics, Université Clermont Auvergne & Stockholm University.

Parrique, T., et al. (2019, July). *Decoupling debunked: Evidence and arguments against green growth as a sole strategy for sustainability*. European Environmental Bureau.

Timothée Parrique is a researcher in ecological economics at Lund University. He holds a PhD in economics from the Centre d'Études et de Recherches sur le Développement (University of Clermont Auvergne) and the Stockholm Resilience Centre (Stockholm University). Titled "The political economy of degrowth" (2019), his dissertation explores the economic implications of the ideas of degrowth. Tim is also the lead author of "Decoupling debunked – Evidence and arguments against green growth" (2019), a report published by the European Environmental Bureau (EEB, and the author of "Ralentir ou périr. L'économie de la décroissance" (2022).

Ecological Economics

Alexandre Berthe

Abstract This article presents the field of ecological economics field which aims to study the interface of humans or society with nature. It examines the relevance of this field to the study of what is at stake in the Anthropocene and how it differs from more standard economic analyses in environmental and resource economics.

Inspired by the work of authors such as Nicholas Georgescu-Roegen, Kenneth E. Boulding, Robert Costanza and Herman Daly, ecological economics is a transdisciplinary field studying the interface between human systems (economy) and ecosystems (ecology). It has a strong internal organisation which was institutionalised by the creation of an international association in 1988 and a journal entitled Ecological Economics first published in 1989 (Røpke, 2004). Many associations at national and continental levels also exist, sometimes with different visions of the field.

In all cases, ecological economics is based on the inclusion of economic relationships within ecosystems, which distinguishes it from environmental and natural resource economics. Environmental and natural resource economics is the branch of mainstream economics — the dominant current of thought in economics based on the analysis of markets and their optimality — which studies in particular how to integrate environmental issues into an optimal functioning of markets. Ecological economics, tangentially, is based on the study of the relationships between social and economic systems and the Earth system that encompasses them. It requires study of the size of the economy, i.e., the total physical volume of material and energy used to be returned in the form of waste, which can be sustainable on this planet (Daly, 1992) without neglecting the efficient allocation of the resources withdrawn and the fair distribution of what is produced. These three objectives are seen as economic by Daly (1992), because they generate costs and benefits. Various trade-offs may arise between these objectives. For example, a trade-off between

A. Berthe (✉)
University Rennes 2, Rennes, France
e-mail: alexandre.berthe@univ-rennes2.fr

© The Author(s), under exclusive license to Springer Nature Switzerland AG 2023

N. Wallenhorst, C. Wulf (eds.), *Handbook of the Anthropocene*,
https://doi.org/10.1007/978-3-031-25910-4_182

allocation and distribution may take the form of an efficient allocation that would not ensure equity of distribution. The trade-off between distribution and the scale of the economy may manifest itself in a situation where one would like to ensure a living minimum for current generations. What then can be done if the provision of a subsistence minimum is not compatible with the respect of a chosen principle of sustainability? For example, what if the need for development in the Global South is in potential contradiction with a given scale of the economy? Ecological economics draws on analyses from a wide variety of disciplines in the human, social and environmental sciences to shed light on all these ecosystem-human system relationships.

The variety of disciplines involved and the different currents of thought in economics make ecological economics a less unified field (Røpke, 2005) than environmental and resource economics. This fragmentation remains one of the major difficulties in its development, since the common identity of this transdiscipline remains weak (Røpke, 2005). In particular, the distinction between a European (ecological socioeconomics) and an American vision should be noted. In contrast to the American vision, the European approach seems less permeable to environmental and resource economics.

Ecological economics therefore has no clear boundaries, but this methodological and disciplinary pluralism is a strength and is even part of the essence of this field of research. In cases of strong uncertainties, which are particularly prominent with regard to environmental and climate issues, the plurality of analyses would provide a better answer than a single analysis claiming to solve the problem as a whole. Spash (2012) calls for preserving the diversity and above all the interdisciplinarity of ecological economics, through epistemological work that distinguishes it from environmental and resource economics and by identifying the theoretical foundations common to this field. Thus, beyond the acceptance of a plurality of approaches, this epistemological work should make it possible to realise a selection among all the approaches and the related knowledge concerning the environment. Spash (2012) therefore calls for the identification of ontological and epistemological bases common to all the methods developed in ecological economics. More than a methodological pluralism, ecological economics would then resemble a pluralistic methodology, i.e., a methodology capable of containing a plurality of approaches. Although transdisciplinarity and the inclusion of pluralism as a principle make the field difficult to circumscribe, issues of justice, intragenerational equity, a long-term approach and the value of nature in itself are more central in this field than in environmental economics.

Environmental and resource economists most often believe that economic science must be objective, notably through a principle of neutrality, whereas ecological economists consider that analyses should not be detached from values (Illge & Schwarze, 2009). The latter therefore argue in favour of a subjectivist methodology. Thus, the method of ecological economics is not based on a particular normativity, but seeks a clarification of ethical positions as proposed by Spash (2012). Ecological economics therefore considers the possibility of a plurality of ethical judgements that can coexist and that are an integral part of scientific analysis in economics.

Taking into account a plurality of ethical judgements leads to a major question: how to consider the human being, the central object of study in this discipline, in this context? In the first place, recourse to an ecological economist's approach leads to the rejection of consumer sovereignty. The individual is no longer solely a homo economicus, but can be perceived through the diversity of his actions. The individual is then not analysed as a purchaser of products, but as a social actor capable of making judgements; a homo politicus, a citizen endowed with a sense of justice. The search for justice no longer takes place solely through the market, but also through the institutions to be built. To illustrate this, we can quote Vatn and Bromley (1994, p.142): 'just as preferences count for consumer choice within constraints, judgments can be used as the driving concept for citizens choosing basic norms or modifying existing constraints'.

By rejecting homo economicus as the only anthropological horizon, ecological economics also rejects the sole recourse to the study of changes in the utility of agents to understand the world (Gowdy & Erickson, 2005). Consequently, ecological economics questions the claim that sustainability issues can be resolved on the basis of a vision of humans centred on their own interests and on an associated global objective of economic growth. Moreover, it is no longer possible to provide a comparison by a single measurement scale, whether cardinal (strong commensurability) or ordinal (weak commensurability). It is then a matter of considering a plurality of incommensurable but comparable value elements by using multi-criteria analyses, for example, or by using the debate among citizens as a tool for identifying solutions for the future by taking ecosystems into account.

More generally, the questioning of homo economicus and the inclusion of human societies in an ecological system that goes beyond them make it possible to rethink the position of the human being and to decentralise it. The analysis may even then include the possibility of an intrinsic value of nature and the need to seriously consider approaches that are no longer anthropocentric but biocentric. Concerning human societies, ecological economics proposes to open up economic analysis to a diversity of possible ways of living and modes of coordination far from the simple triptych — market, producer, consumer. The study of an environmental problem can then involve, for example, the identification and analysis of a common management of environmental resources as proposed by Ostrom (1990). It therefore leaves open possibilities for inventing a radically new economy capable of reconciling human beings with the planet on which they live.

References

Daly, H. E. (1992). Allocation, distribution, and scale: Towards an economics that is efficient, just, and sustainable. *Ecological Economics, 6*, 185–193. https://doi.org/10.1016/0921-8009(92)90024-M

Gowdy, J., & Erickson, J. D. (2005). The approach of ecological economics. *Cambridge Journal of Economics, 29*, 207–222. https://doi.org/10.1093/cje/bei033

Illge, L., & Schwarze, R. (2009). A matter of opinion—How ecological and neoclassical environmental economists and think about sustainability and economics. *Ecological Economics, 68*, 594–604. https://doi.org/10.1016/j.ecolecon.2008.08.010

Ostrom, E. (1990). *Governing the commons: The evolution of institutions for collective action*. Cambridge University Press.

Røpke, I. (2004). The early history of modern ecological economics. *Ecological Economics, 50*, 293–314.

Røpke, I. (2005). Trends in the development of ecological economics from the late 1980s to the early 2000s. *Ecological Economics, 55*, 262–290. https://doi.org/10.1016/j.ecolecon.2004.10.010

Spash, C. L. (2012). New foundations for ecological economics. *Ecological Economics, 77*, 36–47. https://doi.org/10.1016/j.ecolecon.2012.02.004

Vatn, A., & Bromley, D. W. (1994). Choices without prices without apologies. *Journal of Environmental Economics and Management, 26*, 129–148. https://doi.org/10.1006/jeem.1994.1008

Alexandre Berthe is associate professor at the University of Rennes 2 (LiRIS). He is a specialist in ecological and agricultural economics. His work focuses on territorial dynamics in the context of ecological and energy transitions, especially on biogas production in rural territories, and on the inequality-environment nexus.

Growth

Christian Arnsperger

Abstract This article provides an overview of the main concepts needed today to locate the discourse on economic growth within the Anthropocene. Economic growth is built into the economic system that currently dominates. It obeys an outdated, radical imaginary: that of human progress as the triumphant denial of the limits of the biosphere. This imaginary needs to be replaced by a new one. The main task for social science in this day and age is to reflect on and design viable, thriving, non-growing economies.

Economic growth is a polymorphous object. It is a *concept* that underlies the discourse of people who think in a particular way about "the economy" as a complicated machine geared towards maximizing material output (Mitchell, 2011). It is also an *institution*—a set of more or less loosely held principles affirmed through rules, regulations and norms organizing societal interactions and political decisions in order to generate consensus and coordinate policies (Schmelzer, 2016). It is, furthermore, a *system effect*—the emergent result of a multitude of individual and collective decisions concerning such diverse domains as consumption, investment, fertility, saving, production, advertising, innovating, and so on (Raworth, 2017). It is, finally, an *imaginary*—a structuring worldview that sees modern civilization as a project of ever-expanding mastery and well-being under the banner of "progress" (Greer, 2015).

Economic growth was characterized by Simon Kuznets in his 1971 Nobel lecture as follows: "a country's economic growth may be defined as a long-term rise in capacity to supply increasingly diverse economic goods to its population, this growing capacity based on advancing technology and the institutional and ideological adjustments that it demands." Enshrined in this definition are the ideas that humans have potentially limitless wants and needs which cannot legitimately be criticized or

C. Arnsperger (✉)
University of Lausanne, Lausanne, Switzerland
e-mail: christian.arnsperger@unil.ch

© The Author(s), under exclusive license to Springer Nature Switzerland AG 2023
N. Wallenhorst, C. Wulf (eds.), *Handbook of the Anthropocene*,
https://doi.org/10.1007/978-3-031-25910-4_183

questioned, and that basically more of anything and everything is better for anyone and everyone.

Kuznets was, just like his contemporary Walt Rostow, the proponent of the famous "stages of economic growth" approach to development (Rostow, 1960), born in a family that emigrated from the Soviet block to the United States—and this surely influenced both men's neoconservative imaginary in a significant manner. Indeed, as Amadae (2003) and Schmelzer (2016) have shown, modern growth theory emerged within the Cold War era, marked by the conflict between capitalism and socialism around output maximization, well-being through consumption and accumulation as well as geopolitical dominance through militarization.

As Arendt (1958) argued in her early observations about Russian outer-space conquest, within the ambit of the Soviet-US rivalry the space age and the growth age combined to create a new radical imaginary: human progress as the triumphant denial of the limits of the biosphere—or, more precisely, the sleight of hand by which limitlessness could be attained while seeming to remain within the (increasingly irrelevant) physical limits of the biosphere. This inevitably segued into the notion that outer space beyond the earthly biosphere was to be the next frontier. Anthropologically speaking, exponential economic growth is at least as much a result of the Anthropocene as it is one of its causes.

Economic growth is built into the currently dominant economic system, which generates a variety of *growth imperatives* (Richters & Simoneit, 2019). This means that due to various systemic properties of the capitalist market economy, sustained and effectively never-ending growth is necessary if we are to avoid a collapse of the economy and the immense hardships that would come from it. This risk is all the more real because, over the past three centuries, economic growth has in most periods gone hand in hand with colonial violence and increasing inequalities (Hickel, 2018).

One of the main mechanisms propelling economic growth is the *rebound effect*. This effect can take many different shapes, all of which have in common the translation of efficiency gains into output increases and also, in most cases, throughput increases. One particular case, called "technical progress," is when improved technology increases labour productivity, which in a capitalist market economy subsequently allows profit-maximizing firms to expand output, thus usually also expanding both labor use and resource utilization. Therefore, in principle, with profit-driven technological innovation, cheap natural resources and energy, and unchecked rebound effects, aggregate value-added can increase hugely (which is synonymous with macroeconomic GDP growth) and both profits and wages can increase. This is the mutual-interest mechanism that underlies the long-standing "social contract" which ruled capitalist social democracies after World War II and made possible the post-war boom that led headlong into the oil crises of the 1970s and the neoliberal backlash of the 1980s and 1990s.

Given all the foregoing elements, it is perhaps no surprise that economic growth has become a highly contentious concept within the Anthropocene. Clearly, the literatures on the limits to growth (Meadows et al., 1972), degrowth (Latouche, 2006; Hickel, 2020) and post-growth (Jackson, 2021), all anchored within ecological economics (Martínez-Alier & Muradian, 2015), view the waves of the Great Acceleration since the 1850s and the 1950s as radically imperiling the Earth's habitability. They thus follow in the wake of Kenneth Boulding's famous "Spaceship Earth" parable (Boulding, 1966), asserting that open-ended, frontier-pushing "cowboy economics" needs to be replaced by closed-systems, limit-conscious "spaceship economics." Interestingly enough, however, the very same closed-systems thinking has also spawned a whole ideology of "green growth" and human-created artificial biospheres (Allen, 2000), which claims that through appropriate technological progress, humanity will be able at the same time to grow its numbers and to ensure growing material comfort and well-being for all humans.

A key issue in this regard is to what extent the economy's physical throughput and/or output can be *"decoupled"* from its resource consumption and, more broadly, its deleterious environmental impacts. Decoupling can be relative, or it can be absolute. Advocates of green growth adhere to the doctrine of the "environmental Kuznets curve" and suggest that with appropriate technological progress, growth could become a mechanism that actually cleans up the environment: this is the idea of absolute decoupling. In actual fact, absolute decoupling virtually never occurs (Jackson, 2017) and if there is any decoupling at all, it is merely relative: as the economy grows, the associated environmental damages also increase, but they do so at a slower rate.

Relative decoupling is merely a way to delay resource depletion and environmental collapse; it does not solve the fundamental issue—namely, that as long as an economy grows by more than 1 percent annually, any resource efficiency gains through recycling or a "circular economy" are bound to be cancelled out by ever-increasing throughput and output (Grosse, 2010). By the same token, economic growth also makes it impossible for all of humanity to secure a decent living standard while remaining within planetary boundaries (Millward-Hopkins et al., 2020).

The degrowth movement rightly challenges the implicit presupposition of unlimited human wants and desires. Contrary to what classical thinkers such as Thomas Malthus believed, human wants and desires can be reduced without harm (Kallis, 2019). On the contrary, overcoming compulsive consumption and material accumulation can lead to the discovery of more authentic pleasures and freedoms in the form of an "alternative hedonism" (Soper, 2020). If the Anthropocene is to stabilize into lastingly sustainable ways for humans to inhabit the planet, it is inevitable that while basic needs must be satisfied for every earthling (Gough, 2017), substantial reductions in what is considered to be "necessary" will have to occur, and hence the main task for ecological economics in this day and age is to reflect on and design viable, thriving, non-growing economies (Paulson et al., 2020).

Traditional cultures have always known this. Working less with more efficient tools is antidote to the rebound effect. To create more efficient tools and then to increase the duration and/or the arduousness of work is the very incarnation of the rebound effect. Clastres (1974) showed that Indigenous peoples, unlike so-called moderns, do not have the compulsion to exploit all the rebound effects or to constantly generate them. Sufficiency, not efficiency, is the wisdom that will make the Anthropocene from the growth compulsion.

References

Allen, J. (2000). Artificial biospheres as a model for global ecology on planet Earth. *Life Support and Biosphere Science, 7*(3), 273–282.
Amadae, S. M. (2003). *Rationalizing capitalist democracy: The cold-war origins of rational choice liberalism*. University of Chicago Press.
Arendt, H. (1958). *The human condition*. University of Chicago Press.
Boulding, K. (1966). The economics of the coming spaceship Earth. In H. Jarrett (Ed.), *Environmental quality in a growing economy* (pp. 3–14). Resources for the Future/Johns Hopkins University Press.
Clastres, P. (1974). *La société contre l'État*. Minuit.
Gough, I. (2017). *Heat, greed and human need: Climate change, capitalism and sustainable well-being*. Edward Elgar.
Greer, J. M. (2015). *After progress: Reason and religion at the end of the industrial age*. New Society.
Grosse, F. (2010). Is recycling 'part of the solution'? The role of recycling in an expanding society and a world of finite resources. *SAPIENS, 3*(1) Available at https://journals.openedition.org/sapiens/906
Hickel, J. (2018). *The divide: A brief guide to global inequality and its solutions*. Penguin.
Hickel, J. (2020). *Less is more: How degrowth will save the world*. Penguin.
Jackson, T. (2017). *Prosperity without growth: Foundations for the economy of tomorrow*. Routledge.
Jackson, T. (2021). *Post-growth: Life after capitalism*. Polity.
Kallis, G. (2019). *Limits: Why Malthus was wrong and why environmentalists should care*. Princeton University Press.
Latouche, S. (2006). *Le pari de la décroissance*. Fayard.
Martínez-Alier, J., & Muradian, R. (Eds.). (2015). *Handbook of ecological economics*. Edward Elgar.
Meadows, D., Meadows, D., Randers, J., & Behrens, W. (1972). *The limits to growth*. Potomac Associates.
Millward-Hopkins, J., Steinberger, J.K., Rao, N., & Oswald, Y. (2020). Providing decent living with minimum energy: A global scenario. *Global Environmental Change, 65*, 102168.
Mitchell, T. (2011). *Carbon democracy: Political power in the age of oil*. Verso.
Paulson, S., D'Alisa, G., Kallis, G., & Demaria, F. (2020). *The case for degrowth*. Polity.
Raworth, K. (2017). *Doughnut economics: Seven ways to think like a 21st-century economist*. Penguin.
Richters, O., & Simoneit, A. (2019). Growth imperatives: Substantiating a contested concept. *Structural Change and Economic Dynamics, 51*, 126–137.
Rostow, W. W. (1960). *The stages of economic growth: A non-socialist manifesto*. Cambridge University Press.
Schmelzer, M. (2016). *The hegemony of growth: The OECD and the making of the economic growth paradigm*. Cambridge University Press.
Soper, K. (2020). *Post-growth living: For an alternative hedonism*. Verso.

Christian Arnsperger is professor of sustainability and economic anthropology at the University of Lausanne. He holds a PhD in economics from the University of Louvain (Belgium) and has also been active as a scientific advisor to the Alternative Bank Switzerland. He publishes widely in several languages on the critique of economic growth, the existential underpinnings of capitalism, and the links between money and sustainability. His latest books are *Écologie intégrale: Pour une société permacirculaire* (with Dominique Bourg, PUF, 2017) and *L'existence écologique: Critique existentielle de la croissance et anthropologie de l'après-croissance* (Seuil, 2023).

Habits

Mariagrazia Portera

Abstract In the wake of the recent impressive resurgence of interest in the notion of habits among philosophers, social scientists, psychologists and neuroscientists, this article examines the relevance of human habits to the current environmental crisis. The recently established, multidisciplinary matrix known as 'Environmental Humanities' has put major emphasis on habits and their role in, on the one hand, compounding the current ecological disaster (it is the case of those 'environmentally damaging habits' such as using too many plastic bags or using plastic straws) but also, on the other hand, in potentially providing a way out of the crisis. Indeed, new ways of behaving that are less harmful to the environment, can only be formed in a way that is robust enough to endure if they are produced through embodied processes of consolidation, i.e. if they become seated in the form of habits.

What is a habit? Is 'habit' about acting repeatedly and automatically in a certain definite way, i.e. a behavioural pattern hard to give up? Or is a habit a plastic propensity, not innate but acquired, which enables the organism to adapt to its unstable environment? How do habits form, and how and to what extent can 'bad' habits be broken and, conversely, 'good' habits be boosted and improved? What is the difference between 'habit', 'habitus' and 'custom'?

In the last few years, we have witnessed a resurgence of interest in the notion of 'habit', particularly in the English–speaking academic world (Sparrow & Hutchinson, 2013; Carlisle, 2014): philosophers, sociologists, anthropologists, scholars in cognitive sciences, neuroscientists and literary scholars have turned to the concept of habit to investigate its role in social relations, its neural and psychological underpinnings, its value in the history of philosophy and literature and how habits affect and shape the functioning of the human mind. 'Habit', however, is far from a notion of recent coinage, at least in the history of Western philosophy.

M. Portera (✉)
University of Florence, Florence, Italy
e-mail: mariagrazia.portera@unifi.it

© The Author(s), under exclusive license to Springer Nature Switzerland AG 2023

N. Wallenhorst, C. Wulf (eds.), *Handbook of the Anthropocene*,
https://doi.org/10.1007/978-3-031-25910-4_184

Philosophical analysis of habit dates back as far as the work of Aristotle (particularly his *Nicomachean Ethics*) and the notion cuts across virtually all schools and traditions until the present day, from Descartes and Kant, who understood habituation as a stultifying force, to Felix Ravaisson and the American pragmatists, who saw in habits one of the fundamental cornerstones of the human nature, to Pierre Bourdieu (2002; see Krais & Gebauer, 2002), who set clear distinction between 'habit' (conceived of as a purely mechanical tendency to repeat certain actions or ways of thinking) and 'habitus'. Indeed, as Claire Carlisle has shown, "habit" has emerged throughout the history of Western philosophy as a concept with a dual nature: on the one hand, a blind and mindless tendency to repetition (Kant, for instance); on the other hand, a creative, embodied disposition indispensable for life (Aristotle, Hume, Dewey, for instance). Which of these two interpretations is right?

Within the field of psychology, there is much evidence that habituation processes, habits and habitual behaviour pervade our social and mental life to a significant extent (Bargh, 1997): 'during much of our waking lives, we act according to our habits, from the time we rise and go through our morning routines until we fall asleep' (Graybiel, 2008, p. 360). The emergence of the new paradigm of 4E cognition (*e*mbodied, *e*mbedded, *e*xtended, and *e*nactive), along with a growing dissatisfaction with the computational theory of the mind, have led to a re-assessment of the notion of habit as one of the foundations of a new conception of the human mind. As Alva Nöe writes: 'Human beings are creatures of habit. Habits are central to human nature [...]. Only a being with habits could have a mind like ours' (Noë, 2009, pp. 97–98). This means that, if we want to understand how the human mind works, we need to turn the spotlight to our embodied, embedded, partially unconscious (but not utterly impermeable to cognition) habits.

There is an aspect of the pragmatist concept of habit, particularly evident in John Dewey's work, which is worth emphasizing: the 'ecological' or 'environmental' nature of human habits. Indeed, according to Dewey, human habits are not the acquisition of a single individual rather the result of the interaction between individuals and their environment; they cannot be understood apart from the context in which they occur. As Dewey argues in his book *Human Nature and Conduct*, habits are formed and re-formed through the ongoing interactions of bodies and their natural and socio-cultural environment (Dewey, 1922); in this sense, the structure and characteristics of human habits reflect the environment in which they grow out. Let us try to capitalize on this idea: considering the 'ecological' nature of all human habits, and given the current state of our *global environment* – i.e., the current environmental crisis – how and to what extent can habits reveal something of the Anthropocene epoch in which we live? As is known, the Anthropocene is, according to a vast majority of scientists, the new, current phase in the history of the Earth in which humans have become geological agents able to transform with their pervasive activities the structures of the planet itself. Widespread extinction, climate change, and air/water/soil pollution are just a few of the many symptoms of the global environmental changes induced by human activities in the Anthropocene. How and to what extent pre-Anthropocene habits differ from the Anthropocene ones? And, more in general, what is the role of habits in the current environmental crisis?

The recently established, multidisciplinary matrix known as 'Environmental Humanities' has put major emphasis on habits and their role in, on the one hand, compounding the current ecological disaster (it is the case of those 'environmentally damaging habits' such as using too many plastic bags or using plastic straws) but also, on the other hand, in potentially providing a way out of the crisis. Scholars working in the Environmental humanities recognise that 'at the heart of global change in the 21st century there is much more than what the natural sciences, with their argumentative strategies, epistemologies and methodologies, can track and measure: there are 'human choices and actions – questions of human behaviour, habits, motivation that are embedded in individual practices and actions, in institutional and cultural pathways, and in political strategies' (Holm et al., 2015). A growing body of recent literature has addressed the question as to how climate change and the current environmental crisis can be mitigated through personal actions and the acquisition of better environmental habits; there is today 'an urgent need for a robust theory of consumption that addresses how habits form, how they change and how policy can contribute to the formation of new habits that are less environmentally intrusive' (Wilhite, 2015, p. 100). Examples of (very simple) 'environmental (bad) habits' are: leaving the light switched on when nobody's in the room; leaving the tap turned on while brushing the teeth; buying more food than needed (thus increasing the amount of waste) etc. These may look like trivial and ultimately unimportant habitual gestures or actions, but their ecological impact as a whole is far from negligible; moreover, each of these minor actions, which we perform in most cases automatically and without being completely aware of what we are doing while we are doing it, taps into and relies on a more general propensity (acquired, not innate) to think of natural resources as if they were infinite and us, human beings, as if we were the ultimate and only masters of the planet. They tap into a 'neo-liberal' habit of thinking that is specific to the Anthropocene sociocultural environment.

A further point about habits is worth stressing: while it is true that the automatic power of habit can force us to repeat certain entrenched (and environmentally unsustainable) modes of action again and again, like leaving the lights or the tap on, it is also true, on the other hand, that new ways of behaving, less harmful to the environment, can only be formed robust enough to endure if they are produced through embodied processes of consolidation, i.e. if they sediment in the form of habits. Habits are at the same time a curse and a blessing, as Clare Carlisle has noted in her book (Carlisle, 2014): they are one of the main factors of the ecological crisis, but also a possible way out of it. In this sense, a new, ecologically orientated, anthropological theory of habits in the Anthropocene is needed, focusing on their structure, formation processes, role and value for a better and more sustainable relation of humans to their environment.

How do habits form and how bad habits can be broken and changed? Recently, theories of (sustainable) environmental habits revolving around the notions of 'affordance' and 'nudging' have been put forward (Knussen & Yule, 2008; Kaaronen, 2017). According to these theories, to steer people towards more sustainable behaviour, it is not necessary to make them aware of their, largely unconscious and automatic, environmentally unsustainable habits, rather it is enough and much more effective to implement minor changes to the everyday infrastructures and architectures so that people can be nudged into environmentally friendly behaviour. Is nudging the key to better habits? Criticisms have been raised towards this idea (Pedwell, 2017), arguing that nudge theory – which focuses on pernicious habits to get rid of them, instead of addressing the more complex question of how more plastic and more intelligent habits can be formed, and which targets isolated individuals alone – is just another facet of the manipulative, neo-liberal habit of thinking that has been dominating, so far, in the Anthropocene.

References

Bargh, J. A. (1997). The automaticity of everyday life. In R. S. Wyer (Ed.), *Advances in social cognition* (Vol. X, pp. 1–61). Erlbaum.

Bourdieu, P. (2002). Habitus. In H. Jean & E. Rooksby (Eds.), *Habitus: A sense of place* (pp. 27–34). Ashgate Publishing.

Carlisle, C. (2014). *On Habit*. Routledge.

Dewey, J. 1944 (1922). *Human nature and conduct*. Henry Holt and Company.

Graybiel, A. M. (2008). Habits, rituals, and the evaluative brain. *Annual Review of Neuroscience, 31*, 359–387.

Holm, P., et al. (2015). Humanities for the environment. A manifesto for research and action. *Humanities, 4*(4), 977–992.

Kaaronen, R. O. (2017). Affording sustainability: Adopting a theory of affordances as a guiding heuristic for environmental policy. *Frontiers in Psychology, 8*, 1974. https://doi.org/10.3389/fpsyg.2017.0197

Knussen, C., & Yule, F. (2008). 'I'm not in the habit of recycling': The role of habitual behavior in the disposal of household waste. *Environment and Behavior, 40*, 683–702.

Krais, B., & Gebauer, G. (2002). *Habitus*. transcript Verlag.

Noë, A. (2009). *Out of our heads: Why you are not your brain, and other lessons from the biology of consciousness*. Hill and Wang.

Pedwell, C. (2017). Habit and the politics of social change: A comparison of nudge theory and pragmatist philosophy. *Body and Society, 23*(4), 59–94.

Sparrow, T., & Hutchinson, A. (Eds.). (2013). *A history of habit: From Aristotle to Bourdieu*. Lexington Books.

Wilhite, H. L. (2015). The problem of habits for a sustainable transformation. In K. L. Syse & M. L. Mueller (Eds.), *Sustainable consumption and the good life* (pp. 100–110). Routledge.

Mariagrazia Portera is currently a Research Fellow at the University of Florence, Department of Literature and Philosophy, after being a post-doctoral fellow at IASH, Institute for Advanced Studies in the Humanities, University of Edinburgh, at the Free University of Berlin, at the University of Zagreb and at the Centre for Advanced Studies South East Europe, University of Rijeka. She holds a PhD in Philosophy from the University of Florence. Her areas of expertise are the history of Aesthetics between the 18-century and the 19-century in Germany, the history of Darwinism, experimental aesthetics, and environmental and evolutionary aesthetics. She has authored more than 60 publications (articles, book chapters, books) and coauthored book chapters and papers with biologists, geneticists, philosophers of biology.

Habitus and Education

Kathrin Audehm

Abstract This contribution takes up the problem of the social roots of human behaviour. Pierre Bourdieu's (1930–2002) concept of habitus provides the basis for an education-sociological reflection on the inclusion of human behaviour into power structures concerning the genesis and transformation of a system of habitual dispositions through education and socialisation. This concept has become part of social and educational sciences in Europe and beyond. Thus, first there is an explanation of the perspective chosen for this article. Because of the tool-like nature of the concept, then the article focuses on the following aspects: the problem of enforcing legitimate culture through education, the generation of lifestyles by means of distinctive judgements of taste as well as processes of habitus transformation. This is followed by considerations and discussions concerning a transformative way of learning as well as a transformative kind of education and how both may contribute to educating people towards sustainable development.

Pierre Bourdieu's concept of habitus refers to the totality of subtle, matter-of-course dispositions of perception, action and thought. Like Dewey, Bourdieu also understands social agents to be part of their social environments. The social environment confronts human actors not just in the form of an outside world of products, goods and values, of documents and archives of knowledge but urgently calls for action, thus enforcing immediate action and reaction in the context of social situations (Bourdieu, 1970a). From this, results a kind of practical knowledge which manifests itself through judgements of taste, though behaviour and lifestyles and is rooted in bodily dispositions. Practical knowledge shapes itself to become a social sense which enables people to perform socially appropriate practices without having to take a detour via consciousness. Habitus works as a *modus operandi* which generates practices and then aligns, systematises and unifies them.

K. Audehm (✉)
University of Cologne, Cologne, Germany
e-mail: kathrin.audehm@uni-koeln.de

© The Author(s), under exclusive license to Springer Nature Switzerland AG 2023

N. Wallenhorst, C. Wulf (eds.), *Handbook of the Anthropocene*,
https://doi.org/10.1007/978-3-031-25910-4_185

The space of social practice is permeated by structures of power and rule and structured by an unequal distribution of capital. Depending on class and gender, cultural origin and age, there are different opportunities and risks when it comes to the accumulation of economic, cultural, social and symbolic capital through the course of life and within social fields where certain kinds of capital are more relevant than others, and this holds also for entire social spaces or societies. These instances of distribution of capital are quite consolidated, yet still subject to gradual and far-reaching processes of change, or they change as a result of historical breaks, such as the French colonial rule in Algeria which is the starting point for Pierre Bourdieu to develop his concept of habitus, or the peaceful revolution in the GDR. Distribution of capital determines options for action and structures practical meaning, which is why habitus, as a structural system, is at the same time an *opus operatum*.

If both perspectives are brought together, habitus results as a structured and structuring system of schemes of perception, action and thought, in the context of which in Bourdieu's works the human body as a living and feeling, as an acting and mouldable body increasingly takes on the nature of a tool for gaining insight (Bourdieu & Wacquant, 1992) which, through actions, incorporates existing power structures. In Bourdieu's works, habitus frequently functions as a means to explain the transfer of social into symbolic power. In his 'Foundations of a Theory of Symbolic Violence', developed in cooperation with Jean-Claude Passeron (Bourdieu & Passeron, 1970/1990, Book I), education in this context becomes a fundamental tool of symbolic power, as it enforces certain cultural practices as being legitimate while excluding others. Basically, this is true of all kinds of education, in all cultures and at all times.

Symbolic power requires certain meanings to be generally recognized and adds its own power to those social power structures (Bourdieu & Passeron, 1970/1990, p. 4). In contrast to Karl Marx's (and Friedrich Engel's) opinion, it is thus not reduced to a superstructure phenomenon, and it reinforces the social constraint resulting from social facts. However, in contrast to Èmile Durkheim, it is considered to be socially differentiated, in the way that different social groups are to different degrees subject to the symbolic power of educational practice. Educational practice is violent because it is based on a cultural arbitrariness which is of a double nature, on the one hand being part of an already existing and thus pre-given symbolic practice and on the other, in relation to social power structures, only enforcing certain cultural practices – such as ways of speaking and dialects, personal hygiene and health care, judgements of taste and aesthetic ways of expression – as legitimate culture or expressions of education (ibid, Chap. 1). This happens through an irreducible triangle of educational action, educational authority and educational work.

Educational work aims at the incorporation and transformation of habitual dispositions: "*it is a transformative action tending to inculcate a training, a system of durable, transposable dispositions*" (ibid, p. 36, Italics in the original). It is all the more violent, the further away the cultural expressions of each educational addressee is from the idea of cultural perfection which dominates the respective educational discourse, such as those of workers' and farmers' children who are less able to compensate for (allegedly) lacking resources for the accumulation of (allegedly) legitimated cultural capital in the same way as, for example, the offspring of middle class families. These examples illustrate the relations between social and symbolic power structures of (Western) capitalist societies.

In the UNESCO Programme 'Education for Sustainable Development' (ESD), if we consider education as a tool of symbolic power, the following questions arise. What educational actions might be capable of generating cultural practices of sustainability which may achieve (educational) authority both among addressees and professional actors, and how and over what period of time does the imprinting of appropriate habitual dispositions happen through educational work? What cultural practices of what groups of the population of a globalised world are considered legitimate or are excluded as being illegitimate in this context?

As early as in his analysis of the Kabyle house, Bourdieu worked out that the world of objects and the world of symbols are structurally related to each other (Bourdieu, 1970b). Bourdieu analyses this in his empirical study on 'distinction' (Bourdieu, 1979/1987), by the example of the then contemporary French society, and he further develops his habitus concept in regard to the manifestation of a space of lifestyles. As a socially determined judgement of taste, habitus becomes a creation principle and classification system for cultural practices, in the context of which certain practices are systematised and unified to become a particular lifestyle which is recognizable as such and can be classified (ibid, Chap. 3). Now, a socially differentiated homology of the positions of social groups and their judgements of taste becomes obvious, making their lifestyles as the entirety of their kinds of practice systematically different from each other. In this context, at the heart of the analyses there is the body, as the most visible shape of socio-cultural practices as they appear, for example in the way people speak, look and laugh, their common diet, clothing and sports. The study hints at processes of social change, however lifestyles appear to be conspicuously stable and are fixed to a practical logic of distinction and delimitation.

If these considerations are applied to lifelong socialisation processes, the following questions arise concerning the probability of developing sustainable lifestyles. Can a general and globally effective, i.e. cross-social and cross-culture, hegemonial consensus be achieved which breaches the logic of distinction? Bourdieu's study of contemporary history indicates the logical necessity of social-ecological transformation, otherwise the manifestation of sustainable lifestyles would fail because of habitus generating practices which are almost incompatible with them. Although from Bourdieu's point of view habitus, generates certain lifestyles, automatically and not at all coincidentally, how much does a certain habitus conform to social structures?

Anyone who enters a new or massively changed social space, who must act successfully in previously unknown social fields, or who must cope with social climbing or social decline, cannot rely on a well-known system of habitual dispositions but must perform countless acts of gaining insight which, through practice (that is frequently lengthy and painful), are transformed into persistent and transferable dispositions. It is logically necessary that culture-, social- and life phase-specific actions of gaining insight are made congruent with already incorporated experiential knowledge (Krais & Gebauer, 2002, p. 62). Thus habitus can be transformed even without explicit educational efforts or in the course of lifelong socialisation processes.

The relation of genesis or formation and transformation is not clearly explained and remains disputed, however Bourdieu himself points out this possibility. The starting point is the insights gained through collaborative research in Algeria and the relative degree of the non-conformism, crisis-proneness and incoherence of a social-historical habitus which, as a culturally homogeneous habitus of honour, is confronted with a homogenous economic habitus characterising French colonial rule. Algerian social actors are confronted with contradicting classification principles and rules for recognition, resulting in fissions. However, this insight is systematically included only as late as in 'Pascalian Meditations' (Bourdieu, 2000), although in particular transformation in the form of a conversion of habitus, which Bourdieu impressively demonstrates in the example of a Kabyle chef and street peddler in Algiers (Bourdieu, 2013, p. 180ff), providing links to the questions raised before.

Habitus is an embodied system of perception, action and thought schemes, and shows stubbornness, without Bourdieu telling about any existing scope of action (Rehbein & Saalmann, 2014). Being a split habitus which, due to social-historical processes of change, is confronted with several and contradicting classification principles and recognition rules, it may clash with social-historical and socio-cultural demands and may even generate contradicting practices. Thus a habitus-structure conflict arises, which at the same time liberates the social actors from the bonds of their habitual dispositions.

No social law which compulsorily ties social actors to fixed and unchangeable structures of social groups, fields or spaces can be based on Bourdieu's concept of habitus (Audehm, 2017, pp. 175–177). However, it illustrates the difficulties any education towards sustainable development is confronted with, and it points out the inherent power relations of any educational practice, even more so as Bourdieu himself develops considerations concerning a rational kind of education (Bourdieu & Passeron, 1970/1990) which does not treat those actually being unequal as equals, thus symbolically consolidating and reproducing social inequality. In this context, questions arise about the practicability of such a rational pedagogy and its connectivity to non-Western educational practices and discourses. As concerns the manifold debates on transformative education and transformative learning (Laros et al., 2017), which lead us to hope that there will be a successful education for sustainability, we may conclude that these require 'a clear reference to current social-theoretical analyses' as well as 'to concepts of critical citizenship education' (Lingenfelder, 2020).

References

Audehm, K. (2017). Habitus. In A. Kraus, J. Budde, M. Hietzge, & C. Wulf (Eds.), *Handbuch Schweigendes Wissen, Erziehung, Bildung, Sozialisation und Lernen* (pp. 167–178). Beltz Juventa.

Bourdieu, P. (1970a). *The logic of practice*. Stanford University Press.

Bourdieu, P. (1970b). The Berber house or the world reversed. *Social Science Information, 9*(2). https://doi.org/10.1177/053901847000900213

Bourdieu, P. (1979/1987). *Distinction: A social critique of the judgement of taste*. Harvard University Press.

Bourdieu, P. (2000). *Pascalian Meditations*. Stanford University Press.

Bourdieu, P. (2013). *Algerian Sketches*. Polity Press.

Bourdieu, P., & Passeron, J.-C. (1970/1990). *Reproduction in education, society and culture*. Sage Publications.

Bourdieu, P., & Wacquant, L.'c. J. D. (1992). *Reflexive sociology*. Polity Press in association with Blackwell Publishers.

Krais, B., & Gebauer, G. (2002). *Habitus*. transcript.

Laros, A., Fuhr, T., & Taylor, E. E. (Eds.). (2017). *Trans-formative learning meets Bildung: An international exchange*. Sense Publishers.

Lingenfelder, J. (2020). Transformative Bildung Was bedeutet Transformative Bildung im Kontext sozial-ökologischer Krisen? In: *Ausserschulische Bildung* 1/2020, https://www.adb.de/download/publikationen/ab1_2020_jahresthema.pdf

Rehbein, B., & Saalmann, G. (2014). Habitus. In G. Froehlich & B. Rehbein (Eds.), *Bourdieu-Handbuch. Leben – Werk – Wirkung* (pp. 110–118). Verlag J.B. Metzler.

Kathrin Audehm is a professor of education and heterogeneity at the University of Cologne, Department of Educational and Social Sciences at the Faculty of Humanities. She holds a PhD in Philosophy from the Free University of Berlin, where she also qualified as a university lecturer with a theoretical reflection on educational authority. Her fields of expertise are ethnographical and qualitative research concerning symbolic power and performative practices such as rituals in families and at school as well as socio-material aspects of educational practice, processes of habitus formation during adolescence, and gender constructions in popular culture.

Health

Katharine Zywert and Stephen Quilley

Abstract In the Anthropocene, as ecological disruptions intensify and economic growth becomes increasingly untenable, health and health systems must be reimagined. At issue is not only the health of individual human beings but also community wellbeing and planetary health. Promising prefigurative practices that could enable such a transition: (1) create mutual benefits across social-ecological scales; (2) disrupt dominant Anthropocene trajectories; (3) reduce the ecological and economic costs of health and care; and (4) operate within re-embedded networks of community reciprocity.

Human health is dependent upon the health of our finite planet. The trajectories of the Anthropocene, including rapidly accelerating material consumption, climate change, biodiversity loss, soil degradation, and changes to the water cycle, pose substantial risks to the long-term health and wellbeing of our species (Cole, 2019). Ecological disruptions like climate change and land degradation have direct health effects (e.g. the effects of floods and droughts), environmentally-mediated health effects (e.g. changes to the transmission and virulence of infectious diseases), and indirect health effects (e.g. effects related to conflict and migration) (Whitmee et al., 2015). To secure human health as ecological disruptions intensify, paradigmatic change within health systems is needed to shift away from an exclusionary focus on the health of individual human beings toward greater accountability for the health of human and non-human communities, ecosystems, and the planet as a whole. This article will present a set of principles that could enable such a transition along with brief synopses of promising prefigurative practices.

K. Zywert (✉)
Independent Researcher, Waterloo, Canada
e-mail: katharinezywert@gmail.com

S. Quilley
University of Waterloo, Waterloo, ON, Canada
e-mail: squilley@uwaterloo.ca

© The Author(s), under exclusive license to Springer Nature Switzerland AG 2023

N. Wallenhorst, C. Wulf (eds.), *Handbook of the Anthropocene*,
https://doi.org/10.1007/978-3-031-25910-4_186

With the transition to industrial modernity, human societies came to rely on economic growth to achieve social goods including health. However, a substantial body of evidence demonstrates that economic growth cannot be decoupled from ecological destruction when considered at a global scale (Ward et al., 2016). If growth and environmental disruption are inseparable, stepping back from ecological overshoot will require a fundamental transformation of the global political economy to make it possible to achieve health and wellbeing without economic growth (Zywert & Quilley, 2020). Especially in high-income nations with extremely complex, materially and energetically intensive healthcare infrastructures, the transition to a post-growth health system is unprecedented and, for many, unthinkable. Health systems in a post-growth economy will need to contend with: reduced availability of energy and material resources; reductions in the complexity of health systems that mirror broader processes of economic and social decomplexification; the social and economic implications of relocalization and the emergence of more place-bound communities; and less space for growth in healthcare fields as the economy contracts (Hensher & Zywert, 2020).

The most promising approaches for securing human and planetary health in such a context tend not to be found within professionalized healthcare fields, but at the margins of health systems (Zywert & Quilley, 2020). Promising prefigurative alternatives may depart quite significantly from the mindsets, goals, and power dynamics that define the dominant social-ecological regime. As such, they may not uphold the tenets of mainstream sustainability, but may instead embody disruptive ideas about death, ageing, or chronic illness. They may also reject mainstream approaches such as profit-driven healthcare models, the reliance on high technology treatments and pharmaceuticals, or even dependence upon formalized economic markets for professionalized care. Promising prefigurative alternatives instead embody four principles that position them to make a meaningful contribution to health system transformation in the Anthropocene:

Create Mutual Benefits Across Social-Ecological Scales Human civilizations cannot thrive unless the planet's self-regulating biophysical systems remain within parameters that are conducive to the flourishing of the human species (Rockström et al., 2009). As such, ecological disruptions unfolding in the Anthropocene threaten the health of future generations. In modern health systems, the health of individuals is prioritized over the health of social-ecological systems. This prioritization has resulted in emergent problems at higher and lower scales (e.g. antibiotic resistance and climate change) (Zywert & Quilley, 2017). Anthropocene health systems should incorporate initiatives that benefit health across social-ecological scales.

Disrupt Dominant Anthropocene Trajectories The Anthropocene is defined by rapidly rising material consumption, energy use, population growth, and ecological disruption (Steffen et al., 2015). Technological and managerial approaches have proven insufficient to address the scale of the social, ecological, and economic challenges we currently face. In this context, it is important to ensure that social innovations for health and care do not unintentionally reinforce the trajectories of the Anthropocene, but instead create new pathways toward sustainability, health, and wellbeing (Olsson et al., 2017).

Reduce the Ecological and Economic Costs of Health and Care Healthcare expenditures continue to grow substantially faster than GDP and make a significant contribution to resource and energy use in high and middle-income countries (Hensher et al., 2020). If absolute decoupling of economic growth from ecological destruction is impossible (see Ward et al., 2016), health systems in the Anthropocene will need to transition toward structures and practices that can enable health across scales at a much lower ecological and economic cost. This transition will not be easy given that welfare states in most nations are funded by taxation schemes that are dependent upon growth. Healthcare will need to use resources more strategically in future and could be effectively supported by diverse community-based initiatives that address the social and environmental determinants of health and that enable illness prevention and health promotion.

Operate Within Re-embedded Networks of Community Reciprocity A post-growth transition necessarily implies a degree of relocalization of social and economic life. In more place-bound communities, reciprocal networks of support grounded in long-term relationships of obligation take on greater importance within community life. Such networks have a low ecological and economic price tag, often operating informally beyond the domain of market exchange. They also make it possible for individuals to develop stronger connections to place, family, and community, connections that can strengthen health and wellbeing (Quilley & Zywert, 2019).

Diverse initiatives that embody these principles already exist at the margins of health systems. The following prefigurative practices illustrate what the principles described above can look like in practice:

Care Farming integrates health and social care into local agricultural production. Care farms leverage the skills and labour of community members experiencing mental health issues, addictions, and disabilities. They are also working farms that produce crops and animal products, often using organic methods. For participants, working on the farm, taking care of animals, and cultivating vegetable crops has been shown to yield improvements in physical, social, and mental health. At the same time, care farming improves the economic viability of small-scale farms, promotes organic methods, and regenerates local ecosystems (Elings, 2020).

Mutual aid is any informal network that provides reciprocal support within a defined community. Mutual aid takes many diverse forms, from guilds and friendly societies to the hyper-local "caremongering" groups that emerged during the COVID-19 pandemic. Mutual aid networks are uniquely positioned to support community members with issues related to food security, income insecurity, disability, frailty, and mental illness. Existing within the domain of livelihood activities, they do not involve monetary exchange, but instead create new pathways for informal health and care supports that promote community inclusion and long-term caring relationships within communities (Quilley & Zywert, 2019; Steinberger, 2020).

In Geel, Belgium, family foster care for people with serious mental illnesses offers an alternative model of mental healthcare rooted in dignity and community integration. In this unique approach, individuals with serious mental illnesses are welcomed into the homes of ordinary families. "Boarders" become like family, contributing as they are able to the daily work of operating the household while becoming valued members of extended family and community networks. The approach is social rather than therapeutic, with no specific medical regimen to follow. Families are supported by professional healthcare providers who are available at a moment's notice to address crises, and families are compensated by the state to participate (at a much lower cost than hospital or other forms of community care). The model has thrived for hundreds of years and is deeply rooted in Catholic tradition, intergenerational continuity, and a sense of moral obligation (van Bilsen, 2016).

When considered together, prefigurative alternatives that align with the principles described above contribute to health system transformation by demonstrating the potential cross-scale benefits of more sustainable approaches to health and care. There are countless local examples such as those described here, many of which may quickly gain momentum as social-ecological systems transformations intensify. Prefigurative alternatives could take the burden off professionalized health care services, address the social and ecological determinants of health, and potentially respond even more effectively than modern medicine to complex challenges like mental health.

In the broadest terms, the health-dilemmas posed by the Anthropocene highlight a stark divide between two trajectories and ontologies. Mainstream responses are increasingly influenced by a pervasive, and technologically mediated, death-denial, epitomised most acutely by vogue ideas of 'transhumanism' (Porter, 2017). In contrast, the ecological-economic path reframes both (i) the Cartesian subject as a more permeable and nested category of being, and (ii) the ontological meaning of death, illness and suffering as intrinsic to human experience. In short, the dynamics of the Anthropocene require a profound reimagining of what health is and how health systems can support wellbeing across social-ecological scales.

References

Cole, J. (2019). *Planetary health: Human health in an era of global environmental change*. CABI.

Elings, M. (2020). Care farming: Making a meaningful connection between agriculture, health care, and society. In K. Zywert & S. Quilley (Eds.), *Health in the Anthropocene: Living well on a finite planet* (pp. 226–240). University of Toronto Press.

Hensher, M., & Zywert, K. (2020). Can health care adapt to a world of tightening ecological constraints? Challenges on the road to a post-growth future. *BMJ, 371*, 1–4.

Hensher, M., Tisdell, J., Canny, B., & Zimitat, C. (2020). Health care and the future of economic growth: Exploring alternative perspectives. *Health Economics, Policy, and Law, 15*, 419–439. https://doi.org/10.1017/S1744133119000276. PMID: 31685052.

Olsson, P., Moore, M.-L., Westley, F. R., & McCarthy, D. D. P. (2017). The concept of the Anthropocene as a game-changer: A new context for social innovation and transformations to sustainability. *Ecology and Society, 22*(2), 31. https://doi.org/10.5751/ES-09310-220231

Porter, A. (2017). Bioethics and transhumanism. *The Journal of Medicine and Philosophy: A Forum for Bioethics and Philosophy of Medicine, 42*(3), 237–260. https://doi.org/10.1093/jmp/jhx001

Quilley, S., & Zywert, K. (2019). Livelihood, market, and state: What does a political economy predicated on the 'individual-in-group-in-PLACE' actually look like? *Sustainability, 11*(15), 4082.

Rockström, J., Steffen, W., Noone, K., et al. (2009). A safe operating space for humanity. *Nature, 461*, 472–475.

Steffen, W., Broadgate, W., Deutsch, L., et al. (2015). The trajectory of the Anthropocene: The great acceleration. *The Anthropocene Review, 2*(1), 81–98.

Steinberger, J. (2020, April 8th). Pandenomics: A story of life versus growth. Open Democracy. https://www.opendemocracy.net/en/oureconomy/pandenomics-story-life-versus-growth/. Accessed 6th January 2021.

van Bilsen, H. P. J. G. (2016). Lessons to be learned from the oldest community psychiatric service in the world: Geel in Belgium. *BJPsych Bulletin, 40*, 207–211.

Ward, J. D., Sutton, P. C., Werner, A. D., Costanza, R., Mohr, S. H., & Simmons, C. T. (2016). Is decoupling GDP growth from environmental impact possible? *PLoS One, 11*(10), e0164733. https://doi.org/10.1371/journal.pone.0164733

Whitmee, S., Haines, A., Beyrer, C., et al. (2015). Safeguarding human health in the Anthropocene epoch: Report of the Rockefeller Foundation – *Lancet* commission on planetary health. *The Lancet, 386*, 1973–2028.

Zywert, K., & Quilley, S. (2017). Health systems in an era of biophysical limits: The wicked dilemmas of modernity. *Social Theory and Health (Online First Article), 16*, 1–20. https://doi.org/10.1057/s41285-017-0051-4

Zywert, K., & Quilley, S. (Eds.). (2020). *Health in the Anthropocene: Living well on a finite planet* (pp. 285–305). University of Toronto Press.

Katharine Zywert works at the intersection of social-ecological systems change and health. Her research focuses on community responses that enhance the long-term prospects for human health and wellbeing on a planet in ecological crisis. Katharine holds a PhD in Social and Ecological Sustainability from the University of Waterloo, a Master's in Medical Anthropology from the University of Oxford and a Graduate Diploma in Social Innovation from the University of Waterloo.

Stephen Quilley is Associate Professor of Social and Environmental Innovation in the School of Environment, Resources and Sustainability. Trained as a sociologist, Stephen has previously lectured in Environmental Politics at Keele University (2006–2012) Environmental Sociology at University College Dublin (1999–2005) and held a Research Fellowship at the ESRC Centre for Research on Innovation and Competition (CRIC) in Manchester (1997–1999). With academic research interests ranging from the historical sociology of Norbert Elias and urban regeneration, to the long term dynamics of human ecology, Stephen has also worked on policy-related projects relating to sustainability, urban regeneration, food systems, resilience and social-ecological innovation.

Human Reconfiguration of the Biosphere

Mark Williams, Jan Zalasiewicz, and Julia Adeney Thomas

Abstract The biosphere coevolves with the atmosphere, hydrosphere and lithosphere to maintain a habitable space on Earth. Over billions of years – and despite periodic setbacks – it has evolved increasing complexity, from its microbial beginnings to the complex interactions between animals, plants, fungi and unicellular microscopic life that sustain its present state. Recently, the biosphere has been profoundly changed by humans. In part, this includes increased rates of *extinction* that are reminiscent of past fundamental perturbations to life. But the change is even more profound, resulting from a combination of marked *translocations* of species beyond their indigenous ranges, overt *concentration* of biomass in humans and their farm animals, *reconfiguration* of landscape habitats and over-utilisation of ocean life, excessive *appropriation* of energy from the biosphere (including its fossilised component), and increasing *interconnectivity* between technology and life.

The biosphere is a fundamental component of the Earth System, existing for billions of years, and co-evolving with the atmosphere, lithosphere and hydrosphere to maintain a habitable space for life (Vernadsky, 1998). It is ubiquitous at the Earth's surface, extending several kilometres into the lithosphere, where its subsurface mass alone is considered to be ≈15% of the total carbon in the biosphere (Bar-On et al., 2018). It also extends high into the atmosphere, where microbes are important for atmospheric processes such as cloud formation (e.g., DeLeon-Rodriguez et al., 2013). The total mass of carbon in the biosphere is of the order of ≈550 gigatons (Gt, a gigaton = one billion metric tonnes), with plants accounting for ≈450 Gt: by comparison, animals are ≈2 Gt. Before widespread deforestation by humans, the

M. Williams (✉) · J. Zalasiewicz
University of Leicester, Leicester, UK
e-mail: mri@le.ac.uk; jaz1@le.ac.uk

J. A. Thomas
University of Notre Dame, Notre Dame, IN, USA
e-mail: jthomas2@nd.edu

© The Author(s), under exclusive license to Springer Nature Switzerland AG 2023

N. Wallenhorst, C. Wulf (eds.), *Handbook of the Anthropocene*,
https://doi.org/10.1007/978-3-031-25910-4_187

mass of the biosphere may have been approximately double its present size (Bar-On et al., 2018).

A record of the biosphere – as preserved in rocks – extends back over more than 3 billion years (Brasier et al., 2015). For much of that time the biosphere was entirely microbial, and microbes continue to be its most fundamental component (Cavicchioli et al., 2019). A microbial biosphere can survive without animals and plants, but animals and plants cannot survive without microbes. Inside the human body, for example, is a microbiome of several trillion bacterial cells that protect us from harmful pathogens. This microbiome is beneficial to our digestion, and even helps our mental wellbeing. The deep interconnections between microbes and all life exist at many levels and have deep origins; an oxygenated ocean and atmosphere, essential for large multicellular organisms, owes its origin to microbes, and microbes continue to generate a significant component of the available oxygen for life.

The rocks of the Phanerozoic Eon – the past 539 million years – preserve a rich fossil record of complex biogeography and ecoregions that track the position of oceans and continents over time and changes in Earth climate states. It is during this time that complex animal-rich ecosystems became widespread, and plants colonised the land, to form marine and terrestrial ecosystems that are familiar to us. This fossil record also tracks major perturbations of the biosphere, with five mass extinctions in the past 500 million years and the recovery of flora and fauna after each of these. By tracking change, the fossil record provides a comparison and contrast with present patterns of environmental change and biodiversity loss caused by humans.

So, what are humans doing that is fundamentally changing the biosphere? Six, closely inter-related factors can be considered as transformative: increased *extinction* rates, *translocation* of organisms, *concentration* of biomass in certain organisms, *reconfiguration* of ecosystems, overt *appropriation* of energy, and increasing *interconnection* between life and technology.

The first factor – *extinction* – is a repeating theme in the Phanerozoic, caused by major global environmental change, and in one instance at the Cretaceous-Paleogene boundary 66 million years ago, by an asteroid strike. Whilst extinction in the Phanerozoic is not unique, its cause by a single species likely is. Current extinction rates are considerably higher than background levels (Pimm et al., 2014) and have accelerated since 1900 CE (Ceballos et al., 2015). Whilst many species are not *yet* extinct, contractions of their populations and ecological ranges threaten global biodiversity, leading some authors to refer to a sixth mass extinction (Barnosky et al., 2011) and a 'biological annihilation' (Ceballos et al., 2017). As yet, the geological record of this extinction is only partially explored, for example in records of local extirpation and reconfigurations of flora and fauna (see Barnosky et al., 2019).

The second factor, *translocations* of species, has – like extinction – occurred before, as in the Great American Interchange, when North and South America became conjoined through the isthmus of Panama about three million years ago. By contrast, present translocations are caused by humans and not by geographical change. They are also, uniquely in Earth history, global, even to geographically remote areas like oceanic islands, and have accelerated through the past 200 years

(Seebens et al., 2017) to homogenise the global flora and fauna (Samways, 1999). Tens of thousands of introduced species globally have led – in some cases – to wholesale reconfiguration of ecosystems, leaving a recognisable palaeontological signal in sediments accumulating in seas, estuaries, rivers, sinkholes and lakes.

The third factor, *concentration*, is also unique to human intervention. According to Bar-On et al. (2018), 96% of the carbon-biomass of *all* present-day mammals is concentrated in humans (36%, ≈0.06 Gt) and farm mammals (60%, ≈0.1 Gt), with only 4% in wild mammals (≈0.007 Gt). Such patterns – especially the concentration of biomass in one large terrestrial mammal – may be unprecedented in terrestrial vertebrate evolution. Similarly, poultry (mainly chickens) represent 70% of carbon-based biomass of all living birds, with 30% for wild birds. Although not yet systematically examined as a geological archive, these changes – many accelerating in the twentieth century (see Bennett et al., 2018), will likely produce a palaeontological record of domesticated animals, one that is already visible in the archaeological record. To this we can add the two-fold loss of biomass to the biosphere from deforestation during the period of human civilization (Bar-On et al., 2018), which has certainly left a palaeontological record.

Extinction, translocation and concentration are all symptoms of our fourth factor, the global *reconfiguration* of ecosystems by humans. These processes, beginning over 10,000 years ago, involving refashioning of the landscape, have led to plants of the families Poaceae and Gramineae (grasses like wheat, maize and rice) being produced in vast quantities (over 2000 billion tonnes per year; Wrigley, 2010) that feed the livestock that feed and clothe humans. Together, domesticated animals and cultivated plants now utilise half of the habitable surface of the land (Ritchie, 2017 and references therein). This has left a distinctive and global fossil signature too, for example in the pollen record (Mottl et al., 2021) that identifies a worldwide acceleration in rates of vegetation change between 4600 and 2900 years ago that is unprecedented over the past 18,000 years. Uniquely, as a terrestrial species, humans have not only reconfigured the land, but have also caused wholesale changes to ocean ecosystems (Halpern et al., 2008).

The fifth factor is humanity's excessive *appropriation* of energy from living biomass – most of which is in forests, together with that taken from the chemical energy stored by the ancient biosphere over hundreds of millions of years in coal, oil and gas deposits. At present, through photosynthesis, the energy converted from solar energy into biomass each year is 2 ZJ (zeta joules). Overall, this maintains the 19 ZJ of chemical energy stored in all of the living biomass of the biosphere (Schramski et al., 2015). But as we have already noted, this amount is about half what was previously stored (Bar-On et al., 2018). This process of 'reduction of the biosphere' is accelerating, with 10% or more of the loss occurring since 1900 CE (Schramski et al., 2015). Together with the rapid depletion of stored chemical sources of energy (from fossil fuels), some authors have likened this to the Earth battery being run down, the energy being dissipated into space, with ultimately catastrophic consequences for life (Schramski et al., 2015).

The sixth factor is radically different from anything in the previous more than 3 billion years of biosphere evolution and is the *interconnection* between technology and life, that is a major component of what Peter Haff has called the technosphere, the globally emergent system that includes humans, their technology, institutions, governments and associated bureaucracies (Haff, 2019). This interconnection between biology and technology is visible in the engineering of broiler chickens to increase their adult body mass fourfold, and their 'nurture' – over a few weeks – in wholly unnatural ecosystems to enable ~63 billion of them to be consumed each year (Bennett et al., 2018). These broiler chickens – as with the billions of farm mammals – cannot exist without the technosocial ecosystems that support them. Another example here would be the development of chicken cells grown in factories to make chicken meat – and indeed the meat of cattle might be made similarly.

Whether new technologies like growing meat cells in factories can – in due course – be more environmentally friendly and help to foster more sympathetic relationships between humans and animals is a subject of ongoing discussion (Corbyn, 2020). Elsewhere in this book, in the chapter we have written called 'Mutualistic cities', we argue for more sympathetic relationships between the human and non-human components of life. These relationships will have to involve sustainable consumption of the Earth's resources including its biosphere, industrial and domestic processes that more resemble the functioning of natural ecosystems and recycle, and the making of space for other organisms. Only if we mitigate the impacts of our wholesale reconfiguration of the biosphere described here, will we avoid a level of damage to biodiversity that will be catastrophic.

References

Barnosky, A. D., Matzke, N., Tomiya, S., Wogan, G. O. U., Swartz, B., Quental, T. B., et al. (2011). Has the Earth's sixth mass extinction already arrived? *Nature, 471*, 51–57.

Barnosky, A. D., Wilkinson, I., Zalasiewicz, J., & Williams, M. (2019). Chapter 3.2: Late quaternary extinctions. In J. Zalasiewicz, C. N. Waters, M. Williams, & C. Summerhayes (Eds.), *The Anthropocene as a geological time unit*. Cambridge University Press.

Bar-On, Y. M., Phillips, R., & Milo, R. (2018). The biomass distribution on Earth. *PNAS, 115*, 6506–6511.

Bennett, C. E., Thomas, R., Williams, M., Zalasiewicz, J., Edgeworth, M., Miller, H., Coles, B., Foster, A., Burton, E. J., & Marume, E. (2018). The broiler chicken as a signal of a human reconfigured biosphere. *Royal Society Open Science, 5*, 180325. https://doi.org/10.1098/rsos.180325

Brasier, M. D., Antcliffe, J., Saunders, M., & Wacey, D. (2015). Changing the picture of Earth's earliest fossils (3.5–1.9 Ga) with new approaches and new discoveries. *PNAS, 112*, 4859–4864.

Cavicchioli, R., Ripple, W. J., Timmis, K. N., et al. (2019). Scientists' warning to humanity: Microorganisms and climate change. *Nature Reviews. Microbiology, 17*, 569–586. https://doi.org/10.1038/s41579-019-0222-5

Ceballos, G., Ehrlich, P. R., Barnosky, A. D., García, A., Pringle, R. M., & Palmer, T. M. (2015). Accelerated modern human-induced species losses: Entering the sixth mass extinction. *Science Advances, 1*(5), e1400253. https://doi.org/10.1126/sciadv.1400253

Ceballos, G., Ehrlich, P. R., & Dirzo, R. (2017). Biological annihilation via the ongoing sixth mass extinction signaled by vertebrate population losses and declines. *PNAS, 114*, E6089–E6096.

Corbyn, Z. (2020). *Out of the lab and into your frying pan: the advance of cultured meat*. https://www.theguardian.com/food/2020/jan/19/cultured-meat-on-its-way-to-a-table-near-you-cultivated-cells-farming-society-ethics

DeLeon-Rodriguez, N., Latham, T. L., Rodriguez-R, L. M., Barazech, J. M., Anderson, B. E., Beyersdorf, A. J., et al. (2013). Microbiome of the upper troposphere: Species composition and prevalence, effects of tropical storms, and atmospheric implications. *PNAS, 110*, 2575–2580.
Haff, P. K. (2019). Chapter 4.1. The Technosphere and its relation to the Anthropocene. In J. Zalasiewicz, C. N. Waters, M. Williams, & C. Summerhayes (Eds.), *The Anthropocene as a geological time unit*. Cambridge University Press.
Halpern, B. S., Walbridge, S., Selkoe, K. A., Kappel, C. V., Micheli, F., D'Agrosa, C., et al. (2008). A global map of human impact on marine ecosystems. *Science, 319*, 948–952.
Mottl, O., Flantua, S. G. A., Bhatta, K. P., Felde, V. A., Giesecke, T., Goring, S., et al. (2021). Global acceleration in rates of vegetation change over the past 18,000 years. *Science, 372*, 860–864.
Pimm, S. L., Jenkins, C. N., Abell, R., Brooks, T. M., Gittleman, J. L., Joppa, L. N., Raven, P. H., Roberts, C. M., & Sexton, J. O. (2014). The biodiversity of species and their rates of extinction, distribution, and protection. *Science, 344*, 1246752. https://doi.org/10.1126/science.1246752
Ritchie, H. (2017). *How much of the world's land would we need in order to feed the global population with the average diet of a given country?* https://ourworldindata.org/agricultural-land-by-global-diets
Samways, M. (1999). Translocating fauna to foreign lands: Here comes the Homogenocene. *Journal of Insect Conservation, 3*, 65–66.
Schramski, J., Gattie, D. K., & Brown, J. H. (2015). Human domination of the biosphere: Rapid discharge of the earth-space battery foretells the future of humankind. *PNAS, 112*, 9511–9517.
Seebens, H., Blackburn, T. M., Dyer, E. E., Genovesi, P., Hulme, P. E., Jeschke, J. M., et al. (2017). No saturation in the accumulation of alien species worldwide. *Nature Communications, 8*, 14435.
Vernadsky, V. I. (1998). *The biosphere* (complete annotated edition: Forward by Lynn Margulis and colleagues and introduction by Jacques Grinevald). Copernicus (Springer-Verlag). 192 pp.
Wrigley, C. W. (2010). *An introduction to the cereal grains: Major providers for mankind's food needs*. Woodhead Publishing Limited.

Mark Williams is a palaeobiologist at the University of Leicester. He studies the evolution of life on Earth over hundreds of millions of years and has been a long-time member (and former Secretary) of the Anthropocene Working Group. He has co-authored popular science books with Jan Zalasiewicz on climate change, ocean evolution, and the story of life on Earth, and with Jan and Julia Adeney Thomas he co-authored *The Anthropocene: A Multidisciplinary Approach*. His new book with Jan is called *The cosmic oasis: the remarkable story of Earth's biosphere* (Oxford University Press).

Jan Zalasiewicz is a geologist (now retired) at the University of Leicester, and a member (formerly Chair) of the Anthropocene Working Group. He is co-editor of *The Anthropocene as a geological time unit: a guide to the scientific evidence and current debate*, has been involved in other AWG publications, and with Julia Adeney Thomas and Mark Williams has co-written *The Anthropocene: A Multidisciplinary Approach*. His books for a general audience that *inter alia* explore Anthropocene concepts include *The Earth after Us* and (with Mark Williams) *The Goldilocks Planet* and *Ocean Worlds*.

Julia Adeney Thomas is a historian at the University of Notre Dame. She writes about politics, nature, photography, and the Anthropocene in Japan and globally. Her publications include *Reconfiguring Modernity: Concepts of Nature in Japanese Political Ideology*; *The Anthropocene: A Multidisciplinary Approach*, written with Jan Zalasiewicz and Mark Williams, and *Strata and Stories* with Jan Zalasiewicz along with numerous articles and edited volumes. She is at work on *The Historian's Task in the Anthropocene* (under contract with Princeton University Press).

Hunger

Karina Limonta Vieira

Abstract Hunger is a social problem that affects people all over the world. This complex social phenomenon is a biological expression of sociological evils, i.e., of social vulnerability. To understand hunger requires inquiring into the meaning of hunger in the Anthropocene and the challenges that are present in fighting hunger. This leads to looking into the right to food, the violation of human rights, and the right to food security. New forms of human responsibility and new practices for global transformation are required, involving environment, technology and education. Developing education to ensure security of future food systems is the challenge to be met in order to end hunger.

Hunger is represented in the literature as a need to ingest food, i.e., to eat from a biological point of view (Jonsonn, 1981; Biesalski, 2012), as well as being defined as "an uncomfortable or painful physical sensation caused by insufficient consumption of dietary energy" (FAO, 2019, p. 188). The types of hunger, according to Behera et al. (2019), are acute hunger, chronic hunger and hidden hunger. Acute hunger is when a person's inability to consume adequate food is due to crises like droughts, El Niño, wars and disasters. Chronic hunger is the state of long-term undernourishment when the body absorbs less food than it needs. And hidden hunger is the micronutrient (vitamins and minerals) deficiency in the body. Acute, chronic and hidden hunger can provoke mortality and comorbidities, delay in cognitive, motor and language development, and have consequences for both adults and children such as short stature, obesity, poor academic and learning performance, as well as low work capacity and productivity.

In addition to the biological point of view, hunger is also a complex social phenomenon and is linked to poverty and social vulnerability, for it is unevenly distributed among families (Basset & Winter-Nelson, 2010). Specifically, for anthropology, hunger is an issue of poverty, inequality and wealth distribution (Russel, 2006).

K. L. Vieira (✉)
Faculty of Education, Leipzig University, Leipzig, Germany
e-mail: karina_augusta.limonta_vieira@uni-leipzig.de

© The Author(s), under exclusive license to Springer Nature Switzerland AG 2023
N. Wallenhorst, C. Wulf (eds.), *Handbook of the Anthropocene*,
https://doi.org/10.1007/978-3-031-25910-4_188

It can then be said that hunger is the biological expression of sociological evils (de Castro, 1973). According to Behera et al. (2019), factors such as climate change, social exclusion, job instability, food shortages and waste, food price volatility and social discrimination affect hunger and poverty. These factors show that hunger is a global challenge, affecting almost 690 million people who were starving in the year 2019. Indeed, ending hunger by 2030 is an uncertainty, according to The State of Food Security and Nutrition in the World report (2020).

The hunger crisis in the world is a phenomenon that was neglected and ignored for a long time (de Castro, 1973; Jonsonn, 1981; Vernon, 2007). The negligent factors around hunger involve economic and political interests and moral prejudices, i.e., hunger is a prohibited topic and morally considered an inevitable part of the human condition, or a divine retribution for the sins of the human being (Vernon, 2007). However, now a hungry person is a violation of human dignity and incompatible with the universal claim for a decent life (Müller, 2019), and good health and well-being (Santos, 2020). Therefore, an understanding of hunger beyond biological, social, economic and moral concepts is an urgent question.

It must be considered that even in the face of this hunger crisis, the planet Earth has gone through numerous political, social and geological transformations and that the main agent of this transformation is the human being. So, what does hunger mean in the Anthropocene? What are the challenges s in the fight against hunger? The world is experiencing a new geological epoch, the Anthropocene, in which human actions have impacted the planet Earth and, as a consequence of this impact, we have hunger (Janssens et al., 2020). In this sense, hunger is not natural, but is a consequence of human action on the planet (Russel, 2006). Although, for instance, food is being planted, at the same time, it is wasted. Moreover, human actions such as pollution, deforestation and global warming have caused flooding and lack of rain, where many areas are affected by floods and droughts. As a result of this, many crops are lost and many people have no food, causing hunger (Allahyari & Sadeghzadeh, 2020). According to Biesalski (2012), factors such as wars, climate change, migration of people looking for food, global economy and profits can generate a global hunger crisis. In this sense, hunger is one of the most serious manifestations of crisis in the world (Jonsonn, 1981), involving socio-economic and environmental problems, in which human actions cause natural and social disasters (Allahyari & Sadeghzadeh, 2020).

In the current period in which we live we are faced with hunger as a global challenge, which takes into account, on the one hand, environmental issues, such as food waste and the high production of food and its impact on the environment, and on the other hand social issues, such as the right to food or food security, and the end of hunger. Environmental researchers highlight that food waste produces an excess of garbage and tons of greenhouse gases, as well as the fact that 1/3 of the food produced in the world is thrown away, causing a lack of equitable food distribution, as well as the damage to the environment caused by the high production of food (Behera et al., 2019). Humans' impact on the environment has threatened the factors necessary to life, including water, soil, plants, and air to ensure their own livelihood and have created adverse conditions to satisfy their own needs (Allahyari & Sadeghzadeh, 2020, p. 42). Hunger is seen in social terms as the violation of human rights and the right to food security (Schanbacher, 2019; Tobin, 2009). In order to end hunger, a form of agricultural production is required that makes possible equitable food distribution (Santos, 2020).

The current challenges call for societal changes towards hunger. In this case, new forms of human responsibility emerge and new practices for global transformation in relation to hunger, involving the environment, technology and education. Some suggestions proposed here are Zero Hunger, agricultural breeding or Green Revolution, and agroecological education. The responsibility of the human being, in principle, consists of achieving the aims of Sustainable Development Goal 2 (SDG2) – Zero Hunger of United Nations (2019) by 2030, which include ending hunger and ensuring access to food, ending all forms of malnutrition, doubling agricultural productivity, ensuring sustainable food production systems, maintaining genetic diversity. Furthermore, increasing investment in infrastructure, research, services and technology, correcting and preventing trade restrictions in world agricultural markets, and adopting measures to ensure the functioning of food commodity markets. Zero Hunger requires a connection between the two fields of agricultural production and the better distribution of food. Moreover, the praxis for that transformation still includes agricultural breeding or the green revolution and agroecological education. Agricultural breeding and the green revolution have contributed to improving food, increasing food production using methods such as crossing and selection, line breeding, hybrid breeding, cell and tissue culture, cloning, marker technology and genomic selection, digital phenotyping, genetic engineering and genome editing. This makes it possible to increase the harvestable yield, the final quality, resistance to disease and resistance to pests, and to make it more adaptable to environmental extremes which has contributed to the alleviation of chronic hunger in the world (Avana-Tientcheu & Tiambo, 2020). Also new models of agroecological education are required (David & Bell, 2018; Francis et al., 2011) to develop and apply technology to assure productivity, new ideas for agro-production, nature conservation, the protection of traditional agriculture landscapes, and to develop a quality of education that ensures future food systems will be effective in the face of the complexity and uncertainty which are present in the challenge to end hunger.

References

Allahyari, M. S., & Sadeghzadeh, M. (2020). Agricultural extension systems toward SDGs 2030: Zero hunger. In W. L. Filho, A. M. Azul, L. Brandli, P. G. Özuyar, & T. Wall (Eds.), *Zero hunger* (pp. 41–52). Springer Nature.

Avana-Tientcheu, M. L., & Tiambo, C. K. (2020). Breeding and productivity in ending Hunger and achieving food security and nutrition. In W. L. Filho, A. M. Azul, L. Brandli, P. G. Özuyar, & T. Wall (Eds.), *Zero hunger* (pp. 130–146). Springer Nature.

Basset, T., & Winter-Nelson, A. (2010). *The atlas of world Hunger*. The University of Chicago Press.

Behera, B. K., Rout, P. K., & Behera, S. (2019). *Move towards zero Hunger*. Springer Nature.

Biesalski, H. K. (2012). *Der Verborgene Hunger: Satt Sein Ist Nicht Genug*. Springer Spektrum. English edition: Biesalski, Hans Konrad. 2013. Hidden Hunger (trans: O'Mealy, Patrick). Springer.
David, C., & Bell, M. (2018). New challenges for education in agroecology. *Agroecology and Sustainable Food Systems, 42*(6), 612–619. https://doi.org/10.1080/21683565.2018.1426670
de Castro, J. (1973). A fome. *O Correio*, Unesco 1(3). https://www.pjf.mg.gov.br/conselhos/comsea/publicacoes/artigos/arquivos/art_fome.pdf. Accessed 20 Mai 2021.
FAO, IFAD, UNICEF, WFP, WHO. (2019). *The state of food security and nutrition in the world. Safeguarding against economic slowdowns and downturns*. Food and Agriculture Organization of the United Nations. http://www.fao.org/3/ca5162en/ca5162en.pdf. Accessed 25 Mai. 2021.
FAO, IFAD, UNICEF, WFP, WHO. (2020). *The state of food security and nutrition in the world. Transforming food systems for affordable healthy diets*. Food and Agriculture Organization of the United Nations. http://www.fao.org/3/ca9692en/ca9692en.pdf. Accessed 25 Mai. 2021.
Francis, C. A., Jordan, N., Porter, P., Breland, T. A., Lieblein, G., Salomonsson, L., Sriskandarajah, N., Wiedenhoeft, M., DeHaan, R., Braden, I., & Langer, V. (2011). Innovative education in agroecology: Experiential learning for a sustainable agriculture. *Critical Reviews in Plant Sciences, 30*(1–2), 226–237. https://doi.org/10.1080/07352689.2011.554497
Janssens, C., Havlík, P., Krisztin, T., Baker, J., Frank, S., Hasegawa, T., Leclère, D., Ohrel, S., Ragnauth, S., Schmid, E., Valin, H., Van Lipzig, N., & Maertens, M. (2020). Global hunger and climate change adaptation through international trade. *Nature Climate Change, 10*, 829–835. https://doi.org/10.1038/s41558-020-0847-4
Jonsonn, U. (1981). Hunger and society. *Food and Nutrition Bulletin, 3*(2), 1–10.
Müller, J. (2019). *Globaler Hunger als Verletzung der menschlichen Würde. Zu den normativen Grundlagen einer moralischen Herausforderung*. J.B. Metzler.
Russel, S. A. (2006). *Hunger an unnatural history*. Basic Books.
Santos, M. J. P. L. (2020). Equitable food distribution and sustainable development. In W. L. Filho, A. M. Azul, L. Brandli, P. G. Özuyar, & T. Wall (Eds.), *Zero hunger* (pp. 274–282). Springer Nature.
Schanbacher, W. (2019). *Food as human right*. Praeger.
Tobin, J. (2009). *Hunger efforts and food security*. Nova Science Publishers.
United Nations-UN. (2019). *Sustainable development goals*. https://www.un.org/sustainabledevelopment/hunger/. Accessed 27 Mai. 2021.
Vernon, J. (2007). *Hunger. A modern history*. The Belknap Press of Harvard University Press.

Karina Limonta Vieira, Dr., Member of Interdisciplinary Center for Historical Anthropology (Free University of Berlin) and Germany Society for Education Sciences. Research interests are human development, education, pedagogical anthropology and anthropocene. Publications are *Antropologia da Educação: levantamento, análise e reflexão no Brasil* [Anthropology of Education: survey, analysis and reflection in Brazil] and *Cinema e educação nutricional: aprendizagem cultural alimentar na educação infantil* [Cinema and nutrition education: food cultural learning in early childhood education].

Industry

Etienne Maclouf

Abstract This article proposes to tackle our entry into Anthropocene as an observable phenomenon not only through its effects, chiefly on the composition of the atmosphere and the disappearance of biodiversity, but also through its key factor: the process of industrialization of human activities. We owe our entry into this new era to new social entities that we have recently engendered: modern organizations on which we now depend to meet our vital and social needs, and which meanwhile endanger our own survival. Are these organizations the result of an improvement of what previously existed or do they represent a break in our history? Depending on the response to this question, the diagnosis and remedies will be different.

This chapter refers to my recently published essay in organization studies entitled "Why industrial organizations will not save the planet. Or the anti-manual of CSR and sustainable development" (Maclouf, 2020). Here I recall why industrial organizations (IOs) must be held a key factor in ecological and climatic disasters. Then, I explain why we should consider the possibility that, although we are the designers, perpetrators and participants of those organizations, we are actually unable to control their deleterious trajectory. The lesson of this exercise of lucidity is tough: if this is true, to save human species from its own industrial disaster, we should succeed in emancipating ourselves from the social systems that currently protect us from physical and biological threats, and that provide everything we need.

E. Maclouf (✉)
Université Paris Panthéon-Assas (LARGEPA), CESCO (MNHN), Paris, France
e-mail: etienne.maclouf@u-paris2.fr

© The Author(s), under exclusive license to Springer Nature Switzerland AG 2023

N. Wallenhorst, C. Wulf (eds.), *Handbook of the Anthropocene*,
https://doi.org/10.1007/978-3-031-25910-4_189

Industrial Organizations, Key Factor in Our Entry Into Anthropocene

A commonly held view is that markets are the prime movers of industries. Passively, through supply and demand, they would follow and adapt to human needs. As a consequence, if we want to stop the destruction of our planet, we have to change our behaviours as consumers, and thus drive corporate changes. This is the challenge that cognitive approaches propose to take up, by organising workshops aimed at changing our "software", building new imaginations, and triggering effective changes.

However, this view of the modern economy is at best wrong (Simon, 1996), at worst hypocritical. Our industries have been deliberately designed as legal persons, with strategic capabilities (Porter, 1985) for themselves (Coleman, 1990). Although they are "the most intensive and effective destroyers of environment" (Perrow, 1997), there is no empirical evidence that they engage significant changes for climate nor biodiversity. Worse, according to the knowledge accumulated in organizational studies since the end of the nineteenth century, although we are the creators and participants of industries, there is no theoretical certainty that we are able to control them. Let us see how this is possible.

After 12,000 years of relative socio-technological stability in agriculture-based societies, some human communities suddenly started extracting and burning fossil fuels to replace animal labour, including human labour, by machines.

According to standard economics, buying fossil fuel to use machines is an innovation, i.e. a new product or service that meets a solvent demand. Although fossil fuels are traded on markets, we cannot reduce this innovation to the fact that farmers wanted to buy them. In order to grasp this dynamic, we use a much broader definition of innovation, as the introduction of new practices in relation to existing ones.

This understanding refers to social change, which also needs to be defined. As all social animals, human beings build social systems to survive. But unlike other animals whose social forms do not vary very much over time and space, there are many different kinds of anthropo-social systems, associated to what we call cultures.

One obvious way to highlight social structures, thus structural changes, is to observe social interactions. From morning to night, each of us interacts with others, either face to face or through remote means of communication such as mail, phone, email, chat, SMS, voicemail. If the observer notices regularities over time, then these interactions do not just follow a random law. Regularities demonstrate the existence of so-called structures (Jacob, 1970).

Since industrialization unleashed endless processes of social creative destruction (Dupuy, 2012), human-induced social change has become a main source of contingency, until the physical world's responses to our activities manifest themselves in turn, through climate change and biodiversity loss.

Contrary to the rationalist models of diffusion inherited from the French contemporary disciple of Charles Darwin, Gabriel Tarde (Kinnunen, 1996; Tarde, 1895), innovations such as machines and fossil fuels for agriculture cannot be understood as a series of discrete innovations. They were part of larger structural changes. But, moreover, when it comes to our social structures, it is not so easy to characterize what is really changing. For instance, in the early part of the internet's history, up to the 2000s, one counter-intuitive finding was that social interactions had not become free of physical barriers, contrary to what has been widely claimed or hoped for. Simply, those who already had the ability to travel before Internet have a more extensive electronic network than those who cannot travel (Castells, 2001). From this point of view, if we do not take other aspects into account such as the speed or content of exchanges, the Internet lead more to a crystallisation of existing social structures. During the last two decades, the rapid growth of new forms of digital social networks has probably had impacts that should be measured.

To grasp the evolution of social structures in Anthropocene, and thus the true challenge we are facing if we hope to change our industries, we must change our interpretative framework. Industrial agricultural machines and fuels are not discrete innovations, new tools and products at the disposal of farmers to replace and improve traditional techniques. As many authors have demonstrated since the beginning of modern industrialisation (Bateson, 1972; Coleman, 1990; Polanyi, 1944; Simon, 1996), they are elements of recently evolved creatures, large-scale and free-standing: industrial organizations (IOs).

To analyse their role in Anthropocene, we propose the following definition: *as organizations and organizing are closely related concepts, IO also refers to both the process of industrialization and to the result of this process, established industrial organizations.*

How Do IOs Work and Why Do They Lock Humanity Into a Deleterious Trajectory?

We are not dealing here with the history of IOs but with their capacity to encompass us in their own functioning and for their own ends, regardless of the vital interests of humans.

IO as Social Artefacts

As any human social systems, IO exist because individuals "engage in regularized social practices across time and space"(Giddens, 1984, p. 14). This observable reality also refers to cognitive processes, humans using their innate ability to socialize, i.e. to connect to social systems and to behave in accordance to them. IOs as well are artefacts that we partially share through symbolic communication and that partially determine our behaviours (Cooren et al., 2011; Harari & Sapiens, 2014; Luhmann, 1995).

IO as Artificialisation Processes

Admitting that they produce the goods and service we need, protecting or separating us from physical and biological contingencies, IOs differ from traditional organizations. In their material expression, IOs were concrete means to enter into modernity in Arendt's sense of artificialisation, "cutting the last tie through which even man belongs among the children of nature" (Arendt, 1958).

IO's Managers as Social Engineers for Bureaucratic Control of Human Activities

To enact (Daft & Weick, 1984) new social structures, IOs used new kind of social engineers. Smart in the sense of abstraction and reasoning abilities measured by IQ, creative, in the respect of progressive values, trained in engineering, administration or business schools, but also in sciences, including medicine, those modern managers are expected to apply the bureaucratic control to any human activity (Walton, 2005). Although it takes complex forms on the ground, bureaucratic control is based on some simplistic ideas, mainly formal and impersonal procedures attested by objective measures, i.e. figures, meant to improve organization's performance.

From a critical perspective, bureaucratic control has not to do with objective performance but with power-knowledge dominations (Foucault, 1975; Townley, 1993). To take an image, 1 day, an agronomist knocked on one farmer's shoulder: "you are now obsolete; we will explain you how to produce." Then, open and converging collaboration between engineers from all other disciplines, agronomy, mechanical, chemical, construction, food processing, retail, banking, accounting, marketing, human resources, legal, IT, and so on, started the industrialisation of work, and endless process in which we are still involved. Any activity is targeted, as soon as a self-proclaimed industrial organization's manager (IOM) proposes to rationalise workflows (Brunsson et al., 2012).

IO's as Standardization Process (from Cultivars to CVs)

Following Taylor's seminal work (Taylor, 1911), to industrialize human activities, IOMs need to design standardized workflows:

> Nor has any one workman the authority to make other men cooperate with him to do faster work. It is only through enforced standardization of methods, enforced adoption of the best implements and working conditions, and enforced cooperation that this faster work can be assured. And the duty of enforcing the adoption of standards and of enforcing-this cooperation rests with the management alone. (p.83)

Concretely, standardization aims at reducing diversity in processes, products and services. The belief that standardization generates efficiency is related to: (1) the industrial definition of quality, as conform to a standard, *good* or *bad* quality being related to customer's values; (2) the economies of scales, industrial organization reducing unit production's cost; (3) the externalisation of social and environmental costs; (4) the sham of the slogan "producing more with less", which must be replaced by "producing something else".

CSR as IO's Adaptive Responses

According to the original claims by Taylor himself, scientific management was first meant to help protect environment and forests by making our systems more efficient. Some spurious comparisons, for instance with slash and burn practices, maintain a doubt in favour of IOs.

On the contrary, IOs always remained the main destroyers of the environment. Not only are they the sole social entities able to accumulate such efficiency on our material world, but they also show a remarkable capacity to regenerate in the face of institutional pressures for more responsible activities. Pro-environmental managers started to "responsibilize" IOs, i.e. to make them accountable to the interests of stakeholders (Coleman, 1990). However convinced they may be, environmental IOMs face institutional maintenance processes, for instance when circular figure of compromise leads to status quo (Taupin, 2012) or when competing IOs' start to collaborate at meta-organizational level (Raynard et al., 2021).

IOs Follow Directionless Red Queen Effect to the Death

IOs are competing artefacts devoted to enforce the adoption of standardized practices in all human activities. Their survival and development thus depend on their ability to make existing practices become obsolete, in order to implement and maintain their new ones.

As a corollary, following infinite destructive creation processes and infinite expansion goes with infinite risk (Beck, 1986; Fressoz, 2012). Like evolutionary processes do (Sachse, 2011), IOs may be out of control and directionless, no matter if it leads to our own destruction.

The industrialization of human activities exerts destructive dominance over humans and nature. If we wish to get out of the deleterious world of IOs, the most relevant issue is the global phenomenon of artificialization, in the sense of emancipating ourselves from the laws of nature to depend only on our own artifacts. As long as we want to control our physical, biological, and psychic environment, using the reductionist and deductive logics characteristic of the engineering world, we will remain in a war that ignores itself, against nature, including ourselves.

The reverse is also true. Going to war with IOs seems to me counterproductive and dangerous because of our situation of deep vulnerability. We must rather replace IOs by a new artisanal society, based on experimental knowledge, transmission by peers, without domination by experts. Faced with such an unknown, it is absolutely normal that no practitioner or scientist knows how to define the target nor the path to follow.

References

Arendt, H. (1958). *The human condition*. University of Chicago Press.

Bateson, G. (1972). Steps to an ecology of mind. In *Collected essays in anthropology, psychiatry, evolution, and epistemology*. Jason Aronson.

Beck U. (1986). *Risk society: Towards a new modernity* (Ritter, M. Trans). Sage.

Brunsson, N., Rasche, A., & Seidl, D. (2012). The dynamics of standardization: Three perspectives on standards in organization studies. *Organization Studies, 33*(5–6), 613–632.

Castells, M. (2001). *The internet galaxy*. Oxford University Press.

Coleman, J. S. (1990). *Foundations of social theory*. Belknap Press of Harvard University Press.

Cooren, F., Kuhn, T., Cornelissen, J. P., & Clark, T. (2011). Communication, Organizing and Organization: An Overview and Introduction to the Special Issue. *Organization Studies (01708406), 32*(9), 1149–1170.

Daft, R. L., & Weick, K. E. (1984). Toward a model of organizations as interpretation systems. *The Academy of Management Review, 9*(2), 284–295.

Dupuy, J.-P. (2012). *L'avenir de l'économie. Sortir de l'économystification*. Flammarion.

Foucault, M. (1975). *Discipline & Punish: The birth of the prison*, 2d édition, Vintage (First edition in french: 1975).

Fressoz, J.-B. (2012). *L'apocalypse joyeuse. Une histoire du risque technologique*. Seuil.

Giddens, A. (1984). *The constitution of society. Outline of the theory of Structuration*. Polity.

Harari, Y. N., & Sapiens, A. (2014). *A brief history of humankind*. Publish in agreement with The Deborah Harris Agency and the Grayhawk Agency.

Jacob, F. (1970). *La logique du vivant*. Gallimard.

Kinnunen, J. (1996). Gabriel Tarde as a Founding Father of Innovation Diffusion Research. *Acta Sociologica (Taylor & Francis Ltd), 39*(4), 431–442.

Luhmann, N. (1995). *Social systems*. Stanford University Press.

Maclouf, E. (2020). *Pourquoi les organisations industrielles ne sauveront pas la planète. Ou l'anti manuel de la RSE et du développement durable.*, Le Bord de l'Eau, (En Anthropocène).

Perrow, C. (1997). Organizing for environmental destruction, *10*(1).

Polanyi, K. (1944). *The great transformation*. Rinehart.

Porter, M. E. (1985). *Competitive advantage*. Free Press.

Raynard, M., Kodeih, F., & Greenwood, R. (2021). Proudly elitist and undemocratic? The distributed maintenance of contested practices. *Organization Studies, 42*(1), 7–33.

Sachse, C. (2011). *Philosophie de la biologie*. Presses Polytechniques et Univ.

Simon, H. A. (1996). *The sciences of the artificial*. MIT Press.

Tarde, G. (1895). *Les Lois de l'imitation*. Kimé.

Taupin, B. (2012). The more things change… Institutional maintenance as justification work in the credit rating industry. *Management, 15*(5), 529.

Taylor, F. W. (1911). *The principles of scientific management*. Prlimpton Press.

Townley, B. (1993). Foucault, power/knowledge, and its relevance for human resource management. *Academy of Management Review, 18*(3), 518–545.

Walton, E. J. (2005). The persistence of bureaucracy: A meta-analysis of Weber's model of bureaucratic control. *Organization Studies, 26*(4), 569–600.

Etienne Maclouf is full Professor at the University Paris Panthéon-Assas (LARGEPA), and associate researcher at the CESCO (MNHN-CNRS). He is Doctor in Organization Studies and master degree in evolutionary biology. His research is dedicated to human organizations in Anthropocene. He publishes academic articles and summarized his current working propositions in a recent essay: "Why industrial organizations will not save the planet. Or the anti-manual of CSR and sustainable development" (Le Bord de l'Eau. En Anthropocène, ed. N. Wallenhorst, 2020, in French).

Institution

Jean-Yves Robin and Catherine Nafti

Abstract Debates on the Anthropocene do not only raise geological or climatological questions. What is at stake is a dominant institutional discourse; it represents an undeniable threat to the future of humanity. This doctrine is based on the dogma of infinite growth. Yet, scientific experts show that humanity is heading toward a void if we do not break away from this institutional paradigm.

Mentioning the institution means taking the risk of confusing this authority with another dimension which is just as essential, that of the organization. Admittedly, the organizational means deployed reveal institutional choices, but it is important to distinguish these two realities even if they remain inextricably linked, as the current institutional analysis suggests (Remi, 2002). In the first case, the leader, for example, will have the task of setting out a course, indicating a direction, defending values or goals. The latter are by no means objectives to be achieved, they are aims which can possibly be approached, but they always end up resisting any project of influence. These offers of meaning, despite their partly illusory nature, can attract the support of the actors and ultimately mobilize them. It is therefore important that they can identify with the messages and speeches formulated by the hierarchy in charge of the future of a working group in particular. However, nothing works anymore with the institutions. They are struggling to define a mobilizing project. François Dubet (2002) announced their decline and for the past 20 years, it has become obvious, at least in Europe, that this disenchanting and disenchanted prognosis is to some extent legitimate. The end of great integrators, such as family, work or religion, is evoked, particularly in the West. However, this predicted collapse is partly linked to the fact that actors in many professional sectors, but also believers in certain confessions (Pierre-Louis et al., 2017) sense how the outlined "purpose" is out of step with the embodied "purpose" (Jean-Pierre, 2010). This is how idealism rubs shoulders with a caustic cynicism. For example, it is always possible for a

J.-Y. Robin (✉) · C. Nafti
Catholic University of the West, Angers, France
e-mail: jean-yves.robin@uco.fr; catherine.nafti@uco.fr

© The Author(s), under exclusive license to Springer Nature Switzerland AG 2023

N. Wallenhorst, C. Wulf (eds.), *Handbook of the Anthropocene*,
https://doi.org/10.1007/978-3-031-25910-4_190

Human Resources department to promote the integrative culture of a large industrial group while developing management methods (within the organizational framework) that illustrate how the human factor has become an adjustment variable. And as far as religion is concerned, the different monotheisms have generated many deadly conflicts throughout history, so that the prophecy was somehow proven wrong, which is precisely what the institutional analysis means by the Mülhmann effect (René, 1997).

In such a context, these contradictions end up sowing doubt, then emerges an institutional chaos that will facilitate the development of vocational chaos on the part of individuals. The professions involving relationships with others are, notably in France, particularly affected by this endemic social disease, which appears in the form of disengagement and burn-out. These syndromes are excellent analyzers of the institutional decline because what causes this withdrawal or exhaustion is certainly increasing heavy workload but, in the end, the resulting "ethical suffering" (Christophe, 2015) is first and foremost triggered by the fact that the actors concerned no longer identify with what is being asked of them. This shows how these symptoms reveal a "crisis of meaning", as if the subjects no longer knew which way to turn. They are disoriented – why bother to continue and stay in an activity when working conditions have become so difficult that they push both male and female co-workers to mistreat each other, while sometimes mistreating the users and customers they are supposed to take care of. The latter are then shocked, confused and amazed. It is under these circumstances that violence is sometimes unleashed. It is exploding in the open, in the media, following incidents that could not be regulated both at school and in hospitals, but also in certain public services such as the General Treasury, Social Security or family allowance funds offices. These events illustrate how much suffering develops and takes the form of a more or less suicidal or deadly revolt. When the actors have nothing more to lose, they may yield to the devastating charm of an insurrectional adventure.

This phenomenon is made possible precisely because we are witnessing an implosion, that of the great institutional referents that, in the past, were authoritative: the Church, the army, the police, Justice, business enterprises, schools, families, the Republic, have become very fragile institutions. None of these instances is able to contain this potential lifting of inhibitions, especially since the spaces to say, tell each other and be heard have become scarce. This is why the crisis of the narrative, mentioned in particular by Roland (2011), contributes to the weakening of the democratic space. What are the recognized places that can be invested by everyday actors for them to interact and negotiate today? A singular fact will help us understand. When some health professionals in a hospital structure, while carrying out an unpublished research in 2016, showed their bitterness in front of their management precisely because the spaces for dialogue (practice analysis group – wrap-up meetings – transmission time) were shrinking considerably, they were answered, in the wake of an audit carried out by consultants who were steeped in certainties, that the staff did not know how to get organized. An optimization of the way healthcare was organized would make it possible to avoid unjustified recruitment, thus saving time for this type of activity, but to do so, it was important to reduce informal exchanges and in particular coffee breaks.

In the name of principles derived from rules borrowed from the management paradigm, all the organizations are invited to convert and follow the path set by managers. Thus, the figure of "governance" obsessed with figures ends up dismantling any form of power inspired by the "government" of men (Alain, 2015). After all, hospitals, schools and universities are just like any other companies. As a result, the commodification of services allows many spill-over effects, even if this situation generates some "orientation conflicts" (Frederik, 2006). It would obviously be wrong to believe that this institutional problem only concerns the fields of social work, education or health. Producing cars and planes at a time when greenhouse gas emissions have become a threat to the future of humanity also means raising issues within social and environmental jurisdictions. Concealing this data by inserting software in the vehicles themselves, as it was the case with the Volkswagen group, sends out a clear message. In the end, what pays off is the fact of "lying at work" (Duarte, 2015) or ignoring a certain number of abuses. It would be possible to multiply the examples by describing, among other things, certain practices used in the agri-food industry. In all these cases, some actors no longer recognize themselves in the activities imposed on them; if they do, it is at the cost of a renunciation, of a repression against their sensitivity to the environment that they rub shoulders with. The individual thus becomes this "buffer self" (Charles, 2011) in the impossibility of resonating with what surrounds him. How can we survive on a daily basis when every day the staff of an industrial slaughterhouse becomes accomplice to objectively condemnable practices? Without falling into a specist ideology (a doctrine aiming at challenging discrimination between the animal and human species), it is important to understand that this entry into the Anthropocene disrupts patterns of thought. Thus, will the institutions be able to invoke indefinitely the virtues of competitiveness and growth as if they were intangible values? Many models are to be questioned to invent another world.

The main one is the relationship between culture and nature. The former had believed that it could free itself from its dependence on animal and plant species. Obviously, the survival of future generations will depend on our ability to resonate with this nature which humanity belongs to and to think of new ways of functioning (Hartmut, 2018). To this end, it will be important to promote spaces for dialogue. In other words, deliberation and direct democracy are not vain utopias but rather necessities. It is within these collective spaces that reasonable arrangements could be made, both locally and globally. These cooperation links cannot be built in the immediate future. Such a project requires time. It is probably an exercise that also requires citizenship education. But are business schools, such as engineering schools, willing to renew their way of thinking (Jean-Yves & Benoît, 2017)? Nothing could be less certain. Growth remains a dominant paradigm. This process of "dynamic stabilization", as Rosa says (2018), will jeopardise our common humanity and the future of Planet Earth. Man is somehow dispossessed of his destiny. He cannot escape from this mad rush to production. Like a hamster in its cage, he never stops running to make the wheel turning but he gradually discovers that he obtains no result, condemned to keep his balance in this wheel to repeat an identical movement over and over again. In other words, we are only reproducing the same arguments as in the nineteenth century. It is as if the prefabricated categories of meaning remain frozen, inexorably embedded in the marble of the economic law of the market. However, what characterizes any offer of service is its provisional, precarious, plural and partial nature. And yet, the certainties mentioned in the previous lines are shared by a large number of leaders although they are in agony precisely because they do not help to defend the criteria of the just, especially in terms of the redistribution of resources. Thus it is acceptable that a minority, namely barely 1% of the world's population, holds more than 50% of the world's wealth (Vincent & Fabienne, 2015). This time bomb could well ignite minds and spark conflicts between continents. This is already the case with terrorism. And what will Europe do when it is confronted with unprecedented migratory waves linked to global warming? From now on, we can no longer afford the luxury of waiting. Before our eyes is the need to define a basic international right of citizenship, the right to breathe, to eat, to care for ourselves and to educate ourselves without restraint. Politicians are certainly facing unprecedented challenges. Like all citizens of the world, their responsibility is not only "causal" but also "final" (Hans, 1979). Their challenging task will be to identify horizons of meaning, to take actions that can re-enchant the world, all this without any form of indoctrination. Every leader, every citizen on this planet remain marked by values that are often divergent. But beyond these differences and disputes, a crucial question arises: what common values should be promoted? Without doubt, discussion, deliberation and palaver are the only means we can favour, without arrogance or naivety if we want to avoid the worst. Indeed, the desire to live together is always threatened by the unconscious project of definitively resolving it. Eros and Thanatos are fighting an unceasing battle. Let's hope that our common future will not be marred once again by cruel tragedies. The driving force of history throughout the 19th and 20th centuries shows how difficult it was to escape the process of creative destruction as if it were impossible for one category of humanity not to pursue the symbolic or effective death of the other.

References

Boutinet, J.-P. (2010). *Grammaires des conduites à projet*. PUF.

Choquet, P.-L., Elie, J.-V., Guillard, A. (2017). *Plaidoyer pour un nouvel engagement Chrétien*. Les éditions de l'atelier.

De Gaulejac, V., & Hanique, F. (2015). *Le capitalisme paradoxant – un système qui rend fou*. Seuil.

Dejours, C. (2015). *Le choix – Souffrir au travail n'est pas une fatalité*. Bayard.

Dubet, F. (2002). Le déclin de l'institution. La Découverte.

Gorri R. (2011). *La dignité de penser*. Les liens qui libèrent.

Hess, R. (2002). *"Institution" in: Vocabulaire de psychosociologie* (pp. 181–188). Erès.

Jonas, H. (1979). *Le principe responsabilité, une éthique pour la civilisation technologique*. Cerf.

Loureau, R. (1997). *La clé des champs, une introduction à l'analyse institutionnelle*. Anthropos.

Mispelblom-Beyer, F. (2006). *Encadrer, un métier impossible ?* Armand Colin.

Robin, J.-Y. (2022). Chefs d'établissement: le burn out n'est pas une fatalité.
Robin, J-Y., & Ravelau, B. (2017). Fabriquer des managers, des patrons ou des dirigeants. In *Savoirs, Revue internationale de recherche en éducation et formation des adultes* (pp 9–50).
Rolo, D. (2015). *Mentir au travail*. PUF.
Rosa, H. (2018). *Résonance – une sociologie de la relation au monde*. La Découverte.
Supiot, A. (2015). *La gouvernance par les nombres*. Fayard.
Taylor, C. (2011). *L'âge séculier*. Seuil.

Jean-Yves Robin is a psychosociologist. He is a Professor at the Catholic University of West. He is the author of several books on managers and executives. His latest book: Chefs d'établissement –Le burn-out n'est pas une fatalité ! (Robin, 2022).

Catherine Nafti is a sociologist. She is a lecturer at the Catholic University of West. She is the author of several books on the relationship to knowledge. She is the editor of Actor and Knowledge.

Lack

Sebastián Agudelo

Abstract Lack is considered here from the viewpoint of logic, philosophy and anthropology in a conceptual effort to link it to a compensatory human response in the world, which leads us to the Anthropocene.

First of all, it is important to note that when we consider the noun "lack" as a concept, we can only attribute to it a relative value and we see that it alone does not provide a satisfactory definition. Since it is the denying of something, for this very same reason it lacks positivity. It states its absence. In this sense, the positivity, that which the object lacks, is considered as projecting onto it, revealing its negativity or non-occurrence. But by definition, something that is, is not what it is not, i.e. what it lacks. If something that is lacking in something is not constitutive of it, it seems nonetheless that this negative focus can help us to grasp not the object itself, but a longing for something else. There are two aspects to this. If a negative description of an object is not a definition of what it is, but it is rather a subjective or comparative perception of what something is lacking in order to be what it should be under a certain norm, we have the problem of what the object becomes. And there is also the problem of the subjective view of observers which attributes lacks to objects and therefore reveals a particular standpoint from which they see and interact not only with the objects but also with themselves, who are projecting absences onto these objects. An object cannot be what it lacks, it must also contain an identity, an essence or a positivity not (yet) identified by the mere statement that it is not something else. However this lack is considered in view of a process of becoming or by its observers who have not identified its positivity or are awaiting a new one which is favourable to them. If its negativity is relative, it is because it is relational to the positivity that humans present.

Hegel dealt with the logical consequences of such a negative construction in his dialectic, where he concluded the necessity of a positive side to every conceptual definition. A conceptual definition should not introduce external criteria and must

S. Agudelo (✉)
Independent Philosopher, Berlin, Germany

© The Author(s), under exclusive license to Springer Nature Switzerland AG 2023

N. Wallenhorst, C. Wulf (eds.), *Handbook of the Anthropocene*,
https://doi.org/10.1007/978-3-031-25910-4_191

correspond to its object. It must be distinguished from other concepts and objects, the qualities of which it does not possess, but constitute a negative content. The consideration of a lack can only be determined in a judgment based on its own concept. Thus, an object cannot lack, strictly speaking, something that does not constitute part of its normative, positive definition (its absolute) (Hegel, 1969). For example, the sentence "fish lack the capacity to take in oxygen in the atmosphere" is not a definition of the respiration process of fish because it does not restrict itself to their natural realm (the concept of fish), but to that of amphibian, mammals, etc. A fish is not a deficient mammal, it is a fish as far as it, together with its other qualities, can capture the oxygen it needs in the water.

The difficulty that lack introduces when it is treated as a concept, is that the object for which it stands is compared to another that belongs to a realm that does not conform to it. This gives rise to the risk of misleading our judgment about it from the outset.

The lacks that we see and that we experience start with those that prevent us from satisfying our needs and expectations. Since our origins, human activity has been devoted to adapting the natural world to our vital purposes, which cannot be reduced to mere survival, but yearn perpetually for more (Nietzsche, 1968). Lack is rooted in the human absence of natural dispositions prepared to behave in a world in which human beings become aware of their deficiencies and project external solutions by building what they need or repairing what does not work for them in its current state. This is a classical theme that has made its way from ancient mythology to philosophy and human sciences, including the new turn of environmental, climate and geological science towards a branch of humanities. The lack then becomes objective because it permeates all human experience.

Plato's version of the Prometheus myth points out the necessity to compensate for the lack of inherent attributes with external ones, which were not granted to human beings at the moment of creation, because they had all been previously allocated to the animals (Plato, 1967). Human lack appears relative when it is compared to the anatomical and behavioural completion of animals (Gehlen, 1987) and objective when it is considered as a constant human feeling whereby we always find new grounds for discontent in order to set ourselves in motion. As human beings lack the organs and the coordination to defend themselves, predate and move efficiently, they were given, according to the myth, the mastery of fire, without which they would have not attained the necessary conditions for their survival. Fire is the quintessential example of the techniques and cultural knowledge needed to mediate raw nature. It is also a symbol of a deep need to project onto the future and to foretell what it could bring in order to better prepare for it, even if this preparation omits consequences that can only be thought of afterwards (Stiegler, 1998). The myth stands furthermore for a capacity to trick and disrupt fate, that may be natural or divinely imposed, to account for human interests, both natural (eating, heating, keeping wild animals away) and cultural (use and cult of bones, ceremonies of sacrifice, mythological celebration of fire). The feeling of lack triggers the setting of a wide variety of activities in which positivities that fulfil human life are found, at least temporarily. Thanks to the techniques and the projection of action, human beings' lack of specifically determined nature becomes a chance of making something else of themselves, as the Kantian pragmatic stance rooted in the Age of Enlightenment urges. The lack of instincts accounts for the need for moral behaviour. A lack is both a danger and an opportunity. The lacks in the world are relative to our own inherent lacks and these force us to transform the world in processes through which we end up transforming ourselves. These human processes are always going on. This may be because they have not yet attained the positivity they were looking for, or a new positivity is discovered after attaining the precedent one, or the by-products of these processes reveal a whole new set of negativities calling for other positive resolutions. These transformations of the world and of ourselves are direct and indirect, intentional and unintentional, but mostly they are driven by an inextinguishable human impulse to do more, to aspire for more, a perpetual craving and an unending preparation for our repeated needs. Lacks evolve and as such they are never overcome in the historical process. But it can also be that these lacks are just a prejudice introduced into the concept of "human beings" by comparisons with ideals that do not necessarily conform to its domains. The question of what human beings are lacking, if they are actually lacking something objectively as a species, cannot be conclusively resolved and remains open to historical and planetary development.

The Western tradition, shaped by monotheism, confers a flawed original nature to human beings who need to redeem themselves by the practice of what it is good, just or beautiful. At the same time, and despite the defects of creation it bestows on them (moral, gnoseological or biological), it enthrones them as masters of the natural kingdom, which becomes theirs by divine right (Genesis 1:28) and serves them as a framework for their efforts to redeem themselves. On the threshold of modernity, with Descartes and Leibnitz, this right attains the form of a rational pursuit to accomplish both the *mathesis* and the *sapientia universalis*. To know the world means to express, in a given and minimal set of symbols (geometrical or mathematical), all its attributes and their possible combinations, as well as the ability to do something by making use of this knowledge (to predict or improve natural or human behaviour). To be able to foretell and to transform the world into wished for circumstances means to get to know it. Encyclopaedic knowledge, encoded little by little in a non-genetic written form that may remain available to whoever learns to read its alphabet, needs to be proven by the duplication and homogenisation of the phenomena that it describes and manipulates (Leibniz, 1998). The complexity of the phenomena it describes requires a constant actualisation and correction of theories, a belief in the acquired truth of which leads to the correction of failures in practices (Peirce, 1877).

The Anthropocene is the totality of unexpected effects of the rational pursuit of the mastery of nature which is, to say the least, still far away from breaking through the complexity of natural phenomena and influencing its course. If we understand the Anthropocene as a particular moment of Earth's dynamics in which direct or indirect human action has an impact on most living organisms and natural phenomena, it is also to be considered in the light of the same relative relationship in which humans consider objects which lack something to meet their expectations. However, the lacks in Earth's natural systems are different from the primitive lacks that human beings projected and that drove their activity in the first place. They are the result of it. They are an induced lack in natural systems, an acquired negativity, which is the product of the excess of human activity and products to compensate for, overcome and surpass their own (self-assumed) deficiencies by continually meeting and actualizing personal and societal expectations. The Anthropocene is the result of a progressive overloading of the Earth with human activity, products and by-products to the point of disrupting the balance in natural processes that it had gradually attained in the long geophysical and biological evolution.

The human process of satisfying present and future lacks creates a surplus of positivities which, in turn, give rise to new and deeper lacks, both objective and subjective. The results of the progressive transformation of the world actualize the relative human lacks and objectify them as a general loss of nature. The thrill of transforming nature for exclusive human benefits turns into a drama in which most living systems and habitats start lacking the conditions to host life as we know it, including the artificial and compensatory life of humans. Responses to this crisis frequently follow the same rationality that caused the problem. The objective loss can be measured in the quantity and diversity of fauna and flora through overwhelming and indiscriminate exploitation, including the abrupt modification and impoverishment of much of their natural habitats. It goes in parallel with the deregulation of natural phenomena and climate warming. The Anthropocene is a Pyrrhic prize won in the battle of the artificial over the organic in the war for the mastery of the world. The paradox is that this triumph is more of a failure, and the way to reverse it is via a better and more diverse knowledge that gives access to a conception of mastery, not of the world itself but of our relationship with it, that calls for a cross-cultural image of human beings (Wulf, 2022).

In this sense, and even if the conditions that the Anthropocene imposes have an impact on all members of the human (and other living) species around the world, it is important to recognize that the term has been accused of cultural and social class asymmetry and it could be considered as a "climate regime" (Latour, 2017). It imposes a responsibility on all cultural traditions and the different social stratifications among them that they do not necessarily have. In many cases, other cultures and social groups actually endorse a non-anthropocentric or a non-hyper-consumerist knowledge and practice, by deviating from the welfare standard of the modern world while encompassing a different positive concept of the dynamics of life in general (Escobar, 2018; Descola, 2013).

The Anthropocene is thus a concept that actualizes on a planetary level the role of human beings and their burden of responsibility, in order to reconsider lacks and excesses from multiple perspectives that can help in the historical process of their compensation. This is as much techno-economical as psycho-sociological, i.e. it addresses the relationship that humans have with the world and with themselves. The feeling that something is increasingly missing in the world and our lives calls for a response that attacks this misfortune, which is no longer only human but that of most living beings. The opportunity arises to turn a long-standing Western prejudice, which has led to imagining and imposing an overarching mastery of existence out of the feeling of lack, into a new rationality in which human interests are understood to be intertwined with those of other human groups and living species, for whose survival we are now responsible.

References

Descola, P. (2013). *Beyond nature and culture*. Chicago University Press.

Escobar, A. (2018). *Designs for the Pluriverse: Radical interdependence, autonomy, and the making of the worlds*. Duke University Press.

Gehlen, A. (1987). *Man. His nature and his place in the world*. Chicago University Press.

Hegel, G. W. F. (1969). *The science of logic. Allen & Unwin.*

Latour, B. (2017). *Facing Gaia. Eight lectures on the new climate regime*. Polity Press.

Leibniz, G. W. (1998). *Philosophical texts*. Oxford University Press.

Nietzsche, F. (1968). *The will to power*. Random House.

Peirce, C. S. (1877). The fixation of belief. *Popular Science Monthly, 12*(1), 1–15.

Plato. (1967). Protagoras. In *Plato in twelve volumes* (Vol. 3) (W. R. M. Lamb, Trans.). Harvard University Press/William Heinemann Ltd.

Stiegler, B. (1998). *Technics and time, 1: The fault of Epimetheus*. Standord University Press.

Wulf, C. (2022). *Education as human knowledge in the Anthropocene. An anthropological perspective*. Routledge.

Sebastián Agudelo is an independent researcher and translator based in Berlin. He holds a PhD in Philosophy from the University of Paris VIII and his work is mainly concerned with problems of language and technological development from an evolutionary perspective. He uses and reflects on approaches from Historical and Philosophical Anthropology, Pragmatism, Critical Theory and Transcultural Aesthetics.

Poverty

Aasha Kapur Mehta

Abstract This article examines the links between Covid-19, lockdowns, massive job losses and the rise in poverty, all of which are rooted in the negative externalities of human behaviour or the downside of the Anthropocene. The rise in the number of the poor due to the pandemic is a setback for global efforts to achieve SDG1 or "end poverty in all its forms everywhere" by 2030. This article proposes that efforts to end poverty must be understood in the context of its dynamic nature as well as the processes and factors causing movement into and out of it. Since "humans are the main driver of planetary change" in the Anthropocene, unless "human systems" are "targeted to do something about it", achievement of SDG1 is likely to remain elusive.

The Context

The Covid-19 pandemic stopped humanity in its tracks and caused fear, anxiety, lockdowns, closure of businesses and massive job losses that led to an increase in poverty and hunger and a fall in Gross Domestic Product across the globe. This contagious, ill health related shock forced us to recognise the negative externalities of human behaviour or the downside of the Anthropocene. Not only did it highlight the inter-connectedness of our relationship with other human beings, life forms, nature and planetary forces but it also forced us to understand the extreme vulnerability of our existence. This was a shock that affected all countries and people, both rich and poor.

An event that occurred in one country wreaked havoc across the developed and developing world, infected 224.8 million people and killed 4.6 million people (Worldometers, 2021). Hospitals in the richest nations were unable to cope with the massive and sudden escalation in the need for medical care. Shortages of life saving drugs, and even oxygen, led to a rising count of preventable deaths. One estimate is

A. K. Mehta (✉)
Institute for Human Development, New Delhi, India

© The Author(s), under exclusive license to Springer Nature Switzerland AG 2023

N. Wallenhorst, C. Wulf (eds.), *Handbook of the Anthropocene*,
https://doi.org/10.1007/978-3-031-25910-4_192

that this pandemic alone, may have pushed some 100 million people into extreme poverty, the worst setback in a generation (World Bank, 2020 cited in HDR 2020). Another estimate of "COVID-19-induced poor" for 2021 is that it may "rise to between 143 and 163 million" (Lakner et al., 2021).

Drivers, Maintainers, Interrupters of Poverty and the Anthropocene

Global efforts to "end poverty in all its forms everywhere" by 2030 (SDG1) need to be understood in the context of the dynamic nature of poverty as well as the processes and factors causing it and movement into and out of it. In the age of the Anthropocene, human decisions and actions can drive people into poverty as well as perpetuate it across the globe, leading to its transmission to future generations.

The Covid-19 pandemic is only one of many shocks that cause poverty. Analysis of panel data and a review of the literature point to several factors that act as 'drivers' forcing people into poverty. These could be the sudden onset of an expensive illness such as cancer, diabetes, a heart attack, or as recent events have shown, Covid-19. A disaster such as a flood or earthquake, a failed crop, a failed investment or a policy change that leads to a loss of livelihood or reduction in income, a sudden disability, loss of work such as due to the lockdown and economic downturn or high interest debt can also drive people into poverty.

On the other hand, illiteracy, lack of skills, lack of access to basic services, living in a remote geographic location that provides few livelihood opportunities, poor access to healthcare facilities or debt, may result in people getting stuck in poverty or 'maintain' them in poverty. Several factors can 'interrupt' or enable escape from poverty. These include access to diversified income sources, linkages with urban areas, improvements in rural infrastructure, accumulation of human, physical and financial assets, access to water for irrigation, increase in wages and social safety nets (Bhide & Mehta, 2004; Mehta & Shepherd, 2006; Mehta et al., 2011).

Actions to End Poverty in the Age of the Anthropocene

An unacceptably high proportion of the global population remains severely deprived in multiple dimensions. SDG 1, Targets 1.1 and 1.2, require that by 2030, we eradicate extreme poverty for all people everywhere, currently measured as people living on less than $1.25 a day; and reduce at least by half the proportion of men, women and children of all ages living in poverty in all its dimensions according to national definitions. The concerted and coordinated global effort that is needed to address this, is missing. Since "humans are the main driver of planetary change" in the Anthropocene unless "human systems" are "targeted to do something about it", achievement of SDG1 is likely to remain elusive.

The first step that needs to be taken to "end poverty in all its forms everywhere" is to ensure its accurate measurement. We have systematically undercounted the true extent of global poverty by using extremely low thresholds for measuring it. The World Bank estimates of poverty were initially set at PPP $1 a day in 1990 and were the average of the poverty lines of the poorest 15 countries. (Ravallion et al., 2008). This was revised or adjusted each time a new set of PPPs were produced by the International Comparison Program – PPP $ 1.08 in 1993, PPP $1.25 a day in 2005 and PPP $1.90 in 2015 (Mehta et al., 2018 Springer). Yet the SDGs use this "implausibly low threshold" to measure poverty, perhaps because a "more honest approach would force us to face up to the fact that the global economy simply isn't working for the majority of humanity."

As Hickel (2015) points out, "$1.90 is not enough for basic human survival.... this amount of money is inadequate to achieve even the most basic nutrition. The US Department of Agriculture calculates that in 2011 the very minimum necessary to buy sufficient food was $5.04 per day. And that's not taking account of other requirements for survival, such as shelter and clothing."

If poverty lines are raised to realistic levels, the percentage of people in poverty will be much larger than the present estimates. For instance, the number of poor people in the world in March 2021 increases from 696 million with a poverty threshold of $1.90 to 1821 million if $3.2 is used and 3269 million if the poverty threshold is set at $5.5 (Aguilar et al., 2021). It is also important to recognise that a significant proportion of those who are above the poverty line are vulnerable to falling into poverty.

Research shows that there is a 'geography of poverty' since it is concentrated in certain spatial locations, both globally and within any country. There is also a 'sociology of poverty,' since the proportion of those who are poor is higher among certain marginalised groups (Mehta et al., 2011). Indigenous peoples remain notably disempowered, and the 1.3 billion indigenous people living in areas endowed with forests have some of the highest poverty rates in the world. Moreover, they are victims of violence, with several of their leaders killed in connection with their environmental activism (UNDP, 2020: 202).

Poverty is also concentrated among certain occupational groups such as those who depend on casual agricultural labour in rural areas or casual labour in urban areas as well as those who lack assets or have meagre or poor quality assets (Gaiha, 1989; Bhide & Mehta, 2008; Government of India, 2008).

Ending poverty in all its forms everywhere, requires concerted efforts to enable these occupational groups, social groups and geographical locations to move out of poverty. Not only must poverty be measured accurately but the resources allocated to addressing it must be adequate. SDG1 Target 1a requires that we "Ensure significant mobilization of resources from a variety of sources, including through enhanced development cooperation, in order to provide adequate and predictable means for developing countries, in particular least developed countries, to implement programmes and policies to end poverty in all its dimensions". In practice however, resources are far from adequate.

For instance, in countries such as India, casual agricultural labour is the largest occupational group that is stuck in poverty. These are the "working poor", for whom the State has not been able to secure the right to an adequate means of livelihood. The Mahatma Gandhi National Rural Employment Guarantee Act (MGNREGA) was legislated "to provide for the enhancement of livelihood security in rural areas of the country by providing at least one hundred days of guaranteed wage employment in every financial year to every household whose adult members volunteer to do unskilled manual work" (Government of India, 2005). While this right to work legislation is extremely important, it is a truncated right for two reasons. First, the number of days of work that can be demanded is limited to 100 days. Second, even this 100 days of work is given jointly to all adults in a family and not to each adult individually. Additionally, in practice, very few households actually get 100 days of work under the programme. Hence, the financial and physical resources allocated to poverty reduction programmes need to increase substantially, in view of the massive scale on which poverty is experienced.

In this context, SDG 8 Target 8.5 requires that we achieve full and productive employment and decent work for all women and men, including for young people and persons with disabilities, and equal pay for work of equal value by 2030. Target 8.6 requires a substantial reduction in the proportion of youth not in employment, education or training by 2020 but this has not been achieved.

We need to prioritise fair wages and salaries for all work, minimum wages that are not exploitative and decent hours of work for all those who can work to survive. This should apply both within and across countries.

The work from home culture and online work across countries has led to exploitation of labour in developing countries. While those in the corporate sector in developed countries maintain time and space boundaries based on office hours, the workforce in the developing world works beyond office hours in order to meet project needs and participate in online meetings. Not only is no extra payment made for the additional hours of work, but additionally, this has negative consequences for the health of the youth in the developing world. This is neither fair nor decent work. Exploitative work conditions, both between and within countries must stop. Instead, invest in activities that enable good health, in school, at work and in public spaces.

The Covid-19 pandemic is only one of many shocks that affect poverty. Human activity has been a geologically recent, yet profound, influence on the global environment. The pandemic also showed us that curtailment of the negative consequences of daily activities could create positive externalities for the planet. Just a few months of lockdown led to a massive reduction in pollution. Clear skies replaced the smoke-filled haze that we have become accustomed to and stars which had disappeared made their appearance. We have extracted, mined, exploited, degraded, over used and contaminated the land, forests, oceans, rivers and mountains and all the abundant resources that we inherited from earlier generations. Natural resources must be used sustainably.

The pandemic also showed us that cooperation across countries can push the frontiers of science. Vaccines were created in record time. If we can find a vaccine for Covid-19, we can find ways of preventing and curing TB, malaria, diabetes, etc.

Ill health is caused by a large number of diseases that can be prevented but are not because of lack of investment in research that can find a cure. Even if a cure is found there are questions of access and affordability. The unequal distribution of Covid-19 vaccine has made it all too clear that there are haves and have nots, even in vaccine access. This, despite the fact that the survival of humankind depends on ensuring that populations across countries get access to the vaccine to prevent the generation of new mutants to which everyone will be susceptible.

SDG 1 Target 1.3 attempts to provide a safety net for the most vulnerable by requiring the implementation of nationally appropriate social protection systems and measures for all, including floors. It also requires that by 2030 we achieve substantial coverage of the poor and the vulnerable. However, existing social protection measures and floors remain far from adequate. A lot more needs to be done by Governments and communities if we are to translate SDG1 from words to action and reality and end poverty in all its forms everywhere.

References

Aguilar, R. A. C., Fujs, T., Lakner, C., Mahler, D. G., Nguyen, M. C., Schoch, M., & Viveros, M. (2021, March 16). https://blogs.worldbank.org/opendata/march-2021-global-poverty-update-world-bank

Bhide, S., & Mehta, A. K. (2004). *Correlates of incidence and exit from chronic poverty in rural India: Evidence from panel data*. Working Paper 15. Chronic Poverty Research Centre, University of Manchester and Indian Institute of Public Administration.

Bhide, S., & Mehta, A. K. (2008). *Economic growth and poverty dynamics*. Working Paper 36. Chronic Poverty Research Centre, University of Manchester and Indian Institute of Public Administration.

Gaiha, R. (1989). On estimates of rural poverty in India: An assessment. *Asian Survey, 29*(7), 687–697.

Government of India. (2008). *Eleventh five-year plan: Inclusive growth*. Planning Commission.

Government of India. (2005). *The Mahatma Gandhi National Rural Employment Guarantee Act 2005*. Available at: https://rural.nic.in/sites/default/files/nrega/Library/Books/1_MGNREGA_Act.pdf.

Hickel, J. (2015). *The Guardian, international edition*, https://www.theguardian.com/global-development-professionals-network/2015/nov/01/global-poverty-is-worse-than-you-think-could-you-live-on-190-a-day, Downloaded March 25, 2018.

Lakner, C., Yonzan, N., Mahler, D.G., Aguilar, R. A. C., & Wu, H. (2021). https://blogs.worldbank.org/opendata/updated-estimates-impact-covid-19-global-poverty-looking-back-2020-and-outlook-2021

Mehta, A. K., & Shepherd, A. (2006). *Chronic poverty and development policy in India*. Sage Publications.

Mehta, A. K., Shepherd, A., Bhide, S., Shah, A., & Kumar, A. (2011). *India chronic poverty report: Towards solutions and new compacts in a dynamic context*. Indian Institute of Public Administration.

Mehta, A. K., Bhide, S., Kumar, A., & Shah, A (Co-editors) (2018). Poverty, chronic poverty and poverty dynamics: Policy imperatives. Springer Nature.

Ravallion, M., Chen, S., & Sangraula, P. (2008). *Dollar a day revisited* (Policy research working paper series 4620). The World Bank.

United Nations. (2020). *We can end poverty: Millennium development goals and beyond 2015.* https://www.un.org/millenniumgoals/poverty.shtml. Accessed 18 November 2020.
UNDP. (2020). *Human development report 2020: The next frontier: Human development and the Anthropocene*. UNDP.
World Bank. (2020). Poverty and shared prosperity 2020: Reversals of fortune.
Worldometers. (2021). Available at: https://www.worldometers.info/coronavirus/ Last updated: September 11, 2021, 18:12 GMT.

Professor Aasha Kapur Mehta is Chairperson, Centre for Gender Studies at the Institute for Human Development, Delhi. Prior to this, she was Professor of Economics at Indian Institute of Public Administration for many years. She has an MA in Economics from Delhi School of Economics, an M.Phil from Jawaharlal Nehru University and a doctorate from Iowa State University, USA. She has conducted a large number of research and consultancy assignments for the Government of India and for international organisations. She was invited to serve on several Committees constituted by the Government of India. She led the work of the Chronic Poverty Research Centre in India for more than a decade. Her books, articles and working papers are on poverty, poverty dynamics, human and gender development indicators, gender budgeting, deprivation and data gaps. Among the large number of books, monographs and articles that she has authored, edited and co-authored are "Poverty, Chronic Poverty and Poverty Dynamics: Policy Imperatives, Springer 2018; "India Chronic Poverty Report: Towards Solutions and New Compacts in a Dynamic Context", CPRC-IIPA 2011; "Chronic Poverty and Development Policy in India", Sage 2006; and "Gendering the Twelfth Plan: A Feminist Perspective", Economic and Political Weekly, Vol 42 No 17, April 21–27, 2012.

Part XVI
Societies and Habitat

Architecture

Emeline Curien and Mathias Rollot

Abstract Only considering that the building sector generates nearly 40% of annual global CO_2 emissions (UN Environment, IEA. Global Status Report 2017, 2017) may probably be enough to justify the presence of an « architecture » entry in this *Handbook of the Anthropocene*. Yet this article broadens the question to clarify both what is behind the complex concept of « architecture » and in which sense it may be a key point if «our job is to make the Anthropocene as short/thin as possible » (Haraway D. Staying with the trouble: making Kin in the Chthulucene. Duke University Press, 2016: 100). Assuming it is not « too late », the article will then try to show how architecture, despite its disastrous past achievements, could paradoxically become a fantastic way to shape truly ecological societies.

First of all, if architecture should matter today, it is because of its very large *ecological impacts* (if « architecture » is understood in the sense of the building/the built environment), as much as its *ecological responsibilities* (if « architecture » also means the work of architects and the architectural discipline). A good way to understand these issues is probably to consider the *concrete age* we have known for a century. Energivorous to produce, but also almost impossible to recycle, concrete has been recently characterized by *The Guardian* as nothing less than « the most destructive material on Earth », for its CO_2 emissions but also for the water, sand and corruption issues it generated this last decades worldwide (Watts, 2019). Concrete would also be easy to accuse for the architectural, functional and aesthetic uniformization that took place worldwide, often against the « art of dwelling » of vernacular local cultures (Illich, 1992), the artisanal know-how and local bio- and geo-sourced low-tech activities. Yet, as Vaclav Smil noted in his seminal *Making the Modern World: Materials and Dematerialization,* « perhaps no other comparison illustrates the scale of this concretization better than this one: consumption of

E. Curien (✉) · M. Rollot
Ecole nationale supérieure d'architecture de Nancy, Nancy, France
e-mail: emeline.curien@nancy.archi.fr; rollot.m@grenoble.archi.fr

© The Author(s), under exclusive license to Springer Nature Switzerland AG 2023

N. Wallenhorst, C. Wulf (eds.), *Handbook of the Anthropocene*,
https://doi.org/10.1007/978-3-031-25910-4_193

cement in the USA totalled about 4.56 Gt during the entire twentieth century – while China emplaced more cement (4.9 Gt) in new construction in just three years between 2008 and 2010, and in the three years between 2009 and 2011 it used even more, 5.5 Gt (NBSC, 2013). » (Smil, 2014: 132)! A clear comparison that perfectly shows both how the problem isn't the same worldwide, and also how it tends to become more and more problematic as the global acceleration goes on.

Yet concrete isn't the only responsible for the current architectural impacts and responsibilities, as far as the vertiginous set already built by the anthropocenic western cultures is also based on a considerable sum of *junkspaces* – in the sense that architect Rem Koolhaas gave to this term: a short-term, cheap and easily replaceable construction, that configures space as sealed packaging, always interiors and disconnected from the terrestrial multiple realities by technical means such as air conditioning, artificial light or elevators (Koolhaas, 2000). May it be shopping centres, airports, hospitals or night-clubs, *junkspaces* are all built with many « junkmaterials » that are neither sustainable nor durable; nor qualitative nor recyclable. It's quite not a surprise that the « modern-liquid » era (Zygmunt Bauman) transformed *architecture* in such a consumable industrial space-product, at least for who may agree with philosopher Günther Anders remarking that « production lives from the death of the single product (which has to be purchased again and again). In other words, the 'eternity' of production rests on the mortality of its products » (Anders, 1956: 91). Architecture is no exception to the rule, as its now fully following the whole industrial economic system based on the fact that *production design its products as the wastes of tomorrow, production is a waste production*. Yet the main problem with edification being the size and the quantity of the products. As each *junkspace* only serves for a very limited time before being old-fashioned or becoming *obsolete* (Tischleder & Wasserman, 2015), from it birth it already represents tons of pollutant composite chemical materials, whose only destiny is to be « thrown away » – forgetting that, in a biospheric perspective, « you can't throw anything away because there is no « away » » (Sale, 1985: 118)– without any chance to enter a sustainable becoming. So that « there is already more Junkspace under construction in the twenty-first century than survived from the 20th… » (Koolhaas, 2000: 408).

From this point of view, the anthropocene's era not only fulfilled the earth with a huge quantity of buildings and artificialized ground surfaces; it also constructed a huge time bomb, a delayed crystalized pollution that awaits us in a very short future, in the sense that at the end of its short-life terme, this built junk-mass promises to be very difficult either to repair or to be conserved, both impossible to transform and to recycle to fit their new environments. Also built to deploy the full potential of fossil fuels, thesebuildings are quite unable to function without a phenomenal amount of energy to be heated, cooled and ventilated, so that it is difficult to imagine they could resist the different « energy-scarce world » scenarios that emerges today, even from the most serious and official institutions. However, junkspace continues to be built every day, in such a way that the United Nations expects global space cooling to double its energy consumption by 2040 (UN, 2021: 6)…

Architecture – designed or not by architects – also becomes a crucial issue as the world is largely driven by a constant urbanization phenomenon that makes the global population more and more urban each decade – 33% in 1960, 56% in 2020 (UN Population Division, 2018) –, and since predictions announce that, by 2050, more than 68% of the world population projected to live in urban areas (UN Department of Economic and Social Affairs, 2018). In this context, by 2060, « global material use is expected to more than double » and « the floor area of the global buildings sector projected to double » (UN, 2021: 29); mostly in countries that « do not currently have mandatory building energy codes in place » for now (IEA/OECD, 2017). Without ecological education and rules to constrain the markets, it is likely that more urbanization means above all more impervious soils and therefore more floods, more human constructions and therefore less wild ecosystems, more infrastructures and *Large technical systems* (LTS) (Hughes, 1983); and therefore less autonomous and resilient micro-grids of transports, energy and food productions. And finally, more slums. Indeed, despite efforts from the international communities in improving slums and preventing their formation, slums continue to grow worldwide, in the sense that the United Nation itself admits that « absolute numbers continue to grow », so that « the slum challenge remains a critical factor for the persistence of poverty in the world ». In 2016, « one in eight people live in slums », representing « around a billion people » living « in slum conditions » (UN, 2016).

Will societies choose to follow the « solutions » of the *home automation*, *smart cities* and *smart landscape* (Koolhaas, 2015) sold by tech companies to answer those issues – leading to centralized, expert and technology-driven « sustainable » megalopolis for the wealthiest? Or will their choose the path of degrowth, of vernacular & low-tech solutions (Watson, 2020) and of « eco-decentralist design » (Todd & Tukel, 1981) initiated by the « Green Cities » movement (Berg et al., 1989)? Whatever solutions may be taken, a fundamental point lays in the fact that each technical option we choose will also have its certain impact on our mammal bodies and psyches, our cultural local imaginaries and cosmologies.

Indeed for now, modern architecture has been driven by the modern paradigm of the private and the control, the « development » and the economic growth, the idea of a linear constant scientific progress and its blurred link with mystified social progresses, all this being embedded in the overall *paradigm of the new* (Rosenberg, 1960). But many hints tend to prove that the Anthropocene era imposes to shape an architecture that would, on the contrary, help us to slow down, to let things happens without us (what Lurianic Kabbalah called *Tzimtzum*), to reconnect to a conscience and a knowledge of the societies and *milieus* that we inhabit, and above all, to reconsider our deep personal and cultural relationship with the biosphere so that it can really become a shared articulation of more-than-human bioregions (Glotfelty & Quesnel, 2014). In this sense, architecture may even represent a very interesting medium to consciously reshape our ways of life and our mental structures. Far away from staying the perfect tool for extractivism it was, from pursuing the pure anthropocentric views that characterized our modern era, and from reinforcing the structures of social domination with coercive-adapted spatial orders, architecture could rather become the perfect way to follow « The Earth Path », and « ground our spirits in the rhythms of nature » (Starhawk, 2016).

Finally, if the timeless « art of *aedification* » have something to do with our specific *anthropocene* era, and if it may become a solution rather than the main problem, it's probably thanks to the fast-extending scope of competences and interests that transforms the architectural discipline into a field that possess ecological opportunity. Or, to say it with designer Bruce Mau: « If you think about architecture as a methodology – independent from the outcome – you would see that architecture has a deep culture of synthesis informed by civic values. If you have this capacity, that's the most valuable capacity of this time in history » (Mau, 2012: 26). Indeed, « architecture » as discipline already changed, and continues to change so much that few researchers already talk about « post-architecture » (Harriss et al., 2021). This because, as we saw it, tackling the ecological issues of our era will not require to design new labelled industrialized green-constructions, but mostly means to act a very deep restructuring of the building sector and professions, the relationship we nourish with domestic and shared space, the hierarchy we establish between experts and vernacular, the dualism we trace between human and non-human. There is no doubt that, expending its field of interest far from its original « art of *aedification* » ground, architecture as discipline could be considered the perfect tool to address the issues highlighted by the *ecological humanities* (Rose & Robin, 2004). In this sense « the expansion of the discipline and the experimentation with alternate forms of architecture practice is not simply a new trend, but a survival tactic » (Harriss et al., 2021: 14): not only for the architects and architectural discipline, but for the whole of humanity!

References

Anders, G. (1956). On promethean shame. In M. Christopher John (Ed.), *Prometheanism. Technology, digital culture and human obsolescence*. Rowman & Littlefield, 2016.

Berg, P., et al. (1989). *A Green City program for the San Francisco Bay area & beyond*. Planet Drum Foundation, Wingbow Press.

Glotfelty, C., & Quesnel, E. (Eds.). (2014). *The biosphere & the bioregion. Essential writings of Peter Berg*. Routledge.

Harriss, H., et al. (Eds.). (2021). *Architects after architecture. Alternative pathways for practice*. Routledge.

Haraway, D. (2016). *Staying with the trouble: Making Kin in the Chthulucene*. Duke University Press.

Hughes, T. P. (1983). *Networks of power. Electrification in Western society. 1880–1930*. Johns Hopkins University Press.

IEA/OECD. 2017. Energy technology perspectives 2017. .

Illich, I. (1992). « Dwelling. Address to the Royal Institute of British Architects, York, U.K., July 1984 », In *The Mirror of the Past. Lectures and Addresses 1978–1990*. Marion Boyars Publication.

Koolhaas, R. (2000). Junkspace. In Chung et al. (Eds.), *Harvard design school guide to shopping* (Vol. 2). Taschen.

Koolhaas, R. (2015). « Smart landscape: Intelligent architecture ». *Artforum*.

Mau, B. (2012). « The massive changer ». In R. Hyde (Ed.), *Future Practice: Conversations from the Edge of Architecture*. Routledge.
NBSC (National Bureau of Statistics of China). (2013). Statistical Data, http://www.stats.gov.cn/english/
Rosenberg, H. (1960). *The tradition of the new*. Horizon Press.
Sale, K. (1985). *Dwellers in the land. The bioregional vision*. Sierra Books Club.
Smil, V. (2014). *Making the modern world. Materials & dematerialization*. Wiley.
Starhawk 2016. The earth path. Grounding your spirit in the rhythms of nature, .
Todd, J., & Tukel, G. (1981). *Reinhabiting cities and towns: Designing for sustainability*. Planet Drum Foundation.
Tischleder, Babette B., Wasserman, Sarah (eds.). 2015. Cultures of obsolescence. History, materiality, and the digital age, Palgrave Macmillan.
Rose, D. B., & Robin, L. (2004). « The ecological humanities in action: An invitation ». *Australian Humanities Review*.
UN Department of Economic and Social Affairs. (2018). *Population dynamics, World Urbanization Prospects, The 2018 Revision.*
UN Environment, IEA. (2017). *Global Status Report 2017.*
UN Environment. 2021. 2021 *Global status report for buildings and construction: Towards a zero-emission, efficient and resilient buildings and construction sector.*
UN Habitat. (2016). *Slum almanac 2015/2016: Tracking improvement in the lives of slum dwellers.*
UN Population Division. (2018). *World urbanization prospects: 2018 revision.*
Watts, J. (2019). « Concrete: The most destructive material on earth ». *The Guardian*, Mon 25 Feb 2019. https://www.theguardian.com/cities/2019/feb/25/concrete-the-most-destructive-material-on-earth
Watson, J. (2020). *Lo-TEK design by radical Indigenism*. Taschen.

Emeline Curien (1984) teaches at ENSArchitecture of Nancy (France) and is a researcher at the Laboratoire d'histoire de l'architecture contemporaine. She trained as an architect and completed a Ph.D. in Art history. She conducts research on contemporary European architecture and the broader ecological question. She is the author of the books *Gion A. Caminada, S'approcher au plus près des choses* (Actes Sud, 2018) and of *Pensées constructives, Architecture suisse alémanique 1980–2000* (Fourre-Tout, 2019).

Mathias Rollot (1988) teaches at ENSArchitecture of Grenoble (France) and is a researcher at the Centre for research on sound space and urban environment (CRESSON). He trained as an architect and completed a Ph.D. in Architecture. He is the editor and author of more than fifty papers and fifteen books about ecology and contemporary architecture, such as *L'obsolescence* (Métispresses, 2016), *Critique de l'habitabilité* (L&S, 2017), *Qu'est-ce qu'une biorégion?* (Wildproject, 2021) & *Les territoires du vivant* (Wildproject, 2023).

Cities in Transition

Benoît Dugua

Abstract Cities and metropolises, places of human vulnerability, are at the heart of the Anthropocene as much as of its solution. The socio-ecological transition is defined as a profound transformation of territorial metabolism which required to "re-enchant" planning. The big challenge is to involve a large population in the formulation of planning choices and, above all, to link together the alternative proposals arising from the many places of bottom-up initiatives. The implementation of transition policies is conditioned by a new transactional process among stakeholders. From "cities in transition" to "cities in transactions", anthropocenic modalities of urban planning are still to be defined.

In "The trajectory of the Anthropocene", Stephen et al. (2015) refer to the global social and environmental upheavals that have particularly affected the Earth system since the early 1950s as a "great acceleration". The phenomenon is closely associated with the process of global urbanization in diverse ways. Historically, the transition from the medieval city to the industrial age resulted in a major transformation of the urban systems. Cities started relying more and more on fossil fuels (coal, gas and oil) and steadily increased their consumption of material resources and their production of polluting emissions, discharges and waste. As "major places of all accumulations - material, social and economic - cities and metropolises are at the heart of the climate problem as much as of its solution" (Da Cunha & Thomas, 2017, p.19). Today, the 40,000 world largest cities emit more than 70% of the total amount of greenhouse gases and consume between 60% and 80% of the world's energy, even though they occupy less than 1% of the land area (Satterthwaite, 2008). The change takes place in a relatively short period of time (about two hundred years) compared to the evolution of human history and the beginning of sedentarisation and agriculture (end of the ice age, about 10,000 years BC).

B. Dugua (✉)
University of Reims Champagne-Ardenne, Reims, France
e-mail: benoit.dugua@univ-reims.fr

© The Author(s), under exclusive license to Springer Nature Switzerland AG 2023

N. Wallenhorst, C. Wulf (eds.), *Handbook of the Anthropocene*,
https://doi.org/10.1007/978-3-031-25910-4_194

The urban explosion is one of the most spectacular phenomena of the twentieth century, described as the emergence of an "urban civilisation". The publication in 2011 of the UN-Habitat's Global Report on Human Settlements entitled "Cities and Climate Change" shows the interest of international organisations in understanding urban issues. Since 2006, the world population has become predominantly urban. More than half of the world's population (about 7 billion people) now lives in cities (55% in 2015 and 30% in 1950). Global urbanisation has resulted in a significant increase in the number of giant cities: more than 500 agglomerations have now more than one million inhabitants (only 15 in 1900). Urban systems are particularly vulnerable to the effects of climate change: concentration of populations and activities, interdependence of natural and technological risks, dependence on fossil fuels, urban heat islands, urban sprawl, reduction of biodiversity, social inequalities, etc. Almost a quarter of the world's urban population lives in slums. "More than 90% of urban growth is concentrated in developing countries and it is estimated that 70 million new residents move into urban areas in these countries every year. Over the next twenty years, the urban population of the world's poorest regions, South Asia and Sub-Saharan Africa, is expected to double, leading to a substantial increase in informal settlements and their inhabitants in these regions" (Habitat III, 2015, p.4).

Metropolisation can be analysed as the spatial and material realization of globalization; it ensues from a dual process of urban concentration on a global scale and, of the simultaneous diffusion of populations and activities in reticular and newly interconnected territorialities. The term "urban" describes therefore complex systems whose boundaries are blurred and increasingly difficult to define. Urban regions (or metropolitan areas) are embedded in still incomplete system of governance. The relationship between cities and countryside represents an important challenge of the twenty-first century. Outside of the competitive metropolisation process, regions in decline, or "shrinking cities", are also places of innovations (social, environmental, economic, etc.) and alternative models. Like the nature/culture dualism in which Western societies have been locked since the seventeenth century (Descola, 2005), the town/country dualism no longer allows to think and act on territorial interdependencies. Hence, urban regions prove to be the appropriate spatial scale to tackle "anthropocenic" issues, on condition that mechanisms for inter-territorial coordination and reciprocity are accounted for and triggered (Vanier, 2008).

Published in the 1970s, the report of the Club of Rome ("The Limits to Growth" also called "Meadows Report") alerted the international community to the depletion of resources. Following the publication of the Brundtland Report (Our common future 1987), the term sustainable development was first used at the 1992 Earth Summit. Reflecting upon 30 years of existence, many scholars and stakeholders regret that the concept has not had the desired effects. The notion of ecological transition has progressively replaced that of sustainable development (Bourg, 2012). However, there is still a lack of critical analysis of sustainable development to identify the causes of its failure and to ensure that the transition will have better results. Like sustainable development at the end of the 1990s, the necessity of a transition has spread within various institutional, academic, activist and citizen circles. There are many variations of the concept depending on the context in which it is used: ecological transition, energy transition, post-carbon transition, citizen transition, cities and territories in transition, etc. The "energy transition" characterises the shift from the current society based on the abundant consumption of fossil fuels to a more sober and ecological society based on an energy mix that is essentially based on renewable resources (Salomon & Jedliczka, 2013). In the field of urban ecology, the socio-ecological transition is defined as a profound transformation of territorial metabolism, in particular urban metabolism, based on "dematerialization", namely, a reduction in the material intensity of urban societies (Barles, 2017). Others pledge for an "integral transition" calling for "human beings to engage in a civilisational and behavioural conversion in their relationship with the world" (Folléa, 2019). The ecological transition therefore requires a change in behaviour, institutions and techniques, but also a new conceptual framework for the re-organization of space. Sustainable urbanism offers such a framework – guiding public policies and urban planning towards sustainability. This re-design of urban space has proved to help lowering the ecological footprint and the quality of the living environment of its inhabitants (Da Cunha, 2015).

The vulnerability of cities to the effects of climate change has direct and visible local consequences on people's lives: floods, heat waves, urban heat islands, shortages, migration flows, etc. These critical urban situations, shocks and ruptures are gradually putting environmental issues on the international, national and local political agenda. Above all, they lead to the multiplication of citizens' initiatives wishing to act more quickly and on their own scale. Cities in transition also refer to the various citizens' initiatives and experimentations tackling those issues from the bottom up: urban agriculture, shared gardens, proximity exchanges, self-managed spaces, self-building cooperatives, citizen energy, etc. Particularly popular, Transition Towns is a movement initiated in England in 2006 by Rob Hopkins, an environmental activist and permaculture teacher (Krauz, 2014). The awareness of the imminence of peak oil explain the birth and success of the social movement. A "Transition Handbook" was published in 2008 proposing a twelve-step method for starting a transition initiative. The movement brings together local and citizen initiatives and experiments that aim to invent lifestyles less dependent on oil: five hundred official initiatives in some forty countries around the world were recorded in 2013 and labelled by an NGO, the Transition Network. Numerous studies show the multiplication of these innovative actions in search of a shared and more ecological management of urban common goods. The big challenge for the urban planner of the twenty-first century is to involve a large population in the formulation of planning choices and, above all, to link together the alternative proposals arising from the many places of bottom-up initiatives. Italian economist Giacomo Becattini describes these alternative practices as "admirable crumbs" of capitalism with very limited capacity to expand (Becattini & Alberto, 2017). From a more optimistic perspective, they should also be analysed as "weak signals" that could potentially bring about a renewal of planning conception.

The rationale and efficiency of traditional urban planning have been often questioned and criticized. The socio-ecological issues coming along the Anthropocene further demands a renewal of public action. Still, land-use planning and others associated instruments (projects, programs, scenarios, rules) remain the most used in response to climate issues (Burby et al., 2000). Since the first urban settlements in Mesopotamia, about 7500 years BC, humans have never really stopped planning or thinking urban development in prospective and collective approaches. On the contrary, scholars highlight a return to planning on a large territory scale since the end of the 1990s in Europe. Land use planning has therefore been replaced by "strategic spatial planning" or "territorial planning" (Novarina & Zepf, 2009). This new kind of planning gives more importance to coordination between actors and to the materiality of the territory in terms of constraints and resources.

There is a need to "re-enchant" planning in the Anthropocene urban era (Dugua, 2017). Urban vulnerability provides an opportunity to rethink city planning in the light of the inadequacy of previous policies in the face of successive disasters (such as floods). Beyond the technicity of policy innovation, there is also a need to invent new rallying narratives in response to ecological and climatic challenges (Latour, 2015). Interestingly, territorial planning reveals a valuable instrument to describe, organize and stimulate the emergence of new imaginaries articulated to the multiple local places of project and diffuse initiatives (public, private or citizens). Tactical urbanism, transitional urbanism, or temporary urbanism are some of the new terms to describe the ongoing renewal of public action instruments towards experimentations based on assumed uncertainty and the subsequent adoption of trial and error approach.

Besides instrument renewal, it is the very conditions of implementation that are at stake, particularly as a relation between plan and project, while still being insufficiently addressed (Dugua, 2015). This could partly explain the previous failures of sustainability policies (Howes et al., 2017). It seems therefore stimulating to put into perspective the notions of transition (as a passage), interaction (as a relationship) and transaction (as an exchange). The implementation of planning is conditioned by its translation into a transactional process among stakeholders and with the territory, revolving around the current and upcoming social-ecological issues of the Anthropocene. The notion of transaction is often reduced to financial or monetary issues. However, many other forms of socio-spatial transactions - notably between town and country and nature/culture – are associated with the implementation of transition policies (food, energy, land, water, landscape, etc.). From "cities in transition" to "cities in transactions", anthropocenic modalities of urban planning are still to be defined.

References

Barles, S. (2017). Écologie territoriale et métabolisme urbain: quelques enjeux de la transition socioécologique. *Revue d'Économie Régionale & Urbaine, 5*, 819–836. https://doi.org/10.3917/reru.175.0819

Becattini, G., & Alberto, M. (2017). Conscience de classe et conscience du lieu : dialogue entre un économiste et un urbaniste. In A. Magnaghi (Ed.), *La conscience du lieu* (pp. 35–135). Eterotopia France.

Bourg, D. (2012). Transition écologique, plutôt que développement durable. *Entretien avec Dominique Bourg. Vraiment durable, 1*(1), 77–96. https://doi.org/10.3917/vdur.001.0077
Burby, R. J., Deyle, R. E., Godschalk, D. R., & Olshansky, R. B. (2000). Creating Hazard resilient communities through land-use planning. *Natural Hazard Review., 1*(2), 99–106.
Da Cunha, A. (2015). Nouvelle écologie urbaine et urbanisme durable. De l'impératif écologique à la qualité urbaine. *BSGLg* 65(2.)-Varia. https://popups.uliege.be/0770-7576/index.php?id=4116
Da Cunha, A., & Thomas, I. (2017). Introduction. In I. Thomas & A. Da Cunha (Eds.), *La ville résiliente Comment la construire ?* (pp. 15–49). Les Presses de l'Université de Montréal.
Descola, P. (2005). *Par-delà nature et culture*. Editions Gallimard.
Dugua, B. (2015). *Entre mise en œuvre et mise à l'épreuve de la planification territoriale: dynamique des lieux de projets dans l'inter-Scot de l'aire métropolitaine lyonnaise*. Université Grenoble Alpes.
Dugua, B. (2017). *Comment réenchanter la planification territoriale en France ?* Métropolitiques. http://www.metropolitiques.eu/Comment-reenchanter-la-planificationterritoriale-en-France.html
Folléa, B. (2019). *L'archipel des métamorphoses. La transition par le paysage*. Parenthèses.
Habitat III. (2015). *22 – Informal settlements*. Issue papers: https://unhabitat.org/habitat-iii-issue-papers-22-informal-settlements. Accessed 15 Sept 2021.
Howes, M., Wortley, L., Potts, R., Dedekorkut-Howes, A., Serrao-Neumann, S., Davidson, J., Smith, T., & Nunn, P. (2017). Environmental sustainability: A case of policy implementation failure? *Sustainability, 9*(2), 165. https://doi.org/10.3390/su9020165
Krauz, A. (2014). *Les villes en transition, l'ambition d'une alternative urbaine*. Métropolitiques. http://www.metropolitiques.eu/Les-villes-en-transition-l.html
Latour, B. (2015). *Face à Gaïa. Huit conférences sur le nouveau régime climatique*. La Découverte.
Novarina, G., & Zepf, M. (2009). Territorial planning in Europe: News concepts, new experiences. *Disp- The Planning Review, 179*(4), 18–27.
Salomon, T., & Jedliczka, M. (2013). *Changeons d'énergies; transition, mode d'emploi*. Actes Sud.
Satterthwaite, D. (2008). Cities' contribution to global warming: Notes on the allocation of greenhouse gas emissions. *Environment and Urbanization, 20*(2), 539–550. https://doi.org/10.1177/0956247808096127
Stephen, W., Broadgate, W., Deutsch, L., Gaffney, O., & Ludwig, C. (2015). The trajectory of the Anthropocene: The great acceleration. *The Anthropocene Revue, 2*, 81–98. https://doi.org/10.1177/2053019614564785
Vanier, M. (2008). *Le pouvoir des territoires. Essai sur l'interterritorialité*. Economica Anthropos.

Benoît Dugua is associate professor at the University of Reims Champagne-Ardenne (URCA) and within the laboratory EA 2076 Habiter since 2020. At the crossroad between urban planning and landscape, his research focuses on the transformation and implementation of planning and project processes in a transition perspective. He thus contributes to defining the anthropocenic modalities of urban planning as a major challenge of the twenty-first century.

Housing

David R. Cole and Yeganeh Baghi

Abstract This handbook entry examines the issue of housing in the Anthropocene. The issue of housing in the Anthropocene involves many factors and aspects with respect to housing given the facts of climate change. To limit these factors and possible through-lines for this entry, housing in the Anthropocene will be analyzed according to three dimensions to make sense of the future of housing needs alongside climate change: (1) Housing and human population. The fundamental questions with respect to housing in the Anthropocene involve human population, and the concurrent creation of more housing given increases in the human population in the future; (2) Housing and spatial requirements. Following on from the extra requirements for human housing given increases in human population, the critical issue of spatial requirements for this housing will be addressed; (3) Housing and net zero. The third section of this entry examines the possibility of housing in the Anthropocene reaching net zero and thus not contributing further to climate change. For the purposes of this entry, a time ceiling of 2050 will constrain each section, and limit the possibility of this entry becoming embroiled in overly speculative science fiction.

It has been estimated that human housing contributes up to 40% of greenhouse gas emissions worldwide (Huang et al., 2018). Thus, it is clearly an important issue to tackle with respect to reducing emissions in the Anthropocene and helping stave off dangerous climate change. This entry focuses on three dimensions of housing in the Anthropocene to make sense of the complexity and scope of the issue: (1) Population; (2) Space; (3) Net zero. Given these three dimensions, other issues, such as urbanization, infrastructure needs, economics, and the lifestyle choices to help stave off climate change and that are associated with housing may also be addressed.

D. R. Cole (✉)
Western Sydney University, Penrith, NSW, Australia
e-mail: David.cole@westernsydney.edu.au

Y. Baghi
University of Wollongong, Wollongong, NSW, Australia

© The Author(s), under exclusive license to Springer Nature Switzerland AG 2023

N. Wallenhorst, C. Wulf (eds.), *Handbook of the Anthropocene*,
https://doi.org/10.1007/978-3-031-25910-4_195

One may state that the issue of human housing lies at the core of thinking what to do about climate change and human futures, and yet as a coherent subject, the critical and synthetic thought necessary to bring it together can be lacking. This handbook entry tries to enable such thinking and an anthropological approach to living in the Anthropocene given what is known today.

Housing and Human Population

Human population is projected to reach 9.7 billion by 2050 (UN, 2019). Of this 9.7 billion people, it is projected that 68% will live in urban areas (Ritchie & Roser, 2018). Thus, the principal housing landscape that should be figured in terms of the Anthropocene is an urban one, with its concomitant needs and structures. Population growth peaked in 1968 at +2.1%, and by 2050 is projected to be at +0.5%. However, the link between extra housing and population growth, even slowed, is not as simple as every new human being will want or need an extra house in their places of birth. Rather, increases in human population up until 2050 are non-uniform flows, creating extra need for housing in some districts, while others will stabilize and not require large scale new housing plans. For example, continents with higher fertility rates, such as Africa, will continue to feed growth in the human population, yet many of the newly born will eventually look to migrate to other continents such as Europe, thus creating housing demand for newly arrived immigrants in destination countries (Ritchie & Roser, 2018). Further, this demand will be concentrated in urban areas as mentioned above, as the newly arrived migrants look for work and safety in their new countries, as well as being attracted to previous waves of immigrants from their native countries. In sum, less developed countries will continue to support increases in human population, with much of the new inhabitants looking to move to wealthier and more developed nations, whose housing demands will largely be concentrated at the peripheries of large urban zones.

Housing and Spatial Requirements

Housing demands space, and thus the replacement of natural land that could either be used for agriculture or left as wilderness and help, for example, with global warming by retaining tree coverage where possible. However, the spatial demands of human housing are divided along the grounds of wealthier nations, whose urban cities frequently have large suburban or peri urban areas, and poorer nations, whose urban centres are frequently surrounded by slums (Rumbach, 2017). Hence, this contrast in spatial living arrangement sets into motion one of the most contentious aspects of any analysis of the Anthropocene, that the greatest greenhouse polluters are the wealthier, industrial nations, where, for example, in the case of housing spatial arrangements, the expectation will be for continued expansive living spaces, and the congruent high consumption of resources (Armaroli & Balzani, 2007). In developing countries, where the bulk of the population increases to 2050 will come from as we move further into the Anthropocene, the expectation is for basic survival, and hence for food and adequate shelter. Concomitantly, the space that the increased human population will occupy is less in developing countries and will therefore have a lower impact on the natural environment overall, allowing for poor infrastructure to deal with waste and pollution.

The average size for a single-family housing unit in America in 2020 was 2261 square feet (SRD, 2022). Furthermore, in the same year, the average annual electricity consumption for a U.S. residential utility customer was 10,715 kilowatt hours (kWh), or about 893 kWh per month (Ritchie & Roser, 2020). In contrast, the average build size for Hong Kong in 2009 was 484 square feet (Wilson, 2014), and whilst house data is unavailable for poorer countries, per capita electricity consumption is more than 100-fold lower, for example, in 2014, the average person in the Democratic Republic of Congo consumed just over 100kWh (Ritchie & Roser, 2020). Hence, the enormous disparities in house size and energy use, creates differentials in capital and desire between peoples involved with these living arrangements and energy usages. Most people look to the U.S. as possessing a desirable lifestyle that is henceforth copied around the world, and sets up flows of migration and behaviours that are augmented, for example, through the media, and the image of this lifestyle (Cole, 2021), that may be quite distinct from its truth. In sum, the overriding problem for the Anthropocene, is not only to reduce house size and energy consumption per capita, but to limit the effects of wanting these factors as essential in one's life. The desire for large spatial housing and high energy consumption that comes from countries such as the U.S., and its recent media portrayal, is mitigated for and by the Anthropocene as we head for 2050, as searching for lower impact ways in live. In the next section, these improved living arrangements for the Anthropocene will be discussed in terms of 'net zero'.

Housing and Net Zero

Recent climate change directives from governmental and global agency meetings such as COP26 have converged around the slogan of 'net zero by 2050'. In straightforward terms, net zero means that that human activity will no longer in sum add to carbon emissions and/or greenhouse gases that are producing climate change. However, proposals for net zero frequently emanate from companies and capital interests that are trying to hide their real emissions profiles behind abatement and offset schemes to give their operations a green image (Balouktsi, 2020). In contrast, rather than simply critiquing the notion of net zero as an accounting and corporate example of greenwashing, this entry will take seriously the possibility of reaching this goal for housing and given the facts of population and spatial requirements as stated above. This section of the entry will divide the concept of housing and net zero into three sections, that follow Félix Guattari's (2005) social ecology in the three ecologies as an overall logic and rationale. Guattari's approach to climate change and other related actions requires attending to the ecologies of mind, society, and the environment, as three inter-related and inter-dependent spheres, and to bring forward an overall 'ecosophy' (Ibid.), that is used here as a thinking and action pathway to net zero for housing by 2050.

Firstly, the sphere of the mind needs to address net zero by 2050 for housing. What this precisely means for the Anthropocene and the study of the anthropology of housing depends upon how one might figure 'the mind' and its influence in this context. Certainly, in the context of western democracies and its citizens, 'the mind' in terms of sustainable housing and net zero by 2050, determines a set of socio-cultural, affective qualities and attributes that align themselves with green issues and doing something to help with environmental issues (Birkeland, 2020). The cognitive aspects of 'the mind' investigate practical solutions for housing net zero by 2050, such as investing in PV (Photovoltaic) arrays that run HVAC (Heating, Ventilation, Air Conditioning) systems (Elnabawi & Saber, 2021) and cutting down one's carbon footprint through recycling and living a sustainable lifestyle in one's house, e.g., by growing your own food or by reusing your grey water. Responsibility for the greatest changes in housing emissions falls on residents from wealthier nations, and thus they are required to make the biggest adjustments in their thinking and feeling toward environmental matters. In contrast, residents in poorer nations, with far smaller carbon footprints, and housing that does not contribute to global warming in the same way as the wealthy, industrial elite, have less adjustment and thinking differently to do with their minds (Rumbach, 2017). In sum, there is no one set of instrumental, uniform, or top down list of constraints on thinking and action that will absolutely determine the necessary changes to the mind for net zero by 2050 for housing. Rather, there are combinatory and differentiated vectors, or multiple entwined forces combined within the sphere of the mind, that could head towards net zero in housing, and that are contextually more relevant given the real impact of one's current housing lifestyle.

The second sphere according to Guattari's three ecologies (2005), that are being used here as organizing principles to head towards net zero for housing by 2050, is society. Generally speaking, one might say that global human society is in transition, from a global socio-economic system dominated by fossil fuels, to systems with energy generation by renewables, and houses that do not add to the release of CO_2 into the atmosphere, either in their construction or running. However, this transition is not consistent, at the same speed, or even adhered to through regulation and compliance by housing developments in advanced industrial nations, where the transition to a non-carbon mode to build housing and to live in these houses is currently possible (Bataille et al., 2016). The societal sphere of the three ecologies opens up from the thinking and acting sphere of 'the mind', which could be reduced to existential, phenomenological and psychological aspects of the current situation, and understood, for example, through constructs such as climate change resilience (Moritz & Agudo, 2013), and includes becoming actively involved with groups and in political action that move us towards decarbonization. The active elements of the societal changes necessary for net zero to happen by 2050, includes consideration, for example, of 'degrowth' for political change based on deliberately slowing the economy (Kallis, 2011), or Kate Raworth's doughnut economics (2017), as a means to regulate the runaway aspects of the economy, and how they accelerate climate change, for example, through housing.

In terms of housing and net zero, degrowth is a mode of societally questioning the ubiquitous push of economics that runs through the processes of building and buying a house. Rather than the socially determined norm of always wanting more, for example, a larger house with expensive gadgets, the degrowth approach looks in the opposite direction, at downsizing, at different, humbler, and less expensive options. For example, an Italian architect called, Mario Cucinella, has designed 3D printed houses from locally sourced mud, as part of a community showcase project outside Bologna called TECLA (2022). The 3D printed houses are designed in the shape of potter wasp constructions, the animal house aesthetic has the advantage of being closer to natural designs and thus fitting in with a purely ecological notion of a house (Cole & Baghi, 2022). 3D printing for house construction has many advantages over traditional housing building techniques, because it cuts out many aspects of the supply chain, and purely extrudes the house on site, rather than the site acting as a magnet for the different elements of the house to come together for construction, and their combined environmental and emissions impacts. In effect, 3D printed houses from locally sourced mud are a degrowth option for future housing developments in terms of their construction not contributing to emissions. If they are to run at net zero, the 3D printed houses have to incorporate currently available sustainability add-ons such as PV arrays, HVAC systems regulated by active PCM (Phase Change Materials), for example, looped into the systems via water-air coupling (Baghi et al., 2021). Kate Raworth's doughnut economics (2017) can be applied to the TECLA development in terms of regulating and controlling the different systems and aspects of the lifestyles that inhabitants enjoy in the humble, degrowth community. In sum, the combination of 3D printed house made of locally sourced mud, and regulating the living community within ecological boundaries via doughnut economics (inter-linked systems thinking) are a means to transition to a slow economy, and which could act as a beacon of hope in the Anthropocene.

Lastly, the third sphere of the ecologies from Guattari (2005) is nature itself, and how to work within a natural context to purposefully head towards net zero in housing for 2050. Guattari (2005) was concerned to implement an updated and relevant shamanism for the contemporary situation. As such, the religious aspect of Guattari's thought focuses on a connection with nature and in diminishing the human-nature dualism and potential divide that has arguably become entrenched in western thought since Plato. Guattari's (2005) three ecologies work against the barrier of human exceptionalism, and the centering of human consciousness in the world as a self-fulfilling prophecy that determines in advance what will happen from a human-first perspective. Rather, examples such as housing in the Anthropocene, and the necessary work to get to net zero by 2050, require using the three ecologies to bring other non-human elements into the fold, as recent theorizations of posthumanism and the new materialism have attempted to achieve. The difference that the three ecologies brings to the fold is that Guattari (2005), does not lose sight of the ongoing damage that capitalism is doing to the planet, or the infantilizing effects of the global media. Rather, the issue at hand, which is to reach net zero by 2050 for housing sits outside of the negative influences of capitalism and the media, and yet retains agency and human relations with the outside through, as Guattari termed it, 'schizoanalysis' (Cole, 2021). One way to rethink human relations with the outside, and in relation to the environment, is through (re)consideration of indigenous and precolonial modes of acting and being, and as they relate to housing and net zero.

The fundamental problem with housing in the Anthropocene is that it is tied in with contemporary capitalism, and the world financial market in terms of capital assets (Soriano, 2018). Hence, unhinging housing from this connection is paramount to envisaging a natural, or environmentally replenishing mode of living in the Anthropocene, and one that could drive us to net zero. Hence, the move towards pre-colonial and indigenous housing concepts, and houses that are closer to organic modes of design and maintenance, are preferable in the context of the Anthropocene (Cole & Baghi, 2022). The question for this section of the entry is how is this going to happen, given the continuing domination of one world capitalism on all relations? The answer is a type of general mobilization to reform and transform housing, away from a modernist or industrial mode of construction and life that serves capital, and towards one that is in line with the natural context in which one exists. Of course, this is an extraordinarily difficult task given the penetration of capital in every aspect of everyday life, and how housing has become a central part of its functioning as property. The point from Guattari and the three ecologies (2005) is that housing is integral to these ecologies and does not only have value as capital. For example, construction materials for net zero housing are actively involved with ecological and natural systems, and as such, need to be locally sourced and incorporate natural substances such as mud (Cole & Somerville, 2020).

Human population will continue to increase until 2050, despite the pandemic and war. Humans will continue to desire large houses and wish to enjoy the benefits of living a lifestyle that has high daily consumption and is productive of emissions and waste. To reverse these trends and to move towards net zero by 2050 in housing is more than an offset or accountancy trick, or a matter of the piecemeal addition of sustainability-oriented technologies such as PV arrays, PCM units and water recycling units to houses, without addressing wider and more profound issues connected the contemporary socio-cultural complex (Cole, 2022). This entry has suggested that the application of Guattari's three ecologies gives greater scope and could have a more profound impact to housing than, for example, the mere acceleration and implementation of technological solutions to housing and emissions. This is because the three ecologies as applied to housing demands that it is rethought from the perspectives of mind, society, and the environment, and as such, is an integrated system of building and living an ecological life.

References

Armaroli, N., & Balzani, V. (2007). The future of energy supply: Challenges and opportunities. *Angewandte Chemie International Edition, 46*(1–2), 52–66.

Baghi, Y., Ma, Z., Robinson, D., & Boehme, T. (2021). Innovation in sustainable solar-powered net-zero energy solar decathlon houses: A review and showcase. *Buildings, 11*(4), 171.

Balouktsi, M. (2020). Carbon metrics for cities: Production and consumption implications for policies. *Buildings and Cities, 1*(1), 233–259.

Bataille, C., Waisman, H., Colombier, M., Segafredo, L., Williams, J., & Jotzo, F. (2016). The need for national deep decarbonization pathways for effective climate policy. *Climate Policy, 16*(sup1), S7–S26.

Birkeland, J. (2020). *Net-positive design and sustainable urban development*. Routledge.

Cole, D. R. (2021). Caught between the air and earth: A schizoanalytic critique of the role of the education in the development of a new airport. *Educational Philosophy and Theory, 54*, 422–433. https://doi.org/10.1080/00131857.2020.1864322. Accessed 28 Feb 2022

Cole, D. R. (2022). *Education, the Anthropocene, and Deleuze/Guattari*. Brill.

Cole, D. R., & Baghi, Y. (2022). When two worlds collide: Creatively reassessing the concept of a house beyond the human. *Qualitative Inquiry, 28*, 531–540. https://doi.org/10.1177/10778004211065800. Accessed 28 Feb 2022

Cole, D. R., & Somerville, M. (2020). The affect(s) of literacy learning in the mud. *Discourse: Studies in the Cultural Politics of Education, 43*, 188–204. https://doi.org/10.1080/01596306.2020.1818183. Accessed 28 Feb 2022

Elnabawi, M. H., & Saber, E. (2021). Reducing carbon footprint and cooling demand in arid climates using an integrated hybrid ventilation and photovoltaic approach. *Environment, Development and Sustainability, 24*, 3396–3418. https://doi.org/10.1007/s10668-021-01571-1

Guattari, F. (2005). *The three ecologies* (I. Pindar and P. Sutton, Trans.). Bloomsbury Publishing.

Huang, L., Krigsvoll, G., Johansen, F., Liu, Y., & Zhang, X. (2018). Carbon emission of global construction sector. *Renewable and Sustainable Energy Reviews, 81*, 1906–1916.

Kallis, G. (2011). In defence of degrowth. *Ecological economics, 70*(5), 873–880.

Moritz, C., & Agudo, R. (2013). The future of species under climate change: Resilience or decline? *Science, 341*(6145), 504–508.

Raworth, K. (2017). *Doughnut economics: Seven ways to think like a 21st-century economist*. Chelsea Green Publishing.

Ritchie, H., & Roser, M. (2018). *Urbanization*. https://ourworldindata.org/urbanization. Accessed 28 Feb 2022.

Ritchie, H., & Roser, M. (2020). *Energy*. https://ourworldindata.org/energy. Accessed 28 Feb 2022.

Rumbach, A. (2017). At the roots of urban disasters: Planning and uneven geographies of risk in Kolkata, India. *Journal of Urban Affairs, 39*(6), 783–799.

Soriano, C. (2018). The Anthropocene and the production and reproduction of capital. *The Anthropocene Review, 5*(2), 202–213.

SRD (Statistica Research Department). (2022). *Size of floor area in new single-family homes in the U.S. 1975–2020*. https://www.statista.com/statistics/529371/floor-area-size-new-single-family-homes-usa/. Accessed 28 Feb 2022.

TECLA. (2022). *3D printed house*. https://www.3dwasp.com/casa-stampata-in-3d-tecla/. Accessed 28 Feb 2022.

UN (United Nations Department of Economic and Social Affairs). (2019). *Growing at a slower pace, world population is expected to reach 9.7 billion in 2050 and could peak at nearly 11 billion around 2100*. https://www.un.org/development/desa/en/news/population/world-population-prospects-2019.html. Accessed 28 Feb 2022.

Wilson, L. (2014). How big is a house? *Average house size by country*. http://shrinkthatfootprint.com/how-big-is-a-house. Accessed 28 Feb 2022.

David R. Cole is an Associate Professor in Education at Western Sydney University. He has co-founded the knowledge area of 'Deleuze and Education' and has published more than 100 important documents in the field, including: *Education, the Anthropocene, and Deleuze/Guattari* (Brill). He has recently started an Anthropocene web site www.iiraorg.com

Yeganeh Baghi is a lecturer at the University of Wollongong. She is presently working on sustainability oriented innovative technologies and how to apply them in housing. Her passion is to use sustainable engineering principles to fight climate change.

Mutualistic Cities

Mark Williams, Julia Adeney Thomas, and Jan Zalasiewicz

Abstract We discuss the cities of the future, and how they might co-habit with the biosphere in a more mutually beneficial way. Mutualistic cities would blend with their local ecology, co-existing with the immediately available resources of water, life, energy and materials, and enhancing the biosphere so that many species can thrive, including people. Such cities can make a significant contribution to stabilizing the Earth System by sustaining and nurturing life in tune with the evolving local ecology through cyclic economies and values prioritizing mutually supportive communities over economic growth. The term "mutualism" comes from biology, referring to inter-species relationships that are beneficial. Mutualistic cities would benefit human and non-human inhabitants by recycling materials, clean air, and water, and providing habitats for many species.

Cities are transformative engines of the Earth's surface, and their accelerating patterns of consumption are a defining component of the Anthropocene. More than 4 billion of us live in urban areas, and this number is growing. Cities may be responsible for as much as three-quarters of the energy used by humans and produce a similar percentage of our carbon dioxide emissions (Güneralp et al., 2017). They use vast resources of freshwater, much of this piped in over long distances and from extensive land areas (McDonald et al., 2014). Although urban areas cover only 3% of the land, their impact on ecologies reaches far beyond this, to those landscapes converted to cultivated plants and pasture, to overfished oceans, and to the quarrying operations that deliver millions of tonnes of cement and other commodities to them. These patterns of consumption are now so great that they threaten the fabric of the Earth System (Rockström et al., 2009), the integrated relationship between earth,

M. Williams (✉) · J. Zalasiewicz
University of Leicester, Leicester, UK
e-mail: mri@le.ac.uk; jaz1@le.ac.uk

J. A. Thomas
University of Notre Dame, Notre Dame, IN, USA
e-mail: jthomas2@nd.edu

© The Author(s), under exclusive license to Springer Nature Switzerland AG 2023

N. Wallenhorst, C. Wulf (eds.), *Handbook of the Anthropocene*,
https://doi.org/10.1007/978-3-031-25910-4_196

air, water and life that has sustained a habitable planet for billions of years (Vernadsky, 1998). If future cities are to thrive, they *must* emulate natural ecosystems, recycling nutrients, water, the material structures of buildings and other waste products, using the energy available locally, and building inter- and intra-species relationships that are mutually beneficial. This is what we mean by a mutualistic city. It is one that will be less visible to the eye, one that blends into its local ecology, and one that provides wellbeing to all of its inhabitants.

Can a city, or even a single building or utility—like a toilet, washing machine or shower—become truly sustainable and how would you go about achieving this? We begin with the most basic of utilities and work up to the level of the whole city, to show how systems can be designed to reduce consumption and waste, and to provide space for the wellbeing of many species. We also draw the reader's attention to, among others, Ebru Özer's (2014) work on building mutualistic relationships in cities, Paul Chatterton's book "Unlocking sustainable cities" (2019), Kate Raworth's "green doughnut" model (see 'Creating city portraits', 2020), and our own work (Williams et al., 2022).

One of the most basic utilities in a home is a toilet. These need to be clean and safe for their users. But the widespread use of water closets is a visible metaphor of our disconnect from the environment. In most modern houses the water in the toilet is piped in and piped out. Even the grey water that accumulates in the bathroom sink adjacent to the toilet is flushed away without being repurposed for the toilet. Most of us do not think about the environmental implications of flushing. But a quarter of a household's water supply is used in this way, and much more in school or office settings (Anand & Apul, 2014).

A mutualistic city would seek to recycle the water and materials within the closet or avoid using water at all, especially in regions subject to extended drought. Such technologies already exist – and have done so for more than a century, that compost and recycle (Lopez Zavala & Funamizu, 2005), and these have proven utility in settings where piped water is absent, or where an ecological solution has been sought to the problem of human waste. Compostable or bio-toilet systems can function without running water, instead depending on heat—which can be supplied from solar energy—and microbes to convert human faeces and urine to materials that are rich in nitrogen, phosphorus and potassium, and which can be re-used as fertilizer and soil improvers. This is much more akin to how a natural ecosystem works.

So why are such technologies so limited in their use, especially in towns or cities that sit in the middle of deserts? Bio-toilets often require more interaction than a simple flush, and what our bodies deposit there also needs to be mixed with other materials, like moss, leaves or sawdust to encourage biodegradation. Some systems also require an optimum temperature range for microbial activity to reduce odours and kill pathogens. And there are the inevitable issues around peoples' responses to their own bodily waste that need to be overcome (Dimpfl & Moran, 2014). Nevertheless, there are some limited examples of where urban communities have successfully used such technologies, as at Neu-Allermöhe-Ost, Germany, where 36 houses with about 140 inhabitants have been using composting toilets for over two decades (Rauschning et al., 2009). Here the end-member compost is recycled into gardens and common land enriching the soil as fertilizer. To become truly mutualistic, a city must foster this type of recycling for all of its utilities.

If the devices within a house can evolve to recycle materials and reduce their reliance on imported freshwater and energy, can the house itself be fully sustainable? Many technological designs exist for energy-efficient homes, based on the sensible utilization of renewable energy (e.g., Wang et al., 2009), and the refitting of existing properties (Oki et al., 2019). In part these can help solve inefficiencies in current housing stocks. But a mutualistic city would seek to go beyond this, to galvanize its inhabitants to utilize resources that work with the local environment, as a natural ecosystem would. And such homes—with their compostable toilet system of course—should provide living space for humans and other species too.

On a small scale these kinds of homes have been devised, as in The Canelo Project in southeastern Arizona (Canelo Project in 2023). Here buildings are made from straw bales, covered with clay and lime plasters, or made with earth bricks fashioned from clay and straw, and they are durable in their landscape. On the opposite side of the world, on the Japanese island of Hokkaido, a similar house has been built by three students from Waseda University, Tokyo. Their house has a timber frame, into which straw-filled acrylic boxes are inserted. In the summer the straw dries out, the latent heat of evaporation cooling the house. In winter the straw becomes damp and produces heat from microbial fermentation (Testado, 2013). While acrylic is a plastic polymer taking energy to produce, and doesn't recycle easily, this form of building and heating is a significant improvement over building and heating with concrete and fossil fuels. Built from more sustainable and local materials, such projects can help to reconnect people with the landscapes around them, embedding them in the limitations of that land, rather than the 'apparently' limitless supply of global commodities. Straw-based buildings are not applicable in all landscapes, of course, but these and many other innovative designs signal ways in which communities and businesses can build in better ways for the ecologies around them.

It is one thing to build houses from sustainable materials, but quite another to scale this to cities. There are, of course, many challenges, from unfamiliarity with the construction materials and methods of building, to more labour-intensive methods (Seyfang, 2009). These challenges might be overcome by changes to building codes and financing. Policies could encourage builders to recycle on-site materials for new construction, and support ecologically attuned designs (Seyfang, 2009). Already over 80% of the construction and demolition waste in Germany and the Netherlands is recycled, but the UK and the US lag far behind, dumping valuable materials into landfills (Osmani, 2011). New policies might also prompt people (or construction companies) to install composting toilets, to design buildings that capture the energy of sunlight and local water supplies, or to install water systems that reuse grey water. Governments could encourage planning that makes space for non-human life, ensuring, for example, that buildings allow for nesting and migration. When materials can no longer be reused, they should be allowed to decay, as they would in a natural ecology, providing habitats for other organisms (Flyn, 2019). These decaying structures could meld into interconnected green spaces in cities

(Lepczyk et al., 2017), ultimately connecting with the hinterland ecology. Such steps—evolutionary rather than revolutionary—are essential, because without them, patterns of urban consumption will quickly degrade the planet's resources.

Mutualistic cities, in transforming urban-dwellers' relationships with space (localizing consumption) and with materials (reducing consumption and recycling), will also alter peoples' relationship with time and one another. Today, societies place a high value on the speed and efficiency best produced by experts, yet even small acts of mutualism involve slowing down. The few minutes needed to mix leaves or sawdust into one's own waste in a composting toilet would mean longer bathroom breaks. Asking architects to adjust their plans to incorporate old on-site materials into new structures would lengthen construction time. Fostering ecological networks requires knowledge, imagination, and patience over years, not by the next quarterly report. Societies today value linear thinking directed at solving discrete problems; decision-making power is in the hands of a few, and economic growth is the goal. Mutualistic societies would reward systemic thinking and collaborative knowledge, replacing growth with the goal of sustainable, strong communities over generations, and widespread democratic support. To work, mutualism rests on everyone benefitting.

If mutualistic cities can evolve, they will have significant advantages over those that continue to over-use local resources and rely on globally traded commodities. If buildings can be recycled, they could adapt to a quickly changing environment including rising sea levels. As humanity's experiment with the Earth System results in an increasingly precarious global habitat for humans and other life, it is cities that embrace mutualism that offer the greatest chance for survival, whilst those that continue to consume with impunity will only degrade the Earth's biodiversity and themselves. Current cities with their global support networks of non-renewables are far more vulnerable to disaster than would be mutualistic cities with local, adaptable, regenerative systems.

References

Arnand, C. K., & Apul, D. S. (2014). Composting toilets as a sustainable alternative to urban sanitation – A review. *Waste Management, 34*, 329–343.

Canelo Project. (2023). *Connecting people, culture and nature*. https://caneloproject.com/about-us/. Accessed February 2023.

Chatterton, P. (2019). *Unlocking sustainable cities. A manifesto for real change*. Pluto Press, London.

Creating city portraits_Methodological Guide_v.1. (2020). Available at https://www.kateraworth.com/2020/07/16/so-you-want-to-create-a-city-doughnut/

Dimpfl, M., & Moran, S. (2014). Waste matters: Compost, domestic practice, and the transformation of alternative toilet cultures around Skaneateles Lake, New York. *Environment and Planning D: Society and Space, 32*, 721–738.

Flyn, C. (2019). Life amid the ruins: abandoned places as islands of biodiversity. https://harcruins.wixsite.com/ruinationanddecay/post/life-amid-the-ruins-abandoned-places-as-islands-of-biodiversity

Güneralp, B., Zhou, Y.-Y., Ürge-Vorsatz, D., Gupta, M., Yu, S., Patel, P. L., Fragkias, M., Li, X.-M., & Seto, K. C. (2017). Global scenarios of urban density and its impacts on building energy use through 2050. *PNAS, 114*, 8945–8950.

Lepczyk, C. A., Aronson, M. F. J., Evans, K. L., Goddard, M. A., Lerman, S. B., & Macivor, J. S. (2017). Biodiversity in the city: Fundamental questions for understanding the ecology of urban green spaces for biodiversity conservation. *Bioscience, 67*, 799–807.

Lopez Zavala, M. A., & Funamizu, N. (2005). Effect of moisture content on the composting process in a biotoilet system. *Compost Science & Utilization, 13*, 208–216.

McDonald, R. I., Weber, K., Padowski, J., Flörke, M., Schneider, C., Green, P. A., Gleeson, T., Eckman, S., Lehner, B., Balk, D., Boucher, T., Grill, G., & Montgomery, M. (2014). Water on an urban planet: Urbanization and the reach of urban water infrastructure. *Global Environmental Change, 27*, 96–105.

Oki, R., Tsuneoka, Y., Yamaguchi, S., Sugano, S., Watanabe, N., Akimoto, T., Hayashi, Y., Wakao, S., & Tanabe, S.-I. (2019). Renovating a house to aim for net-zero energy, thermal comfort, energy self-consumption and behavioural adaptation: A method proposed for ENEMANE HOUSE 2017. *Energy & Buildings, 201*, 183–193.

Osmani, M. (2011). Chapter 15: Construction waste. In T. M. Letcher & D. A. Vallero (Eds.), *Waste*. Academic Press. https://doi.org/10.1016/B978-0-12-381475-3.10015-4

Özer, E. (2014). Mutualistic relationships versus hyper-efficiencies in the sustainable building and city. *Urban Ecosystem, 17*, 195–204.

Rauschning, G., Berger, W., Ebeling, B., & Schöpe, A. (2009). Ecological settlement in Allermöhe Hamburg, Germany – Case study of sustainable sanitation projects. *Sustainable Sanitation Alliance*. pdf available at: https://www.susana.org/en/knowledge-hub/resources-and-publications/case-studies/details/56

Rockström, J., Steffen, W., Noone, K., Persson, A., Chapin, F. S., III, Lambin, E., Lenton, T. M., Scheffer, M., Folke, C., Schellnhuber, H., Nykvist, B., De Wit, C. A., Hughes, T., van der Leeuw, S., Rodhe, H., Sorlin, S., Snyder, P. K., Costanza, R., Svedin, U., Falkenmark, M., Karlberg, L., Corell, R. W., Fabry, V. J., Hansen, J., Walker, B., Liverman, D., Richardson, K., Crutzen, P., & Foley, J. (2009). Planetary boundaries: Exploring the safe operating space for humanity. *Ecology and Society, 14*(2), 32. http://www.ecologyandsociety.org/vol14/iss2/art32/

Seyfang, G. (2009). Community action for sustainable housing: Building a low carbon future. *Energy Policy, 38*, 7624–7633.

Testado, J. (2013). "A recipe to live": Modern Japanese straw house naturally heated by compost. *Archinect News*. https://archinect.com/news/article/76255489/a-recipe-to-live-modern-japanese-straw-house-naturally-heated-by-compost

Vernadsky, V. I. (1998). *The biosphere (complete annotated edition: Forward by Lynn Margulis and colleagues and introduction by Jacques Grinevald)* (192 pp). Copernicus (Springer-Verlag).

Wang, L., Gwilliam, J., & Jones, P. (2009). Case study of zero energy house design in UK. *Energy & Buildings, 41*, 1215–1222.

Williams, M., Thomas, J. A., Brown, G., Pathak, M., Burns, M., Steffen, W., Clarkson, J., & Zalasiewicz, J. (2022). Mutualistic cities of the near future. In J. A. Thomas (ed.), *Altered Earth. Getting the Anthropocene right.* Cambridge University Press.

Mark Williams is a palaeobiologist at the University of Leicester. He studies the evolution of life on Earth over hundreds of millions of years and has been a long-time member (and former Secretary) of the Anthropocene Working Group. He has co-authored popular science books with Jan Zalasiewicz on climate change, ocean evolution, and the story of life on Earth, and with Jan and Julia Adeney Thomas he co-authored *The Anthropocene: A Multidisciplinary Approach*. His new book with Jan is called *The cosmic oasis: the remarkable story of Earth's biosphere* (Oxford University Press).

Julia Adeney Thomas is a historian at the University of Notre Dame. She writes about politics, nature, photography, and the Anthropocene in Japan and globally. Her publications include *Reconfiguring Modernity: Concepts of Nature in Japanese Political Ideology*; *The Anthropocene: A Multidisciplinary Approach*, written with Jan Zalasiewicz and Mark Williams, and *Strata and Stories* with Jan Zalasiewicz along with numerous articles and edited volumes. She is at work on *The Historian's Task in the Anthropocene* (under contract with Princeton University Press).

Jan Zalasiewicz is a geologist (now retired) at the University of Leicester, and a member (formerly Chair) of the Anthropocene Working Group. He is co-editor of *The Anthropocene as a geological time unit: a guide to the scientific evidence and current debate*, has been involved in other AWG publications, and with Julia Adeney Thomas and Mark Williams has co-written *The Anthropocene: A Multidisciplinary Approach*. His books for a general audience that *inter alia* explore Anthropocene concepts include *The Earth after Us* and (with Mark Williams) *The Goldilocks Planet* and *Ocean Worlds*.

Social Contract Theory

Amentahru Wahlrab

Abstract This article examines social contract theory in the Anthropocene. Past and current social contract theories have a narrowly conceived conception of human that pits humans against nonhumans. This article suggests new parameters for a social contract for the Anthropocene, which has as its focus humans, animals, and nature. Based on the thinking of Brian Elliott, this article suggests that a politics of environmental catastrophe will not encourage the adoption of a post- or more-than human social contract. Instead, political engagement must focus on the relationship between capitalism and climate change in order to create new economic, political and social systems.

Hobbes, Locke, and Rousseau are usually considered the original theorists of social contract theory and they each begin their stories with an imagined state of nature (Hobbes, 1994; Rousseau, 1968; Locke, 1980 [1690]). For Hobbes, the state of nature was solitary, poor, nasty, brutish, and short. To survive, humans needed to band together, submit themselves to the authority of one ruler or "sovereign" and thus give up whatever freedom they possessed in the interest of living communal, rich, pleasant, civilized, and longer lives. Though less pessimistic about the state of nature, Locke and Rousseau too imagined that life could be better if humans agreed to give up whatever freedoms existed in the state of nature to advance, grow, and otherwise evolve into a more sophisticated species. At their core, then, social contracts are exchanges of freedom for security. Both in theory and in practice the perennial questions hinge on how much and what kinds of freedom humans should give up in exchange for how much and what kinds of security. Thus, the Anthropocene thesis only adds intensity to these debates as the fact of human caused climate change requires more and different freedoms to be sacrificed to survive and thrive in the Anthropocene.

A. Wahlrab (✉)
University of Texas, Tyler, TX, USA
e-mail: awahlrab@uttyler.edu

© The Author(s), under exclusive license to Springer Nature Switzerland AG 2023
N. Wallenhorst, C. Wulf (eds.), *Handbook of the Anthropocene*,
https://doi.org/10.1007/978-3-031-25910-4_197

While debates over a new social contract for the Anthropocene are motivated by climate change and related existential threats, agreement exists among those who study Earth Systems that human behaviour must change by 2050 (IPCC, 2021). Since the current social contract has put us on an unstable pathway of roughly three degrees Celsius warmer than preindustrial levels by 2099 (Plumer & Fountain, 2021), prominent Earth Systems scientists conclude that: "the Earth System may be approaching a planetary threshold that could lock in a continuing rapid pathway toward much hotter conditions—Hothouse Earth. This pathway would be propelled by strong, intrinsic, biogeophysical feedbacks difficult to influence by human actions, a pathway that could not be reversed, steered, or substantially slowed" (Steffen et al., 2018, p. 8257). Put differently, past human behaviour, economic systems, state systems, and relationship to nature has led us to a catastrophic tipping point. Much of the blame is directed at modernity and modern ways of thinking, doing, and organizing society, especially modernity's insistence on separating nature and humanity—a view that pits humans against nature. Many critics also point to the way neoliberal capitalism has served as the economic system that contributes most to climate change, especially its faith in unlimited growth, exploitation of and conception of nature as a natural resource.

These and other criticisms of the current social contract should be understood as appropriate mechanisms of change. On this point, Sheldon Wolin notes that a recurring feature in the tradition of contract theory is a justification for the proposed social contract: "what was wrong with current political practices such that a comprehensive recodification of basic principles was urgent" (Wolin, 2004, p. 534). Climate scientists agree that humans are responsible for what some refer to as the rapidly approaching sixth great extinction and, thus, a level of insecurity that justifies a new social contract (Guterl, 2013; Kolbert, 2014). Accordingly, if we are to adopt the necessary political ecology required to re-write the social contract, then we must recognize that humans are a part of natural systems rather than separate from them (Elliott, 2016, p. 84). The Anthropocene thesis itself catalogues how humans have failed to recognize their relationship with nature as fundamentally a part of it. Following from this observation that humans are essentially connected to and a part of nature, it makes sense that new systems of social organization are created that keep this observation centred. Current focus on dire warnings based on sound science attempt to depoliticize the fact of climate change. This has not worked (Sullivan, 2021). Change requires not a focus on dire warnings of impending doom, but on transforming the political and economic systems that have produced climate change.

While stratigraphers and others in the Anthropocene Working Group debate the precise moment when humans began making their mark in the layers of the Earth's surface, "a lasting record in rock" (Ellis, 2018, p. 3), the fundamental question of "are human societies even capable of changing to avoid imminent environmental disaster" (Elliott, 2016, p. 156) may get the problem of the Anthropocene wrong. Framing as natural catastrophe only further emphasizes the modern logic of human versus nature and leads to solutions that rely on more consumption (of green products). Alternatively, if climate change is understood as an outgrowth of capitalist modes of production which rely on endless growth and mass consumption, then solutions include increased political agency and revitalization of democratic practice.

Social contract theory appears to have baked into it an affirmation that human societies are capable of change, whether out of fear of brutish life or hope for a more humane one. Cataloguing the reasons to be fearful is easy enough and is the task set by those who study Earth's systems and the Earth as a system. Their findings, that humans are the primary cause of the current climate chaos, further back-up the Anthropocene thesis. They also beg the question of whether humans can be the central feature of the next social contract. This question follows from the observation that the social contracts that got us where we are today all share the view that humans are the most important and that human communities can overcome the coming catastrophe without considering the Earth as a system. The Anthropocene thesis, in contrast, requires that humans evolve consciously with the planet and shape the next social contract accordingly (Wahlrab, 2019; Steger & Wahlrab, 2017). Unfortunately, according to Wayne Gabardi, modern social contracts "all share a common commitment to the central concern of human progress in the twenty-first century." In other words, they fail to incorporate the view that humans must not be considered in isolation, they fail to see a future that is not simply more modern and technologically sophisticated, and they fail to adopt new logics. They simply ask, "how should we re-engineer the biosphere, society, and human nature so that our species can extend the promise of modernity even further and maintain planetary dominion and stewardship?" (Gabardi, 2017, p. 27). Instead, humans must be reimagined as "humanimals" to include animals and the environments they need to thrive. Past frameworks routinized and legitimized human power in the Age of Humanity, with all its associated catastrophe. Any new social contract, one that seeks justice for the human, nonhuman, and the Earth's systems, must adopt a different frame: one of posthumanism.

This term may dredge up strange images (fantasy or science fiction), but in the context of social contract theory in and for the Anthropocene it means taking seriously the relationships between humans and nonhumans (animals and nature). This includes the view that human and animal traits evolve together over time and that although liberal humanism frames animals and nature as resources that are used by humans, this perspective is ideological and political. Along these lines Brian Elliot explains that politics itself can never be a completely human affair, noting instead that "politics is about how to organize the complex interrelations between humans and non-humans" (Elliott, 2016, p. 20). Central here is the finding that politics is the primary source of climate change, rather than moral failures like greed. Rather than frame our understanding of climate change and the Anthropocene purely in terms of moral crisis and environmental catastrophe, we should understand it as "the result of a certain, historically grounded, system of material production and social organization." In a word, "neoliberalism" (Elliott, 2016, p. 66). Elliot goes on to show that neoliberalism, the ideology that governs the twenty-first century social order, is not negatively impacted by revelations of social harm that results from global capitalism: "the social harm caused by climate change is not intrinsically problematic within the neoliberal paradigm" (Elliott, 2016, p. 74). The blaring statement that the 2021 IPCC report is a "code red for humanity" is both true and insufficient. It is insufficient because its truth is centred in a concern for humanity alone. It is also insufficient because it adopts the depoliticized language of science. Anthropogenic climate change was produced by a faith in endless growth for its own sake. Technological fixes that do not address politics and real social change will merely reproduce the collective behaviours that led us here.

References

Elliott, B. (2016). *Natural catastrophe: Climate change and neoliberal governance*. Edinburgh University Press.

Ellis, E. C. (2018). *Anthropocene: a very short introduction* (1st ed.). Oxford University Press.

Gabardi, W. (2017). *The next social contract: Animals, the Anthropocene, and biopolitics*. Temple University Press.

Guterl, F. (2013). *The fate of the species: Why the human race may cause its own extinction and how we can stop it*. Bloomsbury Publishing USA.

Hobbes, T. (1994). In E. Curley (Ed.), *Leviathan*. Hackett Publishing Company.

IPCC. (2021). *Climate change 2021: The physical science basis. Contribution of working group I to the sixth assessment report of the intergovernmental panel on climate change*. Cambridge University Press.

Kolbert, E. (2014). *The sixth extinction: An unnatural history* (1st ed.). Henry Holt and Company.

Locke, J. (1980 [1690]). *Second treatise of government*. Edited by C. B. Macpherson. Hackett Publishing Company.

Plumer, B., & Fountain, H. (2021). A hotter future is certain, climate panel warns. But how hot is up to us. *New York Times*, August 9, 2021.

Rousseau, J.-J. (1968). *The social contract*. Penguin.

Steffen, W., Rockström, J., Richardson, K., Lenton, T. M., Folke, C., Liverman, D., Summerhayes, C. P., Barnosky, A. D., Cornell, S. E., Crucifix, M., Donges, J. F., Fetzer, I., Lade, S. J., Scheffer, M., Winkelmann, R., & Schellnhuber, H. J. (2018). Trajectories of the earth system in the Anthropocene. *Proceedings of the National Academy of Sciences, 115*(33), 8252. https://doi.org/10.1073/pnas.1810141115. http://www.pnas.org/content/115/33/8252.abstract

Steger, M. B., & Wahlrab, A. (2017). *What is global studies?: Theory & practice*. Routledge.

Sullivan, H. (2021). Code red for humanity': What the papers say about the IPCC report on the climate crisis. *Guardian*, August 9, 2021. https://www.theguardian.com/environment/2021/aug/10/code-red-for-humanity-what-the-papers-say-about-the-ipcc-report-on-the-climate-crisis

Wahlrab, A. (2019). Imagining global non-violent consciousness. In C. Hudson & E. K. Wilson (Eds.), *Revisiting the global imaginary: Theories, ideologies, subjectivities: Essays in honor of Manfred Steger* (pp. 123–140). Springer International Publishing.

Wolin, S. S. (2004). *Politics and vision: Continuity and innovation in Western political thought*. (Expanded edition). Princeton University Press.

Amentahru Wahlrab is an Associate Professor of Political Science at the University of Texas at Tyler. His research interests lie at the intersection of globalization, political economy, political violence, and populism. He is the coauthor of *What is Global Studies? Theory and Practice* (Routledge, 2017). He is also the co-editor of the book series *Global Populisms* published by Brill and the book review editor for the journal *Populism* also published by Brill.

Social Ecology

Helga Weisz, Marina Fischer-Kowalski, and Verena Winiwarter

Abstract Social Ecology is an interdisciplinary research field rooted in the traditions of both the Social Sciences and Natural Sciences. Related research fields include Human Ecology, Industrial Ecology, Ecological Economics and Sustainability Sciences. A core paradigm, shared by most schools of Social Ecology, is the insight that human social and natural systems interact, coevolve over time and have substantial impacts upon one another, with causality working in both directions. Social Ecology offers a conceptual approach to society-nature coevolution in relation to history, current development processes and a future sustainability transition.

The Intellectual Roots

The academic roots of Social Ecology can be traced as far back as the nineteenth century, to the political economies of Adam Smith, David Ricardo, Karl Marx and Thomas Malthus. Another influential field was geography. George Perkins Marsh's book *Man and Nature: or, Physical Geography as Modified by Human Action* (Marsh, 1864) inspired two influential publications: *Man's Role in Changing the Face of the Earth* (Thomas Jr., 1956) and *The earth as transformed by human action*

H. Weisz (✉)
Humboldt-Universität zu Berlin and Potsdam Institute for Climate Impact Research (PIK), Berlin\Potsdam, Germany
e-mail: helga.weisz@hu-berlin.de; weisz@pik-potsdam.de

M. Fischer-Kowalski
University of Natural Resources and Life Sciences, Vienna, Austria
e-mail: marina.fischer-kowalski@boku.ac.at

V. Winiwarter
Austrian Academy of Sciences, Vienna, Austria
e-mail: verena.winiwarter@oeaw.ac.at

© The Author(s), under exclusive license to Springer Nature Switzerland AG 2023

N. Wallenhorst, C. Wulf (eds.), *Handbook of the Anthropocene*,
https://doi.org/10.1007/978-3-031-25910-4_198

(Turner et al., 1990). Already in 1969, the German geographer Neef had explicitly talked about the 'metabolism between society and nature' as a core problem of geography (Neef, 1969).

Functionalist Cultural Anthropology focused on the society-nature interface. Leslie White, an early representative of the functionalist tradition, rekindled interest in energetics. White understood change in societies as social evolution, and the mechanisms propelling it were energy and technology (White, 1943). Julian Steward's 'method of cultural ecology' considered the quality, quantity and distribution of resources within the environment (Murphy & Steward, 1966).

Natural scientists worked on Human Ecology since the 1970s (e.g. Ehrlich & Ehrlich, 1970; Ehrlich et al., 1973), focussing on population growth as main driver of environmental impacts and conceptualizing society as one aggregated actor who creates disturbance in ecosystems. This conceptualization of societies ignores the internal complexity unique to social systems and generates the misleading idea that society can be viewed as analogous to a single rational person. Also the focus on population growth was criticized for ignoring the huge inequality in environmental impacts created by the rich and the poor and for generally understating the importance of material consumption in relation to population growth (e.g. Cohen, 1995; Satterthwaite, 2009; Belmin et al., 2021). In addition, the exclusive focus on 'man changing the environment' prevented an understanding of mutual influences between society and nature.

Historical scholars paved another path toward social ecological reasoning. For example, Braudel (1960) viewed the history of the Mediterranean as an outcome of interaction between social and natural processes, with nature forming the *longue dureé* underlying historical events. Sieferle reconstructed the history of the industrial revolution in terms of energy inputs to society and demonstrated how fundamentally the fossil fuel based energy regime altered social structures and dynamics (Sieferle, 2001).

The Chicago School of Human Ecology (Park, Burgess, Duncan) is often seen as an entry point to the modern reading of Social Ecology. This school used analogues from biological ecology to analyze urban development (e.g., hierarchy, competition, succession). For them, however, the natural environment was reduced to spatial structure. For example, Duncan's POET model (population, organization, environment and technology) for describing social processes, in no way referred to natural processes or conditions other than space (Beus, 1993). The neo-Marxist tradition within Sociology tended to become narrowly focused on theories of capital, class and the state. The influential *World Systems Theory* (cf. Ciccantell & Bunker, 1998; Wallerstein, 1999, Goldfrank et al., 1999; Hornborg & Crumley, 2007;) retained this tradition, and explicitly linked it to ecological concerns. Within Human Geography and Environmental Sociology, the economic requirement of capital accumulation has been linked to both economic growth and the continuing (over)exploitation of natural resources (Harvey, 2014; Schnaiberg & Gould, 1994).

Schools of Social Ecology

In the 1970s two US American traditions of the same name emerged, exemplified in the Institute for Social Ecology in Vermont and in the School of Social Ecology at UC Irvine. The former adheres to the idea of deep ecology (Bookchin, 1984) and continues to be centred on eco-activism and a new environmental ethic (Lejano & Stokols, 2013). The latter has strong roots in psychology and considers Social Ecology as a multidisciplinary approach to study recurring social, behavioural and environmental problems, such as crime, health, wellbeing, or urban planning, by taking into account social, physical, and cultural environments (Oishi & Grahem, 2010; Binder et al., 1975).

The Vienna Social Ecology School, dating back to the 1980s, took a different approach. The name Social Ecology, chosen to be distinguishable from Human Ecology, should stress that it is not the human species that mattered but rather the social (and economic and technical) organization of human societies. Luhmann is by far the most important theoretical basis of the social side of the Vienna School of Social Ecology. Luhmann's social theory builds on interdisciplinary systems theory (Luhmann, 1997). The resulting general definition of systems as entities that emerge through self-referential operations (referred to as operationally closed) implies a notion of social systems as entities that reproduce their own boundaries toward their environment through communication. Luhmann attributes the inability of modern societies to cope with global environmental change to exactly the same social structures that constituted the decisive evolutionary advantage of modern societies over traditional ones (Luhmann, 1986). The functional differentiation of modern societies into operationally closed systems of law, science, policy, economy, art or religion, implies that each system optimizes its operations towards its core function (e.g. winning elections, maximizing profit, distinguishing fact from fiction or distinguishing human behaviour as being within or outside the law). Social Ecology noted that this development to modern society was contingent on an unprecedented ability to utilize energy resources, which in turn mobilized increasing use of mineral raw materials and created environmental change at a planetary scale (Fischer-Kowalski & Weisz, 1999; Krausmann et al., 2009; Weisz, 2011). Any contemporary observer of the international COP negotiations to reach a binding climate mitigation treaty will find ample evidence for Luhmann's analysis from 1986. Luhmann's social theory allows for, but does not directly explore, the relational biophysical conditions of sociocultural evolution. To quantify and analyze these relational biophysical conditions, the Vienna school of Social Ecology (and now many other research fields) employ the concept of *social metabolism* (Haberl et al., 2019; Weisz et al., 2015). In addition, the Vienna school distinguishes itself from other schools of social ecology by a strong historical perspective, which analyses transitions between 'sociometabolic regimes', defined as fundamental changes in core mechanisms of 'society-nature interaction', especially in energy capture mechanisms (Krausmann et al., 2016; Lenton et al., 2016).

The core concept employed by the Frankfurt School of Social Ecology is 'societal nature relations' (*gesellschaftliche Naturverhältnisse*). The focus of the Frankfurt approach is on the relations between society and nature in terms of the various societal regulations that define these relations. Basic societal nature relations, which, being related to basic human needs, are indispensable for individual and societal reproduction and development. The link to the concept of human needs inevitably makes 'societal nature relations' a normative concept: the basic societal nature relations should be regulated in such a way that all humans are able to meet their basic needs (Becker et al., 2011). The Frankfurt approach defines as its 'epistemic object' the 'crisis of societal nature relations' (Becker & Jahn, 2006, p. 19). This definition is normative in that it presupposes the existence of a crisis, that is, a radical deviation of the 'is' state from an 'ought' state of societal nature relations. The purpose of Social Ecology is to generate knowledge necessary to understand this crisis and to react to it, towards establishing the 'ought' state of societal nature relations, pursuing the core research question of Social Ecology: 'How can the crisis-ridden societal nature relations be perceived, understood and actively shaped?' (ibid., p. 12).

In addition to these four explicitly branded research schools of social ecology there are many other scholarly traditions that entertain a similar approach. We here just mention two.

Elinor Ostrom's book *Governing the Commons* (Ostrom, 1996) stimulated a rich strand of research, basically debunking Garret Hardin's influential *Tragedy of the Commons* narrative. Hardin stated that rational individual actors will inevitably overexploit common natural resources unless state law or privatization of common goods will prevent them to do so (Hardin, 1968). Ostrom demonstrated that Hardin's analysis was misleading both theoretically and empirically. Theoretically, Hardin did not account for communication between rational actors (technically: he did not consider repeated games). Empirically, Ostrom and her co-researchers studied many cases where the tragedy of the commons had been solved without referring to state law or to privatization but via communication among actors.

The Dutch societal transitions management school focuses on technical and social systems rather than ecological systems. The core concern is the existence of 'persistent' and 'wicked' problems in social system functioning that can only be overcome by a systemic transition. Hence, a socioecological transition (SET) is a transition between two dynamic equilibria, that is, a shift from one more or less stable state to another (Rotmans et al., 2001). The nature of transitional dynamics is described in terms of a generic pattern that consists of a sequence of mechanisms that result in irreversible changes in the system. The concept distinguishes niches, regimes and landscapes and is therefore also frequently termed 'Multi-Level-Perspective' (MLP) (Geels, 2005). Transition management draws together a selective number of frontrunners in a protected environment while also maintaining links to change-inclined regime agents. Nuanced comparisons between various schools of social ecology have been published, e.g. between the Vienna and the Dutch school (Fischer-Kowalski & Rotmans, 2009), between the Frankfurt and the Vienna school (Kramm et al., 2017), or between several frameworks for analyzing social-ecological systems (Binder et al., 2013). Fischer-Kowalski & Weisz, 2016 provide a comprehensive overview on which this article is based.

Many of the Social Ecology approaches reviewed here are highly visible and feature prominently in international assessment reports and science programs. The once eccentric idea that a Social Ecology must take society-nature interactions seriously and address them in an interdisciplinary manner has become mainstream in the Anthropocene.

References

Becker, E., Hummel, D., & Jahn, T. (2011). Gesellschaftliche Naturverhältnisse als Rahmenkonzept. In M. Groß (Ed.), *Handbuch Umweltsoziologie* (pp. 75–96). VS Verlag für Sozialwissenschaften.

Becker, E., & Jahn, T. (2006). *Soziale Ökologie. Grundzüge einer Wissenschaft von den gesellschaftlichen Naturverhältnissen*. Campus.

Belmin, C., Hoffmann, R., Pichler, P.-P., & Weisz, H. (2021). 'Fertility transition powered by women's access to electricity and modern cooking fuels. *Nature Sustainability, 5*, 245–253. https://doi.org/10.1038/s41893-021-00830-3

Beus, C. E. (1993). Sociology, human ecology, and ecology. *Advances in Human Ecology, 2*, 93–132.

Binder, A., Stokols, D., & Catalano, R. (1975). Social ecology: An emerging multidiscipline. *Journal of Environmental Education, 7*, 32–43.

Binder, C. R., Hinkel, J., Bots, P. W. G., & Pahl-Wostl, C. (2013). Comparison of frameworks for analyzing social-ecological systems. *Ecology and Society, 18*(4), 26.

Bookchin, M. (1984). *The ecology of freedom. The emergence and dissolution of hierarchy*. Cheshire Books.

Braudel, F. (1960). History and the social sciences: The long duration. *American Behavioral Scientist, 3*(6), 3–13.

Ciccantell, P. S., & Bunker, S. G. (1998). *Space and transport in the world-system*. Greenwood Publishing Group.

Cohen, J. E. (1995). *How many people can the earth support?* W. W. Norton & Company.

Ehrlich, P. R., & Ehrlich, A. H. (1970). *Population, resources, environment: Issues in human ecology*. Freeman.

Ehrlich, P. R., Ehrlich, A. H., & Holdren, J. H. (1973). *Human ecology, problems and solutions*. Freeman.

Fischer-Kowalski, M., & Rotmans, J. (2009). Conceptualizing, observing and influencing social-ecological transitions. *Ecology and Society, 14*(2), 1–3.

Fischer-Kowalski, M., & Weisz, H. (1999). Society as a hybrid between material and symbolic realms. Toward a theoretical framework of society-nature interaction. *Advances in Human Ecology, 8*, 215–251.

Fischer-Kowalski, M., & Weisz, H. (2016). The archipelago of social ecology and the Island of the Vienna school. In H. Haberl, M. Fischer-Kowalski, F. Krausmann, & V. Winiwarter (Eds.), *Social ecology. Human-environment interactions, vol 5*. Springer.

Geels, F. W. (2005). The dynamics of transitions in socio-technical systems: A multi-level analysis of the transition pathway from horse-drawn carriages to automobiles (1860–1930). *Technology Analysis & Strategic Management, 17*(4), 445–476.

Goldfrank, W. L., Goodman, D., & Szasz, A. (1999). *Ecology and the world-system*. Greenwood Publishing Group.

Haberl, H., Wiedenhofer, D., Pauliuk, S., Krausmann, F., Müller, D. B., & Fischer-Kowalski, M. (2019). Contributions of Sociometabolic research to sustainability science. *Nature Sustainability, 2*, 73–184.

Hardin, G. (1968). The tragedy of the commons. *Science, 162*, 1243–1248.

Harvey, D. (2014). *Seventeen contradictions and the end of capitalism*. Oxford University Press.

Hornborg, A., & Crumley, C. L. (2007). *The world system and the earth system. Global socioenvironmental change and sustainability since the Neolithic*. Left Coast Press.

Kramm, J., Pichler, M., Schaffartzik, A., & Zimmermann, M. (2017). Societal relations to nature in times of crisis—Social Ecology's contributions to interdisciplinary sustainability studies. *Sustainability, 9*(7), 1042.

Krausmann, F., Gingrich, S., Eisenmenger, N., Erb, K-H., Haberl, H., & Fischer-Kowalski, M. (2009). Growth in global materials use, GDP and population during the 20th century. *Ecological Economics 68*(10), 2696–2705. https://doi.org/10.1016/j.ecolecon.2009.05.007

Krausmann, F., Weisz, H., & Eisenmenger, N. (2016). Transitions in socio-metabolic regimes through human history. In Haberl et al. (Eds.), *Social ecology. Society-nature relations across time and space* (pp. 89–108). Springer.

Lejano, R. P., & Stokols, D. (2013). Social ecology, sustainability, and economics. *Ecological Economics, 89*, 1–6.

Lenton, T. M., Pichler, P.-P., & Weisz, H. (2016). Revolutions in energy input and material cycling in earth history and human history. *Earth System Dynamics, 7*(2), 353–370.

Luhmann, N. (1986). *Ökologische Kommunikation*. Westdeutscher Verlag. English translation 1989. Ecological communication, (J. Bednarz Trans.). University of Chicago Press.

Luhmann, N. (1997). *Die Gesellschaft der Gesellschaft*. 2 Bände. Suhrkamp. English translation 2012, 2013. *Theory of Society* (R. Barrett, Trans.). Standford University Press.

Marsh, G. P. (1864). *Man and nature; or physical geography as modified by human action*. Scribners & Sampson Low.

Murphy, R. F., & Steward, J. H. (1966). Tappers and trappers: Parallel process in acculturation. *Economic Development and Cultural Change, 4*(4), 335–355.

Neef, E. (1969). Der Stoffwechsel zwischen Gesellschaft und Natur als geographisches Problem. *Geographische Rundschau, 21*, 453–459.

Oishi, S., & Graham, J. (2010). Social ecology: Lost and found in psychological science. *Perspectives on Psychological Science 5*(4), 356–377. https://doi.org/10.1177/1745691610374588

Ostrom, E. (1996). *Governing the commons. The evolution of institutions for collective action*. Press Syndicate of the University of Cambridge.

Rotmans, J., Kemp, R., & van M. Asselt. (2001). More evolution than revolution: Transition management in public policy. *Foresight, 3*(1), 15–31.

Satterthwaite, D. (2009). The implications of population growth and urbanization for climate change. *Environment and Urbanization, 21*(2), 545–567.

Schnaiberg, A., & Gould, K. A. (1994). *Environment and society: The enduring conflict*. St. Martin's.

Sieferle, R. P. (2001). *The subterranean Forest. Energy systems and the industrial revolution*. The White Horse Press.

Thomas, W. L., Jr. (1956). *Man's role in changing the face of the earth*. The Chicago University Press.

Turner, B. L. I., Clark, W. C., Kates, R. W., Richards, J. F., Mathews, J. T., & Meyer, W. B. (1990). *The earth as transformed by human action: Global and regional changes in the biosphere over the past 300 years*. Cambridge University Press.

Wallerstein, I. (1999). Ecology and capitalist costs of production: No exit. In W. L. Goldfrank et al. (Eds.), *Ecology and the world-system* (pp. 3–11). Greenwood Press.

Weisz, H. (2011). The probability of the improbable: Society–nature coevolution. *Geografiska Annaler: Series B: Human Geography, 93*(4), 1–12.

Weisz, H., Sangwon, S., & Graedel, T. E. (2015). Industrial ecology: The role of manufactured Capital in Sustainability. *Proceedings of the National Academy of Sciences, 112*(20), 6260–6264.

White, L. A. (1943). Energy and the evolution of culture. *American Anthropologist, 45*(3), 335–356.

Helga Weisz is Professor of Industrial Ecology and Climate Change at Humboldt-Universität zu Berlin and Head of the Future Lab 'Social Metabolism and Impacts' at the Potsdam Institute of Climate Impact Research (PIK). Academic degrees: molecular biology (Mag.rer.nat. University of Vienna), cultural studies (Dr. phil. Humboldt-Universität zu Berlin), social ecology (*venia docendi*. University of Klagenfurt). She serves on various expert boards e.g. the scientific advisory board for UNDP's Human Development Reports, the Jury of the German Environmental Price, The International Resource Panel of UNEP, among others.

Marina Fischer-Kowalski is Professor em. and senior scientist at the Institute for Social Ecology, University of Natural Resources and Life Sciences, Vienna. Director of this Institute 1986–2014. Academic background: PhD in Sociology (Vienna University), post-graduate degree at the Institute of Advanced Studies, Vienna. Associate editor of The Anthropocene Review, member of the editorial board of: International Journal of Industrial Ecology (JIE), Environmental Innovation and Societal Transitions (EIST), GAIA, and BioPhysical Economics and Resource Quality (BERQ). Lead expert in the International Resource Panel of UNEP.

Verena Winiwarter, trained in chemistry, history and communication sciences a retired environmental historian at the Vienna University of Natural Resources and Life Sciences. She is a full member of the Austrian Academy of Sciences (2016) and of Academia Europaea (2020). In 2013, she was elected Scientist of the Year in Austria. In 2019, she received a Honorary Doctorate from Aalborg University (Denmark). She serves on several scientific and policy boards.

Societal Growth

Wiebren Johannes Boonstra

Abstract Concepts and theories of how people are interconnected and interdependent are indispensable for our understanding of the onset, development and problems of the Anthropocene. This chapter discusses the concept of 'society' to denote configurations of people in certain patterns of interdependence. Intensive and extensive growth of these configurations is a central feature of the Anthropocene, a time when human societies started crowding out other forms of life in the biosphere. To understand why growth of societies occurs and why it tends to be persistent, the chapter outlines two ideas – the 'Pareto principle' and the 'power-elite hypothesis' – to theorize the reinforcing feedback between growth of human configurations and power relations.

In this chapter I argue that an understanding of the ways in which humans live together is indispensable for understanding the origin and dynamics of the Anthropocene. To demonstrate the credibility of this argument I will first discuss the concept of 'society' as an idea we often use to denote the different ways in which we live together and relate to one another.

Society is the word we use to denote the greatest configuration of people living together. When we speak, for example, of 'global human society', we refer to the whole human population. In contrast, for smaller human configurations, we use other words, such as 'households', 'families', or 'communities'.

Human configurations are not differentiated by the number of people only. Households or families include configurations based on real or imagined kinship ties, while communities are made up of configurations based on geographical proximity and a shared culture and identity. Households, families and communities are in turn incorporated in societies. But an important feature of societies is that they typically include indirect social ties, while families and communities are mostly based on direct social ties and networks (Berreman, 1978). The difference is

W. J. Boonstra (✉)
Department of Earth Sciences, Uppsala University, Uppsala, Sweden
e-mail: wijnand.boonstra@geo.uu.se

© The Author(s), under exclusive license to Springer Nature Switzerland AG 2023

N. Wallenhorst, C. Wulf (eds.), *Handbook of the Anthropocene*,
https://doi.org/10.1007/978-3-031-25910-4_199

however not clear-cut: we can have kin with whom we no longer meet, as well as neighbours we don't see. Moreover, we also frequently live in configurations which are constituted through a mixture of direct and indirect social ties, such as companies, work associations, etc. And, to complicate matters even more, the word 'society' is sometimes also used to refer to groups of people that are linked together directly, e.g. when we refer to so-called 'secret societies' or 'clubs'.

Societies exist, just as other configurations of people, because we depend on other people for our survival and wellbeing. Through living in societies we create the social conditions needed for our survival. These conditions include of course the provision of basic needs, such as food; shelter; and protection from harm, but also affectional needs, knowledge provision, and the need for others and ourselves to attend impulses, propensities, feelings, and sentiments (de Swaan, 2013: 3). The ways in which people experience these various needs and together fulfil them, is different per society, then and now. To allow for this diversity we can define a society in a broad way as "a configuration of people in certain patterns of interdependence" (de Swaan, 2013: 15).

A consideration of the global and historical cultural diversity and varying sizes of human configurations quickly demonstrates that humans are not (genetically) determined to live together in a group of a certain size or to live together in a certain way. Nevertheless human history is still very often described as a deterministic and generic development that runs from small, egalitarian communities to large, hierarchical societies (cf. Harari, 2014).

A number of sociologists and anthropologists argue that the development of human societies is in reality much more dynamic and diverse with impressively dynamic (often seasonal) variations in the size, complexity and structure of human configurations (Mauss, 2013 [1950]; Scott, 2017; Graeber & Wengrow, 2021). These scholars notice how quantitative changes in the scale of human social life causally interrelate with qualitative changes of the ways in which people relate to nature, each other and themselves (Elias, 2012 [1978]: 251–252).

Changes in the size of human configurations (quantity) together with their forms (quality) are used to understand when and how humans fundamentally altered the synergetic dynamics of climatological, geological and biological processes on Earth. Conventional wisdom has it that the Anthropocene, as this period is now called, started when human societies grew extensively and intensively. Extensive growth refers to growth in numbers and geographical reach, while intensive growth refers to increases in the use of energy and materials (Goudsblom, 2002).

During the Anthropocene the dynamics and diversity that so much characterizes the development of human societies during much of the Holocene (Graeber & Wengrow, 2021; Scott, 2017) declined or disappeared. Considered on a planetary scale and at the level of a species, humans now live in a global society that is characterized by great inequality and persistent overuse of natural environments leading to climate change and overall declines in biodiversity.

Shifts in the scale of human societies are pertinent for the origin and development of the Anthropocene. And, as many others have pointed out (Daly, 2019), the key to a sustainable future lies in limiting the scale of human societies. A brief recapture is in place here. First, reducing scale to a spectre of overpopulation only (Ehrlich, 1991) is too simplistic: as I argued already the scale of human society coincides with qualitative changes in the ways in which humans relate to nature, others, and themselves. These changes in turn – so not only size! (for which: Cafaro et al. 2022) – determine the environmental sustainability of human societies. Second, mass societies, or nation-states, are not the inevitable outcome of an evolutionary process because for most of their history humans have seemed to enjoy the agency to choose living in both small and big societies (Graeber & Wengrow, 2021).

Considering the historical diversity of both small and large-scale societies, and people's habits to oscillate living in both, it is puzzling why it proves hard for us nowadays to control or limit the scale of our societies, or why we find it hard to even admit that growth itself is problematic (Boonstra & Joosse, 2013). In the few concluding paragraphs of this chapter I will offer some theory as to why growth of societies seems pertinent or inevitable, and therefore out of hand.

There is a long scholarly tradition that links growth of the scale of social life to changes in the ways humans exercise power over one another and their natural environments (Bodley, 2003; Elias, 2012 [1978]). Here power refers to the ability of people to influence the conduct and context of other human and non-human beings (Boonstra, 2016). Power thus refers to inter and intra-human social relations and interdependencies. As a very general mechanism power is causally interrelated to growth through reinforcing feedbacks: growth of societies drive changes in power, and through changes in power people drive growth.

The former mechanism, i.e. where growth drives power, is also called "the Pareto principle" after the Italian Vilfredo Pareto who discovered a power law probability for income distribution where 80% of wealth is typically owned by 20% of the population. Because this division remains also when societies and economies grow, it means that the 20% will own more and more and that inequality increases exponentially. Power-law explanations, like the Pareto principle, point out how power becomes more concentrated when societies grow on the basis of random chance and probability (Scheffer et al., 2017). The growing inequalities of power between elites and commoners is thus an effect of the scale of society.

The latter mechanism, i.e. where power drives growth, can be called "the power-elite hypothesis". The hypothesis has been promulgated by John Bodley in his book *"The power of scale"* (2003) where he describes the growth of human societies as an elite-directed process wherein the power balance between elites and commoners becomes more and more unequal as societies grow. Through competition amongst themselves, or with powerful outsiders, elites are pressured to use their power to expand the environments and people they could potentially control.

These two mechanisms are not mutually exclusive or alternative theories. Rather to the contrary, the mechanisms work in tandem and reinforce one another. Taken together they can explain why the growth of societies tend to coincide with larger differentials of power (between people, but also between people and other living beings). It means that growth is hard to stop because it promotes and strengthens the position of powerful elites, which, in turn, enables elites to use their power to continue growth.

Both mechanisms also offer some general clues as to why we now experience that growth is hard to avoid or slow down. Growth is persistent due to the reinforcing mechanisms outlined above. As we have seen these mechanisms are constituted from the interdependencies and power relations that link human groups within and between societies. To manage growth therefore is to manage these interdependencies – these 'structures' or 'orders' of societies. It is not enough to focus exclusively on elites or powerful individuals: the social configurations that make up human societies involve elites as well less powerful groups. Next, managing growth also requires understanding its emergent property. When a configuration of people grows from, let's say, 30 to 300 to 3000 to 30,000, more and more interdependencies are forged. In the process, growth itself becomes harder to control for any of the members of this configuration. Simply because the growth of the configuration in terms of the number of its members comes with a much faster growth in the number of interrelations between the members. The growing social complexity that comes with growth becomes harder to control for any of single member or group of members in the configuration. In this way, growth gains relative autonomy from the plans and intentions of any of the people that create and maintain the configuration through their actions and relations (paraphrasing Elias, 2012 [1978]: 95–96).

Nevertheless, limiting the growth of human societies is what is needed to address the grand challenges of the Anthropocene that we now face. It is therefore essential that we understand human societies as the configurations that are formed by the actions of interdependent people (Elias, 2012 [1978]: 103), and how these configurations link to growth and power.

References

Berreman, G. D. (1978). Scale and social relations. *Current Anthropology, 19*, 225–245.
Bodley, J. (2003). *The power of scale: A global history approach*. Routledge.
Boonstra, W. J. (2016). Conceptualizing power to study social-ecological interactions. *Ecology and Society, 21*(1), 21. https://doi.org/10.5751/ES-07966-210121
Boonstra, W. J., & Joosse, S. (2013). The social dynamics of degrowth. *Environmental Values, 22*, 171–189.
Cafaro, P., Hansson, P., & Götmark, F. (2022). Overpopulation is a major cause of biodiversity loss and smaller human populations are necessary to preserve what is left. *Biological Conservation, 272*, 109646.
Daly, H. (2019). Growthism: Its ecological, economic and ethical limits. *Real-world Economics Review, 87*, 9–22.
de Swaan A. (2013). *Human societies: An introduction*. John Wiley & Sons.
Ehrlich, P. R. (1991). Population diversity and the future of ecosystems. *Science, 254*, 175–175.
Elias, N. 2012 [1978]. *What is sociology?* [Collected works, Vol. 5]. UCD Press
Goudsblom, J. (2002). Introductory overview: The expanding anthroposphere. In J. Goudsblom & B. de Vries (Eds.), *Mappae mundi: Humans and their habitats in a long-term socio-ecological perspective, myths, maps and models* (pp. 21–46). Amsterdam University Press.
Graeber, D., & Wengrow, D. (2021). *The dawn of everything: A new history of humanity*. Penguin.
Harari, Y. N. (2014). *Sapiens: A brief history of humankind*. Random House.

Mauss, M. (2013). *Seasonal variations of the Eskimo: A study in social morphology*. Routledge. (Original edition 1950).
Scheffer, M., van Bavel, B., van de Leemput, I. A., & van Nes, E. H. (2017). Inequality in nature and society. *Proceedings of the National Academy of Sciences, 114*, 13154–13157.
Scott, J. C. (2017). *Against the grain. A deep history of the earliest states*. Yale University Press.

Wiebren Johannes Boonstra is Associate Professor in Environmental Sociology at the Department of Earth Sciences at Uppsala University, Sweden. He has authored and co-authored numerous publications on fisheries, farming, sustainability, and has a general interest in the study of non-strategic motivations of human behaviour, such as habits, emotions, and moods. Wiebren also works as Associate Editor-in-Chief for *Ambio: Journal of Society and Nature*; as member of the editorial board for *Society & Natural Resources* and *Sustainability Science*; and as part-time fireman for Uppsala Municipality, at Skyttorp station.

Urban Ecology

Isabelle Hajek and Jean-Pierre Lévy

Abstract At a time when awareness of the advent of the Anthropocene era once again raises questions about the compatibility between finite resources and globalised urban expansion, the article explores the paradigm shift in our understanding of urban ecology away from a vision centred on metabolism alone towards a plural, multidisciplinary and relational conception.

At a time when the reality of the Anthropocene is reopening questions about the incompatibility between finite resources and globalised urban expansion, is there are different ways to think about the relations between city and ecology? Some cues let believe it. There is a proliferation of initiatives that focus on ecological forms of human habitat and on "nature in the city". Some policies aim to establish natural spaces in and between cities and to reduce urban extraction and waste. The idea that nature can only be found outside and far from the city, or that the city necessarily has a predatory relationship with nature, is now being contested. Nonetheless, urban ecology (UE) is an old notion with multiple and contradictory meanings: it owns several linked strands that reproduce an opposition between the study of the city as a social organisation and a metabolic or even ecosystemic vision of UE. Similarly, making a connection between city and nature is no obvious matter. For many people, it is simply about implementing technical adjustments to resolve urban dysfunctions from a perspective in which the environment is seen as a collection of limits, of problems to resolve by the management of "artificial systems". The representation of nature that prevails in urban areas is often an idealised and homogenising, not to say one that fosters a process of social segregation (Hajek et al., 2015). The utopian vision of an autonomous city that manages all its flows is even

I. Hajek (✉)
University of Strasbourg, Strasbourg, France
e-mail: hajek@unistra.fr

J.-P. Lévy
Marne-la-Vallée University, Ponts ParisTech School, Marne-la-Vallée, France
e-mail: jean-pierre.levy@enpc.fr

© The Author(s), under exclusive license to Springer Nature Switzerland AG 2023

N. Wallenhorst, C. Wulf (eds.), *Handbook of the Anthropocene*,
https://doi.org/10.1007/978-3-031-25910-4_200

consonant with the idea of an opposition between an "urban nature" and a spontaneous nature and a planned, controlled city that the urbanistic aesthetic would help to pacify. In the light of these ambiguities, this article is demonstrating that UE can only provide a lasting solution, whether for people or for built and natural environments, on condition that we move away from an exclusively metabolic conception in favour of a plural, multidisciplinary and relational urban ecology.

With respect to UE, two connections crystallise an opposition between a sociological vision of the emergence of the big city in the Chicago School tradition, and a metabolic vision that made its entrance onto the urban stage through the question of urban *excreta* as approached by urban medicine and chemistry in the 18th and 19th centuries. Borrowing most of their ideas from Darwinism, the researchers of the Chicago School, influenced by Georg Simmel (1988 [1895]), like Park et al. (1925), studied flows of immigrants in Chicago in the 1900s, the areas they occupied, how they coexisted with American populations and how they integrated themselves into modern urban society. They were interested in the formation of cities, in their dynamics of growth, aggregation and expansion, and in the social interactions that took place within them. They saw these unprecedented urban changes as corresponding to an "ecological order" in the sense that they were not dependent on a decisional process but on "natural forces" (such as competition, symbiosis …) which tend to produce an ordered grouping characteristic of its population and its institutions. Above all, these studies showed that the ecological dynamics specific to the city generate original forms of culture: these ecological forces produce "moral areas" and turn the modern city into a cultural melting pot, which in turn feeds its "human ecology"; they also shape a new type of personality, "a distanced urban self" that tries to protect itself from the hardships of urban life while being at the same time more socialised to cultural differences. From this perspective, we can say that this vision is one of a human and cultural ecology, uninterested in the relations between urban societies and their natural environment. Nonetheless, these works were the first to emphasise the relations between ecology and social forms. For its part, the metabolic vision in urban ecology, despite its variable meanings, is rooted in an opposition between city and nature. Whether forged in the nineteenth century in the exchanges of materials between city, industry and agriculture (Barles, 2005) in response to the food supply problems and high mortality rates of the cities, before giving way to an urbanism of sanitation (Choay, 1998), or whether becoming concerned about the ecological impact of cities and borrowing from the ecosystemic conceptions advanced in the international writings of ecologists like Abel Wolman (1965), Paul Duvigneaud (1974), Eugene Odum (1976), it is shaped: sometimes by the plan to build an urban order on nature and by improving the living conditions of human beings through a technical utopia, sometimes by the vision of "protecting nature" and a functionalist utopia. Ultimately, it reflects a dual logic of managing natural urban components and naturalising the city: by the artificial and endogenous replication of the major metabolisms (such as maximising the use of materials through integrated management and recycling of urban waste), and by the production of natural spaces, whether intended to foster biodiversity, urban agriculture or an economic approach to nature. To put it differently, this metabolic conception entails a double negation. First, a negation of the city as a social phenomenon that generates representations, practices that shape its ecological signature, and a social division of labour that makes the idea of the total mastery and autonomy of the urban metabolism an "oxymoron" (Barles, 2017). The city is only possible through externalisation. Second, a negation of an "urban nature" that is inherent to the city and not managed, and yet indispensable to the reproduction of the urban and surrounding ecosystems. More fundamentally, this metabolic conception involves a pathological vision of the city associated with a metabolism that "absorbs nature, transforms it then evacuates it in materials and pollutants that are damaging to the planet, and to the health and quality of life of its inhabitants" (Lévy, 2010, p. 4). The recent transformations of urban reality – less circumscribed within organic borders and within an urban-rural continuum than in complex relations of co-extension/embeddedness – and the crisis of the very idea of "nature" are an encouragement to move away from this metabolic paradigm, whose circularity – illusory and techno-centric – excludes citydwellers in their relations to natural and built environments.

A new conception of UE is emerging today, in which the processes of urbanisation and environmental production can be both a source of damage and of resources for ecosystems (Douglas et al., 2015). This UE, which is broadening its scope to encompass ways of life, concerns not only research, but new expectations and practices amongst populations who, in a context of struggles for day-to-day quality of life and environmental justice, are creating spaces in which a new relationship to nature is coming into being. This "paradigm shift" (Lévy, 2010) is perceptible in different expressions that try to think about the city and nature in their reciprocal relations. To mention a few, the rise of an "ecology of the urban landscape" (Clergeau, 2011), which recognises urban activities and spaces as a historical and modern component of the functioning of natural ecosystems, is combining with the emergence of an "aesthetic of nature" (Blanc, 2017) that includes nature, experience and felt sensations in a way that was previously restricted to art and culture. Both of them emphasise the complexity of urban ecosystems, whose heterogeneity and sensory, imaginary and symbolic components cannot be reduced to a single dimension, whether exclusively technical, metabolic, or even socio-spatial or natural. They suggest other – holistic – modes of action with respect to urban ecology, encompassing questions ranging from the maintenance of biodiversity to the material functions of nature in the city (flood management, food production, etc.) to the ambiences experienced and/or created. This shift in urban ecology covers many problems of the urban environment. "Industrial ecology", for example, with a view to reintegrating urban metabolisms into closed biogeochemical cycles, advocates a "territorial ecology" grounded on social bonds that generate flows of alternative materials. Another notion is "environmental health" (Lawrence, 1999) which extends a biomedical model to an ecological interpretation of health to consider the links between the health of citydwellers, the characteristics of the urban environment, and the impact of "ecological inequalities" (Faburel, 2012). But also the urban ecology of citydwellers themselves: whether, for example, the ecology they look for is a relationship to ordinary nature made up of multiple "sensorialities" (touch, sight, hearing, smell), or the ecology they practise in different gardening groups (Chelkoff & Paris,

2015), or in the invention of new forms of urban life to minimize the waste and consumption of urban metabolisms. While for planners and politicians, UE has become a tool to legitimising their interventions, contributing to the production of standardised urban forms and a single framework of thinking all the stronger for being able to draw on a universal dimension linked with the rise of sustainable development, the history of urban ecology and its appropriations prompts the conclusion that it does not refer to a single rationality. That being the case, its future fruitfulness can only lie in the recognition of its eminently plural character, extending the ecological-metabolic vision to its ordinary, lived, felt components, carried forward by a plurality of actors and proceeding less from top-down political regulation than from the production of a shared world.

References

Barles, S. (2005). *L'invention des déchets urbains. France: 1790–1970*. Champ Vallon.

Barles, S. (2017). Écologie territoriale et métabolisme urbain: quelques enjeux de la transition socioécologique. *Revue d'économie régionale et urbaine, 5*, 819–836.

Blanc, N. (2017). The aesthetics of nature. In A. Chone, I. Hajek, & P. Hamman (Eds.), *Rethinking nature. Challenging disciplinary boundaries* (pp. 67–74). Routledge.

Chelkoff, G., & Paris, M. (2015). Natures d'ambiances en bord de routes: collectifs de jardin, dépendances routières et trame verte urbaine. In I. Hajek, P. Hamman, & J.-P. Lévy (Eds.), *De la ville durable à la nature en ville. Entre homogénéité et contrôle social* (pp. 291–308). Septentrion.

Choay, F. (1998). Pensées sur la ville, arts de la ville. In M. Agulhon (Ed.), *La ville de l'âge industriel. Le cycle haussmannien* (pp. 170–286). Éditions du Seuil.

Clergeau, P. (2011). *Ville et biodiversité. Les enseignements d'une recherche pluridisciplinaire*. PUR.

Douglas, I., Goode, D., Houck, M., & Wang, R. (2015). *The Routledge handbook of urban Ecology*. Routledge.

Duvigneaud, P. (1974). *La synthèse écologique. Populations, communautés, écosystèmes, biosphère, noosphère*. Doin éditeurs.

Faburel, G. (2012). The environment as a factor of spatial injustice: A new challenge for the sustainable development of European regions ? In C. Ghenai (Ed.), *Sustainable development: Policy and urban development-tourism, life science, management and environment* (pp. 431–478). Intech.

Hajek, I., Hamman, P., & Lévy, J.-P. (2015). *De la ville durable à la nature en ville. Entre homogénéité et contrôle social*. Septentrion.

Lawrence, R. J. (1999). Urban health: An ecological perspective. *Reviews on Environmental Health, 14*(1), 1–10.

Lévy, J.-P. (2010). Ville et environnement: pour un changement de paradigme. In O. Coutard & J.-P. Lévy (Eds.), *Écologies urbaines* (pp. 3–14). Anthropos.

Odum, E. (1976). *Écologie. Un lien entre les sciences naturelles et les sciences humaines*. Doin éditeurs.

Park, R. E., Burgess, E., & McKenzie, R. D. (1925). *The City*. University of Chicago Press.
Simmel, G. (1988). Die mode. In G. Simmel (Ed.), *La tragédie de la culture* (pp. 89–127). Rivages. [original edition 1895].
Wolman, A. (1965). The metabolism of cities. *Scientific American, 213*(3), 179–190.

Isabelle Hajek is associate professor at the University of Strasbourg and affiliated with the laboratory Societies, Actors and Government in Europe (SAGE, UMR CNRS 7363). She is a specialist in urban ecology and environmental sociology. She notably published: Lutter contre le gaspillage. Réforme ou révolution. *Écologie & politique*, Ed. Le Bord de l'eau, 60, 2020; *Rethinking Nature. Challenging Disciplinary Boundaries* (eds., with Aurélie Choné and Philippe Hamman), Routledge, 2017; *Guide des Humanités environnementales* (with Aurélie Choné and Philippe Hamman), Ed. Septentrion, 2016; *De la ville durable à la nature en ville* (with Philippe Hamman and Jean-Pierre Lévy), Ed. Septentrion, 2015.

Jean-Pierre Lévy is Research Director at the National Centre for Scientific Research (CNRS). He is a member of the laboratory LATTS UMR 8134, Marne-la-Vallée University and Ponts ParisTech School. He has devoted a large part of his research to issues related to urban housing and population pattern. He notably published: (with Olivier Boisard and Julien Salingue), "The ASHA model: An alternative to the Markovian approach to housing vacancy chains: An application to the study of population in Lille (Nord, France)", *Urban Studies*, Vol. 54 (11), 2017, 2448–2471; (with Alexis Bourgeois and Margot Pellegrino), "Modeling and mapping domestic energy behavior: Insights from a consumer survey in France", *Energy Research & Social Science,* 32, October 2017, 180–192; (with Fateh Belaïd), "The determinants of domestic energy consumption in France: Energy modes, habitat, households and lifes cycles", *Renewable and Sustainable Energy Reviews*, n°81, 2018, 2104–2114.

Part XVII
Societies and Governance

Capability, Capabilities

Jérôme Ballet

Abstract This article discusses the capability approach and the Anthropocene from the perspective of Amartya Sen, Martha Nussbaum and subsequent researchers. The capability approach represents a significant step forward in the study of human well-being, albeit that its anthropocentrism has given it a tendency to sideline nature. There are however various possibilities for rectifying this limitation and indeed reflecting on nature's capabilities.

The capability approach was first introduced as a concept in 1979 by Amartya Sen in the *Tanner Lectures on Human Values*, which were published the following year. The concept was then elucidated further in several publications, such as *Commodities and Capabilities* Sen (1985), *The Standard of living* Sen (1987) and *Inequality Reexamined* Sen (1992). Martha Nussbaum also contributed to the development of the concept, especially in *Women and Human Development* Nussbaum (2000), but her interpretation is slightly different on a number of levels, for example, Sen prefers Capability Approach, whereas Nussbaum prefers Capabilities Approach. Although the two authors are at odds over certain aspects, they do agree on the need to encourage a way of thinking about human well-being which deviates radically from that conveyed by utilitarianism. Here, our focus will be on the capability approach in relation to the Anthropocene rather than on a comparison of human well-being and utilitarianism (see Gasper, 1997; Clark, 2008; Comim et al., 2008 for a detailed discussion).

In essence, the capability approach started out as a reevaluation of inequality: it shifted the emphasis from utility and income to justice and human development. At the same time, it has expanded the size of the information base, i.e. the facts that help us judge the worth of societies. Moreover, this information base should not be totally separated from what people deem to be important. Their self-reflection is

J. Ballet (✉)
University of Bordeaux, Bordeaux, France
e-mail: jerome.ballet@u-bordeaux.fr

© The Author(s), under exclusive license to Springer Nature Switzerland AG 2023

N. Wallenhorst, C. Wulf (eds.), *Handbook of the Anthropocene*,
https://doi.org/10.1007/978-3-031-25910-4_201

vital in making evaluative judgments, and the value they ascribe to one thing relative to another should be incorporated into this evaluative system.

Sen puts forward the idea that we should transfer from a system of evaluation driven by utility and income to a system that takes into account the real freedoms people have to do the things they value. He therefore makes a distinction between functionings and capabilities. Functionings represent a person's accomplishments and achievements during their lifetime. Knowing how to read and write, having a job that protects self-esteem, being treated with respect by others, enjoying good health and participating in a community are all accomplishments which people can value, and can hence become goals they wish to achieve in life. Even though resources, in particular income, have a part to play in some of these accomplishments, this does not mean that accomplishments can be reduced to the availability of resources. Indeed, it may be that resources facilitate our existence, but fundamental to this existence are functionings.

Capabilities represent the functionings that a person can attain during their lifetime. We must however apply a dual meaning to 'can attain': the first relates to what a person has actually attained, i.e. a particular capability has been secured, whereas the second relates to what a person could have attained under different circumstances. This change in meaning either comes down to personal choice, or more fundamentally, to the factors (circumstances) which prevent personal achievement. If we compare what a person actually achieves and what they could have achieved under different circumstances, a capability measures the real freedom a person has to lead the life they choose to lead. The capability approach demonstrates the existence of factors of conversion for each of the two types of achievement, and these multiple factors either enable or prevent the transformation of resources into real achievements. For example, factors may relate to personal barriers, such as disability, social barriers, such as discrimination, or geographical and environmental barriers, all of which limit personal choice. Normatively speaking, the capability approach supports the expansion of real freedoms, in other words, the lessening of barriers that stop people choosing to do the things they value in their lives. Differences in capabilities therefore provide a measure of inequality, and judging the worth of societies is correlated with the possibilities these societies give people to achieve what they want to achieve.

One of the criticisms of the capability approach is the way in which it defines what people value. For example, it includes those who favour a consumerist conception of life, which is not entirely compatible with the conception of a better life on Earth (Giovanola, 2005; Fischer, 2014). In addition, it understates the responsibility people have towards others, and towards nature (Ballet et al., 2007; Pelenc et al., 2013), even though Sen himself acknowledges the latter responsibility (Sen, 2010), and it is too closely bound to anthropocentrism (Watene, 2016).

We must remember that the capability approach was developed as a framework for measuring human well-being, which explains why both Nussbaum and Sen consider nature as having an instrumental value. Nussbaum lists 'Other species' among her ten central human capabilities, and this capability can be summarised as follows: "Being able to live with concern for and in relation to animals, plants, and the world of nature". As nature is one element in human well-being, it cannot be detached from this well-being. According to Sen (2013), whose thinking is similar to that of Nussbaum, nature is often seen as the 'state of nature', and assumes that a pre-existing nature remains intact until human action brings with it impurities and pollutants. Logically, it would seem that nature is best protected if humans interfere with it as little as possible, but Sen would disagree with this for two reasons. The first reason is that nature should not be understood by its substance, but by its opportunities for humanity. Otherwise, eradicating smallpox is a bad thing because this would lead to a loss of nature. The second reason is that we cannot simply consider humanity as a force for destruction: humans are capable of showing a sense of responsibility and dedication towards nature. If we indeed view nature as some sort of unblemished wilderness to be preserved at all costs, and distinct from the actions of humanity, nature becomes a dubious, rather romantic concept.

Notwithstanding that Nussbaum and Sen prioritise the relationship between nature and human well-being over substance, others have attempted to advance a deeper interpretation of nature within the capability approach.

The first attempt establishes a link between human well-being and the ecosystem services provided by nature, but even if this concept of services remains anthropocentric, it is not solely devoted to the benefits people derive from nature. Indeed, *supporting services*, which comprise plant/animal habitats and the genetic diversity of ecosystems, are functions of an ecosystem rather than direct/final ecosystem services that benefit society. They guarantee the production of other services by ensuring species reproduction (*via* their habitat) and by maintaining their diversity (*via* genetic diversity). Therefore, including *supporting services* within the capability approach makes it more compatible with strong sustainability, as human life would no longer be possible on Earth without the guarantee that these services are maintained (Pelenc & Ballet, 2015).

The second attempt stipulates that ecological conditions constitute a meta-capability for other capabilities (Holland, 2008a, b). Ecological systems play a key role in the possibilities we have to lead our lives as human beings. We can adapt to the shifting forms of economic, social and political systems, whereas a degraded ecological system may ultimately transcend certain thresholds into the unliveable. Ecological systems are therefore more vital than any other system and must be sustained. They dominate other capabilities, and this explains why ecological conditions constitute a meta-capability i.e. an 'umbrella' capability. However, we should not confuse ecological conditions with the properties of ecological systems, which allow people to exercise other capabilities, including environment-related capabilities. Ecological conditions control how other capabilities are exercised, for example, to be able to feed oneself or to enjoy good health assumes the efficient operation of ecological systems in the provision of soil, water and atmospheric temperature conditions conducive to the production of crops and the absorption of pollution and waste. Moreover, indulging in nature-related cultural and spiritual activities assumes that nature can preserve itself in a state which is consistent with carrying out these activities. Ecological conditions also enhance how other environment-related capabilities are exercised, for example, the capability of playing or expressing emotion may take place independently of ecological conditions, but a natural environment suitable for leisure and walking augments the capability of playing and expressing emotion. If we are to accept that ecological conditions constitute a meta-capability, we then have to design environmental protection policies which maintain ecological conditions compatible with life on Earth and which comprise a built-in capacity for other capabilities to reach a minimum threshold for everyone.

Nevertheless, as Schlosberg notes, if the capability approach is to find its niche in dealing with the challenges of the Anthropocene, we will have to move it beyond its inherent anthropocentrism by including capabilities for animals and for the whole of nature itself.

References

Ballet, J., Dubois, J.-L., & Mahieu, F.-R. (2007). Responsibility for each other's freedom: Agency as the source of collective capability. *Journal of Human Development, 8*(2), 185–201.

Clark, D. A. (2008). The capability approach: its development, critiques and recent advances. In R. Ghosh (Ed.), *Development studies* (Vol. 2, pp. 105–127). Atlantic Publishers and Dist.

Comim, F., Qizilbash, M., & Alkire, S. (2008). *The capability approach: Concepts, measures and applications*. Cambridge University Press.

Fischer, E. F. (2014). *The good life. Aspiration, dignity, and the anthropology of wellbeing*. Stanford University Press.

Gasper, D. (1997). Sen's capability approach and Nussbaum's capabilities ethic. *Journal of International Development, 9*(2), 281–302.

Giovanola, B. (2005). Personhood and human richness: Good and well-being in the capability approach and beyond. *Review of Social Economy, 63*(2), 249–267.

Holland, B. (2008a). Ecology and the limits of justice: Establishing capability ceilings in Nussbaum's capabilities approach. *Journal of Human Development, 9*(3), 401–425.

Holland, B. (2008b). Justice and the environment in Nussbaum's "Capabilities Approach" why sustainable ecological capacity is a meta-capability. *Political Research Quarterly, 61*(2), 319–332.

Nussbaum, M. C. (2000). *Women and human development: The capabilities approach*. Cambridge University Press.

Pelenc, J., & Ballet, J. (2015). Strong sustainability, critical natural capital and the capability approach. *Ecological Economics, 112*, 36–44.

Pelenc, J., Lompo, M. K., Ballet, J., & Dubois, J.-L. (2013). Sustainable human development and the capability approach: Integrating environment, responsibility and collective agency. *Journal of Human Development and Capabilities, 14*(1), 77–94.

Sen, A. (1985). *Commodities and capabilities*. North-Holland.

Sen, A. (1987). *The standard of living*. Cambridge University Press.

Sen, A. (1992). *Inequality reexamined.* Oxford University Press.

Sen, A. (2010). Sustainable development and our responsibilities. *Notizie di Politeia, 26*(98), 129–137.

Sen, A. (2013). The ends and means of sustainability. *Journal of Human Development and Capabilities, 14*(1), 6–20.

Watene, K. (2016). Valuing nature: Māori philosophy and the capability approach. *Oxford Development Studies, 44*(3), 287–296.

Jérôme Ballet is Associate Professor at the University of Bordeaux. He holds a PhD in economics. He has authored twenty books, such as (*Freedom, Responsibility and the Economics of Person*, New York/London, Routledge, 2013) and more that hundred peer review articles. He is President of the *Fund for Research in Ethics and Economics* and was former editor of *Ethics and Economics*.

Consumption

Phillip D. Th. Knobloch

Abstract Consumption is a key concept for understanding the Anthropocene as well as for developing a prospective anthropology of the Anthropocene. This is because the sources of the phenomena associated with the term Anthropocene – such as climate change, the destruction of biodiversity or environmental pollution – can be understood as the result of unsustainable consumption patterns. Simply put, too much is consumed, or too many goods are consumed, and the production, use and disposal of these goods lead to a variety of serious problems. If consumer behaviour has such serious consequences that it ultimately even has a decisive influence on the geophysical Earth system, it seems anthropologically appropriate and revealing to take a closer look at the human as a consumer.

With regard to the historical location of the Anthropocene (cf. Wulf, 2020, p. 16; Dürbeck, 2018, pp. 12–13) in connection with the history of consumption (cf. Trentmann, 2018; Schramm, 2019), the focus on the period after 1945 makes sense insofar as this phase saw a long-lasting economic upswing that was accompanied by the expansion of consumption and consumption options. Primarily in the West, so-called *affluent societies* emerged during this period, for which *mass consumption* is characteristic and in which people increasingly define themselves through their consumption. The *Western consumption model* is primarily based on consumer sovereignty, i.e. consumers largely decide for themselves what they consume: “In principle, anyone with the necessary money can buy” (Schramm, 2012, p. 17; all translations are by the author). The so-called *Western lifestyle* associated with this consumption model is on the one hand associated with serious environmental problems, but on the other hand increasingly enjoys great appeal worldwide. Due to economic globalisation, especially in the twenty-first century, not only are more and

P. D. T. Knobloch (✉)
TU Dortmund, Dortmund, Germany
e-mail: phillip.knobloch@tu-dortmund.de

© The Author(s), under exclusive license to Springer Nature Switzerland AG 2023

N. Wallenhorst, C. Wulf (eds.), *Handbook of the Anthropocene*,
https://doi.org/10.1007/978-3-031-25910-4_202

more goods produced, especially in so-called non-Western countries such as China; they are also consumed worldwide in large and rapidly increasing quantities. This means that the problems associated with consumption (climate change, destruction of biodiversity, environmental pollution, etc.) are continuously increasing for the Earth system.

In the course of the emergence of so-called *consumer societies,* characterised by mass consumption modelled on US society, a modern understanding of "consumption as the satisfaction of needs by economic means" (Schramm, 2012, p. 3) also developed from the middle of the twentieth century onwards. Previously, in the nineteenth century, consumption had been understood primarily as the "consumption up to destruction and depreciation" (ibid.) of goods. In the meantime, a much broader concept of consumption prevails in recent consumer research (cf. also Knobloch, 2020, 2022), which "includes not only the acquisition, but also the use of goods and services by consumers as well as social discourses on consumption (e.g. advertising, consumer criticism)" (ibid.). Emphasis is placed on the (cultural, aesthetic and symbolic) meanings that consumer goods have for consumers or that are attributed to them. Products are not only for use and should not only fulfil a certain purpose, but are sometimes culturally valorised due to specific meanings, which are then also consumed. With regard to *cultural meanings* and *product messages*, a distinction can be made between the product itself (denotation) and meanings attributed from outside (connotation) (Gekeler, 2012, pp. 23–47). Accordingly, there is a discourse about *aesthetic capitalism* (cf. among others Hutter, 2015; Böhme, 2016; Reckwitz, 2017).

The product meanings attributed from the outside include both affirmative messages, such as advertising, and negative messages, which can include the critique of consumption. The precursor of consumer criticism is considered to be the traditional *Christian critique of luxury,* which, for example, turned against the emphasis on outward appearances (Schramm, 2012, p. 5). Consumer criticism then continued in the *conservative cultural critique* of the nineteenth and twentieth centuries, "which also gave priority to inner values and saw in modern consumption a tendency towards de-individualisation and levelling" (ibid.). It was not until the twentieth century that a *left critique of consumption* emerged that was critical of capitalism and also directed in a sweeping way against consumption, since consumers were considered to be manipulated by advertising and product design, for example, thus creating false needs and a false consciousness (cf. Haug, 2009).

In the meantime, a new type of consumer criticism has become popular and dominant, which is no longer directed against consumption per se. This criticism is often ecologically motivated, directed for example against the high consumption of resources or the throwaway mentality of consumers (Schramm, 2012, p. 5). It can also be directed against the social conditions of production, such as insufficient worker protection or inadequate wages in the countries of production. Sometimes the inappropriate treatment of animals is also at the centre of criticism.

It is characteristic of this form of consumption criticism that only specific consumption products or consumption practices are criticised, while other products and forms of consumption – so-called sustainable, critical or political consumption – are valorised and promoted. Therefore, from this perspective, *sustainable consumption* is considered to be the highest form of consumer critique, because the critique is not only verbally formulated, but is accompanied by concrete (political) action.

Whether a product is sustainable – depending on the definition (cf. Gekeler, 2012, among others) – usually cannot be evaluated just in reference to the product itself. Therefore, *product communication* plays a major role if products are to be staged, advertised and marketed as sustainable. This can be done via advertising, via corresponding texts and seals on the packaging, via test reports, etc. Not to be neglected are also the potential consumers and the way they perceive, evaluate and describe the products. From this consumption theoretic perspective, *sustainability* is therefore first and foremost a *narrative* – relatively independent of how sustainable the products really are – that is associated and consumed with the respective product in one way or another (Gekeler, 2012).

In this context, it is revealing to note that also "the Anthropocene [itself] is often presented in scientific texts and media as a narrative, i.e. as a narrative with protagonists, chain of events and plot with cause-effect relationships as well as a specific spatial and temporal structure that serve to create meaning" (Dürbeck, 2018, p. 13). According to Gabriele Dürbeck, one can distinguish between five central narratives of the Anthropocene, which she has repeatedly encountered in her study of various academic treatments of the same: "the catastrophe or apocalypse narrative, the court narrative, the narrative of the 'Great Transformation', the (bio-)technological narrative, and the interdependence narrative" (ibid.).

In summary, Dürbeck notes that the five narratives she has identified partly overlap, partly contradict each other, and tend to interpret the Anthropocene either pessimistically or optimistically. "While some consider the environmental damage to be so bad and irreversible that a catastrophe can no longer be averted, others emphasise the new ecological, technological, if not eco-modernist possibilities for action to shape a better future for humans and speak of a 'good Anthropocene'" (ibid., p. 17).

It is obvious that these narratives can play a significant role in the *staging* and *aestheticisation* of consumer products and consumption practices, and at least in an abbreviated or popularised form they already do for quite a few consumers. According to the *court narrative*, for example, the 'Western' industrial nations and financial markets and their 'technocratic elites' are the main causes of climate change, from which primarily the poor in the Global South suffer (ibid., pp. 14–15). If one takes up this narrative, the individual renunciation of products from large 'Western' corporations and the purchase of fair trade products, such as coffee from a small Mexican cooperative, may be perceived as a personal reckoning, perhaps even as the implementation of an overarching, higher legal order, giving the consumer the good feeling of having made a small contribution to combating the great injustices of this world.

Buying a watch with solar cells or an electric car, but also consuming vegan chicken meat made from soy protein may, with reference to the *(bio-)technological narrative,* give the consumer the good feeling of belonging at least a little bit to the 'new heroes' of the 'technological elite' who, in the sense of 'biofuturism' and 'ecomodernism', are working to ensure prosperity for all and high energy availability (cf. ibid., p. 16).

However, if one doubts that the purchase of often rather bulky electric cars or the consumption of meat substitutes from the food industry are suitable to sustainably avert the dangers of the Anthropocene, the *narrative of the Great Transformation* may be helpful to orient und legitimise one's own consumption differently. On this basis, it may be obvious to participate in a local solidarity farming project, for example, since the transformation narrative propagates "social and political solution strategies" (ibid., p. 15) that are not only to be imposed from above, but are to be developed and implemented "through the participation of civil society and further democratisation" (ibid.). Every bite into an apple harvested in solidarity may then be accompanied by the good feeling of having done something not only for saving the planet but also for the future of democracy.

In summary, it can be stated that in the Anthropocene, due to industrialisation and then in particular due to economic development after 1945, the economic production power in the world has grown strongly, which has led to the well-known climate, biodiversity and environmental problems. As a result, the consumption possibilities and the actual consumption of goods has also greatly increased, if of course not everywhere in the world and not for all people to the same extent. The emergence of a modern *consumer culture* that goes hand in hand with this development can be understood as a phenomenon of prosperity, which can be characterised by the fact that cultural meanings are attributed to consumer products, and these meanings are becoming increasingly important for consumption and the respective consumption preferences and decisions; so important that the symbolic, affective, narrative or aesthetic value often appears even more significant to consumers than the respective use value.

Culturally charged consumption is anthropologically revealing and pedagogically significant, since culturalised consumer products always convey to consumers an image of themselves and the world. Depending on the interpretation, different images can of course emerge, but regardless of the respective interpretations, it can be stated that consumption decisions, consumption practices and of course also the renunciation of consumption (Schütte, 2022) say something about the respective person, but also about his or her view of the world.

The above-mentioned example of the culturalisation of consumption with the help of Anthropocene narratives makes it easy to see that consumption expresses self and world relations as well as images of humanity, which in turn can also be questioned, reflected upon, shaped and transformed through consumption. Through consumption, it is therefore possible to pursue the question of who I am myself, who the others are, by reflecting in reference to consumer products and consumption practices on what and how we currently consume, on what and how we want to consume in the future and, above all, what we should consume.

Consumption shapes and changes not only the planet, but also people themselves, and must therefore be understood as an important part of their identity and formation, as well as an essential aspect of human nature and culture.

When consumer products and consumption practices are linked to narratives about the Anthropocene, it opens up the possibility of inscribing oneself as a protagonist in the respective story of the Anthropocene, be it, for example, in the role of the environmentalist, the saviour of democracy, the heroic biofuturist, or even as an ecological sinner and climate egoist who courageously defies the Zeitgeist.

The Anthropocene is undoubtedly a grand narrative that can be used to contour, stylise, legitimise, exaggerate, dramatise or even theatricalise the significance of one's own way of life. Whether a lifestyle staged as sustainable and a consumer behaviour underpinned by Anthropocene narratives then actually make a contribution to sustainable development cannot, however, be determined by the quality of the staging, culturalisation and aestheticisation. Consumption should therefore be understood as a complex cultural technique, which on the one hand needs to be learned as well as possible, but which on the other hand always needs to be oriented pedagogically, as consumer-aesthetic educational research has shown (e.g. Knobloch, 2020, 2022). For it is not only about the most successful possible consumer-aesthetic performance, but also about the question of how we interpret these narratives and fictions, indeed the *language of things* (Knobloch, 2018), and how we should then relate to the interpretations – which always also transport an understanding of the self and the world.

Hermeneutic as well as *rhetorical* and *dramaturgical competences* are therefore in demand in the Anthropocene, which must be developed through high-quality general education, especially linguistic and cultural education. In addition to the natural sciences, the humanities, social sciences and cultural studies also play an important role in the Anthropocene, as they provide important foundations for a contemporary general education. This can also be justified by the assumption that, in view of the Anthropocene, it is no longer expedient to adhere to the "categorical boundary between nature and culture" (Dürbeck, 2018, p. 17; cf. Wulf, 2020).

Against this background, however, it is also urgently necessary to develop *anthropological concepts* and *educational programs* that are appropriate to the Anthropocene and that refer to (material and cultural) consumption as well as to modern consumer culture. It would be desirable that in this way the aesthetic potentials of consumer products and practices are also used to advance sustainable development and education and to reduce and defuse as much as possible the dangers and problems for nature and culture associated with consumption. Of great importance in this context are, for example, educational programs that are discussed under terms such as Global Citizenship Education, Education for Sustainable Development, Cultural Education, Post- or Decolonial Education or Consumer Aesthetic Education (cf. among others Wulf, 2021; Knobloch & Zirfas, 2016; Scherrer, 2022; Schütte, 2022; Knobloch, 2020, 2022). Irrespective of future developments in these specialist discourses, it can be stated up to this point that people in the Anthropocene should increasingly and comprehensively deal with their consumption and its effects on the Earth system, on the history of the Earth and on themselves.

References

Böhme, G. (2016). *Ästhetischer Kapitalismus*. Suhrkamp.

Dürbeck, G. (2018). Das Anthropozän erzählen: Fünf Narrative. *APuZ, 68*(21–23), 11–17.

Gekeler, M. (2012). *Konsumgut Nachhaltigkeit. Zur Inszenierung neuer Leitmotive in der Produktkommunikation*. Transcript.

Haug, W. F. (2009). *Kritik der Warenästhetik. Gefolgt von Warenästhetik im High-Tech-Kapitalismus*. Suhrkamp.

Hutter, M. (2015). *Ernste Spiele. Geschichten vom Aufstieg des ästhetischen Kapitalismus*. Fink.

Knobloch, Ph. D. Th. (2018). Die Sprache(n) der Dinge verstehen. Eine kulturwissenschaftliche Revision der Hermeneutik Schleiermachers. *Vierteljahrsschrift für wissenschaftliche Pädagogik, 94*(3), 419–435. https://doi.org/10.30965/25890581-09403006

Knobloch, Ph. D. Th. (2020). Konsumästhetische Bildung zwischen Kunst und Produktkultur. In: Bala, Ch. / Hohensträter, D. / Krankenhagen, S. / Schuldzinski, W. (eds.): *Konsumästhetik zwischen Kunst, Kritik und Kennerschaft*. Beiträge zur Verbraucherforschung. Band 11. Verbraucherzentrale NRW, pp. 149–167. https://doi.org/10.15501/978-3-86336-926-2_9

Knobloch, Ph. D. Th. (2022). Konsumästhetische Bildung und Bildungsforschung. In: Heidkamp-Kergel, B. / Kergel, D. (eds.): *Handbuch interdisziplinäre Bildungsforschung*. Beltz/Juventa, pp. 255–271.

Knobloch, Ph. D. Th. / Zirfas, Jörg (2016). Die Kultur des Konsums und die ästhetische Bildung. Grundlegende Perspektiven am Beispiel nachhaltiger Lebensmittel. In: Fuchs, M. / Braun, T. (eds.): *Die Kulturschule und kulturelle Schulentwicklung. Grundlagen, Analysen, Kritik. Band 2: Zur ästhetischen Dimension von Schule*. Beltz-Juventa, pp. 170–183.

Reckwitz, A. (2017). *Die Gesellschaft der Singularitäten*. Suhrkamp.

Scherrer, M. (2022). Das Anthropozän dekolonialisieren. Perspektiven für die Erziehungswissenschaft. In: Knobloch, Ph. D. Th. / Drerup, J. (eds.): *Bildung in postkolonialen Konstellationen. Erziehungswissenschaftliche Analysen und pädagogische Perspektiven*. Transcript.

Schramm, M. (2012). Konsumgeschichte. *Docupedia-Zeitgeschichte*. https://doi.org/10.14765/zzf.dok.2.262.v2

Schramm, M. (2019). Konsumgeschichte: Aktuelle Trends und Perspektiven. In: Hohensträter, D. / Krankenhagen, S. (eds.): *Konsumkultur. Eine Standortbestimmung*. Kadmos, pp. 59–75.

Schütte, A. (2022). Reisen an den Grenzen des Kapitalismus. Autobiografische Berichte über das Aussteigen und Verweigern. In: Knobloch, Ph. D. Th. / Drerup, J. / Dipcin, D. (eds.): *On the beaten track. Zur Theorie der Bildungsreise im Zeitalter des Massentourismus*. Metzler.

Trentmann, F. (2018). *Herrschaft der Dinge. Die Geschichte des Konsums vom 15. Jahrhundert bis heute*. Deutsche Verlags-Anstalt.

Wulf, C. (2020). Den Menschen neu denken im Anthropozän. Bestandaufnahme und Perspektiven. *Paragrana, 29*(1), 13–35.

Wulf, C. (2021). Global citizenship education. Bildung zu einer planetarischen Weltgemeinschaft im Anthropozän. *Vierteljahrsschrift für wissenschaftliche Pädagogik, 97*, 464–480.

Phillip D. Th. Knobloch is a senior researcher in the Institute of General and Vocational Education at the Technical University of Dortmund. His field of work includes general, historical, comparative and intercultural educational science and pedagogy. His research focuses on decolonial and consumer aesthetic theories of education; a current focus is on global citizenship education.

Courage

François Prouteau

Abstract This article highlights the debate concerning courage in politics, originating in Athens as the theme of *parrhēsia*, i.e., truth in democracy, and how the discourse is both enabled and constantly threatened. How does this debate affect the "drama of truth", in the Anthropocene? From ancient Greece to the present day, philosophical discourse has been seated in necessary interaction between the access to the truth sought after by sciences, political powers and structures, and the formation of *êthopoiêsis* whereby individuals constitute themselves as the moral subject of their conduct. Modern authors and contemporary philosophers such as Tillich and Foucault especially, invite us to do so in their work on the concept of courage. Through Foucault's political concepts, we endeavour to construct two dimensions for the courage of truth, ethical (an ethical *parrhēsia*), and by extension, ecological (an ecological *parrhēsia*) linked with scientific research and growing awareness of the Anthropocene.

Faced with the challenges of the Anthropocene, do we want to act as courageous people and societies to keep global warming and the collapse of biodiversity below reasonable thresholds?

The objective of the Paris Agreement in 2015 was to hold global average temperature increase to well below 2 °C and pursue efforts to limit the temperature increase to 1.5 °C above pre-industrial levels. Future climate is partly determined by the magnitude of greenhouse gases (GHG) emissions from human activities. The last scientific reports assert that they are already responsible for approximately 1.1 °C of warming between 1850 and 1900, and find that averaged over the next 20 years, global temperature is expected to reach or exceed 1.5 °C of warming. "Business as usual" policies would lead to a global mean temperature rise of at least 3.5 °C by 2100 (IPCC, 2021). Even global warming limited to 1.5 °C above pre-industrial levels still leads to more heat waves, more intense storms and severe

F. Prouteau (✉)
LIRFE – Catholic University of the West, Angers, France
e-mail: f.prouteau@iffeurope.org

© The Author(s), under exclusive license to Springer Nature Switzerland AG 2023

N. Wallenhorst, C. Wulf (eds.), *Handbook of the Anthropocene*,
https://doi.org/10.1007/978-3-031-25910-4_203

droughts and floods. Some of these long-term irreversible impacts are already underway.

On the threshold of the third millennium, just before the concept of the Anthropocene first appeared in a scientific publication (Crutzen & Stoermer, 2000), informed observers and scientific experts were warning policy makers and governmental leaders all over the world (Kendall Henry, 1992). Some of them began to express their concern at international summits, such as Al Gore (Climate Change Conference, Kyoto, 8 Dec. 1997) or Jacques Chirac (The World Summit on Sustainable Development, Johannesburg, 2 Sept. 2002).

The Anthropocene as a "boundary concept" "frames critical insights into understanding the drivers, dynamics and specific challenges in responding to the ambition of keeping global temperature increase well below 2 °C" (IPCC, 2019). The hope of achieving this goal is diminishing because carbon dioxide concentration (ppm) grows year on year; current CO_2 concentration has not been experienced for at least 2 million years (IPCC, 2021). It is to be hoped that every country will implement as soon as possible a nationally determined contribution (NDC) submitted to the UN for reducing substantially GHG emissions with a net zero target by mid-century, as it was expected after the Paris Agreement in 2015. Unfortunately, these objectives are not being met, due directly to a shortcoming of political courage at different levels (international, national and individual) of responsibility. Extinction of courage (Fleury, 2010)? One may ask if the bigger danger is cowardliness or lies cultivated by skillful speakers (Plato, 1953, *The Defence of Socrates*), "when companies and politicians are making it look like real action is happening, when in fact almost nothing is being done, apart from clever accounting and creative PR [public relations]" (Thunberg, 2019).

Questions of power, control and command are at the heart of the Anthropocene, which is being shaken up by 'the war of logics' (Wallenhorst, 2021): the logic of profit of *Homo œconomicus*, the individual guided by his own interest; the logic of responsibility of *Homo collectivus*, individuals in solidarity with all humans and more broadly with the biosphere; the logic of hospitality of *Homo religatus*, the human defined by the relationship that posits coexistence, (shared existence) above all else. It is in the perspective of this logic of *Homo religatus* that increasingly responsive and reciprocal relations between the subject and the world unfold, in other words, the force of conviviality. Against the hegemony of the logic of profit, accumulation and totalizing consumption, convivialism appeals to the legitimate policy: the power to be and to act of the human person without harming those of others, in the perspective of equal freedom (Convivialist International, 2020). Achieving this goal requires courage and participation (Internationale convivialiste, 2013): "The courage to be as a part is the courage to affirm one's own being by participation" (Tillich, 1952). Is the courage to lead the fight lacking? Human activity has triggered climate change and environmental disasters that threaten us with global tragedy. In this situation, could the knowledge of "what to fear and what to dare" (Plato, 1953, *Laches*) help to be courageous?

For twenty-five centuries, the concept of courage has been the subject of work based on the reality of Greek thought, and therefore in Western history and philosophy till now (Tillich, 1952; Foucault, 2011). The virtue of courage qualifies the Greek hero who knows how to 'listen, choose himself in the midst of the world and accept the test required of every human being: that of never betraying himself' (Marcolongo, 2019). In *The Iliad*, Book Nine, Homer mentions Achilles's courage who can face the full light of day and appear without reticence to the sight of all, someone straightforward, whereas the ingenious Odysseus is extremely versatile (*polytropos*). Homer opens the *Odyssey* with the Greek word *Andra* to qualify Odysseus, and tells of the epics of "brave warriors" (*agrion andra*). Strong, courageous Odysseus manages to extricate himself from the toughest traps, the deepest abysses. He lives an 'adventure' in the original sense of the word, which contains the idea of the future, with its share of hazards and dangers. To overcome it, he must mobilize his will and show his bravery. What a man! In doing so, he learns something about the world and people (Prouteau, 2021), even if it means suffering and taking risks in relationships with others, at the cost of extreme tensions and even breakdowns that can lead to death.

In a more modern era, *The Adventures of Telemachus*, a reference work for pedagogy but also one of the most widely read and translated works of literature for two centuries, also praises courage, not the fierce and fiery courage of Achilles, but that of the wise and moderate Odysseus. Telemachus, like his father Odysseus, confronts several obstacles along the way, but matures as a man, a warrior and a leader. Worry and discouragement beset Telemachus at the beginning of his adventures, but his masters, from his father to his great-grandfather, and even Mentor (the goddess of wisdom Athena) teach him what courage means:

> And we have now to lean on I will name,
> 'Tis courage. And this lesson, prithee, learn,
> Dangers to come deem great, when they are present spurn.
> [...] Thy courage muster and thyself subdue,
> Control thy passions, to thyself be true (Fénelon, 1850)

In recent decades, research publications have helped to look at the Anthropocene from a different scientific aspect, but we can see the interplay of adhesions and oppositions, resistances or resignations, whether in power relations or in individual or collective principles and norms of conduct. In his lectures at the Collège de France (1983–1984), Michel Foucault (2011) conducted a study on the courage of truth based on the *Laches*, one of Plato's dialogues intended for the education of young people, and otherwise entirely devoted to courage. Foucault highlights how until the end of the dialogue, despite all the efforts made to define the nature of courage, no one, not even Socrates (Plato, 1953, *Laches*), can answer the question: what is courage in its truth, what is the truth of courage? But Socrates makes use of this general failing to show that what is at stake about the nature of courage is neither of *epistemē* nor of *technē* to pass on knowledge from master to disciple; "the schoolteacher to whom one should turn [...] to whom everyone should listen since no one has arrived at the definition of courage, is of course *logos* itself, the discourse which will give access to the truth" (Foucault, 2011).

Having yielded to *logos*, Socrates can guide others along its path, in what Foucault calls the circle of truth and courage. The hallmark of all Socratic maieutic is to lead as by the hand his interlocutor, even in old age, to the point where a person must question the way in which he lives. Throughout one's life, he can give an account of himself, *didonai peri hautou logon (*Plato, 1953, *Laches*): "How do things stand with you and logos, can you justify yourself, can you give the logos of yourself? […] It is not the soul but the way in which one lives (*hontina tropon nun te zē*: how you live now and also how you have lived in the past). It is this domain of existence, of the mode of existence, of the *tropos* of life, on which Socrates' discourse and *parrhēsia* will focus. So, it is neither the chain of rationality, as in technical teaching, nor the soul's ontological mode of being, but the style of life" (Foucault, 2011). The different interlocutors agree to be examined by Socrates because of the harmony between his words, the way of his speaking and the sound of his life. The Socratic courageous truth-telling (*parrhēsia*) aims to help them to take care (*epimeleia*) of themselves correctly and to transform the mode of being in the realm of *bios* (existence). It is truly an ethical *parrhēsia* that can lead to "an aesthetics of existence pursuing the task of giving a visible, harmonious, beautiful form to life (to the *bios*)" (ibid.).

The reciprocity between the concept of courage and a way of life, a Socrates' legacy, appears in Stoicism (Tillich, 1952), in ancient Cynicism, in Christianity, through witnesses of the truth, *marturōn tēs alētheias* (St. Gregory of Nazianzus, 2003), and also in certain revolutionary movements or in artwork and contemporary literature (Foucault, 2011). Moreover, since the Second World War, several contemporary intellectuals and writers (Arendt, Aron, Camus…) have emphasized that radical heroism is about bringing undertone (Birnbaum, 2021) to the darkest of times and "that such illumination might well come less from theories and concepts than from the uncertain, flickering, and often weak light that some men and women, in their lives and their works, will kindle under almost all circumstances and shed over the time span that was given them on earth" (Arendt, 1970).

Growing awareness of the Anthropocene, leads us to ask how we can take advantage of this momentum to reactivate the circle of courage and truth whose otherness is the hallmark (F. Gros in Foucault, 2011). Indeed, "there is no establishment of the truth without an essential position of otherness; the truth is never the same; there can be truth only in the form of the other world and the other life (*l'autre monde et la vie autre*)" (Foucault, 2011). According to Foucault, otherness in Socrates' style can play the role of a touchstone (*basanos* in Greek): by confronting it, we are encouraged to take care of our critical reason or practical wisdom (*phronēsis*), of truth (*alētheia)* and of our soul (*psukhē*), not only for the transformation of our way of being and living, but also for the construction of another world.

But the Anthropocene reminds us that a life devoted to public-political affairs (*bios politikos* according to Aristotle) must take into account what a world is: something that can be inhabited, that can be hospitable but also strange and hostile (Ricoeur, 1995, Prouteau, 2019). In this sense, the courage of truth is not only to be considered in an ethical dimension but also in an ecological dimension, that can be called an ecological *parrhēsia*: with science as the theory of the real (Heidegger,

1977), we can learn not only what nature is but also what is a truly human being immersed in the cosmos and belonging to the Earth.

References

Arendt, H. (1970). *Men in dark times*. Mariner Books.

Birnbaum, J. (2021). *Le courage de la nuance*. Seuil.

Convivialist International. (2020). *The second convivialist manifesto: Towards a post-neoliberal world*. https://doi.org/10.1525/001c.12721

Crutzen, P. J., & Stoermer, E. F. (2000). *The "Anthropocene"*. http://www.igbp.net/download/18.316f18321323470177580001401/1376383088452/NL41.pdf. Accessed 21 Aug 2021.

Fénelon, F. (1699/1850). *Telemachus* (W. E. Hume, Trans.). Walter Spiers.

Fleury, C. (2010). *La fin du courage*. Fayard.

Foucault, M. (2009/2011). *The courage of the truth (The government of Self and Others II). Lectures at the College de France 1983–1984*. Edition established under the direction of François Ewald and Alessandro Fontana, by Frederic Gros (G. Burchell, Trans.). Palgrave Macmillan.

Heidegger, M. (1954/1977). *Question concerning technology, and other essays* (W. Levitt, Trans.). Harper Torchbooks.

Internationale Convivialiste. (2013). *Manifeste convivialiste. Déclaration d'interdépendance*. Le Bord de l'Eau.

IPCC. (2019). *Report 2018: Global warming of 1.5 °C. an IPCC special report on the impacts of global warming of 1.5 °C above pre-industrial levels and related global greenhouse gas emission pathways, in the context of strengthening the global response to the threat of climate change, sustainable development, and efforts to eradicate poverty*. https://www.ipcc.ch/sr15/download/#full. Accessed 21 Aug 2021.

IPCC. (2021). *Climate change 2021 – Sixth assessment report: The physical science basis*. https://www.ipcc.ch/report/ar6/wg1/. Accessed 21 Aug 2021.

Kendall Henry, W. (1992). *World scientists warning to humanity*. http://www.ucsusa.org/about/1992-world-scientists.html. Accessed 21 Aug 2021.

Marcolongo, A. (2019). *La part du héros*. Les Belles Lettres.

Plato. [427 347 BC]. (1953). *The dialogues of Plato* (B. Jowett, Trans.). Clarendon Press-Oxford.

Prouteau, F. (2019). Éduquer à la responsabilité en Anthropocène à la lumière de Paul Ricoeur. In N. Wallenhorst and J.-P. Pierron (dir.). *Éduquer en Anthropocène*. Le Bord de l'eau.

Prouteau, F. (2021). *Odyssée pour une Terre habitable*. Le Pommier.

Ricoeur, P. (1995/1998). *Critique and conviction*. Columbia University Press.

St. Gregory of Nazianzus. (2003). *Select orations* (M. P. Vinson, Trans.). Catholic University of America Press.

Thunberg, G. (2019). *Speech at the National Assembly in Paris*. https://www.fridaysforfuture.org/greta-speeches#greta_speech_july23_2019. Accessed 21 Aug 2021.

Tillich, P. (1952). *The courage to be*. Yale University Press.

Wallenhorst, N. (2021). *Mutation. L'aventure humaine ne fait que commencer*. Le Pommier.

François Prouteau is a graduate engineer from the French Grande école Institute Mines-Telecom Atlantique with a PhD in Educational Sciences from the Lumière University Lyon 2. He is teacher and associated researcher at The Catholic University of the West (LIRFE Laboratory). His current research focuses on education facing contemporary anthropological and cultural challenges. He is cofounder of the Institutes of Formation Fondacio, of which he was the director for Europe from 1989 to 2009, and advisor for Africa, Asia and South America. He is currently the President of Fondacio.

Democracy

Benjamin Lewis Robinson

Abstract The Anthropocene presents a challenge to democracy as both a contemporary political form of government and as a normative ideal for politics in general. This article considers the tensions between democratic practices and ecological exigencies as these became virulent around the turn of the twenty-first century. Democracy in the Anthropocene is not only about grappling with long overlooked ecological concerns, but also recognizing how democratic institutions and imaginaries are themselves political-ecological products implicated in earth system processes that operate on radically different orders of scale. If democracy in its ideal form has been oriented to expanding inclusivity and participation in politics, the Anthropocene radicalizes this democratic impulse extending to agency and collectivity beyond the human.

In an interview with *The Guardian* in 2010, the famous climate scientist and author of the Gaia hypothesis, James Lovelock, suggested that 'it may be necessary to put democracy on hold for a while' to tackle climate change (Lovelock, 2010). He refers to the way that in times of war democracies, for their own security, suspend certain democratic freedoms and procedures by declaring a state of emergency. Politics in the Anthropocene, so the climate scientist, may require similar emergency measures.

The perception that democracies are ill-equipped to respond adequately and urgently to the Anthropocene has re-awakened a long tradition of anti-democratic environmentalism and provoked fears of a resurgence of 'eco-authoritarianism'. The fact that the future of anthropocentric disruption to the earth system is strongly linked to developments in the so-called BRICS countries (Brazil, Russia, India, China, South Africa), ranging from some of the world's largest but perhaps most precarious democracies to distinctly nondemocratic regimes, has stressed the need for diplomatic and technocratic rather than democratic solutions. Especially since

B. L. Robinson (✉)
NYU, New York, NY, USA
e-mail: blr4141@nyu.edu

© The Author(s), under exclusive license to Springer Nature Switzerland AG 2023

N. Wallenhorst, C. Wulf (eds.), *Handbook of the Anthropocene*,
https://doi.org/10.1007/978-3-031-25910-4_204

the turn of the millennium, there have been reservations about the neoimperialism of liberal capitalist democracy and the unsustainable consumerism it has fostered. Recent resurgence in nativist populism often vocalizing anti-environmental sentiment has further unsettled confidence in democracy.

The shortcomings of democracy regarding ecological issues have long been acknowledged. At its most abstract, the problem is that democratic politics operate on different scales from the ecologies in which they are involved. Modern liberal democracies work on short election cycles representing delimited political constituencies and furthering the short-term 'interests' of citizens, a model that structurally occludes the far-reaching effects of lifestyle and political choices and the carbon-intensive infrastructures that undergird these. They are thus ill-disposed to apprehend let alone address the 'slow violence' of ecological devastation which takes place across political borders and generations and often disproportionately affects disenfranchised demographics (Nixon, 2011).

The globalization that accompanied the years of triumphalism of liberal capitalist democracy after 1989 provoked cosmopolitan theories of democracy in tandem with theories of ecological democracy and citizenship (Dobson, 2003; Eckersley, 2020), and ecofeminist conceptions of radical democratic practice (Sandilands, 1999). The unsustainable myopic parochialism of liberal democracy was rethought with a global focus on expanding democratic participation and representation across borders and generations. And these theories arose in the context of proliferating people's initiatives militating for global social and environmental justice. So for advocates and opponents of globalization alike, the idea of democracy was the order of the day.

This would change with 9/11 – and, less immediately but ultimately more lastingly, with the 'shock of the Anthropocene' (Bonneuil & Fressoz, 2016), which entered public discourse after 2000. The temporary suspension of democracy for our own salvation advocated by Lovelock has to be understood in the context of the War on Terror. Lovelock's comments come at the end of a decade in which ostensibly liberal democratic states had declared 'states of emergency' to restrict democratic rights and procedures within their borders, while seeking unilaterally to impose Western-style democratization in distant parts of the world. Democracies, it turned out, exhibited an 'autoimmune' tendency to do constitutional harm to themselves while protecting themselves (Derrida, 2005) – a cautionary consideration in the face of the 'climate emergency'.

The Anthropocene exposed further antagonisms within existing democracies, exacerbated by the Covid-pandemic in 2020: between science and popular opinion, expert knowledge and democratic debate, technocratic policies and deliberative politics. Increasingly a global class of experts is advocating different forms of 'earth system governance' (Biermann, 2014), soliciting what some fear will mean a displacement of democratic politics in favour of a 'geocracy' (Bonneuil & Fressoz, 2016). Some have sought in response to revitalize popular democratic resistance to technocratic 'green governmentality' (Luke, 1999); more institutional approaches have focused on negotiating the tension between experts and lay publics (Dryzek & Pickering, 2019), while sociologists of science have explored what it means to democratize science and technology by incorporating diverse knowledges and practices (Jasanoff & Martello, 2004).

In the 1990s one could make a distinction between 'environmental democracy,' which worked within existing frameworks of liberal democracy towards more environmentally conscious policies, and a more radical 'ecological democracy,' which interrogated the framework of democracy itself (Eckersley, 2020). As a rule, however, both approaches sought to bring environmental concerns *into* the political realm. Since 2000 the problem has rather to be understood as situating democratic institutions and practices within the context of the destabilized earth system. Democracy *in* the Anthropocene has fundamental repercussions for democratic theory. Much of the debate around ecology and democracy in the 1990s departed from the perplexity that democracy, understood as essentially open-ended and procedural, seemed to be at loggerheads with an environmentalism that advocates substantial outcomes (Goodin, 1992). But what happens when the very existence of democracy depends on 'ecological preconditions' that need to be protected (Hammond et al., 2020)?

Rather than bringing ecology into democracy, democracy has now to be considered as part of ecological history. In *Carbon Democracy*, Timothy Mitchell addresses the material and infrastructural conditions of democracy in the form developed in Europe and the United States by attending to the history of the extraction, circulation and consumption of coal and oil arguing: "Fossil fuels helped create both the possibility of twentieth-century democracy and its limits" (2011, p. 1). The concentrated and nodal nature of coal extraction and distribution provided for the development of modern forms of political assembly in the nineteenth century (notably, the workers' strike) which fostered liberal democracy and the welfare state. In the twentieth century, in contrast, oil, the extraction of which is less labour intensive and more fluid in its distribution, has led to a curtailing and undermining of democratic politics as well as contributing to some of liberal democracy's most environmentally egregious ideologies such as limitless growth. Indeed, some of the most cherished values of liberal capitalist democracies turn out to be 'fossil freedoms' (Bergthaller, 2017; Chakrabarty, 2021; Charbonnier, 2021) constitutionally implicated in the destruction of their own ecological conditions. The question of who gets to enjoy fossil freedoms and who, historically and in the future, pays the price for them is at the centre of contemporary debates about 'climate justice'.

The Anthropocene demands an interrogation of the 'demos' in democracy. While globalization in the 1990s provoked a rethinking of citizenship calling for an expansion of the plurality and inclusivity of participation among different groups of humans across the 'globe', in the Anthropocene concern extends to the earthbound in general, including all the non-human entities that constitute the 'planet' (Chakrabarty, 2021). This has solicited a renewed focus on the politics of democratic sensibility (Purdy, 2015). There has been a shift in emphasis from the level of policy to everyday practices (Meyer, 2015), to cultural production responding to the challenge of generating a 'sense of planet' (Heise, 2008), and to an appreciation for the 'fragility of things' (Connolly, 2013) through which new intersectional and more than human solidarities can be fostered.

Within a broadly deliberative framework, efforts have been made to conceptualize broader forms of representation of and for the 'affected' (Eckersley, 2020) and to acknowledge an expanded 'formative agency' in political life by rethinking communication beyond the human voice in a manner attentive to 'signals' from non-human entities (Dryzek & Pickering, 2019). To the extent that the keyword in these debates remains 'reflexivity', these theories largely keep within the humanist principles of traditional democratic theory.

The impulse to envision a posthumanist democratic practice has come from indigenous cosmologies and non-Western decolonial democratic forms insofar as they evince a more holistic political ecology (Satgar, 2018; Kothari et al., 2014; Shiva, 2005). These resonate with recent work in science studies, which likewise aims to overcome the 'modern' distinction between nature and politics. Arguing for a more nuanced understanding of the distribution of human and non-human agency in political life, Bruno Latour has made the case for a 'parliament of things' (Latour, 2004; Stengers, 2011). 'Object-oriented democracy' invites a reconceptualization of political representation in the light of methods calibrated to nonhuman agency in the natural sciences, although it arguably remains beholden to the 'traditional' parliamentary model of politics it invokes (Disch, 2016). 'New materialist' approaches in political theory have sought to tie the findings of science studies to more radical democratic theories. Jane Bennet draws on Jacques Rancière to reflect on the ways that what she calls 'vibrant matter' disrupts the prevailing order to produce a democratic scene in which the contours of what counts as political are cast into contention (Bennet, 2009). Expanding the 'demos' for planetary politics comes with a renewed sense of embeddedness, vulnerability, and interdependence and a very different understanding than the liberal one of political participation and belonging.

If these theories sound abstract, it is important to realize that they are responding to concrete transformations of the political sphere that are increasingly difficult to overlook. It is not that new agents are being introduced into politics, but that their constitutional role is at last being recognized. Movements of diverse sorts, from intersectional activism of affected peoples and protests by children to such 'natural' demonstrations as extreme weather phenomena and emergent pandemic diseases like Covid-19, are all contributing to the changing political climate. In the form felicitously referred to as 'grassroots', democracy in the Anthropocene is already underway.

References

Bennet, J. (2009). *Vibrant matter: A political ecology of things*. Duke University Press.

Bergthaller, J. (2017). Fossil freedoms: The politics of emancipation and the end of oil. In U. Heise, J. Christensen, & M. Niemann (Eds.), *The Routledge companion to the environmental humanities* (pp. 424–432). Routledge.

Biermann, F. (2014). *Earth system governance: World politics in the Anthropocene*. MIT Press.

Bonneuil, C., & Fressoz, J.-B. (2016, original 2013). *The shock of the Anthropocene: The Earth, history and us* (D. Fernbach, Trans.). Verso.

Chakrabarty, D. (2021). *The climate of history in a planetary age*. University of Chicago Press.
Charbonnier, P. (2021, original 2019). *Affluence and freedom: An environmental history of political ideas* (A. Brown, Trans.). Polity.
Connolly, W. (2013). *The fragility of things: Self-organizing processes, neoliberal fantasies, and democratic activism*. Duke University Press.
Derrida, J. (2005, original 2003). *Rogues: Two essays on reason* (P.-A. Brault & M. Naas, Trans.). Stanford University Press.
Disch, L. (2016). Ecological democracy and the co-participation of things. In T. Gabrielson, C. Hall, J. Meyer, & D. Schlosberg (Eds.), *The Oxford handbook of environmental political theory* (pp. 624–640). Oxford University Press.
Dobson, A. (2003). *Citizenship and the environment*. Oxford University Press.
Dryzek, J. S., & Pickering, J. (2019). *The politics of the Anthropocene*. Oxford University Press.
Eckersley, R. (2020). Ecological democracy and the rise and decline of liberal democracy: Looking back, looking forward. *Environmental Politics, 29*(2), 214–234.
Goodin, R. E. (1992). *Green political theory*. Polity.
Hammond, M., Dryzek, J. S., & Pickering, J. (2020). Democracy in the anthropocene. *Contemporary Political Theory, 19*, 127–141.
Heise, U. (2008). *Sense of place and sense of planet: The environmental imagination of the global*. Oxford University Press.
Jasanoff, S., & Martello, M. L. (2004). *Earthly politics: Local and global in environmental governance*. MIT Press.
Kothari, A., Demaria, F., & Acosta, A. (2014). Buen Vivir, degrowth and ecological Swaraj: Alternatives to sustainable development and the green economy. *Development, 57*(3–4), 362–375.
Latour, B. (2004). *Politics of nature: How to bring the sciences into democracy*. Harvard University Press.
Lovelock, J. (2010, March 29). James Lovelock on the value of sceptics and why Copenhagen was doomed, interview with Leo Hickman. *The Guardian*.
Luke, T. (1999). *Capitalism, democracy, and ecology: Departing from Marx*. University of Illinois Press.
Meyer, J. M. (2015). *Engaging the everyday: Environmental social criticism and the resonance dilemma*. MIT Press.
Mitchell, T. (2011). *Carbon democracy: Political power in the age of oil*. Verso.
Nixon, R. (2011). *Slow violence and the environmentalism of the poor*. Harvard University Press.
Purdy, J. (2015). *After nature: A politics for the Anthropocene*. Harvard University Press.
Sandilands, C. (1999). *The good-natured feminist: Ecofeminism and the quest for democracy*. University of Minnesota Press.
Satgar, V. (Ed.). (2018). *The climate crisis: South African and global democratic eco-socialist alternatives*. Wits University Press.
Shiva, V. (2005). *Earth democracy: Justice, sustainability, and peace*. South End Press.
Stengers, I. (2011, original 2003). *Cosmopolitics* II (R. Bononno, Trans.). University of Minnesota Press.

Benjamin Lewis Robinson is Assistant Professor of German at NYU. He is the author of *Bureaucratic Fanatics: Modern Literature and the Passions of Rationalization* (De Gruyter, 2019) and is currently engaged in a project at the intersections of biopolitics and ecopolitics titled *States of Need / States of Emergency*. He co-edited a recent special issue of *The Germanic Review* on "*Schuld* (guilt/debt) in the Anthropocene" and the volume *The Work of World Literature* (2021).

Earth System Governance

Frank Biermann

Abstract Earth system governance is a new paradigm in the social sciences to better understand the functioning and the transformative potential of institutional arrangements dealing with socio-ecological systems. Earth system governance breaks with traditional notions of environmental policy by replacing their dichotomies of human-nature and human-environment with the decisive perspective of socio-ecological systems and a more explicit focus on planetary interdependence and justice. Over the last two decades, the earth system governance paradigm has given rise to the creation of a global network of scholars working in this field, the "Earth System Governance Project."

Earth system governance is a new paradigm in the social sciences to better understand the functioning and transformative potential of institutional arrangements dealing with socio-ecological systems. Earth system governance breaks with traditional notions of environmental policy by replacing their dichotomies of human-nature and human-environment with a socio-ecological systems perspective at planetary scale and a more explicit focus on planetary interdependence and justice. This shift from traditional environmental policy to a novel earth-system perspective is the consequence of a new understanding of global complexity and the rapidly growing planetary impact of the human species. In the Anthropocene, the notion of 'environmental' politics no longer offers a theoretically satisfying understanding of the human predicament; the concept of earth system governance may help here to integrate the expanding research on socio-ecological systems at planetary scale with the perspective of governance, institutions, and politics (Biermann, 2007, 2014).

Earth system governance brings a new perspective that is system-focused as opposed to binary human-environment; integrated across levels instead of being merely inter-governmental or local; and progressive as a research approach by moving from positivist institutional analysis to critical theory and transformative global

F. Biermann (✉)
Utrecht University, Utrecht, The Netherlands
e-mail: f.biermann@uu.nl

© The Author(s), under exclusive license to Springer Nature Switzerland AG 2023

N. Wallenhorst, C. Wulf (eds.), *Handbook of the Anthropocene*,
https://doi.org/10.1007/978-3-031-25910-4_205

change. Earth system governance draws on different disciplines, such as political science, sociology and law, and it is conceptually scalable as it allows the study of local- or regional-scale systems within an earth system perspective.

The key concern of earth system governance is not "governing the earth" or the management of the entire process of planetary evolution. Earth system governance is different from technocratic visions of earth system "management" or "geo-engineering." Earth system governance is about the human impact on planetary systems. It is about the societal steering of human activities regarding the long-term stability of geobiophysical systems and the flourishing of all species. It is about global stewardship for the planet based on non-hierarchical processes of cooperation and coordination.

In the practice of earth system governance research, scholars theorize about it in three ways: earth system governance as analytical practice, earth system governance as normative critique, and earth system governance as transformative visioning.

The *analytical theory* of earth system governance seeks to explain current politics and policies by studying institutions and regimes, their interlinkages and the diagnostics of specific institutional designs. The analytical theory of earth system governance also addresses key political problems such as the role of power, ideas, norms, different claims to legitimacy and the distributive outcomes of governance and their normative evaluation in terms of justice.

The *normative theory* of earth system governance is the critique of the current systems of governance. It does not ask what is but what should be, juxtaposing the findings and insights from analytical theory – for instance on the effectiveness of international or national institutions – with the necessities of earth system stability and the needs of social justice. The normative theory is essentially critical theory, focusing on the reform and reorganization of human activity in a way that guarantees to effectively and fairly "navigate the Anthropocene".

This combination of analytical theory and normative critique turns earth system governance into a *transformative approach* in the social sciences. Business-as-usual will not prevent critical transitions in the earth system, and technological revolutions and efficiency gains alone will not suffice. Instead, earth system governance research directly engages with key concerns of societal change. In a highly divided world, earth system governance research poses fundamental questions of justice within and among nations. It raises important queries about the legitimacy and accountability of public action and about effective and fair mechanisms of democratic earth system governance. Earth system governance is as much about environmental parameters as about social practices and capitalist economic systems. Its normative goal is not purely environmental protection on a planetary scale – this would make earth system governance devoid of its societal context. Planetary targets – such as control of greenhouse gases – could theoretically be reached through hugely different political means with different costs for actors in different geographies; earth system governance research focusses on social welfare as well as environmental protection; on institutional effectiveness as well as global and local justice.

As a research field, the study of earth system governance is *interdisciplinary, global and complexity-oriented.* It needs to transcend traditional concepts and approaches of environmental policy because the anthropogenic perturbation of the earth system raises research problems that are qualitatively different from those that have traditionally been examined as environmental problems. Key questions of earth system governance are, to name a few, the institutional architectures, decision-making procedures and distributive policy impacts in areas as diverse as global adaptation to rising sea levels, the halting of global soil deterioration, the protection of climate migrants or the global implications of carbon dioxide removal technologies – all issues that have been barely covered by traditional environmental research.

Earth system governance research thus *transcends levels of analysis*, bridging a planetary problem perspective with the study of national policies and local governance. Earth system governance is more than a problem of global regulation through international agreements. It is foremost about people who take decisions in their daily lives and their political struggles. Earth system transformations affect individuals as much as they are driven by individual decisions within their social, political, and economic structures. As such, earth system governance theory is informed by planetary interdependencies even when studying local institutions, policies, and contestations.

Because of this complexity, earth system governance research draws on insights *from the full range of the social sciences*, from anthropology to international law. Earth system governance research covers local policies on problems ranging from the preservation of local waters to desertification and soil degradation, but also includes the study of international regimes seeking to regulate governments and corporations. Earth system governance research reaches beyond the social sciences as well. Its problem definition makes it part of the overarching context of integrated sustainability science, where social scientists collaborate with natural scientists to advance the integrated understanding of the coupled socio-ecological system that our planet has become (Clark & Harley, 2020).

Over the last decade, the transformative force of the earth system governance paradigm has given rise to the creation of a global network of scholars working in this field, the *"Earth System Governance Project."* This network originated in 2008 as a core project of the former International Human Dimensions Programme on Global Environmental Change, based on a ten-year science and implementation plan agreed upon in 2008 (Biermann et al., 2009). After over a decade of operation, the Earth System Governance Project has matured into a global, self-sustaining network of over thousand scientists, with conferences, taskforces, affiliated research centres, regional fellow networks, an affiliated foundation, and a lively presence in social media. A new 10-year science plan has been published in 2018 (Earth System Governance Project, 2018; Burch et al., 2019), and the network continues to expand, welcoming scientists and scholar activists from all over the world to join its many research programmes.

References

Biermann, F. (2007). 'Earth system governance' as a crosscutting theme of global change research. *Global Environmental Change: Human and Policy Dimensions, 17*, 326–337.

Biermann, F. (2014). *Earth system governance. World politics in the Anthropocene* (Earth system governance series). MIT Press.

Biermann, F., Betsill, M. M., Gupta, J., Kanie, N., Lebel, L., Liverman, D., Schroeder, H., & Siebenhüner, B., with contributions from Conca, K., Costa Ferreira, L. da, Desai, B., Tay, S., & Zondervan, R. (2009). *Earth system governance: People, places, and the planet (science and implementation plan of the earth system governance project)*. The Earth System Governance Project.

Burch, S., Gupta, A., Inoue, C. Y. A., Kalfagianni, A., Persson, Å., Gerlak, A. K., Ishii, A., Patterson, J., Pickering, J., Scobie, M., Van der Heijden, J., Vervoort, J., Adler, C., Bloomfield, M., Djalante, R., Dryzek, J., Galaz, V., Gordon, C., Harmon, R., Jinnah, S., Kim, R. E., Olsson, L., Van Leeuwen, J., Ramasar, V., Wapner, P., & Zondervan, R. (2019). New directions in earth system governance research. *Earth System Governance, 1*, 100006.

Clark, W. C., & Harley, A. G. (2020). Sustainability science: Toward a synthesis. *Annual Review of Environment and Resources, 45*(1), 331–386.

Earth System Governance Project. (2018). *Earth system governance. Science and implementation plan of the earth system governance project*. Earth System Governance International Project Office, Utrecht University.

Frank Biermann is a professor of Global Sustainability Governance with the Copernicus Institute of Sustainable Development at Utrecht University, The Netherlands. He is an internationally leading scholar of global institutions and organizations in the sustainability domain. Biermann pioneered the "earth system governance" paradigm in 2005 and was the founder and first chair of the Earth System Governance Project. He currently directs a research program on the steering effects of the Sustainable Development Goals, supported by the European Research Council. Biermann has authored or edited 19 books. In 2021, the Environmental Studies Section of the International Studies Association presented him with its Distinguished Scholar Award.

Globalization

Jacques Poulain

Abstract After the collapse of Eastern totalitarianism, American neo-liberalism of social markets and free enterprise is triumphing everywhere, as if it is the only legitimate way to live. It seems that it is claiming for itself the name "globalization". This triumph derives directly from the fact that American democracy is founded on freedom and equality of the social partners. Finally in the twentieth century Alexis de Tocqueville, and more recently Louis Hartz, recognized "The big advantage of the Americans is that they could reach the state of democracy without having to make a democratic revolution: they are born equal, they don't need to become equal" (Rajchman J, West C (ed) Post-analytic philosophy. Columbia University Press, 1985, p. 17). To cross the Atlantic Ocean would have enabled them to realize here, on earth, the Christian concept of producing and sharing salvation without having to cast aside the social structures inherited from feudalism. Furthermore social and democratic consensus would allow everybody today to survive economic and cultural crises as well as the antagonisms caused by their private interests or by the interests of whole groups, thereby avoiding the systemic use of violence. Thanks to neoliberalism, the bowing of our economic democracies to social consensus and to its social benefits would become the law of social and economic progress. As a form of life, it would finally enable individuals to integrate this social and economic progress into their personal lives. In this way neoliberalism would complete the process of rationalization of humankind and of the world. It would lead humanity to its philosophical destination.

J. Poulain (✉)
University of Paris 8, UNESCO Chair of Philosophy of Culture and Institutions from 1996 to 2020, Saint-Denis, France

© The Author(s), under exclusive license to Springer Nature Switzerland AG 2023

N. Wallenhorst, C. Wulf (eds.), *Handbook of the Anthropocene*,
https://doi.org/10.1007/978-3-031-25910-4_206

Neo-liberal Globalization

But, as is well known, this globalization which is experienced by the whole of humankind also generates the disappearance of freedom and of human rights that it should help to develop. The neutralization of civil rights and the proliferation of excluded people is happening today in the neo-liberal context of globalization, through depriving people of economic rights that the Welfare State once used as a way of compensating for the social injustice caused by the development of private capitalism. As Sheldon Wolin has pointed out in the journal *Democracy* (Wolin, 1982, p. 19), this deprivation happened in a paradigmatic way in the development of American industrial society, but it is now completely global because it also characterises the development of all other industrial societies, being enshrined in the capitalistic mind that is programmed to maximize profits with the minimum of effort. What H. P. Martin and H. Schumann (1996) called "The trap of globalization: an attack against democracy and welfare state" transforms this exclusion and this neutralization of civil rights only by viewing it as the best way of benefitting from contemporary scientific and technological progress. This deprivation of civil and financial rights is scandalous, not only because it prevents the redistribution of material goods between rich and poor people and between rich and poor countries, but also because it deprives the excluded from human reason.

But this tragic situation does not need to be, although it appears to be the destiny of the victims of globalization. It is indeed based on a purely negative conception of freedom and of human rights as a pure absence of constraints. Much more, this social illness derives from an old philosophical error: from the belief that human beings have to transform themselves directly into perfectly moral animals of justice that have to master their own desires and interests by means of subjugating their own moral consciousness to social consensus.

The total experience of humankind which is happening blindly through neoliberal capitalism does not need to remain blind and allow only the voices of the economic market to speak. This experience also opens up the possibility of a positive anthropological outcome for the human being: because the anthropology of language teaches us today that the dynamics of the human soul and of human institutions are communicative and that the use of language involves us as speakers and addressees of ourselves towards our social partners, we have learnt that our thinking and our use of language are based upon a verbal and mental pre-harmonisation with our fellow humans, with the world and with ourselves. Since this dialogical imagination is entirely affective (love oriented), cognitive, practical and hedonistic, it does not make us the enemies of our fellow humans and we know today that we do not have to protect ourselves against fellow humans as if they were innately asocial and our enemies by means of a system of rights based on this purely negative conception of freedom.

The philosophical contemporary anthropo-biology of A. Gehlen, of F. Kainz and of A. Tomatis (1939, 1957; Kainz, 1960–1975; Tomatis, 1991) teaches us too that the direct appropriation and transformation of ourselves by ourselves that we are trying to achieve through the liberal maximization of the satisfaction of our desires and interests and by a purely negative juridical and legal respect for the interests of the other people contradicts the communicative and dialogical structure of our minds and of our institutions. The transformation that we are intending to achieve by means of this total experience of ourselves can only occur indirectly, i.e. by making an objective judgment about ethical and political life, by means of a judgment that must be as true as our scientific judgment may be. In order to become as conscious and lucid as it has to be, this experience of ourselves requires that we take into account our use of judgment and develop a positive use and a cognitive conception of human rights: as a collective and private development of this judgment and of its consequences in public and private life. This use and this conception are already working by means of this global experience. They are already expressing themselves there because the social illness which is arising through globalization is only an illness of our reflection and leaves our dialogical imagination untouched. But this use and this consciousness need to be intellectually and philosophically recognized and they need to be applied in order that we may heal ourselves efficaciously from this desire for power that is still blindly deregulating the experience of humankind.

This step constitutes a radical challenge for a philosophical theory of globalization: it is requires that this theory criticizes its own most well-rooted prejudices, namely those that inspired the formation of our occidental societies and our own ways of understanding ourselves. These prejudices are expressed throughout the moral dualism inherited from Plato and Descartes that conceived mind, logos and soul as something that has to master the human body, the desires and the interests which this body inspires in the soul. This philosophical view of a dualistic self-control of the soul, fulfilled by means of the subordination of the body, of the desires and of the interests – to the highest interests of the mind – is a false one and it is based on a false description of the human being. It has generated a false conceptualization of the social and political systems because it was used to repress the necessary individual and collective use of judgment as if the human being would have to be mastered as an animal. The culture of humanity was reduced to a civilisational process where the human being had to become a perfect animal of justice, an animal which could regulate perfectly, once and for all, its thoughts, actions and speech acts in order to produce thoughts, actions and feelings that are consensual. It had to be civilized and transformed into a being that would no longer have to judge the truth and objectivity of what is defined as humanity in ourselves and in our fellow humans.

This injustice, caused by the repression of this judgment of truth and of objectivity about human matters and by a blind submission to an economic consensus in juridical, ethical and political matters has become patently evident through our contemporary globalization of impoverishment. Because this repression is only happening on the level of self-reflection, in other words, on the level of our consciousness, our actual use of our dialogical faculty of imagining and of our capacity of judging these matters remains nonetheless unviolated and enables us to step outside the purely economic neo-liberal consensus, which is used in this globalization as a form of life, by judging that this blind consensus is an impossible and a false way to live for any human being.

The Perversity of Globalized Capitalism

As C.S. Peirce taught us, scientific research has to be understood as a dialogue with the visible world that aims to bring about a consensus with this world. It is through our experimentation of the visible world that scientists ask the visible world to confirm or to disconfirm their hypotheses by answering either "yes" or "no" to the question "Are our hypotheses true?" In the same way, the daily and political experience of the field of human life involves our submission to the consensus that can be obtained from our social partners. Communication is used here as a test of our present and mutual hypotheses of life. By invoking the trans-subjective authority of social consensus in the same way as scientists invoke the consensus of visible world with their hypotheses, we are seeking some objective authority, which could to tell us what to desire and what to do. We trust in the infallibility of this consensual authority insofar as we come to understand that it has been no other authority than this social consensus that was always speaking through our words, thoughts and institutions and that thereby regulated our social and mental life. This social consensus also seems to have the same authority and validity with respect to our "internal" nature as the visible world has with respect to our knowledge of the "external" world.

The social consensus that we obtain by means of communication and find through the consensual successes of the capitalistic experience of the social market, appears to have the authority that we need in order to confront the cognitive experience of our secular salvation. The Weberian explanation of the logical dynamics of this capitalistic experience is also valid in neoliberal globalization, well known, but rarely understood. As the predestined Calvinists could be sure that they were chosen by God for salvation, provided that they were successful in their earthly life, the liberal quest for individual and social happiness is measured by means of the successes of capitalistic enterprises. But it considers these successes as the only source of confirmation of the choice of the actions, which determine the development of the liberal enterprises. The successes of life offered the certainty that Calvinists could be saved if, and only if, they were able to restrain themselves and to prevent themselves from enjoying immediately the fruits of their enterprises. In the same way the capitalists have to reinvest their benefits in their enterprises in order to be able to increase their certainty about their own social salvation, because it was the only means which enabled them to feel as disinterested as they were meant to feel.

This moral consciousness of the liberal was nonetheless necessarily perverse because it subordinated the will for happiness and social welfare of their social partners to an arbitrary and egoistic self-certification of their own personal will for salvation. In the name of the social consensus, they asked their social partners to work in order to be sure that they enjoyed the social happiness that the actions of these workers had brought about. In this way, they enjoyed exclusively their ability to subordinate the welfare of their partners to the satisfaction of their moral consciousness. The maximization of the human desires which is taking place to-day in the era of advanced capitalism and throughout the maximisation of the production of goods is still exclusively due to this maximization of the certainty to be socially saved. By preventing us from judging the social results of this desire for certainty, the revelation of the divine social world market nowadays becomes as autistic as the religious revelation of the past. The answer given by the global neoliberal market seems to us to be as necessary and natural as the answer of the visible world to the scientific experimentation. As is well known, impoverishment, unemployment and exclusion of the poor are the prices that must be paid for this increase of capital and they bring about, in the long run, a radical falsification of the liberal way of life. As Sheldon Wolin wrote in his journal *Democracy*, the deprivation of civil rights followed necessarily from this impoverishment and the neoliberal disappearance of the Welfare State. Although the liberal theory of rights enshrined these rights in the American Constitution itself, conceiving them as special forms of freedom and mutual protection that were to be beyond the ordinary reach of legislative and executive power, and assumed to be "above" politics, "what happened during the twentieth century is that the liberal practice of politics rapidly undermined the liberal conception of rights" (Wolin, 1982, p. 19). "The protection of rights presupposed that government will be their defender, intervening to prevent interest groups from violating the rights of other groups and individuals. For this presupposition to work, government itself would have to withstand effectively the pressures generated by interest-groups politics, by the politics of factions, pressures that were guaranteed to be unrelenting by the system of elections, campaign contributions and lobbying. The presupposition collapsed because North-American politics were reduced to interest-groups, there was no general constituency to support government in its role of impartial defender of rights. Instead of playing the role of defender of rights, the government assumed a function more consistent with the politics of interest-groups, that of "balancing" rights against certain overiding matters of State" (Wolin, 1982, p. 20). It followed, wrote S. Wolin, "that many limitations imposed on rights through legislative or administrative rulings have been inspired by minorities obsessed with single issues. The American society became slowly accustomed to the dangerous notion that rights, like crop subsidies or taxes, are part of the normal give-and-take of politics" (Wolin, 1982, p. 21–22). S. Wolin (2008) and J. Stieglitz (2012) have demonstrated the inability of the Nation-States to brake the capitalistic drive for maximizing the rent of the rich and the social price of inequality that people must pay.

The Dialogical Alternative: Sharing Judgments About Economic and Political Matters

The drive towards the neoliberal experimentation with mankind by means of the consensus and the validation of universal valid contracts is forcing us today to discover that we do not have to believe anymore that the human being is – as body, desires, passions and interests -, his own enemy as mind, soul, or good will. This drive toward an experimental blind consensus has forced the social anthropology of language to recognize that the origins and the dynamics of thought and human reason were constituted by the use of language as the power of emitting and receiving sounds and of linking them to our experience. The priority of human judgment is a condition of the possibility of our human life and must be respected in the formation of our social conditions of life. Reason and thought are generated by our use of language: this means that our use of language obliges us to see ourselves as our own addressees who have to judge about the objectivity of our desires, interests and ethical actions and to share our judgements with our social partners in order to be able to enjoy these experiences, and to recognize them as our objective conditions of life.

Because the human being is not a well-formed biological being, but is born one year too early – if one compares it with mammals endowed with a similar complexity -, it has only intraspecific drives (nutritional, sexual and defensive). Humans need too to invent their visual perceptions and physical actions by projecting the harmony between emitted and received sounds in their relations with the world, with fellow humans and with themselves. They need too to invent their visual perceptions, physical actions and consumatory actions by projecting the harmony between their emitted and received sounds in their relations with the world, with fellow human beingd and with themselves. It is for this reason that primitive people and the human child have (1) – to let speak the world animistically by means of using language as a kind of magical prosopopoie as Vico and W. von Humboldt have remarked that it does, in order to be able to perceive it with their eyes, and (2) – to feel this word of the world as the answer of the world itself, which is invariably as favourable as the voice of their mother.

This harmony, experienced in the use of sounds because of the inability of the child and of primitive man to perceive a difference between their own emitted sounds and their own received sounds, was the source of the sacred in human sensibility, i.e. the source of the sacred prosopopoie of the religions. This harmony imposes its own law on the dynamics of imagination, thought and desires in the following way: every hiatus and disharmony with the world, with one's human fellows and with oneself must be overcome by projecting as a new form the dialogical form of preharmonizing oneself with the world, with one's fellow humans and with oneself; this is done according to the harmonized sound and dialogical model. As we spontaneously harmonize the sounds that we are uttering with the same sounds that we are receiving, we preharmonize our perceptions, our actions and our desires as the best favourable ways by means of which the world and our human fellows could answer to us.

The cognitive and dialogical preharmonization which bears the propositions by means of which we objectivize our perceptions, our actions, our thoughts, our feelings and our desires is always the same: we are unable to think a proposition, to produce it without thinking that this proposition is true. In the terms of C. S. Peirce: "every proposition affirms its own truth" (Peirce, 1935, § 340). We must think of our propositions as true in order to be able to objectivize the visual fact, our physical action and our desire in order to produce the only relation to reality that we can obtain by them, for example, the visual perception of the visual fact or the realization and the perception of our physical action. Therefore, our relations with the world, other human beings and ourselves cannot be produced nor completed without judging whether or not these preharmonized linguistic relations that are shown in our propositions are for ourselves these objective conditions of life that they are presumed to be in order to be able to come into existence. Ours relations with ourselves are therefore necessarily indirect. We cannot judge and transform ourselves without judging the objectivity of our relations to the world, nor without judging the objectivity of the speech or thought experience which ensures these relations and these experiences, i.e. without judging the truth of the propositions expressing our knowledge, our need for action or our desires. This whole dynamic of the soul rests itself upon the self-objectivation of our speech-acts and is made explicit by this. It is submitted to the law of truth – in its creative moment as well as in its reflective moment – the act of judging its actual truth by judging the objectivity of the represented experience.

But the fact that I have to express this speech-act means that I cannot judge the truth of the propositions as a private and solipsist subject: I cannot reach a position which would enable me to judge this truth once and for all from the point of view of God. I need the approval of what I am saying by my social partners and I need their positive and affirmative judgment as the only authority which could confirm or disconfirm the truth of this judgment.

Because this common and spontaneous use of our judgment of truth remains unviolated in spite of our errors concerning our moral competence of mastering ourselves by the magical use of speech-acts – where it is presumed sufficient to designate them in order to bring them about -, this use is nevertheless always working although we are not necessarily conscious of this. The reason is simple: as our own addressees, we are unable not to do it. This constitutes a rational and philosophical use of judgment, in which we are all already involved. What we are used to call "philosophy" is only the intellectual discipline, which has to recognize this use of judgment already working in our use of language. This discipline has to help us to become actually what we already are. We have to realize and to fulfil – by our actions and by the execution of our duties – the objective modes of being that all of us have to judge mutually that we are. We have to reject and to forget all the other modes of being. This is a constitutive law of the formation of human beings and the recognition of their cultures (a *Müssen*) from which it follows that we must share our duties (our *Sollen*), a law that is recognized as such as common to all of us. Philosophy, understood as a discipline, has not only to describe the conditions of the recognition of ourselves in the use of this judgment by means of showing the biological, psychological, sociological and linguistic facts, which are describing this judgment as the *conditio sine qua non* of our existence. Philosophy has also to show how we can be consciously obliged to do what we usually only feel forced to do unconsciously: this need of using our philosophical judgment can and must be given to us as the right to be what we have to be. The anthropological alternative to an irrational globalization that is here at stake, is that this unconscious and conscious philosophical way of using our judgment of truth in order to be able to live, is the only means that we have at our disposal to become completely harmonized with ourselves, with our world and with our social partners.

But how does this philosophical use of judgment work in our life to-day and in the context of an unjust neoliberal globalization although we are not necessarily conscious of this, and although this collective use of judgment is not still completed in a theoretical and in a practical way, i.e. in its social applications?

In order to show that it actually already happens, one could give some examples, which are borrowed from our neo-liberal globalized context (Poulain, 2017). This use of a rational judgment is already working in all bargaining between managers and workers where the redistribution of wages, goods and social values takes place in a critical and objective way in order to overcome advanced capitalistic injustice and exclusion. But it also occurs when we are judging the social injustice that is produced by this deprivation of civil rights and the processes of exclusion, which are becoming an increasing threat in our industrial societies. It occurs when we are judging the exploitation of the Third World by industrially advanced societies and when we are considering this exploitation as a way of stealing the raw materials and the working forces of these countries according to rentability principles. It occurs, when we are judging the speculative attacks against financial values as efficient ways of stealing the Nation States and all of us. The only thing that is changing when we see this condemnation as an effect of our individual and collective use of rational judgment, is that we are afterwards forced to see these scandalous actions as a situation which has to disappear, because they are indeed actions in which we are unable to recognize ourselves. These facts are like other facts: they have no value and no existence by themselves, they are not speaking animistically to us as pragmatists and theoreticians of speech-acts might believe that they do. But they have to be judged as our objective conditions of existence in order to actually exist. In this scandalous situation where these facts are facts of a generalized unjust impoverishment and exclusion, they have to be judged and recognized as our inacceptable social conditions of existence.

But this anthropological constitutive use of rational judgment has its own specific ethics and institutional consequences too. It involves by example a reform of our conceptualization of the Nation-State and of its examples in particular States. Before these States can become the defenders of human rights and the only supporters of the power of using violence against violence, they need to be recognized as legitimate ways of sharing our collective use of judgment when it is already working, and when it can be recognized as such. The false image of the human being as an ethical chaos of values may remain characteristic for the “polytheism of values”, as M. Weber described liberal capitalism. It might still be the basis for a social life inside which we are experiencing fights between factions, interest groups, multinational enterprises and competitive political parties but it must nevertheless be recognized as such and abandoned as such.

If we see elections only as collective means of projecting anticipative ways of life and as means of recognizing their objectivity by adopting them, we become able to see that they allow the parties which are proposing them, to find the best ways of sharing mutually their power of judgment. We are able as well to assume that the description of the objective value of these programmes reveals what human beings have to do in this Nation State context and what they actually are. It is by sharing the power of judging the objectivity of their programme with the opposition that the party which has the executive power is itself controlled in the same way as the human mind is generally controlled: by giving the power to all the addressees of the Nation States, i.e. to all its members. The fight between irreconcilable values can only be avoided this way. But the fight between Nation States and multinational enterprises has not to be conceived as the destiny of humanity because it would be presumed to be something that could not be overcome. International life is not condemned to be more conflictual than the internal life of a nation. At the international level, Nation States can only be legitimized if their proposals and political decisions on this international level can be recognized by an international public opinion in terms of validating objective conditions of interaction between some Nation States and others.

My last example is our use of judgment in our economic democracies. The laws of economics are not only laws securing coherence and consistency between contractual economic agreements and decisions which have to be conducted according to the model of the syntactical and formal laws of language. The judgment of truth which is building our use of language in these democracies has to be recognized in order that one can become able to see the judgment of objectivity which is living in every thought, speech-act and action: without this anthropological recognition of the truth as the internal dynamics of language and of human life, language appears as a phenomenon that can only be reconstructed syntactically, semantically or pragmatically. But its truth dimension is never seen by means of these reconstructions as constitutive of the dynamics of language, communication and human life. In the same way economics is not only a science where we must apply formal and probabilities calculus in order to recognize the specific laws of fuzzy logic which animate the random existence of fluent tendencies of our exchanges of money. We are obliged to think about them as such if, and only if, we assume that the blind laws of the social market, such as the laws of demand and offer, are the only principles of economic movements because they would only infallibly express our mutual interests and the generalized consensus, which is expressed by means of the results of the world market. But the economic judgment operating the translation of goods and of working processes into money and wages is as objective as the judgment that takes place in thought and in our use of language. It has to be reconstructed and redescribed as animating economic human life and economic democracies insofar as they are human.

Economic life itself has therefore to be recognized and reconstructed as a way of recognizing and legitimating objective ways and processes of translating money into objective and necessary concrete conditions of life, the existence of which is conditioned by sharing our objective ethical judgment about the application of human rights. What is usually seen as a better state of the economic conditions of the world has to be judged as such in contrast to the absurd blind criteria of the constant growth of benefits and investments. It has to be judged itself as constituting an actual progress in our way of sharing these objective ethical and social judgments, and not the reverse. It is from this point of view that the Nation States may and must demand a wholly fulfilled transparency from the world banks in order to be able to banish the origins of the most dangerous speculative attacks against the value of their money. If they want to be able to be recognized as the objective institutions that they are and that we need them objectively to be, they have to programme a necessary extension of the civil rights for transparency at this level of our international and cosmopolitan democracy and conceive the legal procedures enabling them to ensure the respect of this right for transparency.

References

Gehlen, A. (1939). *Der Mensch*. Athenaüm Verlag.
Gehlen, A. (1957). *Urmensch und Spätkultur*. Athenaüm Verlag.
Kainz, F. (1960–1975). *Psychologie der Sprache*. Neske, Tome 1–5.
Martin, H. P., & Schumann, H. (1996). *Die Globalisierungsfalle. Der Angriff auf Democratie und Wohlstand*. Rowohlt.
Peirce, C. S. (1935). *Collected papers* (Vol. 5). The Belknap Press of Harvard University Press.
Poulain, P. (2017). *Peut-on guérir de la mondialisation?* Ed. Hermann.
Rajchman, J., & West, C. (Eds.). (1985). *Post-analytic Philosophy*. Columbia University Press.
Stieglitz, J. (2012). *The price of inequality: How today's divided society endangers our future*. Norton Company.
Tomatis, A. (1991). *L'oreille et le langage*. Seuil.
Wolin, S. (1982). What revolutionary action means today. *Democracy, New York, 2*(4), 17–28.
Wolin, S. (2008). *Democracy incorporated: Managed democracy and the Specter of inverted totalitarianism*. Princeton University Press.

Jacques Poulain is Professor Emeritus at the University of Paris 8. UNESCO Chair of Philosophy of Culture and Institutions from 1996 to 2020. Member of the European Academy of Sciences and Arts since 2014. He was Vice-president of the Collège International de philosophie, Paris (1985–92) after having taught at the University of Montreal (1968–85). His publications include: *L'âge pragmatique ou l'expérimentation totale* (1991), *La loi de vérité* (1993), *De l'homme. Eléments d'anthropobiologie philosophique du langage* (2001), *Die neue Moderne* (2012) et *Peut-on guérir de la mondialisation ?* (2017).

Migration

Barbara Pusch

Abstract There are several highly diverse connections between the Anthropocene and migration. If we date the beginning of the Anthropocene to coincide with the Industrial Revolution, the importance of migration in this context becomes immediately evident. The Industrial Revolution not only caused major migration movements (from rural exodus to international labour migration), but also made them possible and fuelled them as well. However, this article does not focus on migration as a factor in the emergence of the Anthropocene, but on Anthropocene-induced migration. In doing so, the article addresses two main questions: (1) How are environmental factors conceptualized in migration research, and (2) how can we understand the impact of environmental factors on migration in the Anthropocene adequately? The paper ends with the formulation of key research desiderata.

There are several highly diverse connections between the Anthropocene and migration. If we date the beginning of the Anthropocene to coincide with Industrial Revolution, the importance of migration in this context becomes immediately evident. The Industrial Revolution not only caused major migration movements (from rural exodus to international labour migration), but also made them possible and fuelled them. However, this article focuses less on migration as a factor in the emergence of the Anthropocene than on Anthropocene-induced human migration as it is related to environmental destruction in general and climate change in particular. There is also an Anthropocene-induced migration of nonhuman species, but it is not discussed in this chapter. Apart from a few notable examples, such as the "Philosophy of Movement" (Nail, 2022), the migration of human beings, animals, and plants in the Anthropocene is not analyzed in a common framework.

Migration is as old as the history of humans, and climate-induced migration is also not new. For instance, the spread of homo sapiens has been explained by "abrupt changes in climate" that might have made large areas of North, East, and

B. Pusch (✉)
Universität Koblenz-Landau, Campus Landau, Arbeitsbereich Heterogenität, Landau, Germany

© The Author(s), under exclusive license to Springer Nature Switzerland AG 2023

N. Wallenhorst, C. Wulf (eds.), *Handbook of the Anthropocene*,
https://doi.org/10.1007/978-3-031-25910-4_207

West Africa unsuitable for hominins (Carto et al., cited in Freire de Mello et al. 2020: 1). Neolithic farmers' migrations also seem to be closely related to climate conditions. However, human interventions in the environment and the resulting intentional and unintentional changes in habitats have also led to migration processes throughout human history. Examples include deforestation and droughts, which led to the migration of peoples in Central Europe around CE 300–500 and to the decline of the Mayan civilization from 800–900. Nevertheless, while humans have historically intervened in the environment at local and regional scales (including "ecological imperialism" (Crosby, 2015)), it is only in the Anthropocene that these interventions have taken a global dimension. Experts predict that global climate change will cause storms, floods, droughts, and desertification to increase, both in severity and frequency, and that sea levels will rise. These changes will affect the entire globe. However, according to the Intergovernmental Panel on Climate Change, the Arctic, parts of Africa, small islands, and Asian and African mega-deltas, and thus people from the Global South, will be affected most.

As Piguet (2013) noted, early migration researchers such as Ratzel, Ravenstein, and Huntington reflected around the turn of the nineteenth century on environmental factors such as climate and soil fertility; however, "migration studies lost sight of the natural environment during the twentieth century" (ibid, 2013:3). This gradually started to change from the mid-1980s. However, it was not until the turn of the millennium that environmental migration factors started to gain prominence under the catchphrase "environmental refugees". This was partly due to heightened sensitivity to environmental issues, as well as forecasts from NGOs, which predicted 200–300 million environmental refugees from the Global South by 2050. El-Hinnawie introduced the term "environmental refugee" in the mid-1980s to describe "people who have been forced to leave their traditional habitat, temporarily or permanently, because of a marked environmental disruption" (El-Hinnawi, 1985: 4). With this buzzword, people from the Global South became the "human face of climate change" (Gemenne, 2011: 1).

There is no doubt that climate change and environmental factors will become increasingly important for migration dynamics in the Anthropocene and will have a disproportionate impact on the Global South. Nevertheless, the concept of environmental refugees and the alarmist predictions of their large numbers are both highly controversial, for several reasons. They both assume (1) a simple "cause-and-effect relationship" following the concept of "the more environmental disasters, the more migration". This suggests not only (2) a monocausal relationship between climate change and migration, which fails to comprehend the manifold and multicausal nature of migration, but that those people affected (3) are helplessly exposed to the environmental factor as a driving force. In line with this perception, these are viewed as "helpless victims" (Baldwin, 2017: 4) of external circumstances, incapable of taking action. Moreover, both the notion of "environmental refugees" and the alarmist predictions and resulting discourses have been criticized for being based on the idea that (4) migration is an exception in crisis situations and (5) an expression of failure to cope with (environmental) risks. Furthermore, it is noted that the environmental refugee concept lacks a legal basis since environmental factors are not recognized reasons for flight in the Geneva Refugee Convention. For these reasons, the term "environmental refugee" is increasingly being replaced by terms such as 'climate-induced migrants" or "climate-related migrants" (Gemenne, 2015: 172). However, as these terms do not adequately reflect the complexity of migration, I argue for the term "environment-(co)induced migration", emphasizing that the environmental factor is only one of several drivers of migration, even in the Anthropocene.

The following question then arises: How can we understand the impact of environmental factors on migration in the Anthropocene? For this purpose, it is worth looking into the social science of migration research. Migration decisions always rely on a variety of social, political, and economic factors and can never be attributed to a single driver; this also means that factors that lead to migration decisions do not automatically trigger migration movements. In the environment-migration nexus, this is most evident in the intersection of socioeconomic status, gender, and migration behaviour of so-called "trapped populations" (GOS, 2011:14), who are unable to adapt locally to the impacts of climate change and do not have the financial resources and/or social networks to migrate. Studies indicating "that migration decisions are affected by *perceptions* of environmental change, rather than environmental change as objectively identified using *scientific* evidence" (Koubi et al., 2016: 441; second italics added), and studies of traditional concepts such as *banua conducted on indigenous (im)mobilities in the context of climate change-induced displacement policies (Suliman et al., 2019) also underline this complexity.

Today, the dominant discourse of migration in the Anthropocene is increasingly being questioned from post-structuralist and power-critical perspectives; in particular, the Mobility Justice concept provides a new perspective on the environmental-migration nexus. By focusing on the conditions of and unequal access to mobility along gender, ethnicity, race, religion, age, and social class, the concept frames what is commonly referred to as environmental migration as the result of colonialism and capitalism. Moreover, if it is stated that, in the Age of Humans, "some [people] have become the main agents of changes on this planet" (Gemenne, 2015: 168), but that "most humans are actually the victims of these changes" (ibid.), and that highly vulnerable groups are hidden the hardest, this highlights global social inequalities in the environment-migration nexus. Against this background, Faist and Schmidt (2020) called for climate change and argued that migration should be considered from the perspective of social inequality research. They argued that this would make it possible to focus on both social inequalities that are exacerbated by climate change and the effects of environmentally (co)induced migration on social inequalities.

All of the above shows the complexity of human migration in the Anthropocene and how difficult it is to define it adequately. However, it is even more difficult to make reliable statements about actual environmentally (co)caused migration. We know that from 2019 to 2020 the number of international migrants increased from 272 million to 281 million, with 89.4 million people living in displacement worldwide (IOM, 2022: 4). Furthermore, 55 million of these people were internally displaced (ibid.). For 2020, the Internal Displacement Monitoring Centre (IDMC, 2021: 7) recorded 30.7 million new disaster-related internal displacements worldwide, 98 percent of which were climate-related. However, all these figures also include environmentally disaster-induced migratory movements in which those affected move back to their places of residence after a short period of time, but do not take into account migrations due to low-onset environmental events and climate changes. Therefore, the figures only provide limited insight into the real environmentally (co)caused migration processes in the Anthropocene. Nevertheless, these statistics show us that environmental displacement is primarily the result of South-to-South migration, which often takes place within national borders. In 2020, more than two-thirds of the new displacements triggered by disasters were recorded in East Asia and Pacific (39.3%) and South Asia (30.1%) (IDMC, 2021: 21). This illustrates that people whose lifestyles contributed least to climate change are the most affected; at the same time, it invalidates the apocalyptic threat scenarios of "climate barbarians" (Bettini, 2013:63) in the Global North and underlines social inequality as a global factor in the context of environment and migration.

The above examples show that migration in the Anthropocene has mainly been discussed in the context of climate change and the associated environmental changes. Despite empirical studies and theoretical considerations from a mobility perspective, the Anthropocene migration discourse largely overlooks the multifaceted aspects of migration. Therefore, "The Age of Migration" (Castles et al., 2014), which overlaps in time with the "Age of Humans", is ignored. With the one-dimensional consideration of migration from an environmental perspective, the Anthropocene migration discourse deprives itself of many starting points that would be fruitful for a profound (theoretical) examination of migration in the Age of Humans. For example, the role of human migration is reflected as a shaping force of the Anthropocene in only a few exceptional cases (e.g., Toomey, 2021), which is why migration is mostly considered as a result and not as a possible contributing factor to human planetary interventions. It would also be worthwhile to explore the relations of human and non-human migration in the Anthropocene.

References

Baldwin, A. (2017). Climate change, migration, and the crisis of humanism. *WIREs Climate Change, 8*, 1–7. https://doi.org/10.1002/wcc.460

Bettini, G. (2013). Climate barbarians at the gate? A critique of apocalyptic narratives on 'climate refugees'. *Geoforum, 45*, 63–72. https://doi.org/10.1016/j.geoforum.2012.09.009

Castles, S., de Haas, H., & Miller, M. J. (2014). The age of migration. In *International population movements in the modern world*. Palgrave Macmillan.

Crosby, A. W. (2015). *Ecological imperialism. The biological expansion of Europe 900–1900*. Cambridge University Press.

de Mello, L. F., et al. (2020). Anthropocene and migration: Challenges in the 21 century. In W. Leal Filho et al. (Eds.), *Encyclopedia of the UN sustainable development goals. Decent work and economic growth* (pp. 1–9). Springer. https://doi.org/10.1007/978-3-319-95867-5_85

El-Hinnawi, E. (1985). *Environmental Refugees*. United Nations Environment Programme.

Faist, T., & Schmidt, K. (2020). Die sozio-ökologische Frage: Klimawandel und die (Re-) Produktion sozialer Ungleichheiten. In T. Faist et al. (Eds.), *Soziologie der Migration: Eine systematische Einführung* (pp. 425–447). De Gruyter Oldenbourg.

Gemenne, F. (2011). How they became the human face of climate change. Research and policy interactions in the birth of the 'environmental migration' concept. In E. Piguet, A. Pécoud, & P. de Guchteneire (Eds.), *Migration and climate change* (pp. 225–260). Cambridge University Press/UNESCO.

Gemenne, F. (2015). The Anthropocene and its victims. In C. Hamilton, C. Bonneuil, & F. Gemenne (Eds.), *The Anthropocene and the global environmental crisis* (pp. 168–174). Routledge.

Internal Displacement Monitoring Centre (IDMC). (2021). *Global report on internal displacement 2021*. https://www.internal-displacement.org/global-report/grid2021/

International Organisation for Migration (IOM). (2022). *World migration report 2022*. IOM. https://publications.iom.int/books/world-migration-report-2022

Koubi, V., Stoll, S., & Spilker, G. (2016). Perceptions of environmental change and migration decisions. *Climatic Change, 138*, 439–451. https://doi.org/10.1007/s10584-016-1767-1

Nail, T. (2022). What is the philosophy of movement? *Mobility Humanities, 1*, 9–22.

Piguet, E. (2013). From "primitive migration" to "climate refugees": The curious fate of the natural environment. *In Migration Studies. Annals of the Association of American Geographers, 103*, 148–162.

Sheller, M., & Urry, J. (2016). Mobilizing the new Mobilities paradigm. *Applied Mobilities, 1*, 10–25. https://doi.org/10.1080/23800127.2016.1151216

Suliman, S., Farbotko, C., Ransan-Cooper, H., McNamara, K. E., Thornton, F., McMichael, C., & Kitara, T. (2019). Indigenous (im)mobilities in the Anthropocene. *Mobilities, 14*, 298–318. https://doi.org/10.1080/17450101.2019.1601828

The Government Office for Science (GOS). (2011). *Foresight: Migration and global environmental change*. Final Project Report.

Toomey, N. (2021). The nexus of (im)mobilities: Hyper, compelled, and forced mobile subjects. *Mobilities, 17*, 269–284. https://doi.org/10.1080/17450101.2021.2000840

Barbara Pusch held a professorship for heterogeneity in education at the University of Koblenz-Landau, Campus Landau, from 2019 to 2022. Her primary fields of research include reconstructive social research, education for sustainable development, migration, and transmigration. Her recent publications include: Our Common Future Today: Umwelt- und Nachhaltigkeitsorientierungen von Jugendlichen in der Pfalz (with Ch. Horne). In: *Pädagogisches Jahrbuch 2021 „Zukunft – heute* "(in print); Pädagogische Ansatzpunkte in der Umweltbildung für Zuwanderer/–innen, *Zeitschrift für internationale Bildungsforschung und Entwicklungspädagogik, ZEP* (44/2), 2021, 17–26; Bildung für nachhaltige Entwicklung in Orientierungs- und Wertekursen? – Ein österreichisch-deutscher Kursbuchvergleich, *Österreichisches Religionspädagogisches Forum* (2/28), 2020, 132–148; and Der transnationale deutsch-türkische Raum in der Migrationsforschung: Schwerpunkte und Perspektiven, in: Sokoll, Anne M. N. (eds.), *TürkeiAlmanya. Migration und Interkulturalität im deutsch-türkischen Kontext*. Barleben: docupoint, 2019, 35–57.

Participation

Jean-Louis Genard

Abstract This article proposes a lens to understand and contextualise the emergence of the participatory claim, to situate its specificity in relation to the model of representative democracy and also with regard to other models based on critique of this model. It highlights the importance of Jurgen Habermas's thinking which questions the tension and articulation between participative and deliberative democracies. It then describes both its successes and critiques but also the spread of the participatory model in spheres extending largely beyond politics, especially its impact on the economic and entrepreneurial realms and that of the new social movements. It then shows how the rise of ecological thinking has found it to be a natural ally but also how it has helped redefine it, especially by redesigning the limits of the community and its legitimate 'participants' and by requiring a renewal of the articulations between participation and representation.

Historically, the participative democracy semantics began emerging in the United States, in the 1960s. It is part of a long tradition of reflexion on the conditions and forms of democracy, dating back to Ancient Greece. More precisely, it also belongs to a tradition of protests and critique of representative democracy models in their differing accentuations: models for 'direct' democracy, championed historically by renowned defenders including Jean-Jacques Rousseau; 'random' models based on selecting representatives via a lottery (Van Reybrouck, 2016) which, in a way, acknowledged the impossibility of direct democracy whilst avoiding the perverse effects of voting and electoral battles; the self-management models popularized by Ivan Illich of the anarchist tradition, but also by theoreticians of workers

Translation by Gail Ann Fagen.

J.-L. Genard (✉)
Université libre de Bruxelles, Brussels, Belgium
e-mail: jean.louis.genard@ulb.be

© The Author(s), under exclusive license to Springer Nature Switzerland AG 2023

N. Wallenhorst, C. Wulf (eds.), *Handbook of the Anthropocene*,
https://doi.org/10.1007/978-3-031-25910-4_208

movements; and lastly the deliberative models, particularly in the thinking of Jurgen Habermas.

From a sociological view, the emergence of participative demands from the 1960s should be seen in relation to a double-sided socio-political context. On one side, there is a 'classical' dimension that stresses the heightened de-legitimation of politics, disconnection of political elites from civil society, and obviously the critique of a political participation restricted to the right to vote, exacerbating of media politics and creating a feeling among citizens that they never really count until election time, when the citizen becomes an 'advertising target' (Manin, 1985). On the other hand, over these same years, but part of a much broader context of 'anthropological mutations'[1] (see the dictionary entry) that marked this period, came the figure of the citizen-actor 'capable and skilled', expert in the fields where they have personal experience, having as consequence the recognition and valuation of citizen's skills and expertise in areas where they had hitherto been excluded (Cantelli & Genard, 2008). The expertise they propose is adequate to compete with if not replace that of recognized 'experts'. Their context is also marked by the proliferation of initiatives, reflecting the example of typical of self-governing mechanisms, where 'ordinary' actors take charge of projects, organizations, which until then seemed capable of being managed only by those holding an authority recognized by others (Genard & Cantelli, 2010). This was also the time of an emergence, first driven in the 1960s by the U.S. feminist and civil rights movements, of calls for empowerment, building the capacity of populations that were excluded, stigmatized and marginalized in a way specifically so they could become 'able and skilled' citizens (Cantelli, 2013). These were also the years when forms of participative apparatus multiplied, undoubtedly peaking with the Zapatista movements or the experience of Porto Alegre (De Sousa Santos, 2002), but also with the evolution of social movements towards constitution around participative settings, 'hybrid forums' (Michel et al., 2009). We could also add, from the social sciences, the advancement of different *studies* in their demand, primarily led by *subaltern* and *postcolonial studies,* to restore a dialogue in the first person to the populations that had suffered exclusion. This context also led to the progressive integration of participative apparatus in public policies, for example with consensus conferences but also the proliferation of participative settings initiated by the political world itself, often in the aim of re-legitimation.

This time from a conceptual angle, the emergence of the participative demand can be explained by the convergence of two demands, actually quite different. On one side, there is the call for each and every individual to be able to intervene in political options and decisions that concern them: this is the emphasis on 'direct democracy'. On the other side, there is the demand that perhaps focuses less on the number of people who participate than on the conditions that ensure the rationality of these options and interest, their conformity with the general interest; this would be the 'deliberative democracy' emphasis, with the underlying suspicion that

[1] See the entry "anthropological mutations".

representation does not constitute a guarantee towards this general interest. These two kinds of demands can seem to converge but also appear contradictory as feared, for example, by John Stuart Mill, for whom the call for democracy relied on conditions for deliberation that could reach good decisions, but which at the same time, did not lead to 'participative' consequences as, in his view, most citizens did not meet the minimal requirements for capability.

Without a doubt, the reflexions of Jurgen Habermas provide a way to move beyond these possible contradictions. More than participation in itself, Habermas was interested above all in the old question of how to build a general will that was founded rationally. Habermas placed his works in the tradition of Kant's thinking on the principle of publicity (Kant, 1996 (1784)) and constitution of a public space, and from his earliest works, he had analysed the weaknesses of this space, especially linked to the rise of the media system. Placing himself in this tradition of the Enlightenment, for Habermas, general interest was formed in a public use of reason (Habermas, 1989). It was in order to address the weakening of public space that Habermas proposed a model of deliberative democracy, a term he preferred to participation, articulated to a normative model of deliberation based on what he called the 'ethics of discussion' (Habermas, 1990, 1996), one criteria of which, beyond submitting to the best argument, was the possibility for access to information and the voice of all those who are interested in the 'objects' of this discussion, which certainly also directly refers to participative demands.

As progression of the participative demand was largely constituted in opposition to the decline of political representation, it is easy to understand both the attraction these participative apparatus may have for the governing authorities who are seeking re-legitimation and also, at the same time, the suspicion and distrust that meets these participative initiatives put forward by the same public authority. This was the case when pressures or injunctions to participate were denounced as consequences of liberal critiques of social policies as 'disempowering', or, more directly, as symptoms of a retreat from the achievements of the welfare state. In a way, the transformation of social policies to render responsible or activate the beneficiaries who are asked to be more involved in conditions for access to the services is not completely foreign to this logic of participation. Neither is it foreign to the multiple calls for citizens to act responsibly in the face of the recent coronavirus pandemic.

Yet perhaps more essentially, empirical analyses of participative mechanisms have highlighted their limits, whether – certainly – in the way these mechanisms become instrumentalized politically, but also in differential possibilities to speak out, in the power left to experts which frustrate participative ambitions, or the failure to take participative actions seriously, etc. And, lastly, this time far from Habermas, whose deliberative model gives an excessive privilege to argumentation, participation can and should be understood as something more than merely exchanging arguments (Berger, 2017). An initial period of euphoria has given way rapidly to an intermingling of enthusiasm and scepticism.

We should point out that the ever more popular figure of the individual who is 'participative, capable, skilled and responsible' is obviously not limited to the world of politics; the call for participation has multiplied in several areas, beyond political participation. Whether, for example, in the economic realm which has seen the rise of 'do it yourself', in the field of home repairs and crafts, 'active' leisure time, self-construction, repairing appliances, all accompanied by a flourishing set of guides enabling individuals to learn skills that they probably never imagined.

Beyond these different limits, it is nevertheless clear that the demand for participation henceforth appears as an essential issue at stake for democracy. This issue has also participated in shaping new forms for social movement strategies and, more generally, for protest practices. For example, we are thinking of the occupation of public spaces, squatters or the Zones to Defend (ZAD) that systematically proclaim the virtues of self-organization, spontaneity and autonomy. And this proliferation is not restricted solely to the political realm. Likely in the wake of experience with and claims for self-management, participation has been seen as a way to make companies more democratic and thus the economic world as well, for example by imposing a bicameral type of model for company management (Ferreras, 2012). Another example in economics, is the proliferation of entrepreneurial, participative, cooperative or participative crowdfunding initiatives, including those found in unexpected areas, such as banking, along the idea that participation in decisions and benefits should not be limited only to owners and/or stockholders.

Lastly, coupled with the decline of politics, the ecological emergency, the horizon of the catastrophe and destruction of the Anthropocene have certainly reinforced the urgency of political changes, which has sought synchrony with participative claims, but have also contributed to their softening.

Ecological thinking has indeed often adopted calls for participation but by adding a new dimension. One strong example is the integration of non-humans, in the perspective of questioning humankind's role of exception (Genard, 2020). Or more precisely, the perspective is taking into account the community of 'destiny' shared by humans and non-humans. How can the voices of non-humans effectively be heard? In its undoubtedly dominant component, political ecology indeed plays out against an ethical backdrop that presupposes an extension of the community to beings that require moral and political recognition, on the notion that beings can suffer from political decisions without having the possibility to express their interests, their suffering. Whether it is a matter of human beings generally ignored by politics, but also – and this is obviously the specificity of this political ecology – future generations and non-humans invested with a value not only in virtue of their interdependency with humans, but also as beings themselves. This dimension approaches, under new costs, the question of participation because, while the initial models relied on the need for each person to be able to defend their interest, here this requirement is completed by, or even makes way for, an extended requirement: for responsibility (Jonas, 1984) towards beings who can only 'participate' through the care of those who are able to participate. It is around this question of extending the 'list' of beings taken into account politically that the idea of ecological democracy has grown (Doherty & de Geus, 1996; Goodin, 1996), defined by a tension between an extension of those who participate, and also at the same time, a call for a reflexion on the how to have them participate, and thus in reality to 'represent' them (Eckersley, 2004), to acknowledge their right.

References

Berger, M. (2017). Vers une théorie du pâtir communicationnel. Sensibiliser Habermas. *Cahiers de recherche sociologique, 62*, 69–108.

Cantelli, F. (2013). Deux conceptions de l'*empowerment*. Politique et Sociétés. *32*(1), 63–87. doi: doi:https://doi.org/10.7202/1018721ar. Accessed 23 February 2020.

Cantelli, F., & Genard, J.-L. (2008). Etres capables et compétents: Lecture anthropologique et pistes pragmatiques, *Sociologie S.* https://journals.openedition.org/sociologies/1943. Accessed 12 May 2019.

Cohen, J. (1996). Procedure and substance in deliberative democracy. In S. Benhabib (Ed.), *Democracy and difference: Contesting the boundaries of the political* (pp. 95–119). Princeton.

De Sousa Santos, B. (2002). *Democratizar a Democracia: Os caminhos da democracia participativa*. Civilização Brasileira.

Doherty, B., & de Geus, M. (Eds.). (1996). *Democracy and green political thought. Sustainability, rights and citizenship*. Routledge.

Eckersley, R. (2004). *The green state: Rethinking democracy and sovereignty*. The MIT Press.

Ferreras, I. (2012). *Gouverner le capitalisme ? Pour le bicamérisme économique*. Presses universitaires de France.

Genard, J.-L. (2020). Un bouleversement radical de nos repères anthropologiques et des conditions de la moralité: le déclin ou la fin de l'exception humaine ?, *Sociologie S.* http://journals.openedition.org/sociologies/13202. Accessed 13 June 2020.

Genard, J.-L., & Cantelli, F. (2010). Pour une sociologie politique des compétences. *Les politiques sociales, 1*(2), 103–120.

Goodin, R. E. (1996). Designing constitutions: The political constitution of a mixed commonwealth. *Political Studies, XLIV*, 635–646.

Habermas, J. (1989). *The structural transformation of the public sphere: An inquiry into a category of bourgeois society*. The MIT Press. (Original German edition 1962).

Habermas, J. (1990). Discourse ethics: Notes on a program of philosophical justification. The MIT Press. (Original German edition 1983).

Habermas, J. (1996). *Between facts and norms: Contributions to a discourse theory of law and democracy*. The MIT Press. (Original German edition 1992).

Jonas, H. (1984). *The imperative of responsibility. In Search of an ethics for the technological age.* The University of Chicago Press. (Original German edition 1979).

Kant, I. (1996). *An answer to the question: What is enlightenment?* ed. Mary J. Gregor. Cambridge University Press. (Original German edition 1784).

Manin, B. (1985). Volonté générale ou délibération. Esquisse d'une théorie de la délibération politique. *Le Débat, 33*(1), 72–94.

Michel, C., Pierre, L., & Barthe, Y. (2009). *Acting in an uncertain world an essay on technical democracy* (G. Burchell, Trans.). MIT Press. (Original French version 2001).

Van Reybrouck, D. (2016). *Against elections: The case for democracy*. Seven Stories Press.

Jean-Louis Genard is a philosopher, doctor in sociology, professor at the Université libre de Bruxelles, co-editor-in-chief of the journal *SociologieS*. He has published numerous books, including *La Grammaire de la responsabilité* (Cerf, 2000), *Action publique et subjectivité* (with F. Cantelli, LGDJ, 2007), *L'éthique de la recherche en sociologie* (with M. Roca i Escoda, 2019), as well as numerous articles on anthropological transformations and developments, responsibility, capacitation and empowerment, ethics, public policies and epistemological and ethical issues specific to the human sciences.

Pathways and Solutions

David C. Eisenhauer

Abstract This entry provides an overview of the need for new pathways and solutions for the Anthropocene that consider ongoing social and environmental transformation and explore the possibilities for vibrant and desirable futures. Designing and implementing such efforts requires both novel governance approaches as well as creative research practices. While great uncertainty characterizes the future, there still remain options for achieving more sustainable and just futures.

The Challenge of the Anthropocene

The design and implementation of pathways and solutions during the Anthropocene is both a challenge to and for governance. A challenge to governance because many governance arrangements are no longer capable of achieving desired and sustainable outcomes (Burch et al., 2019). Moreover, many of the norms, values, and goals embedded within prevailing governance systems have contributed to humanity's interference in the operation of the Earth system (Colloff et al., 2021). A challenge for governance because achieving high level, international human development commitments while staying within planetary boundaries, demands creative and transformative policy interventions. Accomplishing such change will require rigorous, systemic, and socially usable analysis of the emerging, novel conditions of the Anthropocene; the array of plausible alternative governance pathways capable of achieving sustainable outcomes; and the institutional, legal, political, cultural, and economic path dependencies and lock-in effects that make transferring to equitable and sustainable trajectories difficult (Colloff et al., 2021; Werners et al., 2021). In short, effective and equitable pathways and solutions are possible, but will require governance approaches that integrate creativity and cultivate an openness to change (Colloff et al., 2021; Leichenko & O'Brien, 2019).

D. C. Eisenhauer (✉)
Bennington College, Bennington, VT, USA

© The Author(s), under exclusive license to Springer Nature Switzerland AG 2023

N. Wallenhorst, C. Wulf (eds.), *Handbook of the Anthropocene*,
https://doi.org/10.1007/978-3-031-25910-4_209

This entry provides an overview of the design of effective pathways through and solutions to risks and hazards of the Anthropocene. It begins by defining both the terms 'pathways' and 'solutions' in the context of governance during the Anthropocene. Next, the entry provides a brief overview of important global demographic trends associated with the Anthropocene. Following this, the concept of 'Earth System Governance' is introduced along with the notion of a 'safe operating space' for humanity and the need to provide a 'social floor'. A variety of techniques and approaches for designing and implementing pathways through and solutions to the challenges of the Anthropocene are then reviewed and summarized. The entry concludes by highlighting that despite current uncertainty it is important to start acting now.

Defining Pathways and Solutions

The terms pathways and solutions point towards the possibility that it remains possible to both prevent dangerous levels of interference in the Earth System while also effectively adapting to unavoidable change. Thus, while the notion of the Anthropocene suggests that some level of risky and undesirable environmental change is 'locked-in', there are still opportunities for humanity to equitably and sustainably thrive now and into the future. Achieving such outcomes requires careful consideration of key demographic, economic, and environmental trends as well as through collaborative, transdisciplinary, and experimental processes (Werners et al., 2021). Additionally, in the context of the Anthropocene, plans and policies to address key problems—such as climate change, biodiversity loss, poverty, and malnutrition—must consider whether such issues can ever be fully resolved or if they must be constantly managed and monitored.

The concept of 'pathways' signifies that responding to and managing the risks associated with the Anthropocene will be an ongoing process with no clear stopping points (Colloff et al., 2021; Werners et al., 2021). A variety of 'pathways approaches' exist, though they generally share some key attributes—including considering flexibility, path dependency, and sequencing (Werners et al., 2021). The increasing prevalence of the term pathways within practitioner and academic discourse reflects an understanding that responding to the challenges of the Anthropocene will require long-term planning that is constantly being reassessed and reexamined.

'Solutions' are generally more discrete and specific. A solution, though, is likely to only continue to solve an identified problem under certain social, ecological, and climatic conditions. If those conditions change over time—a real possibility during the Anthropocene—then that policy or effort may no longer be adequate to achieve societal objectives. For instance, a restored coastal marsh can help solve issues of flood risk and erosion, but if sea level rise occurs too rapidly that marsh may not be able to persist. Thus, a crucial consideration of any given solution involves questioning the conditions under which it can continue to achieve desired goals and outcomes. Because of the uncertainty regarding how long and well an individual solution might work, recent scholarship has focused on the notion of 'solution space', which can be understood as the "the space within which opportunities and constraints determine why, how, white and who adapts" to environmental change and risks (Haasnoot et al., 2020, p. 37). By understanding how the decision space is shaped, it is possible to expand the range of possible solutions available to a given local.

Key Global Demographic Trends to Consider Within Pathways

The world of 2050 and 2100 will look significantly different than the present due, in part, to shifts in population growth, age, and migration. Understanding where people are likely to reside, what their needs might be, and the resources they could draw upon decades from now is necessary for implementing long-term pathways. The global population is expected to stabilize and, potentially, shrink by 2100. While the trend of slowing population growth rates is occurring in every region of the planet, there are key differences. Sub-Saharan Africa is projected to experience the greatest increase in population during the remainder of the century, while Europe's population is projected to decrease during the same time period. A related trend is an aging global population. For the first time in the history of humanity, there are more people over the age of 65 than there are under the age of 5. Urbanization is a third key demographic trend. Presently, about 60% of the world's population is urban. Projections suggest that basically all future population growth will be urban and, by the end of the century, 80% of the global population will be urban (Norman & Steffen, 2018).

Thus, pathways and solutions must take into consideration that the future will likely include a slowing of population growth, an increasingly older global population, and widespread urbanization. This is especially crucial, as in the next few decades more than a billion new urban residents will require the construction and retrofitting of significant housing, infrastructure, and amenities. The design of these urbanization pathways will have long-lasting implications for the planet, as once built infrastructure will continue to shape urban life and environmental impacts for decades (Norman & Steffen, 2018).

Earth System Governance and Pathways

'Earth System governance' is an emerging area of scholarly and practitioner interest (Burch et al., 2019)—the central notion of which is that governance should aim to maintain a safe operating space for humanity by avoiding dangerous or catastrophic interference within the Earth System (Biermann et al., 2012). The planetary boundaries framework defines nine critical planetary processes that shape the Earth system functioning (Rockström et al., 2009). These processes each have a 'tipping point', beyond which additional change could shift the planet to a new, likely more risky state. Thus, Earth System governance presents a framework in which the objective of pathways and solutions should, be in part, be oriented towards the maintenance of a 'safe operating space' for humanity, which is defined as a Holocene-like planet (Biermann et al., 2012). At the same time, the focus on avoiding crossing planetary boundaries does not prescribe particular solutions (Norman & Steffen, 2018). Recent refinements to Earth System governance have included a focus on achieving a 'social floor' in which all people are able to live a good life (Raworth, 2012).

Principles and Approaches for Designing Pathways and Solutions

Because addressing the causes and consequences of social and environmental challenges during the Anthropocene will be necessarily contextual and contingent, there are no clear and universal solutions and pathways for achieving societal goals (Colloff et al., 2021; Werners et al., 2021). Instead, general principles can be identified as shared goals. Such principles include staying below planetary boundaries (Biermann et al., 2012; Rockström et al., 2009), meeting the needs of the present and future generations (Norman & Steffen, 2018), and principles of justice and equity (Ajibade & Adams, 2019; Colloff et al., 2021; Newell et al., 2021). It is also possible to identify broad design principles and approaches for designing more specific and contextual pathways and solutions. These include, collaboration and co-production, experimentation and learning, a future orientation and positive outlook, fairness and justice, and flexibility and robustness.

Collaboration and co-production at all stages of knowledge production, policy design and implementation, and monitoring allows for the bringing together of a wide array of perspectives, insights, and values into the creation of effective pathways and solutions (Colloff et al., 2021; Werners et al., 2021).

Experimentation and learning point towards a need to devise novel strategies for the Anthropocene and learning from both the successes and failures of such approaches (Werners et al., 2021). How to effectively apply lessons from one location and/or issue to another also requires a willingness to experiment and learn.

A future orientation combined with positive visions is crucial for envisioning plausible and desirable worlds within the Anthropocene (Colloff et al., 2021). By visioning what a desirable future might look like and working backwards, it is possible to identify pathways and solutions that can achieve social and environmental objectives while remaining within planetary boundaries (Werners et al., 2021).

Fairness and justice are important considerations that not only have vital normative dimensions but also directly impact the effectiveness of pathways and solutions

(Ajibade & Adams, 2019; Newell et al., 2021). Policies that are perceived as unfair or unjust are unlikely to receive sufficient support. Justice includes dimensions of distribution, recognition, process, and moral principles (Ajibade & Adams, 2019; Newell et al., 2021).

Flexibility and robustness mean that any proposed pathway and/or solution should take into consideration the deep uncertainty of the future (Haasnoot et al., 2020; Werners et al., 2021). Flexibility entails being able to nimbly respond to surprising social, ecological, and climatic changes; while robustness entails a pathway or solutions performing adequately through a wide range of plausible conditions.

The climatic, ecological, and social conditions of the Anthropocene demand new and effective pathways and solutions. That much is clear. What those pathways and solutions will be, though, is uncertain. There are no clear and easy ways forward that will simultaneously stay within planetary boundaries and meet the needs of the growing global population. At the same time, the pressing nature of the challenges posed by the Anthropocene require starting to act now. The principles and approaches outlined in this entry reflect current understandings of how to start designing novel and effective pathways and solutions. As the Anthropocene progresses, it is a certainty that these techniques will need to be reassessed, tweaked, and expanded. While the future remains uncertain, it is evident that it will be significantly different from the past and present. Opportunities still exist to transform the future in ways that are desirable, sustainable, and just (Leichenko & O'Brien, 2019).

References

Ajibade, I., & Adams, E. A. (2019). Planning principles and assessment of transformational adaptation: Towards a refined ethical approach. *Climate and Development, 11*(10), 850–862.

Biermann, F., Abbott, K., Andresen, S., Bäckstrand, K., Bernstein, S., Betsill, M. M., et al. (2012). Transforming governance and institutions for global sustainability: Key insights from the earth system governance project. *Current Opinion in Environmental Sustainability, 4*(1), 51–60.

Burch, S., Gupta, A., Inoue, C. Y. A., Kalfagianni, A., Persson, Å., Gerlak, A. K., et al. (2019). New directions in earth system governance research. *Earth System Governance, 1*, 100006. https://doi.org/10.1016/j.esg.2019.100006

Colloff, M. J., Gorddard, R., Abel, N., Locatelli, B., Wyborn, C., Butler, J. R. A., et al. (2021). Adapting transformation and transforming adaptation to climate change using a pathways approach. *Environmental Science and Policy, 124*, 163–174. https://doi.org/10.1016/j.envsci.2021.06.014

Haasnoot, M., Biesbroek, R., Lawrence, J., Muccione, V., Lempert, R., & Glavovic, B. (2020). Defining the solution space to accelerate climate change adaptation. *Regional Environmental Change, 20*(2), 1–5.

Leichenko, R., & O'Brien, K. (2019). *Climate and society: Transforming the future*. John Wiley & Sons.

Newell, P., Srivastava, S., Naess, L. O., Torres Contreras, G. A., & Price, R. (2021). *Toward transformative climate justice: An emerging research agenda* (p. e733). Climate Change.

Norman, B., & Steffen, W. (2018). Planning within planetary boundaries. In B. Norman (Ed.), *Sustainable pathways for our cities and regions: Planning within planetary boundaries* (pp. 20–44). Routledge. https://doi.org/10.4324/9781315642482

Raworth, K. (2012). *A safe and just space for humanity: Can we live within the doughnut?* Oxfam.
Rockström, J., Steffen, W., Noone, K., Persson, Å., Chapin, F. S., Lambin, E. F., et al. (2009). A safe operating space for humanity. *Nature, 461*(7263), 472–475.
Werners, S. E., Wise, R. M., Butler, J. R. A., Totin, E., & Vincent, K. (2021). Adaptation pathways: A review of approaches and a learning framework. *Environmental Science and Policy, 116*(November 2020), 266–275. https://doi.org/10.1016/j.envsci.2020.11.003

David C. Eisenhauer is a visiting professor of political economy at Bennington College. He holds a PhD in Geography from Rutgers University. His ongoing research explores the intersections of housing inequity, coastal management, and racial injustice in coastal New Jersey. His research has been published in journals such as *Geoforum, Geography Compass,* and *Weather, Climate, and Society*.

Politics

Ulrich Brand and Alina Brad

Abstract "Politics" is a broad concept which is used in a rather narrow sense to detect processes within and around the state or in a broader sense to look as well at the manifold societal (power) dynamics that constitute the basis and certain corridors of state politics. The different ways in which the term is understood have implications for how the ecological crisis, environmental politics and the Anthropocene are understood. Moreover, a critical understanding of politics helps in countering the de-politicising tendencies of the Anthropocene discourse.

The term *politics* is often vaguely used in political and social sciences as well as by the broader public. An important semantic concept is that something is contested, a decision is needed; politics is often equated with political parties and governments. The original meaning comes from Ancient Greece, where *politics* meant those activities, subjects and objects dealing with affairs of common concern that affected the common living of the *polis*, i.e., the "people", who were at the time largely wealthy, free and male.

We refer to a modern understanding of *politics*, i.e., the organisation of common concerns under conditions of a more or less dominant capitalist mode of (re-)production, of colonialism and a dramatically intensifying appropriation of nature, which leads to the hypothesis of the Anthropocene. And we reflect on how politics needs to be rethought in times of deepening and possibly permanent social-ecological crises. *Politics* is intrinsically linked to concepts such as the *state*, *political power, democracy, society* or *the public* and the realities described and analysed by these concepts (Hay, 2002).

A common understanding of politics refers to the term as an overall concept and, at the same time, distinguishes three dimensions: *Polity* points to the form or structure of the political and refers to institutional aspects (including the state's

U. Brand (✉) · A. Brad
University of Vienna, Vienna, Austria
e-mail: ulrich.brand@univie.ac.at; alina.brad@univie.ac.at

© The Author(s), under exclusive license to Springer Nature Switzerland AG 2023

N. Wallenhorst, C. Wulf (eds.), *Handbook of the Anthropocene*,
https://doi.org/10.1007/978-3-031-25910-4_210

constitution but also ideas about social order, such as the governmental system, political parties, etc.); *policies* are the manifold objects and goals, and particularly state actions (i.e., public policies); *politics* focuses on processes such as political procedures (e.g., elections) and on conflict analysis, the decision-making process of actors as well as power relations around collectively binding decisions.

We can distinguish between narrow and broad interpretations of politics with important implications for how the ecological crisis and environmental politics are conceptualised. A rather narrow understanding of politics refers to the manifold activities for preparing and making binding decisions that are oriented towards the common good of a society at the local, regional and national level but also at an international level (e.g., the European Union or the United Nations). Power is largely attributed to the state, i.e., exercised by legal, financial, military, informational means. Politics as structures and processes should contribute to the fulfilment of the purposes of the state, e.g., to welfare, freedom, peace, an intact environment, security, democracy, that serve all members of a political community.

A classic concept – and quite present in public debates today – is that of Max Weber (1919) and, as early as the sixteenth century, of Niccoló Machiavelli who understood politics as struggle over political, and hence state, power. This perspective focuses on political actors, such as political parties or party leaders and their competition with each other.

More recently, the term *governance* (Bevir, 2012) points out that, since the 1990s, the centrality of the state has been questioned by private enterprises, NGOs, media and others who form an integral part of politics. The state is considered to be only one actor among many. However, governance approaches are criticised for underestimating the particular role of the state in modern capitalist societies and for hiding the fact that other societal actors were already involved in politics before capitalism arose (Brad et al., 2021, see below on critical concepts of politics).

In the approaches outlined, the ecological crisis comes into view mainly as a social and political problem. Environmental politics mean the manifold attempts by governments and other public entities at dealing with the ecological crisis (eventually together with other stakeholders). This takes place, for instance, by establishing environmental ministries (*polity*), environmental *policies* as a result of a variety of conflicts, e.g., between the nuclear power industry and anti-nuclear social movements and NGOs. We can call this *explicit environmental politics*. With the exception of governance approaches, these interpretations of politics consider the state to be a more or less neutral actor "above" society that assures social order and solves problems that arise.

Broader views of politics do also look at the institutional structures, policies and struggles for political power. But additionally, they scrutinise the societal relations of power and domination that have changed throughout history and that are intrinsically linked to politics. These societal structures are reproduced along lines of inequality such as class, gender, *race* or international relations and societal relations to nature. Moreover, power as the ability to impose one's interest on others is complemented by the term domination which points to the socially accepted forms of exercising power.

Particularly, critical interpretations, such as the historical-materialist ones in the tradition of Karl Marx and many feminist approaches, argue that politics and public policies are to be understood in close relationship to their socio-economic and cultural context. Here, the state as a polity is not considered to be a neutral entity above society but a *social relation* (Poulantzas, 2013/1978, Jessop, 1990) which is institutionally, materially and ideologically relatively autonomous and an object of political struggles. At the same time, the structures and actions of the state cannot be explained by themselves but rather by considering social practices and forces, the (changing) social context, the contested functions or tasks of the state in societal reproduction and the process of capital accumulation (for a historical-materialist policy analysis, see Brand et al., 2021a).

When it comes to the ecological crisis and environmental politics, it is crucial to understand politics and the state as the many societal relations to nature that are constantly being organised: health, the use of land, regulations and subsidies for industry, agriculture as well as the production and use of chemicals, etc. Behind modern politics stand the deeply inscribed societal goals of the mastery of nature (Horkheimer & Adorno, 2002; Görg, 2011). Existing states, as well as the European Union, are intrinsically linked via tariffs and taxes to the capitalist economy and a more or less dominant position in the international division of labour.

Ecologically unsustainable structures and processes are deeply rooted in the state apparatus, its personnel and rules, their methods of functioning and their knowledge, and their modes and practices. And the state and its materiality are the result of societal alliances, such as the automotive sector or industrial agriculture and, at the same time, organise and support such alliances.

We can speak here of rather *implicit environmental policies* in the sense that economic and financial policies, agricultural and industrial policies, trade and other international policies are very important for shaping societal relations to nature but do so in an exploitative and destructive sense. Explicit environmental politics tend to remain more or less marginalised.

The state also maps out multiple fields of struggle in the relations of production, through the education process, the assignments of roles to individuals, etc. Therefore, the state is a central site or "strategic field" (Poulantzas, 2013) in which a variety of conflicts are handled and the creation of consensus is facilitated through stabilised and shifting relations of forces and compromises using the means of the state – force, law and regulations, discourses and legitimacy, and material and immaterial resources.

In summary, the polity in the sense of state structures as well as public policies is always contested but, at the same time, contributes to a certain stabilisation of socio-economic, cultural and international relations as well as societal relations to nature – the latter being largely destructive. This is a major reason why politics of social-ecological transformation are so difficult to promote (Brand, 2016).

Reflecting the role of politics for the Anthropocene, various dimensions can be considered.

At a general level, introducing politics into the Anthropocene debate helps to overcome a holistic understanding of an assumed "humanity" which is often considered to be responsible for the deepening ecological crisis. We need to look at political and socio-economic dynamics, interests and conflicts as well as at capitalist and colonial power structures within and beyond the state – including the highly hierarchical knowledge systems. This can counter the de-politicising tendencies within the Anthropocene debate (Bonneuil & Fressoz 2017; Görg et al., 2020; on other problematic assumptions of the Anthropocene concept, cf. Lövbrand et al., 2015).

A narrow understanding of politics helps us recognise why and how the political system finds it difficult to push far-reaching environmental politics. Recent debates on the *green state* (Eckersley, 2020) and the *environmental state* (Hausknost, 2020) argue that politics remain very much oriented towards an ecological modernisation of capitalism which will not lead to dealing effectively with the ecological and climate crises.

A crucial common denominator of critical and broader approaches to politics and the state is to detect the many historical processes that caused and are still causing the ecological crisis (beyond just looking at the problems). Historically, and particularly after World War II and the beginning of the *Great Acceleration*, the state as polity and through its policies contributed to economic growth and the (over-)use of resources and sinks. The "imperial mode of living" (Brand & Wissen, 2021) ensures, in principle, unlimited access to cheap nature and cheap labour in the global North and in the global South and reproduces economic and political structures as well as everyday lives.

In that sense, and here the Anthropocene concept opens interesting modes of thought, the environment itself is a power relation (Bonneuil & Fressoz, 2017; Görg, 2011) and it is more appropriate to speak of societal relations to nature.

Consensus is growing that the politics of an ecological modernisation of capitalism is not sufficient. Much more comprehensive processes of social-ecological transformation are needed. This implies an acknowledgement that politics is crucial for such transformations but that at the same time the state is not the steering centre of society. Crucial social logics such as the capitalist growth imperative and the mastery of nature as well as societal power and property relations lie beyond a mere steering of the state.

Therefore, in critical debates on the Anthropocene, the term Capitalocene was proposed (Moore, 2016) to point to the dominant logics within capitalist societies.

To rethink politics in times of the Anthropocene requires a rethinking and remaking of politics that goes far beyond the state and its policies. It implies the variety of societal and also political conflicts that shape societal power relations, such as capitalist property and class relations, the imperial world market, the capitalist growth imperative and a related imperial mode of living, the restriction of democratic decision-making to the political sphere, to name just a few of the huge challenges.

Challenges ahead are increasing chaos and instability caused by climate and other crises which will likely be accompanied by more social and regional polarisation (some people can better cope with crises than others). Any political orientation towards more or less re-installing the state of the Holocene might miss the severity of the ecological crisis.

Politics under these conditions needs to be rethought: As struggles over and mechanisms for obtaining collectively binding forms of mutual self-limitation and the setting of societal boundaries in order to stay within planetary boundaries (Brand, Muraca, et al., 2021b).

It is a crucial and still an open question as to how and whether the Anthropocene concept and discourse might re-orient politics towards fostering the required social-ecological transformations (Görg et al., 2020).

References

Bevir, M. (2012). *Governance: A very short introduction*. Oxford University Press.

Bonneuil, C., & Jean-Baptista, F. (2017). *The shock of the Anthropocene. The earth, history and us*. Verso.

Brad, A., Brand, U., & Schneider, E. (2021). Environmental governance and the state. In L. Pellizzoni et al. (Eds.), *Elgar handbook of critical environmental politics*. Edward Elgar Publishing. in press.

Brand, U. (2016). How to get out of the multiple crisis? Towards a critical theory of social-ecological transformation. In: *Environmental Values 25*(5), 503–525.

Brand, U., & Wissen, M. (2021). *The imperial mode of living. Everyday life and the ecological crisis of capitalism*. Verso.

Brand, U., Krams, M., Lenikus, V., & Schneider, E. (2021a). Contours of historical-materialist policy analysis. *Critical policy studies, 16*, 279, 296. Published online, July.

Brand, U., Muraca, B., Brad, A., et al. (2021b). From planetary to societal boundaries: An argument for collectively defined self-limitation. *Sustainability. Science, Practice and Policy, 17*, 265–292.

Chakrabarty, D. (2018). Anthropocene time. *History and Theory, 57*, 5–32.

Eckersley, R. (2020). Greening states and societies: From transitions to great transformations. *Environmental Politics, 30*, 245–265.

Görg, C. (2011). Societal relationships with nature – A dialectical approach to environmental politics. In A. Biro (Ed.), *Critical ecologies: The Frankfurt school and contemporary crises* (pp. 43–72). University of Toronto Press.

Görg, C., et al. (2020). Scrutinizing the great acceleration: The Anthropocene and its analytic challenges for social-ecological transformations. *The Anthropocene Review, 7*, 42–61.

Hausknost, D. (2020). The environmental state and the glass ceiling of transformation. *Environmental Politics, 29*, 17–37.

Hay, C. (2002). Divided by a common language? Conceptualising power. *Political analysis. A critical introduction*. Red Globe Press, S. 168–193.

Horkheimer, M., & Adorno, T. W. (2002 [1947]). Dialectic of enlightenment. Stanford University Press.

Jessop, B. (1990). *State theory. Putting the capitalist state in its place*. Polity.

Lövbrand, E., et al. (2015). Who speaks for the future of earth? How critical social science can extend the conversation on the Anthropocene. *Global Environmental Change, 32*, 211–218.

Moore, J. W. (Ed.). (2016). *Anthropocene or Capitalocene? Nature, history, and the crisis of capitalism*. PM Press.

Poulantzas, N. (2013 [1978]). *State, power, socialism*. Verso.

Weber, M. (1919/2013). *Econoy and society*. University of California Press.

Ulrich Brand works as a Professor of International Politics at the University of Vienna. He obtained his doctoral degree at Goethe University Frankfurt and wrote his post-doctoral thesis (second monograph) on the internationalisation of the state at the University of Kassel. He is the author of books and articles on critical international politics, ecological crisis, environmental politics, social-ecological transformations, the imperial mode of living and Latin America. His recent books: *The Imperial Mode of Living. Everyday Life and the Ecological Crisis of Capitalism* (with Markus Wissen, Verso 2021, translated into nine languages) and *Capitalism in Transformation: Movements and Countermovements in the twenty-first Century* (as co-editor, Edward Elgar 2019).

Alina Brad is a Senior Scientist at the Department of Political Sciences at the University of Vienna and author of the book *Palm Oil in Indonesia – Political Economy of a Contested Resource* (published in German by Transcript Publishers). She has published a range of articles dealing with issues of international environmental and resource politics as well as social-ecological transformation. In her current research, she explores the contested integration of negative emissions technologies into EU climate policy.

Scale

Zachary Horton

Abstract This article considers the concept of scale as it is used in several disciplines, from physics to the environmental humanities. It examines the constitutive role scale plays in the original and subsequent framing of the Anthropocene concept and the constellating function of the cosmic zoom, a medial articulation of scale that acts as a vehicle for scalar politics. The article sketches a medial understanding of scale as a reciprocal process by which thinkers make "resolving cuts" that stabilize particular scales for inquiry, revealing nonhuman entities and processes unique to those scales, which in turn challenge the epistemic categories that rendered them visible. The Anthropocene represents a shift in scalar thinking that is both profoundly disorienting and potentially transformative of the human subject.

Scale encompasses all things in the cosmos, but only describes the difference between them. It is not a property; it is a relation. If scale is often conflated with size, this is a reflection of our tendency to treat our perspectives as absolute.

Scale is a spatio-temporal ratio between two or more surfaces. The surfaces can vary topologically from three-dimensional objects to terrains to the flat surfaces of screens and printed maps. In cartography, scale is the ratio between features on a map and features on the surface the map represents. The technology of the map relies upon scalar difference. Similarly, the cinema screen may confront the moviegoer with a face larger than their own, or a phone's screen may render planet Earth the same size as a clumsy thumb. Here the phone, the thumb, and Earth form a scalar relationship. It may be casual or it may be profound. Either way, the relationship is animated by difference: two Earths, one I stand upon and one I pinch with my thumb and forefinger to experience as a whole.

Scale mediates between different forms of knowledge, different ways of knowing the world. All knowledge is scaled, and one of the necessary functions of disciplines is to stabilize a scale of inquiry, to make a resolving cut into manifold

Z. Horton (✉)
University of Pittsburgh, Pittsburgh, PA, USA
e-mail: z.horton@pitt.edu

© The Author(s), under exclusive license to Springer Nature Switzerland AG 2023

N. Wallenhorst, C. Wulf (eds.), *Handbook of the Anthropocene*,
https://doi.org/10.1007/978-3-031-25910-4_211

experience and isolate a scale at which observation can take place (Horton, 2018). Every academic discipline, from physics to geology to anthropology to philosophy, possesses at least one reference scale from which it proceeds (Horton, 2021, p. 124–26). Resolving a particular scale requires establishing a definite relationship between observer and surface of inquiry, and makes knowledge possible.

Different disciplines, even if they share the recursive, self-constituting process of producing resolving cuts, nevertheless frame the question of scale differently. Much of the confusion about scale arises from its differential disciplinary formulations. Not only cartographers, but many others conceive of scale in representational terms as representative ratio. Physicists and members of other natural science disciplines conceive of scale as absolute size domain, a generic measure of the range at which a cluster of inter-related events can be resolved. Biologists and engineers often see scale as a compositional structure of parts to whole. Scale in this conception is a set of relations between components of a system. Finally, mathematicians typically view scale as a process of transformation wherein magnitudes are altered without altering the internal relations of a geometrical form or specified mathematical function. Mathematical scale, then, is conceptually inverse from biological scale, which sees internal compositional relations as directly variant with systemic size. As biologist D'Arcy Thompson puts it in his seminal *On Growth and Form*, "there is an essential difference in kind between the phenomena of form in the larger and the smaller organisms" (Thompson, 1942, p. 36). An insect cannot grow much in size (despite 1950s' cinematic imagination) without its internal structure becoming "an inefficient and inappropriate mechanism" (Thompson, 1942, p. 33). Likewise, "A bacillus lives in a world, or on the borders of a world, far other than or own, and preconceptions drawn from our experience are not valid there" (Thompson, 1942, p. 44).

The concept of the Anthropocene is intimately bound up in dynamics of scale. It heralds a shift in perspective, cause, and effect to larger temporal and spatial scales. When first introducing the concept, Paul Crutzen and Eugene Stoermer evoke human effects at new scales: "Considering these and many other major and still growing impacts of human activities on earth and atmosphere, and at all, including global, scales, it seems to us more than appropriate to emphasize the central role of mankind in geology and ecology by proposing to use the term "anthropocene" for the current geological epoch" (Crutzen & Stoermer, 2000, p. 17).

This saturation of "all" scales by the human is particularly alarming given that we traditionally understand human agency to inhere in a narrow range of scales. The Anthropocene, then, marks a discrepancy between the scales of acknowledged or felt human agency and the scales of human effect. To the extent that the Anthropocene is a problem, it is a problem of scale—specifically, a scalar dysmorphia. Timothy Clark has described this general state as a "derangement of scale," where liberal humanist assumptions of agency are absurdly applied to non-human scalar domains, resulting in an "implosion of scales, implicating seemingly trivial or small actions with enormous stakes while intellectual boundaries and lines of demarcation fold in upon each other" (Clark, 2012). Sensing the urgency of thinking other scales, we nonetheless find that when we attempt to do so within a liberal humanist framework, we produce only scalar derangement. Timothy Morton similarly thinks of the massively large-scale phenomena of the Anthropocene (such as climate change) as "hyperobjects," which "force us to acknowledge the immanence of thinking to the physical" (Morton, 2013, p. 2). Framed this way, the scalar problematics of modernity require a non-modern resolution, a new form of thought rooted in materiality, interdependence, and objecthood.

Whether or not such scalar aporias require a reversal of priority between ontology and epistemology has been the subject of fierce debate within the humanities. Dipesh Chakrabarty has influentially argued for an extension of humanistic thought and practice to geological scales while "Cross-Hatching... species history and the history of capital" (Chakrabarty, 2009, p. 220). Claire Colebrook similarly calls for an enlargement of traditional humanist thought rather than its discretization, emphasizing the need for scaling thought while maintaining collectivities: "rather than express resignation and attribute sublimity to the spectacle that intimates a scale beyond our ken, we need to compose a unity that connects rather than disconnects our affects and actions" (Miller Hillis et al., 2016, p. 119). Donna Haraway has suggested we go further, remodelling thought beyond the human as it scales up by tracing its kinship ties with other species and the forces of the earth that she dubs "the Chthonic ones" (Haraway, 2016). Considerations such as these lead Derek Woods to argue that the human is not properly the subject of the Anthropocene. "Scale critique" leads us instead to follow "terraforming assemblages that draw much of their agency from nonhumans" (Woods, 2014: 134).

These and other scholars grapple with an ineluctable scalar dilemma: We must expand the scale of our thinking, dwelling, and kinship, but every increase of scale—every zoom out—causes difference to become occluded, details to muddy and eventually drop away. The same is actually true in reverse: every zoom in to a smaller scale reveals further details, but in a narrower field of view. Scale, as both technological and conceptual mediation, is fundamentally animated by this drama of resolution (Horton, 2021, p. 44). Scaling up reveals less detail over a wider expanse of time or space, and thus runs the risk of occluding difference, universalizing categories, and re-enabling the symbolic violence of homogenization that the humanities and social sciences have worked so hard in the past 50 years to denaturalize and parse as difference. And yet we know that we must find a way to widen our view. The problematics of scale are the problematics of resolution.

Scale is both a verb and a noun. Processes that "scale" transform along a scalar vector, expanding or shrinking, altering perspective, adapting to larger or smaller spans. Meanwhile, "scales" are particular milieus that have been isolated from the full scalar spectrum—the range of all potential scales—and stabilized for observation, habitation, or exploitation. They are bounded, and function as conceptual or medial tools of access. Scaling phenomena, however, pass through many scales and are thus particularly confounding to track. This is why energy, data, capital, and affect tend to elude and overwhelm us: they move fluidly between scales, while our perspectives remain largely scale-bound. Media produce scales, while scaling is a form of mediation. Scaling media enabled the conditions of modernity and are now our only hope to take its measure.

The most self-reflexive construction of scalar relationships takes the form of the cosmic zoom, a medial project that arrays multiple scales onto a single plane, rendering them epistemically equidistant from a subject (Horton, 2021, p. 92). Though cosmic zooms have existed in some form for eons and are now ubiquitous, two of the most influential were Kees Boeke's illustrated book, *Cosmic View* and Charles and Ray Eames's 1977 film, *Powers of Ten* (Boeke, 1957) (Eames Office, 1977). The cosmic zoom is a method of organizing the world, and embeds a politics. Each instantiation of the cosmic zoom stabilizes a particular set of scales and constellates them into a particular relational configuration of relative value, visibility, and hierarchical structure. Certain scales, or regions of the scalar spectrum, are privileged over others and positioned as more or less central to the overall network. Scalar politics consist of the processes by which such trans-scalar constellations are constructed, reinforced, and reproduced, as well as the particular characterizations of the world they constitute.

One salient difference between competing trans-scalar constellations is the degree to which they highlight scalar difference or collapse it. This can be tactical, as when individual and national scales are conflated by fascism, or strategic, as when technoscientific discourses render all scales as similarly accessible and exploitable by human subjects, through the visage of speculative, scale-collapsing technologies from nanotechnology to geoengineering. At this far end of scalar collapse lies an "essentially imperial" figuration of the world through what Denis Cosgrove identifies as the "Apollonian gaze," which renders all scales as a single homogenous milieu, "proclaiming disinterested and rationally objective consideration across its surface" (Cosgrove, 2003, p. x). This scalar structure animated the imperial logic of the Roman Empire, but also characterizes, for example, the logic of neoliberalism, which seeks "the financialization of everything" (Harvey, 2006, p. 33). We can extend this insight to racism and colonial practices of many stripes.

If scale offers a way to conceptualize the conjoined ideological, infrastructural, and geological forces that characterize the present, it also offers myriad trailheads from which to chart new paths through the scalar spectrum of our world. Just as the human subject can morph and evolve when viewed from the radically different scales of species-being and database-driven media, an apprehension of scale can enable new encounters across scale, speculative or otherwise (Horton, 2021). Every trans-scalar encounter is an encounter with difference. The vitality of such medial meetings, whether they confront us with a "posthuman comedy" of nature's return that punctures our inflated scalar pretensions (McGurl, 2012, p. 550) or a self-reflexive mediation that enables us to "to reflect on *the difference of difference* across various scales" (Kember & Zylinska, 2016, p. 233) has the potential to re-sensitize us to a post-human world. A sensitivity to scalar difference can change the human. This is the potential of the Anthropocene, which must be arrayed against its scalar horrors.

References

Boeke, K. (1957). *Cosmic view: The universe in forty jumps*. John Day Co.
Chakrabarty, D. (2009). The climate of history: Four theses. *Critical Inquiry, 35*, 197–222.
Clark, T. (2012). Derangements of scale. In T. Cohen (Ed.), *Telemorphosis: Theory in the era of climate change* (pp. 148–166). Open Humanities Press.
Cosgrove, D. (2003). *Apollo's eye: A cartographic genealogy of the earth in the western imagination*. Johns Hopkins University Press.
Crutzen, P., & Stoermer, E. (2000). The "Anthropocene". *IGBP Newsletter, 41*, 17–18.
Eames Office. (1977). *Powers of ten (1977)*. Eames Office.
Haraway, D. J. (2016). *Staying with the trouble: Making kin in the Chthulucene*. Duke University Press Books.
Harvey, D. (2006). *A brief history of neoliberalism*. Oxford University Press.
Horton, Z. (2018). The trans-scalar challenge of ecology. *ISLE: Interdisciplinary Studies in Literature and Environment, 26*, 1–22. https://doi.org/10.1093/isle/isy079
Horton, Z. (2021). *The cosmic zoom: Scale, knowledge, and mediation*. University Of Chicago Press.
Kember, S., & Zylinska, J. (2016). Media always and everywhere: A cosmic approach. In *Ubiquitous computing, Complexity and culture* (pp. 226–236). Routledge. https://doi.org/10.4324/9781315781129-31
McGurl, M. (2012). The Posthuman comedy. *Critical Inquiry, 38*, 533–553. https://doi.org/10.1086/664550
Miller Hillis, J., Colebrook, C., & Cohen, T. (2016). *Twilight of the Anthropocene idols*. Open Humanities Press.
Morton, T. (2013). *Hyperobjects: Philosophy and ecology after the end of the world*. University Of Minnesota Press.
Thompson, D. W. (1992). *On growth and form*, Abridged edition (Ed.). Cambridge University Press. (Original edition 1942).
Woods, D. (2014). Scale critique for the Anthropocene. *The Minnesota Review, 2014*, 133–142. https://doi.org/10.1215/00265667-2782327

Zachary Horton is Associate Professor of English and Media Studies at the University of Pittsburgh and director of the Vibrant Media Lab. His research focuses on the intersection of technological mediation, ecology, and scale. His book, *The Cosmic Zoom: Scale, Knowledge, and Mediation* was published in 2021 by the University of Chicago Press.

Social Performance

Sebastian Helgenberger and Grace Mbungu

Abstract Sustainability transformations in the energy industry, a key sector in the race to a zero-carbon economy in the Anthropocene, are still being largely understood from a technological standpoint. We propose that the question of how energy transitions can perform for societies and local communities should move to the centre of the debate, in order to unlock untapped social opportunities of transitioning to renewable energy and to speed up climate action. This article draws on four inter-related research perspectives which have made a considerable contribution to conceptualising social performance: sustainability, climate action, business, and development. Understanding and connecting these four perspectives in a transdisciplinary manner is important in our understanding of what Social Performance means in the real world. It also helps to make the concept applicable and transformational. Against this background we propose a community-centred and participatory social performance approach to energy transitions.

The Paris Climate Agreement and the 2030 Agenda for Sustainable Development, with its Sustainable Development Goals (SDGs), have been rightfully celebrated as global milestones for securing livelihoods and opportunities now and in the future. They, however, lack the societal ownership and traction, necessary to supporting and speeding up transformational processes to prevent a global climate breakdown. Transformations in the energy industry, as key sector in the race to a zero-carbon economy, are still being largely understood from a technological standpoint, focussing on technological innovation and change in supply sources (Renn et al., 2020; Setton, 2020). In many cases the social dimension in energy system transformations is reduced to the question, to what extent people and societies would accept these technological transformations or not, rather than generating broad agency. Examples such as Germany's energy transition to renewables suggest that political and economic ownership of citizens, communities, municipalities, and local businesses in

S. Helgenberger (✉) · G. Mbungu
Institute for Advanced Sustainability Studies (IASS), Potsdam, Germany
e-mail: sebastian.helgenberger@iass-potsdam.de; grace.mbungu@iass-potsdam.de

© The Author(s), under exclusive license to Springer Nature Switzerland AG 2023
N. Wallenhorst, C. Wulf (eds.), *Handbook of the Anthropocene*,
https://doi.org/10.1007/978-3-031-25910-4_212

renewable energy, by participating financially and by participating in planning of local energy solutions, can considerably accelerate energy transitions (IASS, 2013).

We conclude that the question how energy transitions can perform for societies and local communities should move to the centre of the debate to unlock untapped social opportunities of transitioning to renewable energy and to speed up climate action in the energy sector. With the global climate and sustainability agendas, referred to above, this implies increasing the social opportunities for present generations without compromising the abilities of future generation to thrive. Against this background we have proposed a social performance approach to energy transitions and energy system transformations: "Social performance of energy sector investments refers to direct and positive social impacts on the well-being of individuals and communities during the development and implementation of energy projects and the usage of locally generated energy, in both monetary and non-monetary ways. In essence, the social performance approach in energy-sector investments and energy-project development puts the needs and well-being of people – both current and future generations – at the centre of energy development and related investments and activities" (Mbungu & Helgenberger, 2021).

In this article we introduce the social performance approach to energy transitions by portraying different origins of the concepts. We present options to operationalize the concept and discuss how the social performance approach contributes to an opportunity-oriented communication in driving sustainability transformations in the Anthropocene. The question how policy interventions, project development processes, investments, and organisations perform for societies, communities, and individuals, i.e., positively contribute to meeting their needs and aspirations, has been addressed through different strands of research. We highlight four inter-related research perspectives which have considerably contributed to conceptualising social performance: sustainability, climate action, business and development. Understanding and connecting these four perspectives in a transdisciplinary manner will be important to understanding what *social performance* means in the real world and to make the concept applicable and transformational.

The proposed social performance approach builds on a human-centred sustainability perspective, where the social, natural-environmental and economic dimensions act as support systems to facilitate well-being for individuals and communities and opportunity to thrive. As such it is closely connected to the 2030 Agenda on Sustainable Development and the SDGs. Against this background the understanding of social performance emanates from the social sustainability dimension, which we have defined as follows in view of energy sector interventions: "The Social Sustainability of a policy intervention, project development, or investment allows for continuity and long-term perspective by identifying and harnessing social opportunities, and by preventing and mitigating social conflicts and community unrest. Social sustainability is facilitated through prioritising the well-being of people and communities for current and future generations, and by pursuing inclusivity and broad political and economic ownership in the development process and its results" (Mbungu & Helgenberger, 2021). This understanding of social sustainability connects to a strand of scientific literature, which centres upon the societal support for an intervention (cf. Setton, 2020), preserving specific societal values, such as intra- and intergenerational equity or human rights (Widok, 2009) and cooperation (summarized in: European Parliament, 2020).

Addressing the social performance of climate action connects to the policy and research discourse on socioeconomic *co-benefits* or sustainable development (SD) benefits of measures to reduce greenhouse gas emissions: "The term 'co-benefits' refers to simultaneously meeting several interests or objectives resulting from a political intervention, private-sector investment, or a mix thereof" (Helgenberger et al., 2019). While climate policies and action have been debated in view of fairly distributing resulting socio-economic burdens within and across generations and world regions, the co-benefits discourse has contributed to changing the narrative "from burden-sharing to opportunity-sharing" (ibid.). Multiple studies have quantified country-specific opportunities, such as employment opportunities or health cost savings, particularly in the energy sector transformation, which allow to connect to specific political agendas and socioeconomic interests (e.g., IASS and TERI, 2019; IASS et al., 2020). Corresponding to the co-benefits approach in the climate action discourse, the social performance approach calls for quantifiable and policy-directed assessments to reconcile socioeconomic interests and priorities with climate action and induce societal traction for decarbonisation pathways (Helgenberger et al., 2021).

With a view on sustainable development, the social performance of business activities is being addressed along with the environmental performance in the context of *Corporate Social Responsibility* (CSR) activities (Halme et al., 2018). More broadly social performance from a business perspective can be defined building on Wood (1991) in terms of principles of social responsibility of a company to motivate action through socially responsive processes, in managing the firm's societal relationships, and the social impacts as observable outcomes of a firm's action, programs and policies. This broader conceptualisation as Corporate Social Performance (CSP) goes back to a long tradition of management scholarship and has entered the management and organisational theory literature in the 1970s (Wood, 1991). The implementation of social performance principles and practices different industries and companies indicate that social performance is not necessarily attached to sustainable practices: in the energy sector for example, the strive for social performance can be found both in the wind energy industry (e.g., Imbali, 2020) and the mining industry (e.g., ICMM, 2020). Recently social performance categories have also been associated with companies' economic performance and innovation, such as gender-inclusive employment (Hakhu & Helgenberger, 2021).

In earlier works we have insinuated that development interventions for a long time did tend to pursue one-dimensional theories of change and social progress, such as providing technical assistance to facilitate access to electricity but ignoring the wider community needs and social opportunities around an intervention, such as community revenues (Mbungu & Helgenberger, 2021). This perspective builds on important groundwork on putting community well-being, empowerment, social justice, and capabilities in the center of development efforts (Nussbaum & Sen, 1993) and we have suggested for the energy sector that "energy development practices can move from unintended social co-benefits toward intentionally fostering the social performance of local energy projects. Therefore, explicit consideration of the well-being of individuals and communities should form the core of the design, development, and implementation of renewable development policies and projects." (Mbungu & Helgenberger, 2021). Recently, the development perspective has been influenced by the discourse on *just transitions* in the context of decarbonizing industries. Besides giving attention to communities, impacted by changing development pathways – particularly employees in the fossil energy industry – the just transition paradigm strengthens the social performance perspective in development by calling for procedural fairness and participation of impacted communities (Błachowicz et al., 2021).

The social performance approach can be used to compare how different energy options (e.g., a renewable wind park, decentralized energy services, such as solar mini-grids, or a coal-mining site) effectively and comprehensibly improve the lives of people and local communities. An important aspect of the social performance approach is its focus on the direct and local contributions of renewable energy projects in fostering the well-being of individuals, in a manner that reflects their aspirations for a good life (Helgenberger et al., 2021). The approach can be used to assess social progress through local energy projects and foster accountability of related investments by ensuring that projects perform both for the people and the planet. The social performance approach helps to identify concrete intervention points or enablers to ensure and increase the positive contributions of energy-sector investments to the well-being of individuals and communities either in monetary ways, such as local economic value creation and employment, or in non-monetary ways, such as community cohesion and social inclusion.

We have developed the Social Performance Index (SPI) as a community-oriented and participatory tool for assessing investments and policy interventions connected to electricity generation in view of the contributions of energy project development, implementation, and use of energy services: "The Social Performance Index (SPI) for energy sector investments is a tool to systematically assess, monitor, compare and communicate the social performance of energy projects for communities, in terms of the (positive/negative) social impact on the well-being of communities and their members" (Helgenberger et al., 2021). Contrasting the community-centered and participatory co-design approach (ibid) of the presented SPI tool, global assessment approaches and indicator lists of social impact and social sustainability have been developed, which can be used to operationalizing social performance on a general level (for an overview see European Parliament, 2020) or surveying a larger population (such as IASS, 2020). Altogether these approaches contribute to identify and communicate actual and potential social opportunities, which have proved to be important to rally societal support and traction for energy transitions and decarbonizing industries (Nagel, 2021).

References

Błachowicz, A., et al. (2021). I*ncorporating just transition strategies into developing countries NDCs and Covid-19 responses.* Climate Strategies https://climatestrategies.org/wp-content/uploads/2021/07/Incorporating-just-transition-strategies-into-developing-countries-NDCs-and-Covid-19-responses.pdf. Accessed 05 Nov 2021.

European Parliament. (2020). *Social sustainability. Concepts and benchmarks*. Study for the European Parliament's Committee on Employment and Social Affairs.

Hakhu, A., & Helgenberger, S. (2021). Green employment for women. Towards gender-inclusive renewable energy careers. *COBENEFITS Impulse*, IASS Potsdam. https://www.cobenefits.info/resources/green-employment-for-women/. Accessed 05 Nov 2021.

Halme, M., et al. (2018). *W*hen is there a sustainability case for CSR? Pathways to environmental and social performance improvements. *Business and Society, 59*(6), 1181–1227.

Helgenberger, S., Jänicke, M., & Gürtler, K. (2019). Co-benefits of climate change mitigation. In W. L. Filho, A. M. Azul, L. Brandli, P. G. Özuyar, & T. Wall (Eds.), *Climate action: Encyclopedia of the UN sustainable development goals* (pp. 1–13).

Helgenberger, S., Mbungu, G.K., Rodriguez, H., & Nunez, A. 2021. The social performance index (SPI). Assessing and monitoring community well-being through energy-sector investments. *IASS discussion paper*. www.cobenefits.info. Accessed 05 Nov 2021.

IASS. (2013). Beiträge zur sozialen Bilanzierung der Energiewende". *Transdisciplinary Panel on Energy Change* https://www.iass-potsdam.de/sites/default/files/files/report_beitraege_zur_sozialen_bilanzierung_der_energiewende_0.pdf. Accessed 05 Nov 2021.

IASS. (2020). *Social sustainability barometer of the energy transition*. https://www.iass-potsdam.de/en/barometer. Accessed 05 Nov 2021.

IASS and TERI. (2019). *Improving health and reducing costs through renewable energy in India. Assessing the co-benefits of decarbonising the power sector. COBENEFITS Study*. IASS/TERI.

IASS, UfU, IET, and CSIR. (2020). Making the Paris agreement a success for the planet and the people of South Africa. Unlocking the co-benefits of decarbonising South Africa's power sector. *COBENEFITS policy* Report Potsdam/Pretoria. www.cobenefits.info. Accessed 05 Nov 2021.

Imbali, F. (2020). A different wind of change – Harnessing Africa's largest wind project for climate action. *Rapid Transition Alliance*. https://www.rapidtransition.org/stories/a-different-wind-of-change-harnessing-africas-largest-wind-project-for-climate-action/. Accessed 05 Nov 2021.

International Council on Mining and Metals (ICMM). (2020). *Social performance. Mining principles*. https://www.icmm.com/mining-principles/9. Accessed 05 Nov 2021.

Mbungu, G. K., & Helgenberger, S. (2021). The social performance approach: Fostering community Well-being through energy-sector investments. *IASS discussion paper*. https://publications.iass-potsdam.de/pubman/item/item_6001015_1/component/file_6001016/IASS_Discussion_Paper_6001015.pdf. Accessed 05 Nov 2021.

Nagel, L. (2021). Tales of doom, tales of opportunity. How climate communication can help to overcome psychological barriers to sustainable action. *IASS discussion paper*. www.cobenefits.info. Accessed 05 Nov 2021.

Nussbaum, M., & Sen, A. (Eds.). (1993). *The quality of life*. Clarendon Press.

Renn, O., Ulmer, F., & Deckert, A. (2020). Introduction. In O. Renn et al. (Eds.), *The role of public participation in energy transitions* (pp. 201–221). Academic.

Setton, D. (2020). Social sustainability: Making energy transitions fair to the people. In O. Renn et al. (Eds.), *The role of public participation in energy transitions* (pp. 201–221). Academic.

United Nations Environmental Programme (UNEP). (2020). *Guidelines for a social life cycle assessment of products and organizations*.

Widok, A. (2009). Social sustainability: Theories, concepts, practicability. In: *Environmental informatics and industrial environmental protection: Concepts, methods and tools* (pp. 43–51).

Wood, D. J. (1991). Corporate social performance revisited. *Academy of Management Review, 14*, 691–718.

Sebastian Helgenberger leads research projects on the social and economic dimensions of climate action and renewable energy and on the social sustainability of energy transitions at the Institute for Advanced Sustainability Studies (IASS), Potsdam. He studied environmental sciences at Leuphana University Lüneburg and ETH Zurich and holds a PhD in socioeconomics for his work on the relevance of global warming on investment decisions in small and medium enterprises.

Grace Mbungu is a research associate at the IASS Potsdam. Her research focuses on the social dimensions of energy access and transitions towards sustainable energy for all, with a focus on the Global South. She holds a BA in Political Science and Gender Studies as well as a Master's in Public Administration (MPA) with a focus on human rights and international development from Bowling Green State University, Ohio, USA. She holds a PhD from the University of Stuttgart.

Society

Christoph Antweiler

Abstract The discovery of man-made irreversible changes in the geosphere and biosphere has enormous significance for scientific as well as societal debates. It is also relevant for the quick answers needed to the problem from the fields of politics and economics. Talk of the Anthropocene has led modern societies to describe themselves and their dynamics in a new way. As a new world observation formula, the Anthropocene often stands for a diagnosis of a comprehensive crisis of society, culture and humanity. The image of the Anthropocene in modern societies is strongly associated with dystopias, narratives of disease, and largely misanthropic images of humanity. A major concern of social science research is the inequalities in driving forces and differing vulnerability to adverse anthropogenic impacts across humanity. Many social-science works are overly normative. The central causative question in social science is what constitutes Anthropocene societies so far and how socioeconomic formations of the future would have to be designed to produce less permanent changes in the surface spheres of the earth.

As a glance at the list of entries in this handbook will show Anthropocene is a topic eminently relevant to the social sciences in many respects. Thinking in terms of the Anthropocene has demonstrated to natural scientists that humans are not merely external factors disrupting "natural processes". Iit has also shown scholars within the humanities, social scientists, and historians that they can play an important role in addressing the current environmental crisis (Magni, 2019). Early on, individual sociologists sought to make sociology fruitful for Earth systemic issues (McHale, 1970). Sociology, however, has embraced the term Anthropocene more hesitantly and sceptically than scholars in anthropology, cultural studies, and literary studies. Similarly to geology and palaeontology to date, even recent textbooks and handbooks in sociology treat the topic stepmotherly. For example, the Anthropocene appears on only six pages of text in the otherwise comprehensive "Handbook of

C. Antweiler (✉)
Institute of Oriental and Asian Studies, University of Bonn, Bonn, Germany
e-mail: christoph.antweiler@uni-bonn.de

© The Author(s), under exclusive license to Springer Nature Switzerland AG 2023

N. Wallenhorst, C. Wulf (eds.), *Handbook of the Anthropocene*,
https://doi.org/10.1007/978-3-031-25910-4_213

Environmental Sociology" (Caniglia et al., 2021). The inviting call from the natural sciences to historians and the humanities and cultural studies is often viewed critically. Some sociologists emphasize that the talk of the Anthropocene is in danger of a reductive "naturalization" of global problems.

The Anthropocene challenges the social sciences to rethink central concepts. The question arises as to how far the social sciences should or even must open up to the Earth system analysis. Adloff and Neckel illustrate this with a central social science term: structure (Adloff & Neckel, 2020: 10). In sociology, structure is usually understood as permanent sets of rules and resources of human practices. The various theories of practice and materiality have already made clear that social structure is also related to material infrastructure, for example, in that some objects suggest certain actions (affordances). In view of the Anthropocene, however, it becomes clear that not only social and nature interact, but that social structures are also related to structures of the Earth system. So the question arises whether and to what extent biophysical conditions of existence should be part of the structural description of a society. Accordingly, we should no longer speak of social formations or socioeconomic formations, as in Marxism, but of "geosocial formations" or "geosocial classes" (Clark & Yusoff, 2017).

From a cultural studies and sociological perspective, the Anthropocene is both: as much a fact-based real concept as it is a momentous fundamental metaphor of permanent crisis perception. Anthropocene functions as a means of raising attention. From a social science perspective, Anthropocene is a self-description mode of society and a world observation formula similar to the time diagnoses "late modern society", "knowledge society" or "risk society" for individual societies. The theoretical challenge of the Anthropocene for the social sciences can be summed up with a double question: How can the genuinely geological category of the Anthropocene be socialized and anthropologized, and how can the social sciences be geologized (Irvine, 2020, Clark & Szerszynski, 2021: 38–41,46–49)?

In terms of social theory, it is still largely unclear what it means that we are in the Anthropocene. In favor of including at least the living world in the sociological analysis is the fact that the habitability of the planet everywhere depends on other organisms that animate the same habitat. It is only through the cooperation of other organisms that the framework conditions are formed that grant habitability, that is, that make a habitat habitable for humans. If we want to "socialize the Anthropocene" from a cultural and social science perspective, then the findings on extractive environmental use, on the dynamics of complex systems and non-linear dynamics call for "geologizing the social" in reverse, as some sociologists or geographers demand. This can be further theoretically grounded if the theorization of the Anthropocene is informed by ethnological concepts of culture, which explicitly do not reduce culture to specific materialities. Given the Anthropocene, we can thus ask about the anthropogenesis of those social formations which impact heavily on earth's surface.

Since societies' dealings with nature, their respective "nature relations", are constitutive of any society, environmental transformations in the form of socio-ecological crises reverberate on globalized societies. Nature relations exist in socio-politically shaped interactions between societies and nature, insofar as they are constitutive for these very societies and their history. Capitalist societies have a special relationship to nature. In particular, today's variant of globalized capitalism can only reproduce itself by means of nature, the exploitation of raw materials as well as the bodies of people. Nature is thus devalued, as are labor, money, food, energy, care, and even living beings (Moore, 2015). Thinking in terms of social relations of nature follows thinking in terms of metaphors of social metabolism. Using the term "social metabolism," Marxist and other materialist approaches concede that there is no simple opposition between humans and nature. However, in contrast to common cultural studies approaches to the Anthropocene, such as those from the "more-than-human" direction, they maintain the distinction between humans as natural beings, but who are precisely a particular part of nature, and the rest of nature (Foster & Angus, 2016). Moreover, they make a different analytical distinction than those between humans and nature or cultures and nature, namely that between society and nature.

When interpreting the Anthropocene as a self-descriptive formula of society, it is noticeable that graphs and figures serve to support certain self-descriptions by evidence over others, such as the diagnosis that the social system or the planet is "sick." This makes us forget the contingency of the respective descriptive formula or leaves it deliberately hidden. The numbers in dominant narratives about the Anthropocene are not just any numbers, but the partial numbers of Earth system scientists and geologists (Will, 2021). The different ways of socially manufacturing evidence are especially important because the Anthropocene is, after all, placed alongsid other epochs in Earth history, not as a period of human history. While the Anthropocene, according to the currently dominant view, has only the duration of a human lifetime, the other geological periods last tens of thousands of years, such as the Holocene, or many millions of years, such as the Pleistocene.

In many respects, the talk of the Anthropocene has so far been unproductive for social analysis. In his major work on social theory, on "modes of societal existence", Latour suspends the concept of "society" in favor of the "collective" and distinguishes 15 special modes of existence that produce specific world relations among modern collectives (Latour, 2012). The exuberant theoretical language makes connections to established social science or cultural theory approaches difficult. Some sociologists have become thoroughly disillusioned, at least in view of the quite few concrete findings and especially analytical statements on the phenomenon of the Anthropocene.

Some consider Latour's farewell to modernity to be premature and the "flight to Gaia" to be little more than an empty formula. The social-theoretical treatment of the Anthropocene is being transferred into unitary terms such as "Gaia" and hybrid composites such as "NatureCultures." This runs the risk of losing any order-creating categories and thus surrendering the analytical discernment of social theory that would be so important for an understanding of global anthropogenic environmental change (Neckel, 2020: 157, 166). A desideratum is to use the whole sociological set of concepts to analyze the social construction of anthropocenic realities. In view of the manifold hybridization of nature and culture, the effects and side effects, the intended and unintended consequences of social action for the environment and society should be at the center of sociological and anthropological Anthropocene research.

A separate sociological and anthropological problem is the strong normative stance of many texts on the Anthropocene. On the other hand, this normative overstretch in research on the Anthropocene constitutes as such a social fact that would have to be analyzed in social science terms. It is precisely the value orientation that often remains implicit in research on the Anthropocene, which is particularly evident in climate research. From a critical anthropological point of view, it is worth examining how biotopes and sociotopes are often implicitly conflated in the debate about social responses to the Anthropocene.

References

Adloff, F., & Neckel, S. (Eds.). (2020). *Gesellschaftstheorie im Anthropozän*. Campus. (Zukünfte der Nachhaltigkeit, 1).

Caniglia, B. S., Jorgenson, A., Malin, S. A., Peek, L., Pellow, D. N., & Huang, X. (Eds.). (2021). *Handbook of environmental sociology*. Springer Nature (Handbooks of Sociology and Social Research).

Clark, N., & Szerszynski, B. (2021). *Planetary social thought. The anthropocene challenge to the social sciences*. Polity Press.

Clark, T., & Yusoff, K. (2017). Geosocial formations in the anthropocene. *Theory, Culture and Society, 34*(2–3), 3–23.

Foster, J. B., & Angus, I. (2016). Marxism in the anthropocene: Dialectical rifts on the left. *International Critical Thought, 6*(3), 393–421.

Irvine, R. D. G. (2020). *An anthropology of deep time. Geological temporality and social life*. Cambridge University Press. (New Departures in Anthropology).

Latour, B. (2012). *Enquêtes sur les modes d'existence. Une anthropologie des modernes*. Éditions La Decouverte.

Magni, M. (2019). *Au racines de l'Anthropocène. Une crise écologique reflet d'une crise de l'homme*. Le Bord de l'Eau.

McHale, J. (1970). *The ecological context*. George Braziller.

Moore, J. W. (2015). Capitalism in the web of life. *Ecology and the accumulation of capital*. London: Verso.

Neckel, S. (2020). Scholastische Irrtümer? Rückfragen an das Antrropozän. In F. Adloff & S. Neckel (Eds.), *Gesellschaftstheorie im Anthropozän* (pp. 157–168). Campus. (Zukünfte der Nachhaltigkeit, 1).

Will, F. (2021). *Evidenz für das Anthropozän. Wissensbildung und Aushandlungsprozesse an der Schnittstelle von Natur-, Geistes- und Sozialwissenschaften*. Vandenhoeck & Ruprecht. (Umwelt und Gesellschaft, 24).

Christoph Antweiler is a cultural anthropologist with a background in geology and paleontology. His main theoretical interests are socio-cultural evolution, pan-cultural patterns (human universals) and vernacular cosmopolitanism. His main research region is Southeast Asia. Among related book publications are *Inclusive Humanism. Anthropological Basics for a Realistic Cosmopolitanism* (Taipei: National Taiwan University Press, 2012), *Our Common Denominator. Human Universals Revisited* (New York and Oxford: Berghahn 2018) and a new comprehensive 650-pages treatise *Anthropologie im Anthropozän* (Darmstadt: WBG, 2022).

Sovereignty

Pierre-Yves Cadalen

Abstract This article discusses the evolution of sovereignty from the Anthropocene momentum. Most of the ecology-related literature finds the Sovereign State scale unfit to solve the environmental problems; however, States do persist, and it is quite impossible to ignore their central dimension both to international relations (IR) and internal politics. Their sovereignty is strongly limited by ecological dynamics, which paradoxically could need sovereign intervention to be mitigated. Political and social changes relating to the Anthropocene will have to deal with States, as they deal with the entire global society. It is worth wondering if the State/civil society separation can survive the Anthropocene, and the conclusion of this article tackles this important challenge for social sciences as well as political thinking and practices.

Sovereignty has long been a central object of political science and political theory academic discussions. It is generally associated to the European political modernity and the Statist organization defended by Thomas Hobbes in the *Leviathan*, an evolution quite concomitant with the recognition of both internal and external sovereignty through the Westphalia Treaties signed in 1648 (Linklater, 2010). Hence, the definition of the State and its expansion as the main political structure throughout the 19th and 20th centuries were also presented as an expression of the general modernization of the world. Within the IR studies, it created on one side a certain tendency to discriminate "real" States, which fit the European model, from "failed" States, unable to master acquire the monopoly over legitimate violence (Acemoglu & Robinson, 2013). On another side, a package can be made of modernization, capitalist expansion, generalization of the State as a political structure, and the Anthropocene (Dardot & Laval, 2020; Grove, 2019).

P.-Y. Cadalen (✉)
CERI Sciences Po, CRBC UBO, AMURE UBO, Brest, France
e-mail: pierreyves.cadalen@sciencespo.fr

© The Author(s), under exclusive license to Springer Nature Switzerland AG 2023

N. Wallenhorst, C. Wulf (eds.), *Handbook of the Anthropocene*,
https://doi.org/10.1007/978-3-031-25910-4_214

Both perspectives have some interest, but each one presents a major obstacle which impedes their full adoption, particularly while thinking about the links between Anthropocene and sovereignty. The perspective which distinguishes firmly between successful States and failed ones tends to forget that some States endure even if their monopoly over security is permanently contested – one might for instance think about the relation between the Lebanese State and Hezbollah. As for the second perspective, which leaves a choice between cultural, political, economic, or social factors to determine the prevalent causality chain of the Anthropocene, it risks missing fundamental points of articulation between political and economic dynamics. The universalization of the State is a real political phenomenon which implies sovereignty, whether its origins are Westphalian or not (Badie, 1992; Teschke, 2009), cannot be ignored from Anthropocene politics.

Ernst Gellner offers, according to us, a much more productive definition of the State, when he writes the following: "Where there is no division of labour, one cannot even begin to speak of the state. But not every specialism makes a state: the state is the specialization and concentration of order maintenance. The "state" is that institution or set of institutions specifically concerned with the enforcement of order (whatever else they may also be concerned with). The state exists where specialized order-enforcing agencies, such as police forces and courts, have separated out from the rest of social life. They *are* the state" (Gellner, 1983, p. 4). The division of labour is both a precondition for capitalist development and State existence, which is the first advantage of Gellner's definition. The second one is that, despite the existence of competitors, the security and order specialization itself makes a State concrete and consistent. Along the lines drawn by Gellner, we can start by saying the State structures are almost universal today and must be considered in the Anthropocene politics, no matter their concomitance with the capitalist development and the industrial age.

In a stimulating book about the Neolithic period, wondering about the origins of centralized power, James Scott defends the first developments of such a power might have been linked to moving climate conditions in the Mesopotamian alluvial plains, which induced the possibility for the cities to control the peasantry and enslave a labour force necessary to produce a surplus production previously unneeded (Scott, 2019). Anthropocene politics urge us to cross middle-run history with *longue durée* and rethinking sovereignty in our time supposes to integrate Westphalia, the twentieth century State expansion, as well as this perspective coming from the Neolithic age analysis. Sovereignty might be weakened by the Anthropocene, but there can also emerge a social need for political control and collectivization, as a reaction to liberal policies and free-trade globalization, just as Karl Polanyi depicted these resorts in the case of the European 1930's (Polanyi, 2001).

Under the Anthropocene, sovereignty matters. The ecological limits exist as an obvious frontier to capital accumulation, unlimited collective will or the full exercise of reason of State (O'Connor, 1997; Pottier, 2017). It is quite clear sovereignty – as a specific *locus* of power – has never concentrated the entire capacities of human self-regulation, or political power. In other words, it has always been quite limited by the complexities power relations present. It is also the case for property rights, which were never absolute, despite the theoretical and formal insistence of liberal theory and liberal politics to make them so (Crétois, 2020). However, property relations as well as States do have an important role in the play of Anthropocene politics. Murray Bookchin invites us to consider sovereignty in its contemporary limitations. The social ecology he praises for implicates to inscribe sovereignty in these very limitations, and to consider human pretentions to master the course of history cannot ignore bioregional characteristics and the ecosystems' qualities (Bookchin, 2012). This limitation of sovereignty offers a thrilling perspective which can work on the articulation between real democracy, attention to the *milieu*'s reproduction, and sovereignty.

This door is, nevertheless, not the only one our contemporary history can open. "Sovereign is he who decides on the exception" (Schmitt, 2005, p. 1), Carl Schmitt wrote as a famous definition of sovereignty. The exception is a regime particularly fit to times of crisis and scarcity, which is quite well shown by the echo coming from the Mesopotamian alluvial plains. In this first chapter, Carl Schmitt basically argues that the liberal tradition does not integrate the moment when political power is founded, necessarily outside of the constitutional order which is posterior to those moments. This is quite coherent with a realist position adopted before by Nicolas Machiavel and Denis Diderot (Diderot, 1751; Faraklas, 1997; Machiavel, 1998). The point is Carl Schmitt, as a supporter of dictatorship, does not consider this exception in another perspective than an unlimited power exercised by a minority over the people in the State-building process. There is another realist approach of sovereignty which is worth saving under the Anthropocene, which presupposes the possibility of a democratic exercise of sovereignty. This democratic exercise of sovereignty can belong to a state of exception and keeps a democratic character if it inserts itself in the very limitations posed by the social and ecological situation.

This opposition between two realist perspectives on sovereignty, the dictatorial one and the democratic one, draws two distinct paths for humanity in the times to come. The former can take the form of climate emergency and permanent measures of exception which would basically aim at preserving the current conditions of economic accumulation, by force and violence if necessary (Amster, 2015; Angus, 2018). The latter can claim for popular sovereignty and the limitations, by the people, of market freedom, as envisaged by Geoff Mann and Joel Wainwright (Wainwright & Mann, 2018). Those conceptions of sovereignty are opposed and might cross the classical opposition between oligarchy and democracy. Mann and Wainwright expose this opposition by quoting one Marx's relevant statement in the *Contribution to the Critique of Hegel's Philosophy of Law*: "in the juxtaposition of sovereignty of the people and monarchical sovereignty "we are not discussing one and the same sovereignty with its existence in two spheres, but two wholly opposed conceptions of sovereignty … One of the two must be false, even though an existing falsehood." Hegel and Schmitt are right – democracy undoes the very possibility of rule. For them this is democracy's great failure; for Marx and us, however, it is its great promise" (*Ibid.*, p. 178).

This opposition between Schmittian sovereignty and democratic sovereignty is far from anecdotic. It is, on the contrary, a core element of Anthropocene politics. The two paths drawn by this opposition are to be covered by cattle if all is left to an emergency dictatorial dimension of sovereignty (Sloterdijk, 2008), or by humans and citizens if democratic sovereignty can firmly articulate the historical projection of our societies with the social and ecological limitations raised by the Anthropocene. This last conception of sovereignty invites us to overcome the traditional division between State and civil society, as the self-organization it bears as a principle cannot leave the application of equality to the sole political sphere – the recent delirious projects advocated by Ellon Musk and other billionaires, whether it be an autonomous island, a crowd of private satellites in extra-atmospheric space or trips to the moon, definitely urge for an exercise of democratic sovereignty that would reinsert these private actions in defined collective perspectives. Thinking sovereignty at the Anthropocene age actualizes the "re-embedment" advocated by Karl Polanyi, which would strongly nuance the abstract division between State and society, inherited from liberal political theory before all (Constant, 2014). Though there is some truth to the inadaptation of the States – as a scale of power – to the global crisis Humanity currently faces (Latour, 2015; Strange, 2011), it is still a relevant level of action – precisely because it stages this opposition between oligarchical sovereignty and popular sovereignty.

The determinist approach along which no salvation can be found from national policies is to be abandoned for the benefit of a more politics-centred approach. A same circumspection can cease us regarding the "collapse" literature, which seems to forget humans can organize themselves politically and have been doing so for millenaries. In James Scott's last book, already mentioned here, one might find this glimpse of optimism, when he discusses the "barbarian ages" as times that are perceived as dark because human societies were not submitted to a centralized power that needed scriptures and administration to reproduce itself. According to him, those times might nevertheless have been the freest of human history. The rethinking of sovereignty through the Anthropocene can take a great advantage of this barbarian times reconsideration. Democratic sovereignty might be found barbaric for power holders, but Humanity being threatened might find more relevant to consider barbaric the current destructive dynamics and renew a rather old slogan: *Popular Sovereignty or Barbary*.

References

Acemoglu, D., & Robinson, J. A. (2013). *Why nations fail: The origins of power, prosperity, and poverty* (Vol. 32, paperback ed., pp. 154–156). Profile Books.

Amster, R. (2015). *Peace ecology*. Paradigm Publishers.

Angus, I. (2018). *Face à l'Anthropocène: le capitalisme fossile et la crise du système terrestre*. Écosociété.

Badie, B. (1992). *L'État importé. Essai sur l'occidentalisation de l'ordre politique*. Fayard.

Bookchin, M. (2012). *Qu'est-ce que l'écologie sociale ?* Atelier de la création libertaire.

Constant, B. (2014). *De la liberté des Anciens comparée à celle des Modernes*. Berg international.
Crétois, P. (2020). *La part commune: Critique de la propriété privée*. Paris.
Dardot, P., & Laval, C. (2020). *Dominer: enquête sur la souveraineté de l'État en Occident*. La Découverte.
Diderot, D., 1751. Autorité politique.
Faraklas, G. (1997). *Machiavel: le pouvoir du prince*. Presses universitaires de France.
Gellner, E. (1983). *Nations and nationalism. New perspectives on the past*. Cornell University Press.
Grove, J. V. (2019). *Savage ecology: War and geopolitics at the end of the world*. Duke University Press.
Latour, B. (2015). *Face à Gaïa: huit conférences sur le nouveau régime climatique*, Les Empêcheurs de penser en rond. La Découverte : Les Empêcheurs de penser en rond.
Linklater, A. (2010). The English school conception of international society. Reflections on Western and non-Western perspectives. *Ritsumeikan Annual Review of International Studies, 9*, 1–14.
Machiavel, N. (1998). *Œuvres*. Laffont.
O'Connor, J. (1997). *Natural causes*. The Guilford Press.
Polanyi, K. (2001). *The great transformation*. Beacon Press.
Pottier, A., 2017. « Le capitalisme est-il compatible avec les limites écologiques? ». Prix Veblen du jeune chercheur.
Schmitt, C. (2005). *Political theology: Four chapters on the concept of sovereignty*. University of Chicago Press.
Scott, J. C. (2019). *Homo Domesticus: une histoire profonde des premiers États*. La Découverte.
Sloterdijk, P. (2008). *Règles pour le parc humain: une lettre en réponse à la "Lettre sur l'humanisme" de Heidegger*. Éd. Mille et une nuits.
Strange, S. (2011). L'échec des Etats face à la mondialisation. *Esprit, 12*, 62–75.
Teschke, B. (2009). *The myth of 1648: Class, geopolitics, and the making of modern international relations* (paperback edition. ed.). Verso.
Wainwright, J., & Mann, G. (2018). *Climate leviathan: A political theory of our planetary future*. Verso.

Pierre-Yves Cadalen is currently post-doc researcher at the CRBC – Université Bretagne Occidentale and associate to the CERI-Sciences Po. He recently obtained his PhD in Political Science and International Relations. His works are mainly related to the relations of power around the Environmental Commons. His former studies, which also included a Bachelor in Philosophy, led him to theorize the new forms of power related to the Anthropocene era. Recent publications: « L'Amazonie et le vivant à l'épreuve de l'écopouvoir », *Raisons politiques*, n° 80, 2020, pp. 77–90; « Le populisme écologique comme stratégie internationale », *Critique internationale*, n° 89, 2020, pp. 165–183; « Republican populism and Marxist Populism: Perspectives from Ecuador and Bolivia », in *Discursive Approaches to Populism Across Disciplines – The Return of Populists and the People*, Palgrave MacMilan, 2020.

Sustainable Development

Daniel Curnier

Abstract This article briefly describes the historical and political emergence of the idea of sustainable development, as well as its conceptual roots. It then criticizes the ideological assumptions that drove its adoption by the international community, its dissemination and its retrieval by civil society and the private sector. This critical review relies particularly on the half-century-old question of limits to growth and on academic debates distinguishing strong and weak sustainability. It finally explores the anthropological and ontological consequences of a development model embedded in a strong interpretation of sustainability, founded on a non-dualistic approach of human-nature relationships.

The most commonly encountered definition of sustainable development is a short quotation from the 1987 World Commission on Environment and Development report to the UN: "Sustainable development is development that meets the needs of the present without compromising the ability of future generations to meet their own needs" (WCED, 1987, Chap. 2). Named after the chairwoman of the Commission, the *Bruntland Report* extends considerations on environmental challenges that had emerged during the 1960s. Some studies had already then demonstrated the existence of ecological limits to human activities and the impossibility of an infinite economic growth (Boulding, 1966; Meadows et al., 1972). The report also followed the idea of "ecodevelopment" (Sachs, 1974) that had been discussed during the first United Nations Conference on the Human Environment held in Stockholm in 1972. But the US administration had in the meantime buried this idea in the context of the rise of neoliberal ideology (Berr, 2009). The *Bruntland Report* therefore emerged from the pragmatic efforts of Scandinavian countries to bring socio-ecological challenges back to the negotiation table.

D. Curnier (✉)
University of Lausanne, Lausanne, Switzerland
e-mail: daniel.curnier@eduvaud.ch

© The Author(s), under exclusive license to Springer Nature Switzerland AG 2023

N. Wallenhorst, C. Wulf (eds.), *Handbook of the Anthropocene*,
https://doi.org/10.1007/978-3-031-25910-4_215

Adopted by the United Nations members at the 1992 Rio Earth Summit, the report's intra- and intergenerational ethics became a global but specific understanding of public good. Indeed, this understanding aims at squeezing environmental constraints into the box of neoclassical economic theories while pursuing reduction of inequalities. The public sector, companies and civil society enthusiastically adopted the concept.

The dissemination of the idea of sustainable development following the Summit has had several positive effects. First, it enabled the propagation of environmental concerns. Second, it helped the implementation of measures to reduce the environmental impacts per unit produced. Third, the notion of needs of future generations made the long-term issue increasingly important. Finally, it stimulated academic research on the concept of sustainability. Conducting research notably allowed for the distinction between two categories of interpretations of sustainability: weak sustainability (which maintains the course of productivism) and strong sustainability (which advocates for a need for a paradigm shift) (Norton, 2005).

However, almost 30 years after its adoption by the United Nations member states there is no choice but to conclude that sustainable development is a failure: inequalities have not ceased to increase and environmental degradation has accelerated. The causes of this failure are rooted both in the *Bruntland Report* and in its political retrieval. The WCED's pragmatic approach resulted in a lack of conceptual clarity, especially when suggesting a possible balance between economic efficiency, social equity and environmental compliance. By refusing to set priorities between economic means, social goals and biophysical boundaries, the idea of sustainable development led to the marginalization of social issues and the subordination of environmental measures to the imperative of a continuously larger production.

Some decision makers and entrepreneurs took advantage of this lack of clarity to divert the concept of sustainable development and melt it into the hypercapitalist discourse. As a matter of fact, greenwashing strategies that misuse adjectives such as "sustainable", "green" or "eco" have helped reproduce the exploitation of the environment and human beings. Moreover, these strategies have hidden the fact that these backward trends are increasing. Multilateral organizations later inflected their discourse towards a "green economy". Preferred to "sustainable development" during the Rio + 20 Earth Summit in 2012, this idea exemplifies the marginalization of social issues and ecological constraints for the benefit of the pursuit of economic growth, hardly altered by slight environment-inspired modifications.

Yet the very economic structures of the model of productivist development are unsustainable: sustainable development is an oxymoron (Latouche, 1999). In fact, the dominant understanding of development, that relies on Rostow's model (1960), defines a path determined by subsequent "stages" that can be reached thanks to economic growth (Rist, 1996). However, an unlimited growth in a finite world is, intrinsically, unsustainable.

The operationalization of sustainable development has been focused on neoclassical economics tools and technical innovation, following the weak interpretation of sustainability that does not challenge the indefinite growth of income (Solow, 1997). Forgetting about social issues, the myth of progress has been mobilized to allegedly solve the question of ecological limits. Thanks to its technical power, the human species is supposed to be able to perpetuate an economic growth decoupled from proportional energetic and material flows, while developing hypothetical innovations to solve resource depletion and ecological unbalance. At its most extreme, the ideology of progress now envisions manipulating the functioning of the Earth system through geoengineering (Barrett, 2008) or setting human beings free from biophysical constraints through transhumanism (Tirosh-Samuelson, 2012).

Sustainable development thus belongs to an "anthropology of extraction" of human beings from nature (Papaux, 2015), strengthening the "naturalist posture" based on the domination of nature specific to modern European thought (Descola, 2013). These Western foundations have moreover been extensively criticized for the reproduction of asymmetric international structures inherited from the process of colonization. This point lead French economist Serge Latouche to call sustainable development a "sham" (2003). But it is indeed the foundations of modernity that explain sustainable development's incapacity to initiate the necessary social transformations. By choosing hypermodernization over demodernization, advocates of the concept chose to reproduce a linear development model that is productivist, technicist and dualist.

This critical analysis of sustainable development raises a number of questions regarding the future. What should be done with the political program embedded in the concept of sustainable development, which has now been assimilated on a global scale? Is it possible to consider the concept of development in a different manner than through the lens of growth? And what would be the consequences of a development approach based on strong sustainability?

A critical discussion of the concept of sustainable development should first help rule out the weak interpretation of sustainability. Not only because of its theoretical fragility, but also because of its incapacity to fulfil its original goal: preventing the crossing into an unstable and uncertain biogeochemical regime that threatens human activities (and perhaps existence as a species). The idea of the Anthropocene (Crutzen, 2002), which seals the end of weak sustainability, did not exist at the time of the publishing of the *Bruntland report* nor of the 1992 Earth Summit in Rio. Yet, if development is to be considered as a social transformation process rather than a quest for an ultimate or superior stage on a linear path, the search for a postmodern development model is still relevant.

Taking the Anthropocene into account requires considering the future in a totally different way from what the paradigm of modernity has been offering and calls for proposing, in the present case, radically new principles for the organization of society. Questioning the dominant development model and imagining alternatives should henceforth stem from the theoretical framework of strong sustainability. This position leads to the definition of priorities that are lacking in the definition of sustainable development: economic activities rejoin their function as a means for serving fair social goals, all of which are integrated in the binding and irreducible biogeochemical context of the Biosphere (Vernadsky, 1926). But what about the anthropological and metaphysical consequences of such a shift?

It may be more interesting to address this issue the other way around. If entering the Anthropocene leads to redefining a human being's relationship to the *milieu* (Berque, 2013), it is hard to imagine how that would not transform its relationship to its body, interiority and metaphysical world, as well as to other human beings. Yet the contemporary issue of development is first and foremost a political question, namely the collective definition of a common social project – at the global scale and enshrined within planetary boundaries (Steffen et al., 2015). A metamorphosis of the relationship to other human and non-human beings should therefore transform individual and collective answers to the fundamental questions that direct development choices which humanity has been facing for a very long time: what is a meaningful life at the individual level? What is a meaningful life at the collective level? How does this articulate with the living conditions of all human dwellers of the Earth today and tomorrow? How does this articulate with the living conditions of all non-human dwellers of the Earth today and tomorrow? And how does all this articulate with planetary boundaries and ecological balances?

The chances are that a system of answers rooted in an "anthropology of immersion" rather than an "anthropology of extraction" (Papaux, 2015) will finally result in the definition of a development model that explicitly reunites personal, social and environmental goals, linked to a holistic relationship to alterity, whether tangible or metaphysical, human or non-human, alive or not.

References

Barrett, S. (2008). The incredible economics of geoengineering. *Environmental and Resource Economics, 39*(1), 45–54.

Berque, A. (2013). *Thinking through landscape*. Routledge.

Berr, E. (2009). Le dévelopemment soutenable dans une perspective post keynésienne: retour aux sources de l'écodéveloppement. *Économie appliquée, 602*(3), 221–244.

Boulding, K. E. (1966). The economics of the coming spaceship earth. In H. Jarrett (Ed.), *Environmental quality in a growing economy* (pp. 3–14). Johns Hopkins University.

Crutzen, P. J. (2002). Geology of mankind. *Nature, 415*, 23.

Descola, P. (2013). *Beyond nature and culture*. University of Chicago Press.

Latouche, S. (1999). The paradox of ecological economics and sustainable development. *Democracy and Nature, 5*(3), 501–509.

Latouche, S. (2003). L'imposture du développement durable ou les habits neufs du développement. *Mondes en développement, 121*, 23–30.

Meadows, D. H., Meadows, D. L., Randers, J., & Behrens, W. W. (1972). *The limits to growth*. Universe Books.

Norton, B. G. (2005). *Sustainability. A philosophy of adaptive ecosystem management*. University of Chicago.

Papaux, A. (2015). Homo faber. In D. Bourg & A. Papaux (Eds.), *Dictionnaire de la pensée écologique* (pp. 536–540). Presses universitaires de France.

Rist, G. (2014). *The history of development. From western origins to global faith*. Zed Books. (Original edition 1996).

Rostow, W. W. (1960). *The stages of growth. A non-communist manifesto*. Cambridge University Press.

Sachs, I. (1974). Environment and styles of development. *Economic and Political Weekly, 9*, 828–837.
Solow, R. M. (1997). Georgescu-Roegen versus Solow/Stiglitz (reply). *Ecological Economics, 22*, 267–268.
Steffen, W., Richardson, K., Rockström, J., et al. (2015). Planetary boundaries: Guiding human development on a changing planet. *Science, 347*(6223), 1259855.
Tirosh-Samuelson, H. (2012). Transhumanism as a secularist faith. *Zygon, 47*(4), 710–734.
Vernadsky, V. I. (1986). *The biosphere*. Synergetic Press. (Original edition 1926).
World Commission on Environment and Development. (1987). *Our Common Future*. Oxford University Press.

Daniel Curnier holds a PhD in environmental science from the University of Lausanne and a MA in development studies with a specialization in global ecology. He teaches geography at a high school for adults, is a freelance researcher and advises Swiss policy makers at the cantonal and federal level. He is specifically interested in ecological economics, ecological taxation, environmental philosophy, ecopsychology, agroecology, education for sustainability and sociology of environmental movements. He is the author of *Vers une école éco-logique* (Le Bord de l'eau, 2021) and *Le Monde d'aujourd'hui, expliqué à mon fils né ce matin* (Metispresses, 2022).

Transactions (Social and Democratic)

Philippe Hamman

Abstract This entry discusses social transactions as modes of "conflictual cooperation" resulting in hybrid products along time sequences where successive adjustments between social groups and actors take place. They relate to conflicts of interests *and* values, which are especially difficult to resolve (conflicts between tradition and modernity, identity and alterity, etc.). In this sense, transactions are also a matrix for action, as in representative democracy. Here bipolar transactions between elected representatives and technicians play a structuring role in the daily decision-making process and cannot be separated from calls for reassessing the citizens' role by having them participate, resulting in tripolar transactions. From a prospective point of view, the production of compromise is a central concern regarding the pressing calls for sustainability. Mobilizing social transactions is a way to approach possible hybridizations in the absence of a single overarching principle of legitimacy and depending on the relative (political, technical, social, economic) capacities of the different actors.

In 1978, Jean Remy, Liliane Voyé and Émile Servais published *Produire ou reproduire?*, a seminal work for the French-language sociology of social transactions that promoted a "third way" between the sociology of action and holism. The authors examined the elaboration of practical compromises in concrete situations where conflict is not limited to an economic opposition, and is characterized instead by modes of "conflictual cooperation", embedded in a dialectic of autonomy and interdependences, resulting in hybrid products (Remy et al., 2020: 19–98). Social life is seen as "a confrontation of multiple actors in partly conflictual relationships, negotiating to establish areas of agreement on the basis of their respective capacities to exert pressure" (Remy et al., 1978: 89).

In this sense, social transactions are temporal sequences of successive adjustments, forming a process of socialization and learning to adjust to the Other, in a

P. Hamman (✉)
University of Strasbourg, Strasbourg, France
e-mail: phamman@unistra.fr

© The Author(s), under exclusive license to Springer Nature Switzerland AG 2023

N. Wallenhorst, C. Wulf (eds.), *Handbook of the Anthropocene*,
https://doi.org/10.1007/978-3-031-25910-4_216

universe structured by opposed couples. This echoes the view developed by Georg Simmel (1950 [1917], Part 1) of society as made of unstable balances and reciprocal actions combining opposite but complementary demands, at odds with the irenic vision of the "end of History".

Tensions exist between a variety of opposed couples: tradition and modernity, identity and alterity, etc., which refer to equally legitimate principles that tend to be mutually exclusive. Social transactions precisely relate to conflicts of interests *and* values, which are especially difficult to resolve when internalized conceptions come into play – all the more so considering these conceptions that vary between actors and social groups. Lastly, intercultural dynamics involve playing with formality and informality – including written or oral agreements depending on the tradition, contracts (with various legal values), and practices that may exist without institutionalized regulation. This is combined with a dialectic of trust and mistrust, both between individuals and institutions and between different groups of actors (Blanc, 2009; Remy et al., 2020).

In this sense, social transactions are also a matrix for action. Jean Remy was involved from the start in the conception and planning of the new university town of Louvain-la-Neuve (Belgium) (Remy et al., 2020: 121–124). In a play on words, Michel Marié suggested a distinction between *aménager* [planning] and *ménager* [accommodating]: "The idea of accommodating was first developed through contacts with planners, in response to the often autocratic and undemocratic character of their methods. (…) Accommodating (…) is taking the greatest care of the subject-object, the one for whom planning is done" (1989: 199). What is at stake in the transaction is to combine the production of space (which entails the risk of the *tabula rasa*) and the management of space (which entails the risk of a heritage-oriented preservation, preventing the implementation of new standards). This goes beyond the bounds of negotiation, which reflects the analysis of a modern society where decisions are taken following set, recognized rules. Olgierd Kuty (2004), for instance, argues that man in democratic societies is a negotiator. The growing rise of uncertainties and the diversity of competing registers of justification now force policy-makers to compromise beyond negotiation, if not beyond what is publicly negotiable, for instance in matters of public health or security (Hamman, 2015).

For the purposes of understanding and action, the concept of social transaction allows us to jointly consider the two types of structuring dimensions of social phenomena, pertaining respectively to the production of objective constraints and possibilities, and to the production of meaning and the perception of what is normal and possible. It takes into consideration the actors' positions both in terms of limitations (social structures, relations of domination) and possibilities (the individual's leeway). The diversity of the actors' rationales explains the indeterminacy of these limitations, in a dialectic of shadow and transparency (Blanc, 2006).

This precisely characterizes contemporary democratic transitions: ambiguity is sometimes required for coexistence compromises to be found (Remy et al., 1978; Blanc, 2006). The interplay of identity and alterity emphasized by Simmel (1950 [1917], Part 1) is a defining feature of citizenship: debating with the Other means recognizing both commonalities and genuine differences. Representative democracy fosters the illusion that only elected representatives, who have a mandate to do so by universal suffrage, can take decisions in the name of the public interest (Schumpeter, 1942). Yet, decisions can be analyzed as the outcome of a bipolar transaction between two principles of legitimacy: representation and technical skill. This is where "experts" come in and form the state and the local authorities' administrative apparatus. This transaction, however, is only valid if it remains "informal and tacit" to citizens (Blanc, 2006: 29). Michel Dobry (2009) defines transactions that are both informal and inter-sectoral as "collusive transactions", which establish a "consolidation network". He sees this as one of the main characteristics of most contemporary polities, especially democratic systems.

The bipolar transactions between elected representatives and technicians cannot be separated from forms of local democracy that call for reassessing the citizens' role in decision-making processes by having them participate. This results in tripolar transactions between legitimacies at different levels, based on representation through suffrage for the elected officials, expertise and competence for the technicians, and direct participation for the citizens, a purportedly "more democratic" legitimacy, sometimes involving mobilization (Blanc, 2006). Unstable compromises ensue, with fluctuating conflicts and alliances both between the three groups of actors and within these groups: for instance, for elected representatives, depending on their political affiliations. This is why these transactions can be called multipolar, even though it is worth noting that interaction does not necessarily have a positive outcome: in terms of reciprocity (Thuderoz, 2017), the process can fail, decisions can be made without mutual concessions, or new options that radically transform each side's positions can emerge.

From a prospective point of view, the production of compromise should be a central concern in any analysis of the increasingly pressing calls for sustainability voiced in our Western, urban societies. For the "sustainable city" (Hamman, 2019b), transactional challenges include "mild" but enduring forms of conflict, an example of which is separate waste collection: collection dates can be set at different intervals depending on types of waste; some types may require depositing at specific collection points, and people who may have inadvertently failed to comply with the proper schedule can have their bags turned down, etc. Residents are targeted by information campaigns and often mobilized by policy-makers to become "eco-citizens" with a commitment to a green energy transition, but they are rarely considered as genuine stakeholders in policymaking (Hamman, 2019a).

Sustainability-related issues, as previously with urban and social development, bring to the fore the inertia of space in the face of the transformations man seeks to impose in space, which binds the socio-spatial and the temporal dimensions together. This suggests a distinction between transactions in continuity and disruptive transactions. The former reflect a managerial rationalization based on three assumptions: trust in technology and science to resolve issues; trust in the economic markets' capacity of self-regulation; trust in the intervention of governments, which are believed to retain control over the issues (Harrison, 2000). Conversely, the latter are based on the premise that only breaking away from the currently prevailing model of development and growth will open avenues for sustainability. In practice, a tension manifests itself between innovations in continuity, for instance to reduce the nuisance and impact of car traffic without fighting car traffic itself (by building ring roads, or delineating 30 km/h speed zones in central neighborhoods, etc.) and disruptive innovations, which propose a change of model (developing alternatives to car use, promoting bicycle use, etc.) (Hamman, 2015, 2019a, 2019b).

The "sustainable city" is therefore both a *transitional* concept – following a dialectic of adaptation/disruption that is characteristic of socioecological change (Lockie et al., 2014: 95–105), differing according to uses and contexts – and a *transactional* concept, relating to coexistence compromises that are always partial and potentially challenged, but also renewed (Hamman, 2019a, 2019b).

Divergences on the scope of actions that should be undertaken have emerged, along with a trend towards the technicization of ecological issues, even though they are ultimately always political (Swyngedouw, 2009). Mobilizing social transactions is a way to approach possible hybridizations in the absence of a single overarching principle of legitimacy and depending on the relative (political, technical, social, economic) capacities of the protagonists. Behind discourses on the global emergency, sustainability does not prevail from above, but is introduced into a variety of "transaction domains" (Frödin, 2015: 453). This also applies at the individual level: reducing one's waste, building a bioclimatic house and giving up on using a car, for instance, are not decisions that are mechanically linked. The hierarchization of values is a delicate matter, with the coexistence of autonomous "spheres of justice" (economy, politics, family, leisure, etc.) that have no equivalence principle. We now find ourselves in a regime of negotiated "complex equality" (Walzer, 1983), pointing to the importance of the role of barriers that are not set in stone, but whose constant readjustments are, conversely, a reflection of the democratic process.

This article received support from the Maison Interuniversitaire des Sciences de l'Homme – Alsace (MISHA) and the Excellence Initiative of the University of Strasbourg, France. The author wishes to thank Jean-Yves Bart for translating from the original French.

References

Blanc, M. (2006). Conflits et transactions sociales: la démocratie participative n'est pas un long fleuve tranquille. *Sciences de la société, 69*, 25–37.

Blanc, M. (2009). L'avenir de la sociologie de la transaction sociale. *Recherches Sociologiques et Anthropologiques, 40*(2), 165–178.

Dobry, M. (2009). Critical processes and political fluidity: A theoretical appraisal. *International Political Anthropology, 2*(1), 74–90.

Frödin, O. (2015). Researching governance for sustainable development: Some conceptual clarifications. *Journal of Developing Societies, 31*(4), 447–466.

Hamman, P. (2015). Negotiation and social transactions in urban policies: The case of the tramway projects in France. *Urban Research and Practice, 8*(2), 196–217.

Hamman, P. (2019a). Local governance of energy transition: Sustainability, transactions and social ties. A case study in north East France. *International Journal of Sustainable Development and World Ecology, 26*(1), 1–10.

Hamman, P. (2019b). Urban sustainability governance and social transactions in France: A social sciences perspective. In P. Hamman (Ed.), *Sustainability governance and hierarchy* (pp. 199–215). Routledge.
Harrison, N. E. (2000). *Constructing sustainable development*. State University of New York Press.
Kuty, O. (2004). A conceptual matrix for negotiation. From bargaining to value negotiation. *Négociations, 1*(1), 45–62.
Lockie, S., Sonnenfeld, D. A., & Fisher, D. R. (Eds.). (2014). *The Routledge international handbook of social and environmental change*. Routledge.
Marié, M. (1989). *Les terres et les mots*. Klincksieck.
Remy, J., Voyé, L., & Servais, É. (1978). *Produire ou reproduire?* Vie ouvrière.
Remy, J., Blanc, M., Foucart, J., Stoessel-Ritz, J., & Van Campenhoudt, L. (2020). *La transaction sociale*. Érès.
Schumpeter, J. A. (1942). *Capitalism, Socialism, and Democracy*. Harper Perennial (Partially republished in 2009: *Can Capitalism Survive? Creative Destruction and the Global Economy*. New York: Harper Perennial Modern Classics).
Simmel, G. (1950). In Wolff Kurt H. (Ed. & Trans.), *The sociology of Georg Simmel* (pp. 1–84). The Free Press. Part 1: Fundamentals problems of sociology. (Original edition 1917).
Swyngedouw, E. (2009). The antinomies of the postpolitical cities: In search of democratic politics of environmental production. *International Journal of Urban and Regional Research, 33*(3), 601–620.
Thuderoz, C. (2017). Why do we respond to a concession with another concession? Reciprocity and Compromise. *Negotiation Journal, 33*(1), 71–83.
Walzer, M. (1983). *Spheres of justice: A defense of pluralism and equality*. Basic Books.

Philippe Hamman is a Professor of urban and environmental sociology, affiliated with the Institute for urbanism and regional planning (IUAR) and the laboratory Societies, Actors and Government in Europe (SAGE) at the University of Strasbourg. His recent publications include: *Cross-border Renewable Energy Transitions: Lessons from Europe's Upper Rhine Region* (ed., Routledge 2021). *Sustainability Governance and Hierarchy* (ed., Routledge 2019). *Sustainability Research in the Upper Rhine Region. Concepts and Case Studies* (ed. with S. Vuilleumier, Presses Universitaires de Strasbourg 2019). *Rethinking Nature. Challenging Disciplinary Boundaries* (ed. with A. Choné, I. Hajek, Routledge 2017).

Transformation

Cécile Renouard

Abstract We highlight three complementary levels of transformation, corresponding to three complementary ways of considering intellectual, practical, moral and spiritual resources in order to achieve the ecological and social transition: (a) every day, (b) structural or institutional, (c) inner, psychological and spiritual. These transformations can only be envisaged if we also reflect on the strategies and postures that will enable them to be implemented, in societies marked by institutional and collective inertia. The three horizons approach (Sharpe B (2013) Three horizons. The patterning of hope. Triarchy Press, London) makes it possible to identify the routes of change by highlighting the dominant aspects of the present situation with the desirable outcome, and by inviting the drawing up of scenarios that transform effectively. Thinking about such scenarios goes hand in hand with questioning the different positions adopted by those seeking strategies for scaling up. (Olin (2010) Real utopias, Verso, London) proposed a grid of three main strategies – rupture, action in the interstices, and symbiotic (or reformist) action – for the establishment of democracies driven by a concern for social and environmental justice and an exit from financialized and extractivist capitalist economic models.

The Anthropocene era places humanity in the face of climatic, environmental, social and economic upheaval. Since the 1960s, engineering sciences, scientific ecology and developmental psychology have used the term resilience to refer to the capacity of people and societies to bounce back from shocks: in fact it is possible to measure only clearly identified disturbances in a short time span, and therefore impossible to study the underlying processes underway in a long time span in climate change. Consequently, the notion of transformability (Renouard et al., 2020) has partially replaced that of resilience, to designate the reorganization of complex systems after strong impacts causing an internal change of their own functions. It is a question of describing the emerging properties of new socio-ecosystems. Such studies are also

C. Renouard (✉)
Campus de la Transition, Forges, France
e-mail: cecile.renouard@campus-transition.org

© The Author(s), under exclusive license to Springer Nature Switzerland AG 2023
N. Wallenhorst, C. Wulf (eds.), *Handbook of the Anthropocene*,
https://doi.org/10.1007/978-3-031-25910-4_217

prospective, because they invite us to identify the levers of change to ensure the conditions of the planet's habitability. They force societies to specify the contours of "maintaining an authentically human life on earth" (Jonas, 1979), of "living well with and for others in just institutions" (Ricoeur, 1992). This research implies defining the different individual, collective, theoretical and practical transformations related to this aim.

We highlight three complementary levels of transformation, corresponding to three complementary ways of considering intellectual, practical, moral and spiritual resources in order to achieve the ecological and social transition.

Everyday' transformations are linked to personal or collective modifications of certain aspects of daily life in order to reduce the carbon and ecological footprint, to fight against pollution and waste. According to a recent report (Dugast & Soyeux, 2019), between 25 and 40% of the changes will be due to modifications in the ways of heating, moving, eating, spending leisure time and consuming. These changes will involve new practices (carpooling and soft mobility, renting services rather than buying goods, eating less meat, locally and in season, lower heating in homes) that can be linked to the exercise of virtues by focusing on the dispositions, representations, emotions and motivations of the actors relating to ways of being and acting. It is about guiding desire through access to both data and the moral imagination that allows one to put oneself in the place of others, and to cultivate democratic emotions, a source of voluntary and unconstrained choices for justice and fraternity (Nussbaum, 1995, 2013).

Structural or institutional transformations correspond to choices in terms of public policies, regulations, and strategic orientations. They can be associated with the normative principles of an eco-justice, concerned with defining a framework (economic, legal, etc.) contributing to economic models (Giraud & Renouard, 2012; Raworth, 2017) and lifestyles that respect planetary limits. These principles concern, in particular, the creation and sustainable and equitable sharing of wealth, by establishing rules that favour a reduction in inequalities that are a source of social and environmental degradation (Laurent, 2020). They also relate to the conditions of participation and representation of affected beings in decisions that concern them, as defined by the all affected principle (Fraser, 2009). In a way, the recognition of such principles corresponds to the perspective developed by thinkers such as Simone Weil, insisting on the obligations towards each human person, as opposed to a rights-centred approach (Weil, 1943). The adoption of such regulatory principles would reinforce the need to transform the rules of the game by identifying those areas in which incentives alone are insufficient.

Inner, psychological and spiritual transformations (Macy, 2009) shape a certain type of relationship to the world, to others, to nature, to oneself and to greater than oneself, which nurtures a culture of disempowered engagement. This perspective goes hand in hand with the exercise of detachment as a condition for self-limitation (Castoriadis, 1983) and an openness to otherness, the source of an adjusted culture.

These transformations can only be envisaged if we also reflect on the strategies and postures that will enable them to be implemented, in societies marked by institutional and collective inertia. The three horizons approach (Sharpe, 2013) makes it possible to situate the routes of change by articulating the present situation in its dominant aspects with the desirable horizon by inviting the drawing of scenarios that are effectively transformative and not cosmetic or rhetorical. The objective is threefold, by creating a space for deliberation and reflective practice: to better understand complex situations; to put oneself in the picture; to arrive at better disagreements (which is, for the author, a way of strengthening the very exercise of democratic life).

It is a question of defining a vision concerning an ideal situation to be reached at a certain horizon (H3), starting from an initial situation (H1), to draw paths to reach it (H2). The path goes from 1 to 3 through 2. We can understand these horizons as different types of postures, mentalities, and different ways of approaching realities: H1 is more the perspective of the manager, the leader, the one who must manage the present situation, business as usual. H3 is the posture of the visionary, the one who does not accept the impasses or failures of the present world, the one who anticipates a better future and draws its features. H2 is the way in which the entrepreneur draws paths, which can be conceived mainly from the old world (H2-) or from the objectives of the desirable horizon H3 (H2+).

Developing a collective approach, on subjects that can be extremely varied, is necessary. The idea is to bring together people with a certain level of knowledge on the subject, so as not to draw up maps that have nothing to do with the possible paths. The method consists in creating a space for dialogue, listening, and shared creativity, starting from the different points of view and successively placing ourselves in each horizon and according to the views of different stakeholders.

The perspective also insists on the epistemology underlying our understanding of the world, on the ways in which we look at reality. Reality is made of both flows and structures. One of the challenges is to be attentive to these flows, to these energies that orient our actions and can contribute to reorient and reshape our structures and institutions. Looking primarily at the flows, at the networks of relationships, allows us to look more freely at the structures and at the way in which they can be reorganized. Knowing that these structures also inform the ways in which we are connected to each other and how we perceive the world.

Thinking about H2+ paths goes hand in hand with questioning the different postures adopted by those seeking strategies for scaling up. Three postures described by Olin (2010) correspond to intellectual, existential and political sensibilities. The American sociologist proposed a grid of three main strategies for the establishment of democracies driven by a concern for social and environmental justice and an exit from financialized and extractivist capitalist economic models. It analyzes five declinations of a project to leave capitalism, mobilizing in different ways the social power of civil society, economic power and state power: state socialism, social-democratic regulation, associative democracy, social capitalism, and social economy. With regard to the transformations implied by such alternative projects to the dominant economic model, the three strategies are rupture, action in the interstices and symbiotic (or reformist) action.

The transformations by rupture seek to get out quickly of the social structures, favouring their destruction and the advent of new ones, thanks to a social empowerment. The perspective of rupture is the one inspired by the conception of Marx, according to which the capitalist system will collapse of itself, thanks to a transformation of the class structures towards the deployment of a proletarian class more and more wide and homogeneous, in which the collective capacities of action grow. According to the author, historically, no transformation by rupture has succeeded in obtaining the construction of perennial democratic institutions. Interstitial transformations correspond to strategies strongly rooted in initiatives coming from civil society. They take place on the margins of capitalist societies and are only slightly threatening, hence a certain acceptability by the elites. Symbiotic transformations are linked to strategies that show the articulation between transformations initiated by popular movements and the resolution of concrete problems for the elites. "The democratization of the capitalist state, for example, was the result of joint pressures and struggles from below that were initially seen as a serious threat to the stability of capitalist rule." (2011) This reading grid is valuable for analyzing social movements and citizen initiatives today, even though their inspirers and actors do not necessarily refer to a Marxian, anarchist or social democratic vision. This is what the sociologist defends, pointing out the insufficiency of each of the options taken in isolation and the need to link these different strategies, while knowing that they can be contradictory in their objectives and their effects.

Thus, the three ways in which Olin Wright situates transformative action can be understood not only as the expression of different ideological options that are understood as irreducible to each other, but also as the expression of different sensibilities and postures that can enter into dialogue and, perhaps, invent synergies.

The transformational challenges to be met are immense. One of the complex aspects concerns the relationship between the particular and the universal in a world marked by the tendency to withdraw into one's own identity, even though the global ecological challenges require an additional level of planetary fraternity. Can we nurture conceptions and practices anchored in territories and contexts, and open to the universal?

References

Castoriadis, C. (1983). The Greek polis and the creation of democracy. In D. A. Curtis (Ed.), *Philosophy, politics, autonomy* (pp. 81–123). Oxford University Press.

Dugast, C., & Soyeux, A. (2019). *Faire sa part. Pouvoir et responsabilité des individus, des entreprises et de l'État face au changement climatique*. Carbone 4.

Fraser, N. (2009). *Scales of justice. Reimaging political space in a globalizing world*. Columbia University Press.

Giraud, C., & Renouard, C. (Eds.). (2012). *20 Propositions pour réformer le capitalisme*. Champs-Flammarion.

Jonas, H. (1979). *The imperative of responsibility: In search of ethics for the technological age*. University of Chicago Press.

Laurent, E. (2020). *The new environmental economics: Sustainability and justice*. Polity Press.

Macy, J. (2009). *Coming Back to life: The updated guide to the work that reconnects*. New Society Publishers.

Olin, W. E. (2010). *Real utopias* (p. 2011). Verso.
Nussbaum, M. (1995). *Poetic justice – The literary imagination and public life*. Beacon Press.
Nussbaum, M. (2013). *Political emotions: Why love matters for justice*. Harvard University Press.
Raworth, K. (2017). *Doughnut economics: Seven ways to think like a twenty-first century economist*. Chelsea Green Publishing.
Renouard, C., Beau, R., Goupil, C., & Koenig, C. (2020). *Manuel de la Grande Transition. Former pour transformer*. Les Liens qui Libèrent.
Ricoeur, P. (1992). *Oneself as one another* (K. Blamey, Trans.). University of Chicago Press.
Sharpe, B. (2013). *Three horizons. The patterning of hope*. Triarchy Press.
Weil, S. (1943). *The need for roots: Prelude towards a declaration of duties towards mankind*. Routledge.

Cécile Renouard is president of the Campus de la Transition, a place of teaching, research and experimentation, which she co-founded in 2017, which intends to promote an ecological, economic and humanistic transition. She is a professor of philosophy at the Centre Sèvres (Jesuit faculty of Paris) and teaches at the Ecole des Mines de Paris and ESSEC Business School. She is the scientific director of the research program "CODEV – Companies and Development", at ESSEC. She is a member of the Catholic congregation of the Religious of the Assumption. Latest publication: *Manuel de la Grande Transition*, (with Rémi Beau, Christophe Goupil and Christian Koenig, dir.), LLL, 2020.

Section V
The Profound Changes in the Long Term: Education, Learning, and Socialization

Nathanaël Wallenhorst and Christoph Wulf

Perfecting the Imperfectable: Historical Perspectives

In modern times in the West the formation of the Anthropocene is closely linked to processes of education and learning, the goal of which is to perfect the individual. This is clear if we look at the history of education in Europe (Wulf, 2022).

Comenius expresses this aspiration clearly in his dictum that "Everyone should learn everything in its entirety". Rousseau takes this further. He emphasises the right of each child to find self-fulfilment through education. With industrialisation and the growth of the middle classes the following become particularly important: the access of concrete individuals to *universal norms*, the *rationalisation* of ways of life, and the principle of *representation.* Education contributes to people's self-empowerment, their increased autonomy, and their perfection of themselves. Instrumental reason governs our emotions and demands that we fit into ways of conducting our lives that are increasingly based on economic principles. This development extends to the externalization of emotions and the training of the body.

Education plays a role in disciplining people and the planet to make them economically more useful. It promotes people's functionality. New modes of behaviour are acquired. People learn to fit into superior spatial and time plans. Along with the development and prevalence of rational modes of behaviour there is an increase in the division of labour, with wage labour becoming more widespread. Moderation and rationality are becoming what society rewards. As we become increasingly dependent on how society functions as a whole, there is a growing pressure on us to

N. Wallenhorst (✉)
Catholic University of the West, Angers, France
e-mail: nathanael.wallenhorst@uco.fr

C. Wulf
Freie Universität Berlin, Berlin, Germany
e-mail: christoph.wulf@fu-berlin.de

exercise self-control. With the internal structure of human beings becoming more differentiated as processes of modernisation take place, the contradictions in society become part of people themselves, so that they must grapple with these internal conflicts more and more. At the centre of this development is the subject, what constitutes him or her, their education, their position in the world and with regard to the self.

The self-empowerment of the individual is becoming a key characteristic of the modern age. The subject is formed by the structures of the Anthropocene and by means of his or her own actions. The right to individual freedom, criticism and autonomous action is promoted in the subject's name. As knowledge becomes more and more rationalised and generalised, the subject's power over nature gets entrapped in pitfalls of his or her own making as he or she exercises their power. Subjects cannot assert their power over nature and society without themselves becoming imprisoned as they exercise this power. This process also inevitably leads to the isolation and self-alienation of the subject. Hopes that the conciliatory power of reason might succeed in resolving these antimonies have been only partially fulfilled. The forming of identity as a task of education rests on the repression of nature both inside and outside us. The self-identical self can only be acquired through the sacrifice of nature, i.e. through breaking off communication with nature that has become anonymous to the self as a result of these processes. The need to master nature rationally results in the need to expose oneself to an educational process which produces the isolated, abstract subject who is focussed on the accumulation of power (Kamper & Wulf 1994).

These thoughts illustrate the importance of education, learning and socialization in the developments of the Anthropocene. This is equally true of both negative and positive results. Corrections of the undesirable developments, the reforms associated with this, the implementation of sustainable development and Global Citizenship Education are also dependent on socialization and educational processes. Learning is essential for all Sustainable Development Goals. None of them can be achieved without appropriate educational processes. Extensive learning processes are taking place in all areas. Their focus on sustainability and global citizenship is resulting in the emergence of a new normative orientation of society and culture (Wulf, 2021, 2022).

The Combining of Local and Global Perspectives

Ever since the end of the First World War and the League of Nations and then after the Second World War and the founding of UNESCO there have been several attempts to recognise education as being a global task. However, it soon became evident that global education schemes do not really meet the needs of all the various parts of the world. Education must also pay full consideration to what is going on locally, nationally, and regionally. It is clear today that in education the desired results can only be achieved by combining global and local aspects.

In addition, UNESCO also recognises the value of the educational and socializing effects of culture, in the concrete sense (literature, theatre, performances, visual arts, music). Such a concept of education is therefore different from purely utilitarian ideas that are dominated by economic factors, and it is this that makes it so complex. This is also seen in the three seminal UNESCO reports: the Faure Report "Learning to be" (1972), the Delors Report "Learning – the treasure within" (1996) and "Rethinking education. Towards a global common good?" (2015). What they all have in common is that, in the global society in the Anthropocene epoch, they are attempting to develop a framework for education that has human rights at its core.

The Faure Report deals with the development of two interconnected concepts – *"learning society"* and *"lifelong education"*. It underlines the right of all individuals to learn for their own personal, social, economic, political, and cultural development. The Delors Report picks up on these thoughts and develops the four pillars of learning – *learning to know, learning to do, learning to live together, and learning to be*. It also stresses the close connection between the organisation of the society in which people live and the way that people do and can learn. Since the 1970s considerable changes have taken place in global living conditions (UNESCO, 2015). The problematic developments and the negative effects of the Anthropocene have become irrefutable and have made clear the extent to which all people and societies are becoming increasingly interdependent. New kinds of complexity, insecurity and tension have arisen that can only be dealt with globally. Television, internet, mobile technologies and other digital media offer opportunities for culture and education. About 50% of people are connected to each other via the internet, although this can vary greatly depending on region. Increase in mobility results in new abilities and skills. Sustainability and social development result in global transformations and new learning opportunities. This development has resulted in a new diversity of human images, world views and knowledge systems (Wulf, 2013, 2022; Wallenhorst, 2020, 2021, 2022; Wallenhorst & Wulf, 2021, 2023).

From a conceptual angle, some of these challenges are addressed in the 2015 UNESCO report "Rethinking Education". This aims to determine how education and learning are to be understood in the globalised world of the Anthropocene. On the one hand it presents an update of the values, concepts and practices developed by UNESCO. It also develops ideas and strategies for meeting the challenges presented by changes to our planet. Here *sustainability* and *peace* come to take a central place in education. We must understand *education as a common good* and have a corresponding *planning strategy.*

Education for Sustainable Development

A new phase of Education for Sustainable Development began with the World Education Forum in May 2015 in Incheon in Korea, which dealt with a declaration and action programme from 2015 to 2030. The post 2015 agenda, Education for Sustainable Development and Global Citizenship Education, was then decided at the world summit in autumn 2015 in New York. For the first time, the international community set universal goals, which were valid for all nations. The fourth of the 17 Sustainable Development Goals (SDGs) developed the following new vision of education and learning in the world community: "*Ensure inclusive and equitable quality education and promote life-long learning opportunities for all.*" The programme is a "vision of education and development" that is based on human rights, human dignity, social justice, security, cultural variety, and common responsibility. Education is seen as being a "public good" and a basic human right, which must be ensured to achieve peace, human fulfilment and sustainable development. (UNESCO, 2015).

The Action Programme recommends a public school system that lasts12 years. There should be 9 years of compulsory schooling with free, high-quality teaching, covering at least the primary and middle school stages. It is also recommended that there should be free, obligatory pre schooling, lasting at least a year, and that vocational education should be developed and also literacy classes for adults. Education should be inclusive and equitable. "Inclusive" does not merely mean the inclusion of children with disabilities but is broader and is directed against all forms of exclusion and marginalisation. Equality of access and treatment in education automatically follow. In many regions of the world there is still much to be done in the case of girls and women. The quality of education needs to be improved to promote the creativity of children and young people. A reform of teacher training is needed to raise the quality of education. Finally, it is important to see the improvement of education not only in terms of the school system. Vocational education and lifelong learning must be developed as well as informal and non-formal education. 4–6% of the GDP or 15–20% of public spending should be used on education.

In Goal 4 of the Sustainable Development Goals (SDGs), Education for Sustainable Development is expressly mentioned, to "ensure that all learners acquire the knowledge and skills needed to promote sustainable development". A further task is "Global Citizenship Education" (GCED), which is about developing a *common feeling* of global responsibility for the fate of our planet (Wulf, 2021). Furthermore, education in Human Rights, Peace Education, intercultural learning, and education for tolerance and democracy are promoted.

Perspectives for Education in the Anthropocene

Learning, education, and the raising of awareness are important ways of achieving critical and constructive perspectives in the Anthropocene. This is true of sustainability, Global Citizenship and many other cultural and social areas that are important in how we co-exist on our planet. For this reason, the range of areas where education and learning are important is vast. It ranges from birth, childhood, and youth to social, aesthetic and inter-cultural education, to educational processes in institutions and non-institutional and informal contexts. It embraces cognitive and emotional, theoretical, and practical, mimetic and ritual aspects. Embodiment and the formation of habits also play an important role. (Kraus & Wulf, 2022) Having to create oneself through very different and sometimes contradictory or paradoxical learning processes is an ineluctable condition of being human that comes up against many new challenges in the Anthropocene. Processes of learning and learning things differently become intermeshed and define the diversity and quality of our daily lives (Göhlich et al., 2014).

It was the stability and predictability of the Earth's climatic conditions (following the entry into a new geological epoch, the Holocene, 11,700 years ago) that subsequently made agriculture possible in different parts of the world. Then, our mastery of ecosystems made it possible to manage agricultural surpluses, which led to the emergence of civilizations. We then organized modes of transmission and learning, and then our education systems emerged. The change of geological epoch with the transition from the Holocene to the Anthropocene calls for unprecedented challenges for educational communities. The challenge of the twenty-first century is to re-found our education systems. This is so that we learn to work for the sustainability of human life in society and ensure peace and justice in a sustainable world (Leal Filho, 2019; Hétier & Wallenhorst, 2021, 2022a, b, c; Jagodzinski, 2018; Padilla, 2021; Paulsen et al., 2022; Priyadharshini, 2021; Summers & Cutting, 2016; Wallace et al., 2022; Wallenhorst & Pierron, 2019).

References

Göhlich, M., Wulf, C., & Zirfas, J. (Eds.). (2014). *Pädagogische Theorien des Lernens* (2nd ed.). Beltz Juventa.

Hétier, R., Wallenhorst, N. (Eds). (2021). L'éducation au politique en Anthropocène, *Le Télémaque, 58*.

Hétier, R., & Wallenhorst, N. (2022a). The COVID-19 pandemic: A reflection of the human adventure in the Anthropocene. *Paragrana, 30*, 41–52.

Hétier, R., & Wallenhorst, N. (2022b). *Enseigner à l'époque de l'Anthropocène*. Le Bord de l'eau.

Hétier, R., Wallenhorst, N. (2022c). Promoting embodiment through education in the Anthropocene. In D. A. Kraus & C. Wulf (dir.), *Learning bodies – Tact, emotion, performance. A European handbook*. Palgrave MacMillan.

Jagodzinski, J. (2018) (dir.). *Interrogating the anthropocene. Ecology, aesthetics, pedagogy, and the future in question*. Palgrave Macmillan.

Kamper, D., & Wulf, C. (Eds.). (1994). *Anthropologie nach dem Tode des Menschen. Vervollkommnung und Unverbesserlichkeit*. Suhrkamp.

Kraus, A., & Wulf, C. (Eds.). (2022). *Palgrave handbook of embodiment and learning*. Palgrave Macmillan.

Leal Filho, W. (2019) (dir.). *Encyclopedia of sustainability in higher education*. Springer.

Padilla, L.-A. (2021). *Sustainable development in the Anthropocene*. Springer.

Paulsen, M., Jagodzinski J., & Hawke S. M. (2022) (dir.). *Pedagogy in the Anthropocene. Re-wilding education for a new earth*. Palgrave Macmillan.

Priyadharshini, E. (2021). *Pedagogies for the post-Anthropocene – Lessons from apocalypse, revolution and utopia*. Springer.

Summers, D., & Cutting, R. (2016) (dir.). *Education for sustainable development in further education*. Palgrave Macmillan.

UNESCO. (1972). *Learning to be*. UNESCO.

UNESCO. (1996). *Learning. The treasure within*. UNESCO.

UNESCO. (2015). *Rethinking education. Towards a global common good?* UNESCO.

Wallace, M. F. G., Bazzul, J., Higgins, M., & Tolbert, S. (2022) (dir.). *Reimagining science education in the Anthropocene*. Palgrave Macmillan.

Wallenhorst, N. (2020). *La vérité sur l'Anthropocène*. Le Pommier.

Wallenhorst, N. (2021). *Mutation. L'aventure humaine ne fait que commencer*. Le Pommier.

Wallenhorst, N. (2022). *Qui sauvera la planète ?* Actes Sud.

Wallenhorst, N., & Pierron, J.-P. (Eds.). (2019). *Eduquer en Anthropocène*. Le Bord de l'eau.

Wallenhorst, N., & Wulf, C. (Eds.). (2021). *Humains. Un dictionnaired'anthropologie prospective*. Vrin.

Wallenhorst, N., & Wulf, C. (Eds.). (2023). *Handbook of the Anthropocene* (p. xxx). Springer.

Wulf, C. (2013). *Anthropology. A continental perspective*. University of Chicago Press.

Wulf, C. (2021). Global Citizenship Education. Bildung zu einer planetarischen Weltgemeinschaft im Anthropozän. *Vierteljahreszeitschrift für Wissenschaftliche Pädagogik, 97*(2021), 463–480.

Wulf, C. (2022). *Education as human knowledge in the Anthropocene. An anthropological perspective*. Routledge.

Nathanaël Wallenhorst is Professor at the Catholic University of the West (UCO). He is Doctor of Educational Sciences and Doktor der Philosophie (first international co-supervision PhD), and Doctor of Environmental Sciences and Doctor in Political Science (second international co-supervision PhD). He is the author of twenty books on politics, education, and anthropology in the Anthropocene. Books (selection): *The Anthropocene decoded for humans* (Le Pommier, 2019, *in French). Educate in Anthropocene (*ed. with Pierron, Le Bord de l'eau 2019, *in French*). *The Truth about the Anthropocene* (Le Pommier, 2020, *in French*). *Mutation. The human adventure is just beginning* (Le Pommier, 2021, *in French*). *Who will save the planet?* (Actes Sud, 2022, *in French*). *Vortex. Facing the Anthropocene* (with Testot, Payot, 2023, *in French*). *Political education in the Anthropocene* (ed. with Hétier, Pierron and Wulf, Springer, 2023, *in English*). *A critical theory for the Anthropocene* (Springer, 2023, *in English*).

Christoph Wulf is Professor of Anthropology and Education and a member of the Interdisciplinary Centre for Historical Anthropology, the Collaborative Research Centre (SFB, 1999–2012) "Cultures of Performance," the Cluster of Excellence (2007–2012) "Languages of Emotion," and the Graduate School "InterArts" (2006–2015) at the Freie Universität Berlin. His books have been translated into 20 languages. For his research in anthropology and anthropology of education, he received the title "*professor honoris causa*" from the University of Bucharest. He is Vice-President of the German Commission for UNESCO. *Major research areas:* historical and cultural anthropology, educational anthropology, imagination, intercultural communication, mimesis, aesthetics, epistemology, Anthropocene. Research stays and invited professorships have included the following locations, among others: Stanford, Tokyo, Kyoto, Beijing, Shanghai, Mysore, Delhi, Paris, Lille, Strasbourg, Modena, Amsterdam, Stockholm, Copenhagen, London, Vienna, Rome, Lisbon, Basel, Saint Petersburg, Moscow, Kazan, Sao Paulo.

Part XVIII
Developing Subjectification

Bringing Life

Renaud Hétier

Abstract The transmission of knowledge and stimulation of intelligence are, seemingly, not enough to motivate individuals to act to safeguard the future. Thus, the education system must sensitize them, give them a grounding in the living world, and encourage them to feel solidarity with all that lives in the world. In this way, the living can be experienced as vulnerable, and consequently be protected. However, we can go further still and give children the opportunity to participate in life – to sustain it and make it grow. The abstract sense of solidarity then becomes concrete, and the wonder they feel at the life they have helped create serves as the foundation of moral concern for the preservation of that life.

When we look at the matter of education in today's world, we must inevitably look at it in the broader context: that world has entered the Anthropocene, and there is no way back. No longer is education simply a question of pouring information into the receptacles that are children's minds. Rather, it is a matter of preparing them for a most uncertain future, but in view of that uncertainty, how can we prepare them? What are we to prepare our children *for*? How can they live in an environment where the balance is threatened?

Those most affected by the threat of the meltdown of the living world are, of course, future generations. They are *part* of the living world. This potential collapse (Diamond, 2005) can be viewed as a singular event: i.e. the huge-scale collapse of the living, which will push humanity to – or beyond – the brink of extinction. Alternatively, it can be viewed as a multitude of different types of collapse: political, ecological, and so forth. In any case, though, as soon as children are old enough to digest the information presented to them, they will understand that it is their future which is at issue.

R. Hétier (✉)
UCO, Angers, France
e-mail: Renaud.hetier@uco.fr

© The Author(s), under exclusive license to Springer Nature Switzerland AG 2023

N. Wallenhorst, C. Wulf (eds.), *Handbook of the Anthropocene*,
https://doi.org/10.1007/978-3-031-25910-4_218

Today, we speak of solastalgia (Albrecht, 2005), which is the distress and anxiety caused by environmental devastation, or the prospect thereof. So powerful, in fact, is solastalgia, that some people question whether there is any sense in continuing to procreate. This ability to anticipate the future (whether in a positive or a negative light), though understandable, nevertheless presents a problem. From a certain point of view, our civilisation could be said to be *lacking* in anticipation (failing to alter the disastrous course we are on). However, at the same time, we appear to suffer from *excessive* anticipation (tending even to consume the future). Today, 'adults', whose activities have eaten away at the future available to the next and subsequent generations, are in a position to consume *all* the resources available to those future generations, and in so doing, denying their descendants the right to life.

If children are still being born – and will continue to be – then we must ask ourselves how best to raise them to face the impending doom of humanity. There are two main approaches: one based on rationality, and the other based on a way of coexisting harmoniously. The former can notably be described as 'educating towards' various goals; it consists of giving information, inculcating knowledge, and perhaps 'raising awareness' about the environmental issues facing humanity. This approach has a major flaw, which can be seen in the world even today: the adults today shirking their responsibility for the planet, refusing to change their way of life, are doing so in full knowledge of the consequences of their action (or rather, of their inaction); their attitude is not attributable to any *lack* of information. The latter may take a variety of forms (Éducation et Socialisation, 60): meditation (Hagège, 2017), creativity, resonance (Rosa, 2018), or convivialism. It consists of in-depth examination, engaging people to coexist in relation to the world and to others – this is a subject which we shall revisit in this compendium.

We are facing what truly is an unprecedented threat (though civilisations and societies have collapsed in the past, civilisation as a whole has never broken down entirely – the threat we face is on a scale never seen before). In this position, ought we to rely entirely on innovation (in the often-quoted words of René Char, 'our heritage was left to us without a testament')? Could we not draw upon certain preexisting wisdom? Ancient philosophy teaches us to separate body and soul (Platon, 1999), encouraging us to engage in spiritual exercises (Hadot, 2002). Montaigne (2009) surmised that to philosophise is to learn how to die. At least, in the western world, philosophical practice (Charbonnier, 2013) has broadly addressed the question of finitude, but as a matter of individual fate. However, we could cite Heraclitus' idea of singularity, when he points out that 'no man can step into the same river twice'. Oriental wisdom – in particular, the idea of the transience of all things, in Buddhism – underline the idea of wider-reaching finitude.

However, though finitude is the focus of philosophical debate, ought it to be something on which our children's thinking focuses? The question needs to be asked, because engaging the young in philosophical discussion in schools has become legitimate practice (Tozzi, 2015; Chirouter, 2015). After all, children are aware, as adults are, that their lives will ultimately come to an end, eventually (Hétier, 2012). However, ought we not to learn how to live before we think about dying? 'Nature (…) calls upon [the child] to be a man', writes Rousseau (1966, 42), concluding that: 'How to live is the business I wish to teach him' (idem). Thus, to live is prioritised over all other things – in particular, it is more important to live than to be a responsible citizen. In mental terms, it is no simple matter to feel oneself to be alive. Above all, it requires that the desire (to live) not be damaged. Recently, Edgar Morin (2014) has underlined the importance of 'teaching [individuals] how to live', advocating an overarching approach to knowledge – one that is less compartmentalised than in current teaching – so as to give meaning to the effort of acquiring knowledge: thus, being in, or entering into, a relationship becomes a fundamental part of knowing how to live.

This relationship with life – life understood as a contiguous whole, as a web of relationships – poses significant questions about education. The way in which our practices and sensitivities are evolving, in an individualist, consumerist society, is placing the child at the 'centre' of the educational universe. France's official educational texts, published in 1989, spoke of child-centred education. The attention paid to a child by educators, and in particular by parents, risks doing away with meaning. Durkheim wrote that 'the man who cares only for himself, in fact, cares little for himself'. It is the process of the child's heterogenesis that is at issue: heterogenesis which is not necessarily a loss of self-identity. However, we must make the point, as Morin does: the relationship with alterity that the child needs to build is not solely a (social) relationship with other people; it is a way of relating to the universe as a whole.

Certainly, it is important to 'mirror' a child, even in very early infancy ('What does a baby see, when he looks at his mother's face? Generally, what the baby sees is himself' – Winnicott, 1971, 155)). However, once a sufficiently solid sense of self has been instilled, it is essential that a child be shown the world, and also be shown *to* the world. From this point of view, the information conveyed relates to the baby's rights, as a person. At least, this is the case if it is understood as relational articulation (Chalier, 2008). Finally, a sense of life itself – rather than merely an impression of the world – is conveyed. A child who receives – mentally as well as physically – becomes alive, and as such, becomes a part of the living world. Indubitably, children are intuitively aware of this living fabric, which can be seen in their spontaneous interest in the world, and particularly the animal world. Through its constant motion, the world a child sees is indicative of the power of life.

By focusing on the child, ensuring that s/he is sheltered, and in so doing, not allowing them to have formative experiences, we run the risk of preventing them from feeling a part of the larger fabric of the living world. By their very nature, children cannot 'give life' to others, and will not be able to for a long time to come, and as children, they are not responsible for the lives of others, but are nonetheless interested in the lives of other people. Today, in the Anthropocene, might it be possible to go a step further, and explore a vital power which can counteract the dominant trend towards destruction? In other words, at a time when we are destroying the living world on a massive scale, and hesitate to imagine what the future holds for humanity, and to pass on the gift of life by procreating, education could endow children with a new kind of power: the power to bring life through their immediate actions.

It is tricky to present life and destruction as two opposite ends of a scale. We could, in line with Nietzsche and Schumpeter, speak of the dynamic of 'creative destruction'. Death is necessary, for the continuous renewal of life. In terms of mental dynamics, to destroy is a way of being sure of reality, and having the ability to influence that reality (we cannot destroy something that is a mere figment of our imagination – Winnicott, 1971). It is a similar logic which gives creative activities such importance, even from infancy: by creating things, it is possible to feel truly alive, because in doing so, we engage with the living world. Creating an 'indestructible' environment, furnished with unbreakable or replaceable things, does a disservice to a child's sense of aliveness. An indestructible world is an inert world, and could even be a lethal one (in the same way that a cancer cell is lethal).

The priority, in the current climate of emergency, is to allow children to become aware of how vulnerable the world is (Tronto, 1993), and the 'interdependence' between the world and its inhabitants (thus, *The Convivialist Manifesto* claims to be a 'declaration of interdependence'). This involves encouraging situations where the child has power over things (rather than over other people). The term 'power' should be understood in contrast to responsibility – responsibility is something that children do not yet have. It is also more closely related to '*potentia*' (force and strength of the multitude) than to '*potestas*' (the power of authority) (Mairet, in Rousseau, 1996). Francis Imbert praises the educational situations that Rousseau describes in *Émile* as 'a sufficiently stable and consistent environment', in which the child 'can exercise his own control and, by applying it, regulate his own 'power'' (Imbert, 1989, 83). Ultimately, it is a matter of having 'limited but real power' (idem., 84).

The environmental background upon which our educative thinking is based supports this idea. We know, for example, that 'reforesting the earth on a massive scale would be one of the most effective solutions to attenuate climate change, according to a study published in the journal *Science*' (Le Monde, 5 July 2019). Thus, it is a matter of creating life in order to be able to continue to live. At the child's level, this means that all activities by which they can take care of life, and encourage it, are constructive (Montessori, 1958, 57–61). One of the fictional scenarios that Rousseau proposes in *Émile* is telling: planting broad beans (Rousseau, 1966, 119–121). The child and his governor invest in a plot of ground that seems conducive. They take great care of their crop, until 1 day, they discover that it has been destroyed (Robert, the gardener, has reclaimed the plot of land, where he had – unbeknownst to Émile – planted melons). There are many ways to interpret this story, by which the author presents an experience of ownership. However, we could also use it for reflection about care and vulnerability.

The quality of an experience of life is contingent on the interdependence of these two dimensions. To watch something (or someone) grow and thrive as a result of one's care and attention is, of course, a transcendent experience, which cultivates the wonder of life. However, it is also a question of helping the child to break away from the illusion of indestructible and inexhaustible life, beginning with that of 'omnipotent' parents. In that, education is necessary, as it is apparent that many adults do not acknowledge that fragility of life. For the child to experience the death of something of which he has taken care – be it animal or vegetable – gives true meaning to that care, in that only something that is vulnerable requires care. Today, children are widely (and fortunately) acknowledged to be vulnerable creatures needing to be cared for. However, it is vital now that they learn to take care of others even in childhood, leaving behind the sense of their own vulnerability to feel the vulnerability of others (Derycke & Foray, 2018).

References

Albrecht, G. (2005). Solastalgia: A new concept in human health and identity. *Philosophy, Activism, Nature, 3*, 41–55.

Chalier, C. (2008). *Transmettre de génération en génération*. Buschet/Chastel.

Charbonnier, S. (2013). *Que peut la philosophie ?* Seuil.

Chirouter, E. (2015). *L'enfant, la littérature et la philosophie*. L'Harmattan.

Derycke, M., & Foray, P. (2018). *Care et éducation*. Presses universitaires de Lorraine.

Diamond, J. (2005). *Collapse: How societies choose to fail or succeed*. Penguin.

Hadot, P. (2002). *Exercices spirituels et philosophie antique*. Albin Michel.

Hagège, H. (2017). Des modèles du sujet pour éduquer à la responsabilité. Rôles de la conscience et de la méditation. *Éducation et socialisation, 36*, 1–17.

Hétier, R. (2012). La mort à portée d'un enfant. In A. Kerlan & L. Loeffel (Eds.), *Repenser l'enfance* (pp. 315–329). Hermann.

Imbert, F. (1989). *Émile ou l'interdit de la jouissance*. Armand Colin.

De Montaigne, M. (2009). *Essais*. Pocket.

Montessori, M. (1958). *Pédagogie scientifique*. Desclée de Brouwer.

Morin, E. (2014). *Enseigner à vivre*. Actes sud.

Platon. (1999). *Phédon*. Garnier-Flammarion.

Rosa, H. (2018). *Résonance: Une sociologie de la relation au monde*. La Découverte.

Rousseau, J.-J. (1966). *Émile ou de l'Éducation*. Garnier.

Rousseau, J.-J. (1996). *Discours sur l'origine et les fondements de l'inégalité parmi les hommes*. Librairie Générale Française. Introduction de Gérard Mairet.

Tozzi, M. (2015). *Apprendre à philosopher par la discussion: Pourquoi ? Comment ?* De Boeck.

Tronto, J. (1993). *Moral boudaries. A political Argumnt for en ethic of care*. Routledge.

Winnicott, D. (1971). *Playing and reality*. Routledge.

Renaud Hétier is Professor at the Catholic University of the West (UCO). He is Doctor of Educational Sciences. He is the author of books on education, and anthropology in the Anthropocene. Books (selection): *Humanity versus Anthropocene* (PUF, 2021, *in French*); *Cultivating attention and care in education* (PURH, 2020, in French); *Create an educational space with the fairy tales* (Chronique sociale, 2017, *in French*); Education between presence and mediation (L'Harmattan, 2017, *in French*).

Childhood

Renaud Hétier

Abstract Childhood is, strictly speaking, not a biological phenomenon but a cultural one. Historically, the concept appears when society makes a cultural decision to allow a long time for young people to mature, protecting and stimulating them. To achieve this, society must renounce the desire for immediate labour. It was with Jean-Jacques Rousseau, in the mid-eighteenth century, that childhood became the entire focus of an education: protection, a stimulating environment, renunciation of authority, and – no less – the defeat of the temptation to succumb to child tyranny. In today's world, childhood must be thought of in a new way: though our children are often overprotected, they are also overexposed to digital technologies, the stranglehold of which seems to compensate for the constraint of school. It is therefore important to establish a cultural third space, free for all children to inhabit, and from adolescence, to offer them the option to take on genuine responsibility.

Where do the boundaries of childhood lie? Are they contingent upon nature (notably, reaching physical maturity), or culture (the place assigned to the 'child' in this or that society)? 'Nature' itself is a fairly fuzzy criterion, as in developed countries, puberty tends to occur earlier in girls (younger than 10) than in boys. Generally, 'the category of "the child" remains an extremely slippery one' (Buckingham, 2010, 67).

From a cultural standpoint, we see enormous variety in how children are treated between cultures – at both ends of childhood. Beyond what point is the child considered to be a part of the community (*in utero*, when the mother feels her baby growing inside her? When it can be seen on a scan? When it is given a first name? When it is baptised?)? Rouche points out that in Ancient Rome, a father was required to lift his newborn child from the ground and hold them aloft, presenting them to the whole family (known as *Tollere Liberum*, thought to symbolise paternity and acceptance of the child). Failing that, the child would be 'left on a pile of rubbish outside

R. Hétier (✉)
UCO, Angers, France
e-mail: Renaud.hetier@uco.fr

© The Author(s), under exclusive license to Springer Nature Switzerland AG 2023

N. Wallenhorst, C. Wulf (eds.), *Handbook of the Anthropocene*,
https://doi.org/10.1007/978-3-031-25910-4_219

the city gates, soon to perish for lack of care' and exposure (Rouche, 2003, 98). On the other hand, when can childhood be said to be over? Does it include puberty and adolescence? Is it dependent on a legal criterion, such as the age of legal majority (which varies from culture to culture)? Is a 12-year-old still a child if he/she is working, fighting in an armed conflict, or having sex?

Historically, discussion on the matter has been based on Ariès' pioneering work (2014), which raises the issue of representation (in portraits, children are depicted as miniature adults). Above all, though, Ariès raises the question of the place attributed to the child; in very early childhood, attitudes may be less entrenched, given the very high child mortality rate at the time when Ariès was writing. Thereafter, though, children are catapulted into adult life. Lebrun, Venard and Quéniart (1981, 111) relate how 'A child of two, three, four or older is viewed as a person, already on a par with adults in many regards'. Ariès' conclusions have been debated – particularly with respect to parents' attachment to their children. However, those conclusions invite a discussion of three essential issues: representation, attachment, and the child's place.

Postman (1982), notably drawing upon the work of historians, stresses the contrast between the Middle Ages, when what we now view as children were fully embroiled in the intrigues of adult life, and the more contemporary period, when children are separated from adults by the institution of schools. From this perspective, childhood is a stage of life that is characterised by sheltering young people from adult life – in particular, from the sexual aspect of adulthood. In Postman's view (and Postman was writing even before the dawn of the internet) modern media are jeopardising that separation – hence the title of his book: 'The Disappearance of Childhood'. If childhood indeed is that stage of life when the child is unfamiliar with, and protected from, adult life, we still need to consider how to protect children from themselves. School plays a crucial role in sheltering our young, which is aptly underlined by Buckingham: 'Schooling, for example, is a social institution that effectively constructs and defines what it means to be a child – and indeed a child of a particular age' (Buckingham, 2010, 7).

From the eighteenth Century, mentalities began to shift drastically: 'no longer is it a matter of redressing the harshness characteristic of childhood in centuries past; now, we must safeguard that part of [the child] which is spontaneous, innocent and unique' (Morel). Rousseau obviously plays a crucial role in the discussion, with his seminal work *Emile, or on education*, in which he develops a veritable psychology of childhood, recognising that it is a time unlike any other in a person's life. Rousseau helped reshape the way in which children were represented, with detailed descriptions of their needs from infancy onward (e.g. their freedom of movement), by the importance he gives to attachment ('I would like [the governor] himself to be a child, were it possible; that he should become his pupil's companion, and win his trust while sharing his amusement' (Rousseau, 1966, 55), and by his recognition of the need for a separate place: certainly, Rousseau does not advocate schooling, but favours a type of 'governance' whereby the child and his innocence are long sheltered from the realities of adulthood.

The danger comes not only from the child's environment, but also from what Rousseau terms the child's 'fantasy' – his desires (which go beyond his basic needs, and may cause him to become tyrannical). On the basis of *Emile*, Imbert rightly offers psychoanalytical reflection on the child's impulse-driven existence, and what must restrain it. He describes '*mediation systems* by means of which the "natural" tendency toward unlimited desire and imagined omnipotence can be transformed. Thus, the goal of education is for the pupil to learn to exercise power which is *limited but real*. This involves suffering frustration, and also, for the educator, the ability to "say no to a certain regression but say yes to all the rest" (Oury & Pain, 1972, 277)' (Imbert, 1989, 84). In addition: 'It is up to the educative practices to determine the intensity and quality of the "stimuli" offered, so that the child is not simply at the mercy of his impulses, but can find his place as a desirous being. This requires the construction of a sufficiently coherent framework, regulated by unspoken obligations' (Imbert, 1989, 143). We then see the question of authority: 'abandoned by social discourse that would assert his legitimacy and thereby his authority, [the father] seeks, from his own child, the support which has been taken away from him. The result is obvious: the child is protected, by the father, from ever having to test the boundaries. This merely makes it more difficult for the child to grasp those boundaries as part of his psyche' (Lebrun, 2016, 272).

Today, in the age of media and the internet, the issue of whether and how to protect children is again becoming topical. In particular, this is the case because cultural industries may tend more to encourage impulses than to contain them. In Stiegler's analysis, 'with stratified audience capture systems, the aim is to *replace the mental apparatus* of the ego and the id, and the circuits which form – such as transindividuation circuits where the transindividual is identified as the object and product of desire – *with psychotechnologies* which are responsible for attention control, and deal no longer with desire but with impulses' (Stiegler, 2008, 31). The media and the internet pose different problems. The media – chief among them the television – provides content of potentially questionable quality (and quantity). This point is made by Stiegler. Those responsible in media organisations are ultimately accountable for the content they convey, but the anonymity offered by the internet can encourage irresponsible conduct. With the internet, the issue is its very boundlessness, in two senses: there is no limitation on the number of websites, for example, but above all, an unlimited number of users can register. Free access to pornographic content represents a historical regression, as children are, once more, privy to 'adult secrets'. Thus, there is a risk that childhood will be thrown out of balance. In view of this limitlessness, the school (as an institution) is doubly important in its role to create a sufficiently rigid framework. It is still needed, although as Meirieu notes, 'it does not have rituals capable of triggering what Gabriel Madinier calls the "inversion of dispersion" (1953, p. 109). It lacks rituals to help long-term focus on cultural objects. It lacks rituals to impose deferment on our impulses and, in the assumed constraint of temporality, allow the emergence – always possible but never certain – of thought' (Kambouchner et al., 2012, 171).

We live in an age where children are, on the one hand, more protected than ever (particularly in terms of physical and emotional safety), and on the other, more exposed than ever (think of the impulsive pressures of the consumer society and the omnipresence of temptation). We begin to wonder what the future holds for childhood and, correlatively, what is to become of an institution such as the schooling system, which can continue to support childhood to a greater or lesser extent.

To begin with, we must address the apparent paradox of overprotectiveness in combination with 'overexposure' during childhood. From the standpoint of impulse control, we can appreciate the issue. Indeed, overprotectiveness itself lays the foundations for a degree of '*laisser-faire*', because we wish to protect the child from displeasure (and the resulting 'disaffection'). The individualism of our culture is extending to include childhood – evidence of this can be found in the fact that children are given their own private room, and also their own private hardware (e.g. games console, smartphone, television, tablet or computer). As soon as the child has a space of their own and such resources at their disposal, they are free to do whatever they 'want', although they are not yet capable of wanting sensibly. What children 'want', left to the mercy of their impulses, is to avoid stress, and avoid anything which looks like work, and the frustration it brings. What they 'want' is to immediately satiate their impulses – to play, to let off steam with a violent video game, to excite themselves with pornography, to eat, etc. From the standpoint of construction of the psyche, the very purpose of education is to establish guiding frameworks, mediation and cultural objects to channel and transform impulsive energy. This 'liberation of childhood' (Renaut, 2002) has not come about by accident. It is the result of a neoliberal ideology that aims to destroy institutions, which are decried as enemies of freedom.

Whilst families themselves are cogs in the consumerist machine, striving to prevent children from experiencing frustration, the role of school – as an institution charged with supporting patience, development and sublimation – is more important. Today, the tenets of this role are entirely new, as the school must play it in unprecedented conditions – outside of school, children are left to their own impulses. Kambouchner et al., (2012, 175) describes a '*joy of intelligibility* that the school must try to substitute for the *joy of impulse satisfaction*'. As a step in the right direction, we have two additional recommendations.

The first relates to citizenship. As David Buckingham did, we can advocate a media-based form of citizenship, which takes account of children's media exposure. 'As Richard Lindley suggests, a number of the restrictions imposed on the youngest children are no longer necessary for adolescents; and might even hinder their efforts to take control of their own lives' (Buckingham, 2010, 209). Particularly in view of their access to information, adolescents could be included in a number of civic decision-making processes (which would also encourage them, in turn, to engage more fully with the information available to them). Such an exercise in citizenship would be worthwhile running at school, as has been argued by Dewey and the proponents of institutional pedagogy.

The second recommendation has to do with culture. Culture should not be understood (solely) as a set of content, but above all, as a set of content vehicles. We see this idea in the writing of Imbert and Meirieu (mediations, rituals), and Winnicott points to the notion of an 'intermediary space'. Children need to be able to play and exercise their imaginations, interacting with cultural objects which they appropriate as their own in the process, developing their sensitivity to and intelligence in relation to the world. For this idea to work, school must be an arena for the transmission of these objects, and the way in which they are conveyed must be independent of the learning objectives, which are separate. These cultural content vehicles are all the more important if the alternative is to simply surrender to impulse. However, rather than repelling impulsive energy and motivation, the content vehicles must absorb and transform them – in much the same way as traditional tales make room for anthropological archaisms, give shape to them, and then transcend them (Lafforgue, 1995; Boimare, 1999).

References

Ariès, P. (2014, [1975]). L'enfant et la vie familiale sous l'Ancien régime. Points.

Boimare, S. (1999). *L'enfant et la peur d'apprendre*. Dunod.

Buckingham, J. (2010). *Chilhood and consumer culture*. Palgrave Macmillan.

Imbert, F. (1989). *Émile ou l'interdit de la jouissance*. Armand Colin.

Kambouchner, D., Meirieu, P., & Stiegler, B. (2012). *L'école, le numérique et la société qui vient*. Mille-et-une-nuits.

Lafforgue, P. (1995). *Petit Poucet deviendra grand*. Mollat.

Lebrun, J.-P. (2016, [2009]). Un monde sans limites. Érès.

Lebrun, F., Venard, M., & Quéniart, J. (1981). *Histoire générale de l'enseignement et de l'éducation en France*. Nouvelle Librairie de France.

Oury, F., & Pain, J. (1972). *Chronique de l'école caserne*. Maspéro.

Postman, N. (1982). *The disappearence of Chilhood*. Vintage Books.

Renaut, A. (2002). *La libération des enfants. Contribution philosophique à une histoire de l'enfance*. Bayard.

Rouche, M. (2003, [1981]). Histoire de l'enseignement et de l'éducation. Perrin.

Rousseau, J.-J. (1966, [1762]). Émile ou de l'éducation. Garnier-Flammarion.

Stiegler, B. (2008). *Prendre soin de la jeunesse et des générations*. Flammarion.

Renaud Hétier is Professor at the Catholic University of the West (UCO). He is Doctor of Educational Sciences. He is the author of books on education, and anthropology in the Anthropocene. Books (selection): *Humanity versus Anthropocene* (PUF, 2021, *in French*); *Cultivating attention and care in education* (PURH, 2020, in French); *Create an educational space with the fairy tales* (Chronique sociale, 2017, *in French*); Education between presence and mediation (L'Harmattan, 2017, *in French*).

Competence

Gwénola Réto

Abstract The growth of the notion of competence makes it necessary to clarify its meaning, as well as the anthropological conceptions it conveys. After defining the term, we analyze the importance it has gained in the professional field, outlining an individual capable of autonomy, mobility, productivity and entrepreneurship, serving an economic objective, fully meeting the demands of neoliberalism. A critical approach to the notion leads us to highlight the need to consider vulnerability. We thus suggest three different viewpoints of the notion of competence in face of the challenges posed by the Anthropocene.

Competence, a Prolific Polysemy

Since the 1980s, the notion of competence has generated a great deal of literature. Etymologically, the word, stemming from the Latin word *competentia,* signifies a fair relationship, a decent proportionality, a pursuit of adaptation. Its various uses make it difficult to give the notion an unequivocal definition, whilst also, in part, explaining its success. We summarize its different meanings by building on its semantic polarization. Competence refers to a pole of knowledge, of literacies linked to education. It is in fact the recognized skill in a field that allows its holder to decide or make a value judgment, competence being a person's ability to judge in their area of expertise. Competence, then, refers to a pole of power, because it gives its holder the power to decide and influence, and a hardly questionable authority. In law, it designates in fact the allocation of power and its evaluation, and more largely the legally recognized ability held by an authority to address an issue, pronounce a judgment on it, carry out an act according to pre-determined conditions. Competence, then, logically refers to a pole of professional expertise, mastery, qualification,

---

G. Réto (✉)
Catholic University of the West, Angers, France
e-mail: gwenola.reto@uco.fr

© The Author(s), under exclusive license to Springer Nature Switzerland AG 2023

N. Wallenhorst, C. Wulf (eds.), *Handbook of the Anthropocene*,
https://doi.org/10.1007/978-3-031-25910-4_220

skillfulness, and it is this third pole which contributed to impose the notion in numerous areas of human activity. These three poles probably explain its increase in the economic and political fields, and even in the field of education and training.

For Professionally Increased Individuals

The meaning of the word competence has considerably evolved in relation to the evolution of work, to the demand for increased competitiveness caused by the globalization of the economy, and to the growing complexification of professional situations. In the professional field, starting in the 1980s, it has gradually replaced the notion of qualification, linked to a Taylorist view of work organization, in favor of a greater consideration of people's role in work situations. It allows a reflection of the evolution of professional duties which are less and less linked to the work position, faced with the prospect of an accelerated obsolescence of professional competences. From the 1990s, the competitiveness challenge, in an increasingly complex and competing context, has made it necessary to articulate quality, reactivity, service and innovation. Competence then becomes a key concept of organizational development. Its administration and management are a challenge for organizations and men, and a concern for businesses. It is inescapable at each stage of human resources management (recruitment, training, assessment, performance expectations, pay, career management, mobility, etc.) and becomes the very purpose of the contract. The employee is managed as the holder of a capital, with a "portfolio of competences" of which it is necessary to take care, in order to keep on responding to economic issues.

Because it is popular, competence is the subject of many studies in various fields. Barbier (2017) defines it as a "property given to individual and/or collective subjects by attributing characteristics built by inference from their engagement in activities which are situated, finalized, leading to the attribution of value" (p. 63). Competence generally refers to tasks, human activities performed in identifiable circumstances, with an expectation of efficiency when those tasks, activities or problems are to be solved, a structured nature of the processes of mobilizing knowledge, know-how and behavioral attitudes which ensures efficiency, and the possibility of making predictions on these specificities (Aubret & Gilbert, 2003). According to Le Boterf (2008), it is the implementation of a relevant professional practice whilst mobilizing an appropriate combination of resources. A real ability to mobilize in a situation, competence, according to Zarifian (2009), is about taking initiative, assuming responsibility by the individual who must take charge of him or herself in the work situation. Personal commitment is then inescapable because he or she must come up with a suitable answer in order to successfully deal with events. Therefore, even though it is not directly manifested, competence is inseparable from activity and the context in which it takes place. Structured in a combinatorial and dynamic manner, it is designed, evolving and includes a metacognitive dimension.

Since the 2000s, workers are expected to simultaneously be efficient and flexible so they can adapt to new situations and maintain their employability. A culture of lifelong learning has developed, with a goal of making careers more dynamic, personalized, multifaceted, in order to secure these careers. This is a profound change: the professional must be fully involved in his or her career path, responsible for its sustainability and capable of living all professional transitions, wished for or not. This tendency has only been strengthened in the last twenty years. The uncertain control of many professionals on their daily life and the stable and clearly delineated boundaries are blurred. For many, the spatial and temporal frame of work is becoming less clear, with multiple workplaces and activity period blending with personal time, requiring the extension of the range of professional competences. Transferable and transversal professional competences are just as expected as employment- and company- related competences. More importantly, in this context of diverging and destabilizing movements, professionals must implement permanent coping strategies which demand that they develop a greater autonomy in shaping their career paths. They must demonstrate their ability to learn to learn in order to evolve within changing environments. Relying solely on knowledge, mastered practices and favorable behaviors is limiting, if not completed by a now valuable ability to evolve. Professionals must take responsibility for their employability, broaden and demonstrate their worth on the labor market, develop their ability to leverage their competences and to communicate on these by building on a true personal branding. But at what cost, for what individual and collective risks?

The notion of competence has therefore established itself in the professional field. It also has a strong presence in the context of adult training which values knowledge gained from practice and the required learning abilities. Activity and training reference systems nowadays are built in connection with that notion, serving its development. Expressing an economic and social expectation, competence has also entered the field of education, up to educational purposes. The need for life-long learning goes hand in hand with a concern to instigate, starting at school, the development of complex skills which will be essential to the future adaptation of the individual. It is a question of considering, very early on, how to adapt to a changing environment, which requires the development of flexible intellectual tools, able to adjust to transformations and reinforce knowledge (Legendre, 2008). Integrating competences in school *curricula* reflects a renewed view of learning and suggests considering it in its cognitive, social and affective dimensions. It is a true wave which attracts not only political authorities and educational officials but also educational researchers. Under the influence of the OECD, competence now guides reforms of programs, subjects, teacher training, purposes of education, at the risk of the domination of a frenzied neoliberalism. Faced with the changes of the Anthropocene, is a different perspective desirable, possible?

The Shift Towards Taking Vulnerability into Account

Several voices have raised concerns about the supremacy of competence. The climate of uncertainty, strengthened by the various economic, health and social crises, weakens many individuals, causing social disqualification and the need for greater competitiveness. Individuals find themselves immersed in a ruthless world which is subject to the demands of productivity, competitiveness, urgency and excess which invites them to show mobility, responsiveness, efficiency, urging them to become ever more responsible in a context of generalized fragmentation. Castel warned against the scission between the "visibles", the "individuals who are active, mobile, full of subjectivity and driven by great care for themselves" (2004, p. 121) and the "individuals by default" who lack resources. The first find, in the injunctions made by society, new ways to excel themselves in terms of performance, of social success, required to better themselves in an entrepreneurial adventure at the risk of burning themselves in hyperactivity, stress or burn-out. The latter suffer from the injunction to be self-fulfilled and feel anguish about insecurity, solitude. Poché (2016) moreover denounces the fragmentation of temporality, identity, community and values which contribute to fragilizing lives. Doesn't the polarization of competence on knowledge, power and expertise make it a tool of domination which should be counterbalanced? Which competences would be needed to respond to the changes of the Anthropocene? Following Wallenhorst (2021), we wonder whether a different balance of power is possible in order to counter the depletion of Earth and human resources. Which levers would enable us to consider and assume fragility, vulnerability? Which competences are necessary for us to live together tomorrow, face the challenges posed by the Anthropocene in a fairer manner and tackle a worrying neoliberal determinism? What alternative forms of coexistence should be investigated? We suggest three perspectives to meet the challenges of change, developing perspectives which extend the notion of competence beyond the ratio of productivity and allow the contemplation of ties of interdependency and vulnerability. The first two take on an educational perspective and the third, a more political approach.

From an educational perspective, UNESCO has provided many studies inquiring about the notion of competence, from a critical perspective as well as to promote further mutual comprehension and democracy. Recognizing the changes in cultural landscapes and their diversity, even though peoples, communities and individuals are brought increasingly closer to each other, Leeds-Hurwitz and Stenou (2013) have highlighted the importance of developing intercultural competences which "are abilities to adeptly navigate complex environments marked by a growing diversity of peoples, cultures and lifestyles" (p. 10). Presented as the keystone to understanding the world, this new basic competence "can be seen as part of a broad toolkit of worldviews, attitudes and competences that young people acquire for their lifelong journey". It "enable(s) them to interact with cultural 'others' with a view to bridging differences, defusing conflicts and setting the foundations of peaceful coexistence. (…). Intercultural competences complement human rights as a catalyst for promoting a culture of peaceful and harmonious coexistence". This competence relies on respect, self-awareness/identity, seeing from other perspectives/world views, listening, adaptation, relationship building, cultural humility. Though this definition is somewhat vague concerning competences, skills and attitudes, it does outline potential answers.

In pursuing this clarification effort, UNESCO has also defined the competences understood as the minimal requirements to attain intercultural competences. Thus sets of competences are put forward such as Systems thinking competency, "the abilities to recognize and understand relationships; to analyze complex systems; to think of how systems are embedded within different domains and different scales; and to deal with uncertainty" (UNESCO, 2017, p. 10) or anticipatory competency, "the abilities to understand and evaluate multiple futures – possible, probable and desirable; to create one's own visions for the future; to apply the precautionary principle; to assess the consequences of actions; and to deal with risks and changes" (id.). To these two competences others are added, also useful for thinking collectively about the future: normative, strategic, collaboration, critical thinking, self-awareness, integrated problem-solving competences. Educational issues are therefore environmental, economic and societal, and education appears to be a powerful lever to increase individual and collective well-being, and to live together in greater harmony. These two ranges of competences seem able to contribute to transforming our society and shape our future, in a fairer way than when we focus on professional competence.

Another approach seems equally able to respond to the changes related to the Anthropocene, by taking another look at the very nature of the competences, the development of which is a priority: the path paved by care ethics, and that of individuals' capabilities. Though it is not fully equivalent to the notion of competence, the notion of capability ensures meeting the concern for the most vulnerable individuals, who cannot be productive regarding modern economic expectations. With this notion, Sen (1992) draws attention to the problem of political, economic and social inequalities and offers a political vision for greater social justice, building on an ethical finality: that of living well. Following Sen, Nussbaum defines ten capabilities designed to create the conditions for a fairer world, working towards supporting and reinforcing the effective and potential capacities of humans so they can choose a fulfilling way of life (2011). In contrast to dominant models which consider the quality of life from their standards, it offers to support and reinforce the effective and potential capacities of the human being so he or she can choose a fulfilling way of life. These perspectives meet those of care ethics, which involve a thought that highlights responsiveness to specific situations of vulnerability, by making care for human lives its chief concern, far from unbridled economic productivity. Tronto (1993) defines it as a four-phase active process: caring about, which involves accepting the need for care; taking care of, which is recognizing one's responsibility in action; care giving, the concrete moment of looking after someone; and care receiving, acknowledging the impact of care. The third phase of this process, care giving, requires thinking about care as a practice and due to this effective looking after, it paves the way for the recognition of a precious and often undervalued range of competences, borne by people doing low-profile, low-profit, yet essential work. Far from transhumanism, this approach seems able to increase our humanity.

If the notion of professional competence is not to be excluded in order to live the Anthropocene, we do see its limits and risks of human depletion and weakening. We suggest taking a complementary approach by considering ranges of competences more likely to meet these challenges.

References

Aubret, J., & Gilbert, P. (2003). *L'évaluation des compétences*. Mardaga.

Barbier, J.-M. (2017). *Vocabulaire d'analyse des activités: penser les conceptualisations ordinaires*. PUF.

Castel, R. (2004). La face cachée de l'individu hypermoderne: l'individu par défaut. In *L'individu hypermoderne* (pp. 117–128). Érès.

Le Boterf, G. (2008). *Repenser la compétence. Pour dépasser les idées reçues*. Éditions d'Organisation.

Leeds-Hurwitz, W., & Stenou, K. (2013). *Intercultural competences: Conceptual and operational framework*. Unesco.

Legendre, M.-F. (2008). La notion de compétence au cœur des réformes curriculaires: effet de mode ou moteur de changements en profondeur ? In *Compétences et contenus: Les curriculums en questions* (pp. 27–50). De Boeck.

Nussbaum, M. (2011). *Creating capabilities. The human development approach*. Press of Harvard.

Poché, F. (2016). De l'hyper-vulnérabilité. Diagnostic du présent et clarification conceptuelle. *Revue des sciences religieuses, 90*, 51–61.

Sen, A. (1992). *Inequality reexamined*. Harvard University Press.

Tronto, J. (1993). *Moral boundaries. A political argument for an ethic of care*. Routledge.

UNESCO. (2017). *Education for sustainable development goals: Learning objectives*. UNESCO.

Wallenhorst, N. (2021). *Mutation. L'aventure humaine ne fait que commencer*. Le Pommier.

Zarifian, P. (2009). *Le travail et la compétence: entre puissance et contrôle*. PUF.

Gwénola Réto works at the Catholic University of the West (UCO). She is a Doctor in Education, Careerology and Ethics, and holds a Ph.d in education. She is the author of *La bienveillance à l'Ecole* (2019, Presses Universitaires de Nancy). She focuses on issues relating to ethics, professional development and management, of school organizations.

Empowerment

Cécile Redondo

Abstract This article examines the concept of empowerment in the context of the Anthropocene. It draws on various translations, definitions and interpretations and also on the historical approach to the concept in order to reassess such an approach at the dawn of the Anthropocene, particularly in the field of education and training. The concept of empowerment is present in the discourse of scientists who seek on the one hand to understand, explain and shed light on the educational issues and processes at work, and on the other hand to develop and propose solutions or ways of improving the existing situation. The Anglo-Saxon concept of empowerment, which originated in the field of sociology, is thus employed today in a cross-disciplinary context, including education and pedagogy.

Among the many translations existing in the literature, the term empowerment, of Anglo-Saxon origin, can be defined as a dynamic process consisting, for the individual, in developing or reinforcing his or her "power to act" with a view to the "self-empowerment", "autonomizing" and self-emancipation of individuals (Bihl, 2020; Chapeau, 2020; Grandval, 2020; Mahmoudi, 2020; Maury & Hedjerassi, 2020). For Julian Rappaport (1987), the concept of the "power to act" refers to the possibility for an individual to carry out a desired change, which concerns him/her and which he/she has defined. Yann Le Bossé (2008, p. 138) defines empowerment as "the concrete capacity of people (individually or collectively) to exercise greater control over what is important to them, their loved ones or the community with which they identify". This concept designates a process, a state and an intervention approach aimed at supporting the development of this capacity for action (agency). For Edward Bernstein et al. (1994) and Marc Zimmerman (2000), it is not a question of teaching, promoting or stimulating the power to act (individual or collective) of the individuals concerned, but of contributing to the emergence of the conditions necessary for its manifestation.

C. Redondo (✉)
Aix-Marseille University, Marseille, France

University Jean Monnet, Saint-Étienne, France
e-mail: cecile.redondo@univ-st-etienne.fr

© The Author(s), under exclusive license to Springer Nature Switzerland AG 2023

N. Wallenhorst, C. Wulf (eds.), *Handbook of the Anthropocene*,
https://doi.org/10.1007/978-3-031-25910-4_221

Empowerment thus concerns the development of certain skills/capacities/abilities and values in the individual, such as action and autonomy, with a view to the emancipation of the individual, and the transformation of society at a collective level (social and societal change). The dimensions (of the capacity) of action in the world and of emancipation are thus central to the acquisition of collective power and transformation of the world.

The historical approach (Le Saux-Pénault, 2020; Maury & Hedjerassi, 2020; Roy, 2020) places the appearance of the concept in the 1960s–1970s in the USA within the framework of the civil rights movements, in particular the feminist and anti-racist struggles and the struggles of gays and lesbians, and then in the 1980s in Latin America with the Brazilian philosopher and pedagogue Paulo Freire within the framework of the development of the popular education movements. The dimension of political power and social conflict was thus present from the birth of the concept of empowerment and was developed through the thinking of Paolo Freire who was described as a liberation theorist.

In the field of education and training, the concept has an important place from the moment we consider that education is an essential vector for strengthening individuals' power to act and "empowering" them. At this level, empowerment refers to the learning process that enables individuals to gain more power (Carton & Tréhondart, 2020, p. 79). For Freire, it is a matter of providing learners with "empowering" experiences (Maury & Hedjerassi, 2020). The concept has thus taken on a growing importance with the development of the *éducations à* teaching modules: education for sustainable development (Redondo, 2020), citizenship education, media and information education (Bihl, 2020; Carton & Tréhondart, 2020; Frisch, 2020; Mahmoudi, 2020), parenting education (Chapeau, 2020; Grandval, 2020), and gender equality education (Le Saux-Pénault, 2020), etc.

With the theory of "conscientization", Freire addresses the pedagogical dimension of the empowerment process. Indeed, for him, all pedagogical work begins with an awareness of the forms of oppression integrated into the social structure and internalized by the subject (Carton & Tréhondart, 2020; Pont, 2020). Conscientization is defined as "the process by which man engages with reality and practice, that is, the set of activities that enable him to transform his environment and the social world." (Roy, 2020, p. 25) As such, the concept of empowerment could be pertinently articulated with that of "praxeology" in the sense of Chevallard (2007): a permanent back and forth between action and reflection oriented towards social transformation. Returning to Freire, his "pedagogy of the oppressed" finds its foundation in the process of conscientization, which consists of "articulating a reflective component with a 'transformative action' (Freire, 1980)." (Roy, 2020, p. 32) This dimension of action aimed at social transformation is exercised, for example, in the context of participatory approaches and collective actions of citizen participation.

Normand (2017) clearly identifies the pedagogical deployment – active, participative and cooperative – that can be employed in educational situations based on the concept of empowerment. It is a matter of learners building (or acquiring), strengthening and expressing their capacities (at a collective level), developing or even rehabilitating their "power to act", which is expressed in the classroom on the one hand through the "active" attitude of the students, and on the other hand through "strategies of cooperation and mutualization" (Normand, 2017, p. 419). For this author, the central concepts of "student activity" and cooperation/collaboration in learning thus seem to be associated with – or even inherent in – the concept of empowerment transposed into the educational field. For Simonneaux and Simonneaux (2017, pp. 426-431), the concept of empowerment is to be compared with the concept of "engagement", the modalities of which are many. Three levels of commitment are possible, starting with minimal individual involvement of the participants, up to commitment envisaged on a collective scale. According to Simonneaux and Simonneaux (2017), the concept of empowerment is expressed in this – more evolved – form of "collective commitment" which also incorporates the concept of the "empowerment" of individuals. Other practices, such as "activist initiation" and "educ-action" involve a greater degree of commitment which Simonneaux and Simonneaux (2017, p. 426) compare to the concept of "activism", i.e., the militant involvement of learners in actions to promote change (social, environmental, etc.). The challenge is for students to "engage in activist action in the school setting" (Simonneaux, 2011, p. 375) since "activism", "in a more radical sense, is aimed at preparing to take part in an active and informed way in political action for social justice and environmental preservation" (Simonneaux & Simonneaux, 2017, p. 426). On the lines of "educ-action" and "activism", we will finally retain the triplet of concepts proposed by Pagoni and Tutiaux-Guillon (2012, p. 9) – "empowerment (appropriation of one's power of control), accountability (responsibility) and commitment (engagement)" – constituting, according to them, a founding theoretical basis for the whole of the "*educations à*" modules.

In the field of education and training, we have therefore identified a set of concepts that are relatively close to empowerment, which include nuances in its understanding, and which contribute to its theoretical and practical deployment. The concept of empowerment is thus multi-faceted with multiple and non-univocal translations, but with some guiding principles that we have tried to highlight: the importance of action, collective mobilization, cooperation/collaboration, etc.

In our opinion, the concept of empowerment is particularly interesting, relevant and applicable in the context of the Anthropocene. It provides a key to understanding the Anthropocene through an anthropological approach by containing arguments from the human and social sciences and by being part of a forward-looking dynamic.

The social, economic, cultural, political, environmental and anthropological dynamics and issues that are expressed in the Anthropocene and that are also present in the different interpretations of the concept of empowerment make it possible to counter the oversimplification of many scientific works (reduction to one field). The dimension of interdisciplinarity is therefore central here. The principles of decentering and questioning at the heart of social pedagogy (Le Breton & Sabin, 2020), which for Freire are conditions for emancipation and triggers for action, are also at the heart of the Anthropocene.

From the point of view of the changes needed at the dawn of the Anthropocene, the concept of empowerment is thus a fully appropriate one. It allows us to develop perspectives that can contribute to a transformation and improvement of the global situation by indicating the most relevant and effective way to act collectively in the Anthropocene (participatory democracy, citizen participation, etc.). This collective action that has been lacking throughout the modern era and this lack of action that has led us into the Anthropocene.

In a dynamic of identification of the anthropological changes necessary for the human adventure to continue in the Anthropocene, the concept of empowerment thus constitutes, according to us, a lever to prepare for this new geological period, including in the educational field, associated with a study of the pedagogical and didactic issues and the problems involved.

References

Bernstein, E., Wallerstein, N., Braithwaite, I., Guttierrez, L., Labonte, R., & Zimmerman, M. (1994). Empowerment forum: A dialogue between guest editorial board members. *Health Education Quarterly, 21*(3), 281–294.

Bihl, J. (2020). La littératie médiatique par et pour l'*empowerment*: La place du pouvoir d'agir dans un dispositif évaluatif de compétences. *Spirale – Revue de recherches en éducation, 66*, 65–75. https://doi.org/10.3917/spir.066.0065

Chapeau, J. (2020). Éduquer à la protection: Les pratiques professionnelles de développement du pouvoir d'agir à l'épreuve du mandat de protection de l'enfance. *Spirale – Revue de recherches en éducation, 66*, 119–136. https://doi.org/10.3917/spir.066.0119

Chevallard, Y. (2007). Passé et présent de la théorie anthropologique du didactique. In Ruiz-Higueras L., Estepa, A., & Javier García, F. (Dirs.). *Sociedad, Escuela y Mathemáticas: Aportaciones de la Teoría Antropológica de la Didáctico*, Universidad de Jaén, pp. 705–746.

Carton, T., & Tréhondart, N. (2020). La plateformisation de l'éducation aux médias et à la citoyenneté: Regards critiques et enjeux d'émancipation. *Spirale – Revue de recherches en éducation, 66*, 77–94. https://doi.org/10.3917/spir.066.0077

Freire, P. (1980). *La pédagogie des opprimés*. Libertalia.

Frisch, M. (2020). Engagement, développement professionnel autonome et complexité en classes inversées. *Spirale – Revue de recherches en éducation, 66*, 95–104. https://doi.org/10.3917/spir.066.0095

Grandval, M. (2020). Développer le pouvoir d'agir des parents en contexte contraint: Entre place symbolique et place physique. *Spirale – Revue de recherches en éducation, 66*, 137–150. https://doi.org/10.3917/spir.066.0137

Le Bossé, Y. (2008). L'*empowerment*. De quel pouvoir s'agit-il ? Changer le monde (le petit et le grand) au quotidien. *Nouvelles Pratiques Sociales, 211*, 137–149.

Le Breton, H., & Sabin, G. (2020). Pédagogie sociale: Le quotidien comme source d'émancipation. *Spirale – Revue de recherches en éducation, 66*, 37–49. https://doi.org/10.3917/spir.066.0037

Le Saux-Pénault, E. (2020). Légitimité et pouvoir d'agir des enseignants qui éduquent à l'égalité des sexes: Une recherche-action à l'école primaire. *Spirale – Revue de recherches en éducation, 66*, 165–182. https://doi.org/10.3917/spir.066.0165

Mahmoudi, K. (2020). Esprit critique et pouvoir d'agir: Vers le développement d'une « attitude critique » ? *Spirale – Revue de recherches en éducation, 66*, 51–63. https://doi.org/10.3917/spir.066.0051

Maury, Y., & Hedjerassi, N. (2020). *Empowerment*, pouvoir d'agir en éducation: À la croisée entre théorie(s), discours et pratique(s). *Spirale – Revue de recherches en éducation, 66*, 3–13. https://doi.org/10.3917/spir.066.0003

Normand, R. (2017). *Empowerment*/capacitation. In A. Barthes, J.-M. Lange, & N. Tutiaux-Guillon (Eds.), *Dictionnaire critique des enjeux et concepts des « éducations à »* (pp. 419–425). L'Harmattan.

Pagoni, M., & Tutiaux-Guillon, N. (2012). Présentation: Les *éducations à*… Quelles recherches, quels questionnements ? *Spirale, 50*, 3–10.

Pont, E. (2020). La réhabilitation professionnelle des personnes paraplégiques: Des instruments de pairémulation au prisme de la pédagogie freirienne. *Spirale – Revue de recherches en éducation, 66*, 195–204. https://doi.org/10.3917/spir.066.0195

Rappaport, J. (1987). Terms of empowerment/exemplars of prevention. Toward a Theory for Community Psychology. *American Journal of Community Psychology, 15*, 121–148.

Redondo, C. (2020). La place de la notion d'*empowerment* dans le champ de l'éducation au développement durable: Quels transferts entre développements théoriques et traductions pratiques? *Spirale – Revue de recherches en éducation, 66*, 151–164. https://doi.org/10.3917/spir.066.0151

Redondo, C., & Ladage, C. (2019). Quand l'enjeu pédagogique dispute la place de l'enjeu didactique et réciproquement : Le cas de l'éducation au développement durable. *Éducation & Formation, e-312*, 109–121. http://revueeducationformation.be/index.php?revue=32&page=3

Roy, A. (2020). D'acteurs affaiblis à acteurs politiques: La conscientisation permanente des militants quart-monde. *Spirale – Revue de recherches en éducation, 66*, 25–35. https://doi.org/10.3917/spir.066.0025

Simonneaux, J. (2011). Légitimité des savoirs et des expertises : L'exemple du développement durable. In A. Legardez, & L. Simonneaux (Éds.), *Développement durable et autres questions d'actualité : Questions socialement vives dans l'enseignement et la formation* (pp. 365–382). Éducagri.

Simonneaux, J., & Simonneaux, L. (2017). Engagement. In A. Barthes, J.-M. Lange, & N. Tutiaux-Guillon (Eds.), *Dictionnaire critique des enjeux et concepts des « éducations à »* (pp. 426–431). L'Harmattan.

Zimmerman, M. (2000). Empowerment theory: Psychological, organizational and community levels of analysis. In J. Rappaport, & E. Seidman (Eds.), *The handbook of community psychology* (pp. 43–63). Plenum Press.

Cécile Redondo is a teacher-researcher at the University of Saint-Étienne. She has a doctorate in educational sciences. Her thesis focuses on the theme of education for sustainable development (ESD) through the prism of the pedagogical and didactic practices of the professionals concerned. She is the author of several publications in the field of education and training. Publications (selection): *Quand l'enjeu pédagogique dispute la place de l'enjeu didactique et réciproquement* (Ladage & Redondo, 2019); La place de la notion d'*empowerment* dans le champ de l'éducation au développement durable (Redondo, 2020).

Health Education

Annette Miriam Stroß

Abstract In view of the worldwide developments in the fields of medicine, new technologies and natural sciences during the last decades, "health" as one of the central human issues needs to be radically reconsidered. Human consciousness and human action can be described as one of the most important pivotal points in the new era of the Anthropocene (cf. Wulf, Internationale Zeitschrift für Historische Anthropologie 29(1):13–35, 2020; United Nations (UN). Resolution der Generalversammlung, verabschiedet am 25. September 2015. Transformation unserer Welt: die Agenda 2030 für nachhaltige Entwicklung. www.euro.who.int/__data/assets/pdf_file/0006/129534/Ottawa_Charter_G.pdf. Zugegriffen am 10.10.2021, 2015). On the threshold of the self-destruction of large sections of humanity or even of humanity as a whole, we need to examine what considerations and what actions can be described as sustainable in order to bring a new understanding of health into the world and new capacities of dealing with health.

The Problematic Status Quo of Existing High-Tech Medicine

The ambivalences of a specialized medicine that works with empirical-experimental methods involve a definition of functionality of the body based on defined parameters and measurable values. People are brought into a state of potential "sickness" through comprehensive preventive measures supported by an enterprising industry. And the (inevitable) dying with dignity is countered by a high-tech medicine that intervenes 'at all costs' to prevent death and seek to prolong life (cf. Beck, 1986).

The perhaps unintended result is a continuous process of incapacitation. People are denied their own way of perceiving things; it is considered to be inadequate or interpreted as a psychopathological phenomenon on the basis of clinical diagnostics. When reduced to the status of a *homo patiens*, the individual cannot recognize

A. M. Stroß (✉)
Karlsruhe University of Education (Baden-Wurttemberg), Karlsruhe, Germany
e-mail: annette.stross@ph-karlsruhe.de

© The Author(s), under exclusive license to Springer Nature Switzerland AG 2023

N. Wallenhorst, C. Wulf (eds.), *Handbook of the Anthropocene*,
https://doi.org/10.1007/978-3-031-25910-4_222

the physical experience of a dynamic and lasting sense of wellbeing any more than consciously be able to accept or read the symbolism of occurring symptoms as an expression of his/her individual way of life (cf. Rainville, 2004).

How to Find Alternatives and New Approaches

A radically new view of "health" means opening the door to new perspectives. (The transition to a new understanding of health took place in the 1990s, as a contrast to the increasing technological requirements in many fields of human interaction in industrialized societies.) This would include, for example: the statement by the theologian and philosopher Ivan Illich that in our high-tech, polluted world "health" and "responsibility" (in the sense of control over the ecosystem) belong to an idealized past, and the talk of "personal health" (combined with the idea of isolated individuals in competition with one another) has long since degenerated into a farce (Illich, 1991); the point of view of the philosopher Hans-Georg Gadamer, who speaks of the non-controllability of health that follows an overarching rhythm of life (Gadamer, 1993); the analysis of the sociologist Hagen Kühn, which is critical of restraint and connects the term "healthism" with the collective exaggerated search (addiction) for "health" (Kühn, 1993).

Radically new perspectives also include the willingness and courage to be open to new visions of human existence in the awareness of the dialectic of the Enlightenment and the tendency to dogmatic narrowing of positive utopias. In our opinion, the statements of the 2030 Agenda can only be read adequately against this background. Here it is stated: "We are determined to take the bold and transformative steps which are urgently needed to shift the world on to a sustainable and resilient path [...] so that all human beings can fulfil their potential in dignity and equality and in a healthy environment" (UN, 2015, 1/2).

According to this, human beings are called upon to recreate themselves and 'their' future, locally and regionally as well as relating to the earth and the cosmos as a whole, towards more humanity and care, in vibration with all life (cf. Weber, 2017). (In this context, the origin of the word "health" should be remembered, which can be derived from Old English terms *hælþ* (wholeness, being whole, sound or well), *halig* (holy, sacred) and *hælan* (to heal).) That does not mean entering into blanket resistance to new technological developments in natural science, medicine and technology, but rather recognizing and promoting the functionality that serves people and their consciousness development – instead of trying to overcome the essence of being human by the transhumanist agenda.

In order to strengthen the chances of people increasingly determining their own lives in freedom and dignity (cf. WHO, 1986; UN, 2015), it is necessary to argue counterfactually against a medically one-sided and politicized health understanding of the (late) modern age. In contrast to the notions of the human being as an 'unpredictable disruptive factor', whose digital monitoring and control is considered a new goal of a global health policy agenda, the new view of people reconsiders them as a sensitive, feeling, human and community-oriented, historically-biographically anchored in different collectives and stories, (self-)reflective as well as being capable of consciousness.

For the ongoing existence of the planet and humanity it seems furthermore necessary to leave behind the 'great' narrative that human beings rule over nature, animals, plants, etc. as well as their own kind. In this process it is important to find new, culture-specific forms of expression interested in a different, sensitive way in dealing with animate and inanimate nature (such as air, water, minerals), i.e. the entire ecological system on this planet.

The variety of 'little' narratives that may redefine the topic of health in the future can neither fall back on the naive self-image of a supposedly pedagogical habitus, which is still characterized by the powerful mechanisms of control and self-control that have emerged in the course of social modernization processes. Nor can those narratives follow other programmes of hybrid self-overestimation of humankind, which for many decades have been characteristic of the functioning of numerous institutions in the highly industrialized societies, especially in the (north) western hemisphere, and which have become even more obvious in many of these countries since the corona pandemic was announced.

Reconstructing Existing Individual and Collective Patterns

The basis for a radical new perspective on the discussions about maintaining and promoting health, which goes 'to the roots', is provided by an approach that encompasses (self-)educational processes of the individual as well as the recognition of self-conceptions that are historically and biographically set in stone on various levels (political, cultural, religious). At the centre of such an approach are mindfulness (reflexivity, meditativeness) and attention (alertness, awareness) to existing – culturally anchored and individually effective – health-related fixations, which so far have found expression in dogmatic ways of thinking and behaving related to an understanding of health that is governed by imperatives. (The terms "self-education" and "education" used in this context refer to the German term "Bildung"; cf. Stroß, 2018b.)

In our own research work we speak of myths – anchored in the tradition of the Christian West – which are related to personal ideas (patterns) and which can be reconstructed on the basis of individual histories of health and disease. Compared to the psychological concept of "subjective ideas" (cf. Flick, 1998), the myths approach takes a more comprehensive, cultural and socio-critical perspective. In this way, the far-reaching character of the individual entanglements involved in a quasi-religious search for salvation as well as in secular and socially mediated patterns (e.g. "having to function") can be emphasized (cf. Oevermann, 1995; Böhme, 2000; Stroß, 2018a). Lateral approaches to the historically and socially determined fixations of health, the awareness of metaphors and myths anchored in personal history, coincide with the emergence of multiple 'little' narratives that can be expressed as being process-like, poetological-aesthetic and in resonance with life ("everything that exists”). In this way, the previously abstract understanding of health, which was collectively and biographically subliminally powerful, can be transformed into sensitive processes of self-education that engage with what is found and resonate with other people as well as nature, thus gaining new experiences and opening up scope for creativity (cf. Rosa, 2019).

Perception-Based Self-Education and Education

The simple desire to stay healthy and prevent illness, accompanied by instructions and rules of conduct can no longer be the goal of health-related efforts today. Any form of health instruction that strictly insists on compliance with prescribed norms in dealing with oneself and with other people can no longer be the target in a society that is oriented towards responsible and self-determined human beings in the twenty-first century. Even in kindergartens and elementary schools the imparting of simple moral rules, generating fear and deterrence – to which a security policy tends in times of crisis even in the twenty-first century – has to be seen as a kind of relapse into the concepts of virtue and morality of health education in the 18th and 19th centuries (cf. Stroß, 1996, 2021b).

The new way of dealing with health as being something individual and something that can be initiated in others in a non-technological and non-hierarchical way, refers to the potential inherent in every human being. This involves changing ways of seeing and acting and thus overcoming boundaries, in order to “grow beyond oneself” for the benefit of humanity and the planet (cf. Stroß, 2018c, 2021a). By questioning what is taken for granted in the so-called normalcy of everyday life, the thought and perception patterns that prevail in our society, which are often very one-sidedly geared towards performance and personal strengths, can be reconstructed, in order to recognize the mechanisms working in the background.

From this point of view, health education in the new era of the Anthropocene includes a transformation process that not only addresses people intellectually, but also touches them deeply and is able to change people into communal, community-oriented beings. In this way, it becomes possible to experience one’s own subjectivity (as being individual and transpersonal) and lead a more self-determined life. The willingness to change and of let this happen, by getting involved in transforming experiences is just as central to this as looking at the mental patterns and the changeability or plasticity of one’s own perception. Existing health concepts can therefore be contrasted with a concept of “soft resilience”, which is geared towards recognizing one-sided and technological understandings of health and leaving them behind.

Lateral Approach and Success Factors

As recent research results show, the (self-)educational processes that lead to a far-reaching reorientation with regard to a new perception-based understanding of health and illness are linked to certain factors that promote success (cf. Stroß, 2021c, d). These can be summarized as follows:

The process itself, the permanent readiness to open up to new perspectives, to recognize other realities, is more important than the search for new 'truths' and the alignment with new certainties.

Only the willingness to accept what is perceived through "one's own glasses" enables far-reaching changes in thinking and acting instead of remaining resistant to a reality that has been perceived as unacceptable.

Resistance to existing patterns of thought and behaviour (through dogmatic interpretations, projections in the form of "enemy images") goes hand in hand with the maintenance or establishment of new dogmas, which in turn again are characterized by exclusion, the absolutization of truth and a renewed search for security (by limitations, restrictions, etc.).

The creation of new perspectives and points of view, the opening up to "stories" that can be written one way or another, often only takes place in a way that is in the mind only. Only changing and realigning one's own perception allows new perspectives to come effortlessly. Then, in the awareness of the contingency of stories, rewriting can be done sensitively, metaphors and myths can be perceived as relating to both body and mind and a new kind of action that is free of reactive mechanisms (restrictive beliefs, influences) can follow.

Conclusion: In the new era of the Anthropocene it is important and also possible to create a lateral (implicit, embedded) and process-oriented understanding both of health and health promotion and also of dealing with diseases and suffering. An anthropologically based trust in the still untapped potential of humanity is essential for this. This trust does not contain any naive ideas or images about the existing state of the planet and humanity. 'Little' narratives about health emerge from this trust – stories of unimagined life opportunities for human beings on this planet that come to life through deep awareness (*anamnesis*) being shaped by conscious reversal processes (*metanoia*) and embedded in the certainty of a constant dynamic creation of which humans – as sensitive and contemplating actors – *can* be part.

References

Beck, U. (1986). *Risikogesellschaft. Auf dem Weg in eine andere Moderne*. Suhrkamp.

Böhme, J. (2000). *Schulmythen und ihre imaginäre Verbürgung durch oppositionelle Schüler. Ein Beitrag zur Etablierung erziehungswissenschaftlicher Mythosforschung*. Klinkhardt.

Flick, U. (Ed.). (1998). *Wann fühlen wir uns gesund? Subjektive Vorstellungen von Gesundheit und Krankheit*. Juventa.

Gadamer, H.-G. (1993). *Über die Verborgenheit der Gesundheit*. Suhrkamp.

Illich, I. (1991). Gesundheit in eigener Verantwortung: Danke, nein! *Universitas, 46*(5), 490–496.
Kühn, H. (1993). *Healthismus. Eine Analyse der Präventionspolitik und Gesundheitsförderung in den USA*. Edition Sigma.
Oevermann, U. (1995). Ein Modell der Struktur von Religiosität. Zugleich ein Strukturmodell von Lebenspraxis und von sozialer Zeit. In M. Wohlrab-Sahr (Ed.), *Biographie und Religion. Zwischen Ritual und Selbstsuche* (pp. 27–102). Campus.
Rainville, C. (2004). *Metamedizin. Jedes Symptom ist eine Botschaft*. Amrita.
Rosa, H. (2019). *Resonanz. Eine Soziologie der Weltbeziehung. Frankfurt a.* Suhrkamp.
Stroß, A. M. (1996). "Gesundheitserziehung" als Norm? Historische Stationen eines pädagogischen Praxisfeldes in der Moderne. *In Zeitschrift für Gesundheitswissenschaften, 4*, 102–110.
Stroß, A. M. (2018a). Bildungspotentiale bei Studierenden und Professionalisierungsperspektiven durch gesundheitsbezogene Mythenarbeit. In A. M. Stroß (Ed.), *Gesundheit und Bildung. Reflexionsansprüche und Professionalisierungsperspektiven* (pp. 85–108). Springer VS.
Stroß, A. M. (2018b). Bildungsprozesse erster und zweiter Ordnung – eine pragmatische Annäherung. In A. M. Stroß (Ed.), *Gesundheit und Bildung. Reflexionsansprüche und Professionalisierungsperspektiven* (pp. 109–130). Springer VS.
Stroß, A. M. (2018c). Schulische Gesundheitsförderung und Prävention im frühen 21. Jahrhundert: Perspektiven einer Gesundheitsbildung. In Theo-Web (Ed.), *Zeitschrift für Religionspädagogik, 17*(1), 3–23. www.theo-web.de/ausgaben/2018/17-jahrgang-2018-heft-1/news/schulische-gesundheitsfoerderung-und-praevention-im-fruehen-21-jahrhundert-perspektiven-einer-gesundh-1/?tx_news_pi1%5Bcontroller%5D=News & tx_news_pi1%5Baction%5D=detail & cHash=669e6ef90cf1d4ed31e50df697519b02. Zugegriffen am 10.10.2021.
Stroß, A. M. (2021a). Gesundheitsbildung als politische Vision: Ein 'anderer' Zugang zur RAVA-Strategie. In G.-B. von Carlsburg & A. M. Stroß (Eds.), *(Un)pädagogische Visionen für das 21. Jahrhundert. (Non-)Educational Visions for the 21st Century. Reihe: Baltische Studien zur Erziehungs- und Sozialwissenschaft* (pp. 453–474). Peter Lang.
Stroß, A. M. (2021b). Curriculare Grundlagen für eine "Gesundheitspädagogik und –didaktik". Ein neues Studienfach für angehende Grundschullehrerinnen und -lehrer? In M. Goldfriedrich & K. Hurrelmann (Eds.), *Gesundheitsdidaktik* (pp. 65–83). Beltz.
Stroß, A. M. (2021c). "Als das kleine Virus tanzte ...". Ein bildungswissenschaftlicher Essay, In Paragrana (Ed.), *Internationale Zeitschrift für Historische Anthropologie, 30*(2), 106–120.
Stroß, A. M. (2021d). Gesundheitsorientierte Erwachsenenbildung unter bildungswissenschaftlichem Gesichtspunkt. In forum erwachsenenbildung. *Die evangelische Zeitschrift für Erwachsenenbildung, 54*(2), 15–21.
United Nations (UN). (2015). Resolution der Generalversammlung, verabschiedet am 25. September 2015. Transformation unserer Welt: die Agenda 2030 für nachhaltige Entwicklung. www.un.org/Depts/german/gv-70/band1/ar70001.pdf. Zugegriffen am 10.10.2021.
Weber, A. (2017). *Sein und Teilen. Eine Praxis schöpferischer Existenz*. transcript.
World Health Organization (WHO). (1986). Ottawa-Charta zu Gesundheitsförderung. www.euro.who.int/__data/assets/pdf_file/0006/129534/Ottawa_Charter_G.pdf. Zugegriffen am 10.10.2021.

Prof. Dr. phil. habil. Annette Miriam Stroß M.A., 2000–2008 Professor of Educational Science at the University of Vechta (Lower Saxony), since 2008 Professor of Educational Science with a focus on health education at the Karlsruhe University of Education (Baden-Wurttemberg), Faculty of Humanities and Human Sciences, Institute for Educational Science with a focus on extracurricular fields, Bismarckstrasse 10, D-76133 Karlsruhe. Research fields: educational theories; health education & health sciences; history of educational science and professionalization; educational historiography. Selected monographs: Pädagogik und Medizin, Weinheim: Deutscher Studien Verlag 2000; Reflexive Gesundheitspädagogik, Berlin: Lit 2009; Bildung und Gesundheit. Wiesbaden: Springer 2018.

Integral Development

Tanguy Marie Pouliquen

Abstract This article examines the meaning of the expression "integral personification" of the person as the process by which any individual becomes more and more a unique person by their capacity to assert their uniqueness in an intersubjective dynamic: to receive, to integrate, to give back, to share. This integral personification opens up to the "integrality" of the human being, it operates by "integration" (the process of interiorisation), it aims at the "integrity" of the subject (their unity: thought, being, action), it ends in the capacity "to integrate" socially the difference (of all human beings).

The term "personification", as an act of becoming oneself, which M. Nédoncelle first used, attributing it to *The reciprocity of consciences* (1942), describes the process by which any individual becomes more and more a unique person by their capacity to assert their uniqueness in an intersubjective dynamic. Following J. Maritain, who in *Integral Humanism* (1936) recognised in the spirit the intangible and transcendent dimension that surpasses and fulfils the human being, the process of personification becomes "integral" when it involves all the dimensions of the person (understood as body, soul, spirit and relationships), open to "the gift of life at its root" (Bergson, 1907), a vitality which, inseparable from the person (Henry, 1963), commits that person to a lively interpersonal communion. Integral personification reflects the permanent way of being of *docibilitas*, an availability to "learn life from life" (Cencini, 2014), in order to expand it by giving it back.

The historical study of personification follows that of the very concept of "person" that E. Housset (2007) develops into four categories (ontico-categorical, ontologico-categorical, ontico-existential, onticologico-existential) while adhering to the last one as the guideline of "vocation". The person is called by life to be

---

translation Wilfrid Shaker

T. M. Pouliquen (✉)
Catholic University of Toulouse, Toulouse, France

© The Author(s), under exclusive license to Springer Nature Switzerland AG 2023

N. Wallenhorst, C. Wulf (eds.), *Handbook of the Anthropocene*,
https://doi.org/10.1007/978-3-031-25910-4_223

always more moved by relationship, which is "received by that which commits them personally" (p. 499). What the person becomes, from the spirit itself, is that of an exodus where they grow by existing in and for alterity, according to the triple modality of time, transcendence, but also exile: paradoxically they have to exit from themselves in order to become… themselves.

From the spirit where the human being is the "shepherd of the being" (M. Heidegger, *Question II*, 1968) the inner conversion (Kehre) receives the manifestation of the being, the unveiling (a-lètheia) of life. It is because the capacity of relationship must find its interior form of existence. Following G. Marcel in *Being and Having* (1935), the person, challenged by relationships, is called to the locus of being itself, and passes from the mode of having to the mode of being which establishes a living relationship to the interiority of the phenomena. The authentic existence, opening up to the being, develops the attitudes that allow this openness: the willingness to receive from the outside, the commitment that unifies life and finds its peak in hope, the capacity to be in harmony with oneself in the long term, the taking into account of others, etc. It is through these openings that the passage from the individual to the person takes place in a process of personification that brings alive the growing subject by their consent to an infinity of life, of being (Stein, 1998).

This challenge of life, carried by an "onto-don-logy" – the logic of the gift of the being – open to the transcendence of the being, evolves into a vital process of unification in the growing subject. Far from Sartre's self-referential freedom (*Being and Nothingness,* 1943), the unification is that of the person to the fullness of life that calls them to renounce their ego – the "hateful ego" of B. Pascal. Finite freedom is carried by two movements open to universality, the energy of desire – always looking for more – and by a language (*logos*) which gives it an open meaning. Finite freedom aspires to an infinite freedom that can only be satisfied by a truly transcendental and inherent personal and relational infinite finite. This return to the principle of life, while remaining in its cause, opens the integral personification to the religious question: the human being aspires to be fully connected. More than a result to be achieved, it is a process to be created personally and socially, always evolving towards its accomplishment.

People, being in constant development of themselves, have, as emphasised by M. Blondel, "to undergo a growth" (*The Being and the Beings,* 1935, p. 204), through the quality of the relationships with others which invites them to "surpass themselves" (p. 286). We are not self-sufficient, we "must act for others, with others and through others" (*Action*, 1893, p. 229). Blondel brings to an end individualism, "by attaching us elsewhere than to ourselves" (p. 230). The alterity is constitutive of the identity, the personification becoming tangible by the opening to "oneself as another" (Ricœur, 1990).

Integral personification is understood by the intensification of the dynamic of the being and existentially according to the gift of life – to receive, to integrate, to give back, to share – a dynamic of permanent formation of the person (*docibilitas*) as a result of a good vulnerability (Ide, "*Positive Vulnerability*"), or malleability of persona who develop by unifying themselves in order to unify (Pouliquen, 2014). In this structuring perspective, the gift of life refers to four dimensions.

Firstly
the gift of life is primarily to be received. The gift of life precedes us, it is the transcendent level, open to an origin, to a given life. We receive our life freely, initially from the origin of life, from our parents, from society, from the environment. Here the person is made for welcoming life and to be thankful for it: the origin of life urges gratitude for such a gift.

Secondly
the gift of life should be united with the total existence of the person. Here everyone is called upon to build up themselves, to develop their own abilities, to fortify their life by strengthening their freedom with the help of the gifts received, there where the search for fulfilment (the last level Maslow, 1971) drives the dynamism of integration (Wojtyla, 1983). It is the anthropological level. The subject as a "person", becomes what he or she decides to be. They are what they are by committing themselves from the given life, therefore without promoting *the virtue of selfishness* (Rand, 2008).

Thirdly
the gift of life is there to be given back. Life "fills" people when they give it. It is specifically the ethical dimension. Without needing a "counter-gift" (Mauss, 2007), and therefore without needing a contract based on reciprocity, but rather with the necessity of a relational experience of two persons, human beings become themselves by coming out of themselves by giving themselves sincerely to others. They become more and more alive by giving that which they have received. Giving life gives… life. The free gift is transformed into a gratifying dynamic because it is relational, that is to say into a free gift of oneself "for" others, by expropriation of oneself for others. Supported by a double gratitude, we make ourselves "available" to others thereby transforming the anonymity of our "me" into a personal and mutually dependent "I", which is a commitment "for us".

Fourthly
to serve communion between people finalises the gift of life. From this perspective, we find our perfection in the common good through circular relationships, exchanges of quality gifts that we inspire. All true integral personification is communal. Communion is the goal of the personification, by building an "I-you-we" (Ulrich, 1974) which requires practical criteria. The "communal we" is not the sum of anonymous individualities, but rather of unique persons who are responsible together for the building of the common unity. The community is formed by bonds of love, exchanges of freely given gifts, that are not only congenial bonds sharing a natural affinity, but bonds formed by action. The more life is received at its roots – welcomed as a freely given gift –, the more it will be shared freely by circulation of gifts generating a genuine sisterhood or brotherhood. This is why the community itself "aspires to set itself up as a person" (Mounier, 1934), one that is a "person of persons."

Does considering the primacy of integral personification not make it possible to base social relationships in an accompanying mode: that is to serve the growth of the persons, their unification, rather than to exert a hold on them?

Homo viator human beings – like beings on the road – become unified more and more by enlightening their conscience. Conscience is the receptacle of this integration process. Unlike J. Locke (1690) who encloses it within itself – "the person is an intelligent and thinking being endowed with reason and reflection aware of his identity and of his permanence in time" (*Essay concerning Human Understanding*, 1690) – the conscience, driven by the continuously renewed gift of life, is an opening, it accompanies the process of integration by the recognition of the goodness of existence. The conscience acts freely by a *Grammar of Assent* (Newman, 2010), by *Donation and consent* (Tourpe, 2000).

Within this approach, a "unified man" is set in motion by a "dialectic of the recognition" of the difference (Fessard, 1959). The existential awakening takes place by the emergence of consciousness of being, by the awareness of positive feelings of communion, but also by negative feelings (anguish, nausea, absurdity, despair), which are so many ways to the depths of existence. However, there is alienation when a reductive, possessive relationship with people develops, which is completely the opposite of a process of interiority as would be a desire desired from above (*de superioribus*), in oneself (*ad interioribus*) and towards others (*ad exterioribus*), according to the Augustinian triad.

The invasion of technology (technocratic paradigm) leads to a loss of interiority. A unified person protects and cultivates their "interior life" (Thibon, 2012). Integral personification is an antidote to practical atheism promoted by the technocratic paradigm of which J.-C. Larchet (2016) underlines the effects on intimate life: substitution of verticality by horizontality, rushing ahead (Finkielkraut, 2001), continuous changing caused by continuous dissatisfaction, replacement of the psychic stability by the unceasing changes, agitation of passions (violence, sexualism, greediness, money-cult, emotions, vainglory), superficial communication replacing communion, etc.

A "unified person" seeks to live in harmony with their environment, they are neither a predator of others, nor of natural resources: they lead their life soberly in harmony with the surrounding nature (*For an ecological commitment,* 2014). To this end everything is linked. The ecology of simple life makes sobriety its main benchmark by aiming for less technique (low technology), intensity of life (slow life), unnecessary food (slow food). A certain logic of reduction of consumption can therefore be associated with it. Nevertheless, it is not simply a question of doing less, but rather more a question of recovering the coherence of personal life with the environment by placing the human person (human ecology) at the heart of the ecological reflection.

Integral personification, that unifies the human being in the gift, is positively utopian. The solution (*topos*) to be found is in front of us as it has to be integrated. The gift is always to be invented; it thus invites us to constantly move forward. Integral personification involves cultivating our desire not by exalting ourselves, but by expressing the echo of the gift of life, primary and freely given.

As opposed to any reduction of others which dissolves them (Lévinas, 1961), the relationship between two unique entities, our own and that of another, calls for an attitude of poverty synonymous with shared non-violence: to serve others and not use them. This mutual service is marked by a vulnerability which leads to an unconditional need of the other, as a result of their uniqueness (Grieux, 2012). The acceptance of the transcendence of others enlarged to the transcendence of the relationships definitively buries any hold on them. The hand that wants to take is turned into a hand that welcomes. In the inner reversal that opens up to the transcendence of others, the offered flower can become fruit (fertility) by dying to its sufficiency. My brother, my sister, "no I will not use you; I will always honour you" underlines the poet C.-K Norwid (1863).

Through conscience, life receives itself from its origin, integrates with the irreducible "I", gives itself back to the "you" of others and builds the unity of the "we" that gathers. The integral personification of the person is ultimately a beautiful four-legged table where it is good to be together: it opens up to the "integrality" of the human being, it operates by "integration" (process of interiorisation), it aims at the "integrity" of the subject (their unity: thought, being, action), it ends in the capacity "to integrate" socially the difference of all humankind.

References

Balthasar, Hans Urs von Balthasar. (1973). *Die Personen des Spiels, 1. Teil: Der Mensch in Gott.* Johannes Verlag.

Bergson, H. (1907). *L'évolution créatrice.* Presses universitaires de France.

Blondel, M. (1935). *L'Être et les êtres*. Alcan.
Cencini, A. (2014). *La formation permanente… Y croyons-nous ?* Lessius.
Collectif. (2014). *Pour un engagement écologique: Simplicité et Justice*. Parole et Silence – Diocèse de Nantes.
Fessard, G. (1959). *Correspondances*. Archives de Philosophies.
Finkielkraut, A. (2001). *Internet, l'inquiétante extase*. Mille et une nuits.
Grieux, E. (2012/4). *Salut et fraternité*. Revue Projet 329.
Henry, M. (1963). *L'essence de la manifestation*. PUF, collection "Epiméthée".
Housset, E. (2007). *La vocation de la personne*. Presses Universitaires de France.
Ide, P. (1997). *Et bien dîtes don: don*. Editions de l'Emmanuel.
Larchet, J. C. (2016). *Malade des nouveaux médias*. Cerf.
Lévinas, A. (1961). *Totalité et infini*. Martinus Nijhoff.
Locke, J. (1690). *Essay concerning human understanding*. Oxford.
Lubac, H. (1938). *Catholicisme. Les aspects sociaux du dogme*. Cerf.
Maritain, J. (1936). *Humanisme intégral*. Aubier.
Mauss, M. (2007). *Essai sur le don. Forme et raison de l'échange dans les sociétés archaïques*. Presses universitaires de France.
Mounier, E. (1934). *Révolution personnaliste et communautaire*. Montaigne.
Nédoncelle, M. (1942). *La réciprocité des consciences*. Aubier.
Newman, J. H. (2010). *Grammaire de l'assentiment*. Ad Solem.
Pouliquen, T. M. (2014). *Devenir vraiment soi-même. Itinéraire d'un développement personnel chrétien*. Editions des Béatitudes.
Rand, A. (2008). *La vertu d'égoïsme*. Les Belles lettres.
Ricoeur, P. (1990). *Soi-même comme un autre*. Seuil.
Stein, E. (1998). *Être fini et l'Être éternel*. Nauwelaerts.
Thibon, G. (2012). *Hommes de l'éternel, conférences au grand public (1940–1985)*. Mame.
Tourpe, E. (2000). *Donation et consentement*. Lessius.
Ulrich, F. (1974). *Gegenwart der Freitheit*. Johannes Verlag.
Wojtyla, K. (1983). *Personne et acte*. Editions du Centurion.

Tanguy Marie Pouliquen is a full professor of ethics at the Catholic University of Toulouse, religious priest of the Community of the Beatitudes, teacher-researcher (CERES, Jean Rodhain chair) and radio columnist, He studies the concrete meaning of integral humanism and has published fifteen books on ethics and spirituality, including recently: *Becoming truly oneself. Itinerary of Christian personal development* (Editions des Béatitudes, 2014, *in French*); *Fascination of new technologies and transhumanism* (Editions des Béatitudes, 2017, *in French*); *Hyperconnected and free. Better living in the digital age without falling back to the Stone Age* (Editions des Béatitudes, 2020, *in French*); *Make the right decision* (Editions des Béatitudes, 2020, *in French*).

Knowledge

Angela Barthes

Abstract As a form of capital, knowledge is symbolic. Mastery of knowledge is one of the key stakes in a power struggle. Entering the Anthropocene, and becoming aware of both planetary limits and the potential for political socialisation, are all forms of knowledge. Knowledge and education in the Anthropocene are uncertain, problematic, systemic, complex, multi-referential, prospective, coherent, and also political in the sense that they make it possible to manage public affairs in a collective and enlightened manner. They differ in how they emerge and in their educative purposes concerning other metanarratives, such as ecological transition or sustainable development.

The circulation and retention and diffusion of knowledge is one of the cornerstones of societies, as much in their internal, socio-economic structuring as in their relations with others. With the entry into a knowledge-based economy, the circulation of knowledge increases under the effect of transnationalisation, and first state, then local and public education policies have to adapt. Yet knowledge is the result on the one hand of voluntary actions by the social key players (parents, peer groups, etc.) and, on the other, of social life, or more precisely the socialisation processes that continue throughout life in a variety of forms. This diverse knowledge only exists because it is formed socially (Barthes & Alpe, 2016). It transcends the individuals or groups that produce it, use it, and circulate it. This is what distinguishes it from understanding, which is attached to an individual or a group of individuals. There are effectively many forms available in society, and they are the basis of culture, forming part of a continuous movement between theory and practice, with each nourishing the other. Defining the knowledge to be transmitted in a society is a political issue, where power is at stake. Knowledge has a hierarchy: not all pieces of knowledge have the same social value, and they do not all provide the same social advantages. The capacity for mastering knowledge (assimilating it, producing it,

A. Barthes (✉)
Aix-Marseille University, Digne, France
e-mail: angela.barthes@univ-amu.fr

© The Author(s), under exclusive license to Springer Nature Switzerland AG 2023

N. Wallenhorst, C. Wulf (eds.), *Handbook of the Anthropocene*,
https://doi.org/10.1007/978-3-031-25910-4_224

validating it) determines the power conferred by this social position. Not all knowledge has the same validity depending on the way in which it is constructed. Knowledge is thus a symbolic form of capital, and mastery of it is a key issue in incessant struggles. These hierarchies also evolve over time and are both the object and the result of power strategies. Certain social groups also tend to claim that knowledge is specific to them because it is the foundation of their identity and their place in society. This comes with two consequences: implementation of strategies for transmitting the knowledge essential for socialisation; and the coexistence within a social group of a wide range of different types of knowledge, included within the power relationships. This makes conflicts inevitable. For example, of all the knowledge available in contemporary societies, that said to be traditional (or local) and informal has progressively lost its importance. For many years, it has been banished by education systems because it interfered with the principle of universality and ran the risk of countering the political project for education and maintaining particularisms.

Two major systems for producing knowledge have thus acquired a particular place in society: the sciences, because modernity has been built on a rationalist premise, and because technology and the resulting technical progress has long been tried and tested (even if this notion of progress is being reviewed today…); and formal education, because it has been entrusted with the mission of transmitting in a systematic manner all the validated and hierarchised knowledge, which both completes and contests other forms of knowledge transmitted by the family, the milieu… in a mechanism of acculturation. But these knowledge systems are based on systems of selection and exclusion (at least in principle): not all knowledge is scientific, not all knowledge is considered transmissible by formal education. Only a small part of this knowledge belongs simultaneously to both sub-sets. Today, the link between these two forms of knowledge has become both a key issue and a source of tension. Scientific knowledge is the upstream guarantor of the validity of a significant part of the knowledge transmitted through formal education, and formal education is generally considered to be the gateway to scientific knowledge. In contemporary societies with highly structured educational systems, the sub-set of academic knowledge (in the broadest sense) is relatively well-defined. As soon as we move into non-formal modes of education, the outlines become less clear and significantly more variable. As proof of this, all that needs to be done is to examine the content of the knowledge passed on through family education: the proximity or distance of this content, in relation to academic and socially valued knowledge, is a very strong determiner for inequalities in academic success. In all cases, however, the knowledge concerned by formal education necessarily maintains complex relations with the other forms of knowledge available in society.

Generally, education on environment and development matters is stretched between both poles: that of weak education, which contents itself with erasing the most harmful effects of development, essentially considered from an economic point of view, and that of strong education, which aims to transform society and develop fair, collective arrangements (Lange & Kebaili, 2019; Lange & Barthes, 2021). In the case of weak education, the knowledge dispensed is often of a technical nature (Slimani et al., 2020) or refers to the authority of experts and the accumulation of knowledge – the *deficit model* used by English-speaking countries (Levinson, 2012), – anomie, or political neutralisation (Barthes, 2018; Barthes et al., 2021). Strong education aims to understand what is at stake in the transformation of society and the development of fair, collective social and societal arrangements (in the sense relating to social justice). The project is then conceived as education in policy that is emancipatory, critical, creative, and mobilising, and the knowledge becomes complex.

Entering the world of the Anthropocene places knowledge and transmission strategies at the heart of new issues related to power and new questions, while education is increasingly obliged to manage the overall stakes affecting society (Barthes et al., 2017). According to Charbonnier (2017), the Anthropocene corresponds to a new way of both organising knowledge with regard to the relationships that Man has with nature, and thinking about their political scope. It accelerates the process of raising awareness of planetary limitations and the need to build new relationships with knowledge and how it is transmitted. As a reminder, global trends in environmental and development education are organised into three major periods. Schematically, environment-related education (ERE) from the 1960s (Reid & Dillon, 2015; Sauvé, 2017) gave way to education in sustainable development (ESD) in the 1980s (Jickling & Wals, 2008; Barthes & Lange, 2018), which was in turn supplanted by metanarratives, including education in the Anthropocene (EA) in the 2010s (Hétier & Wallendorst, 2020). The respective weight of the activist, citizen, and experimental spheres, the institutional and political spheres, and the spheres that produce "learned" knowledge, differ depending on the period, while the dominant educational endpoints that emerge are dependent on the power relationships within society. Thus, and still speaking very schematically, education relating to the environment comes for the most part from an activist model with emancipatory educational aims, while education in sustainable development is more part of a first professional, and then managerial, model that comes from the institutional sphere with behavioural then developmental aims, and education in the Anthropocene corresponds more to a reformulation of citizen resistance with political citizenship aims. We should also mention that several metanarratives – the Anthropocene, ecological transition, climate urgency…etc – are the follow-on to education in sustainable development and are not equivalent in how they emerged, their political aims, or the knowledge they dispense. Knowledge of the Anthropocene is uncertain (Alpe & Barthes, 2013), problematic (Fabre, 2017, 2022), systemic (Giordan, 2017), complex (Morin & Le moigne, 1999; Favre, 2017), multi-referential (Ardoino, 1988), and prospective (Barthes et al., 2019). It aims to be coherent, in Audigier's sense of the term (2017), and political in the sense that it makes it possible to manage public affairs in a collective and enlightened manner (Barthes, 2017). The knowledge taught thus tends to exonerate any risk of absolutism, the consequences of which at the educational level are dogmatism and indoctrination, as well as certain forms of commitment. Far from normative drift (lessons in good moral behaviour – sorting waste properly, for example), it thus also avoids current relativist drift (Roth et al., 2020). The relativism of the value of truth can lead to confusion between knowledge, whose construction is robust, and opinions with little basis, and conclude that all distance between knowledge and opinions has disappeared, while considering that all knowledge is of equal value, or more generally that the distance between academic knowledge and social practices has been negated.

Education in the Anthropocene presupposes that there are questions about values, endpoints and stakes, and relationships with the world, truth, and power. If we consider that the Anthropocene corresponds on the one hand to reaffirmation of scientifically constructed knowledge and, on the other, reformulation of citizen resistance to normative liberal orthodoxy, in that case education in the Anthropocene is scientific, but it is also part of political citizenship. The knowledge dispensed is chosen with the aim of increasing both individual and collective capacities, and the values of social justice. We should also mention that the metanarrative of ecological transition refers more willingly to a normative dimension of weak education, the value of which is resilience and whose purpose is adaptive management of crises while maintaining the social structures in place and essentially technocratic knowledge.

References

Alpe, Y., & Barthes, A. (2013). De la question socialement vive à l'objet d'enseignement: Comment légitimer des savoirs incertains? In *Dossiers des sciences de l'éducation* (n° 29, Presses Universitaires du Mirail, pp. 33–44). Toulouse.

Ardoino, J. (1988). *Vers la multiréférentialité. Perspectives de l'analyse institutionnelle*. Méridiens-Klinksieck.

Audigier, F. (2017). *Cohérence. Dans Dictionnaire critique des enjeux et concepts des éducations à Paris* (pp. 565–566). L'Harmattan.

Barthes, A., et Alpe, Y. (2016). *Utiliser les représentations sociales en éducation*, Collection Logiques Sociales, Paris, L'Harmattan.

Barthes, A. (2017). Quels outils curriculaires pour des « éducations à » vers une citoyenneté politique? *Educations, 17–1*, International sciences and technical edition, ISTE, p. 25-40, n°1-327. Londres.

Barthes, A., Lange, J.-M., et Tutiaux-Guillon, N. (dir.) (2017). *Dictionnaire critique des enjeux et concepts de « éducations à »*. L'Harmattan.

Barthes, A. (2018). The Hidden curriculum of sustainable development: the case of curriculum analysis in France. *Journal of Sustainability Education*, *18*, Mars, Prescott.

Barthes, A., et Lange, J.-M. (2018). Researchers' positions and construction of curricula of education for sustainable development in France. *Journal of Curriculum Studies, 50*(1). Routledge, p. 96–112. London

Barthes, A., Blanc-Maximin, S., & Dorier, E. (2019). Quelles balises curriculaires en éducation à la prospective territoriale durable? Valeurs d'émancipation et finalités d'implications politiques des jeunes dans les études de cas en géographie. *Education et socialisation* n° 51. Montpellier.

Barthes, A., Sauvé, L., et Torterat, F. (2021). Questions environnementales et éducation au politique ? Numéro 63, *Éducation & Socialisation* (A paraitre).

Charbonnier, P. (2017). Généalogie de l'Anthropocène. La fin du risque et des limites. *Annales: Histoire, Sciences Sociales, 72*(2), 301 3028.

Favre, D. (2017). *Complexité. Dans Dictionnaire critique des enjeux et concepts de « éducations à »* (pp. 317–322). L'Harmattan.

Fabre, M. (2017). *Qu'est-ce que problématiser ?* Vrin.
Fabre, M. (2022). L'éducation au politique. Les problèmes pernicieux. *Série Education, ISTE Edition Ltd*, Londres pp. 241.
Hétier, R., et Wallendorst, N. (2020). *L'éducation politique en Anthropocène* (Le Télémaque). Presse universitaire de Caen.
Jickling, B., & Wals, A. (2008). Globalization and environmental education: looking beyond sustainable development. *Journal of Curriculum Studies, 40*(1), 1–21.
Giordan, A. (2017). *Systémique. Dans Dictionnaire critique des enjeux et concepts de « éducations à »* (pp. 565–566). L'Harmattan.
Lange, J.-M., et Kebaïli, S. (2019). Penser l'éducation au temps de l'anthropocène: Conditions de possibilités d'une culture de l'engagement. *Éducation et Socialisation, 51*(51).
Lange, J.-M., et Barthes, A. (2021). *« Education à » et « questions socialement vives »: Eduquer en contexte d'anthropocène*. Carrefour de l'éducation.
Levinson, R. (2012). A perspective on knowing about global warming and a critical comment about schools and curriculum in relation to socio-scientific issues. *Cultural Studies of Science Education, 7*(3), 693–701.
Morin, E., et Le Moigne, J.-L. (1999). *L'Intelligence de la complexité (avec)*. L'Harmattan.
Reid, A., & Dillon, J. (2015). *Major works in environmental education*. Routledge.
Roth, X., Barthes, A., et Cohen, J. (2020). Plaidoyer pour les antivaleurs? In *Penser l'éducation* (n°47, pp. 89–108). Rouen.
Sauvé, L. (2017). Education relative à l'environnement. In *Dictionnaire critique des enjeux et concepts de « éducations à »* (pp. 565–566). L'Harmattan.
Slimani, M., Barthes, A., et Lange, J.-M. (2020). *Les questions environnementales au miroir de l'évènement Anthropocène: tendance politique et hétérotopie éducative* (Le Télémaque. n° 58.2020-2, pp. 75–88). Caen.

Angela Barthes is Full professor at Aix-Marseille University, France. Holder of a double Ph.D./ HDR (geography and educational sciences), she is recognized as a speciaslist in education, political education, sustainability, and development of territories. She is the author of a dozen books.

Resonance

Nathanaël Wallenhorst

Abstract In April 2016, the German sociologist Hartmut Rosa published his book *Resonanz* (since translated into a variety of languages). To date, it has enjoyed even more success than Rosa's previous bestseller, *Acceleration*. These two books reflect a two-pronged movement in the Frankfurt School's Critical Theory, to which Rosa regularly reminds readers that his work belongs. Firstly, we see reflection on the processes of alienation in today's capitalist society, and the worsening of conditions for coexistence on our planet. Secondly, theorists are proposing a way (albeit a conceptual one, yet to be put into practice) to prevent such alienation, which could ultimately destroy the human adventure if left unchecked.

Rosa's intellectual work identifies *acceleration* as the main cause of alienation in today's world (Rosa, 2013). The imperative for growth (inherent in capitalism) drives a process of perpetual, unsustainable acceleration; caught up in the current, we gradually lose our humanity. For Rosa, the crux of the problem in modern society is this increasingly frenetic pace of relationships with other people, with ourselves, with objects and with the material world. This process of dynamisation is characteristic of modern social acceleration. Expanding upon Georg Simmel's analyses of the accelerating pace of life, in the late nineteenth and early twentieth centuries, Rosa sees three forms of acceleration at work simultaneously in today's world (technological acceleration, increasingly rapid social changes, and the increasing pace of life). To remain stable, societies must keep up with the growth and accelerating innovation. Therefore, competition and competitiveness are central features of the interactions between individuals and organisations. However, they have become central features of our mental processes too. The way in which we relate to the world suffers as a result of this dynamic stabilisation of contemporary societies; that relationship is at risk of becoming dysfunctional.

N. Wallenhorst (✉)
Catholic University of the West, Angers, France
e-mail: nathanael.wallenhorst@uco.fr

© The Author(s), under exclusive license to Springer Nature Switzerland AG 2023

N. Wallenhorst, C. Wulf (eds.), *Handbook of the Anthropocene*,
https://doi.org/10.1007/978-3-031-25910-4_225

Rosa proposes a kind of alternative to this intrinsic way of modern life. As a counterweight to the destructive alienation, he posits the concept of resonance (Rosa, 2018). The idea is that we should learn to expand our sense of being in the world (currently, our sense of being is shrunken, so we are barely aware of anything outside of ourselves). Resonance occurs outside the mechanisms of acceleration, and could help to build a new, postcapitalist world for all to share. Resonance aims to end the deadly *hubris* of maximising individual interests, which is so characteristic of how we live today. What is important is the relationship between the subject and the world, and resonance is defined as the process of uniting the two. It means a profound relationship with the world, which will have a major and lasting effect on the person in question. In concrete terms, resonance is experienced through different types of activities in daily life: 'Politics and democracy are types of resonance, as is work. Resonance is partly linked to the very foundation of our existence. I believe humans need to be sure of their relationship with the world. However, we do not have the means to provide an answer on this. Albert Camus (and, to a lesser extent, Friedrich Nietzsche) wrote that, at the root of our existence, there is a silence which we cannot comprehend, and to which we cannot be indifferent. The work of James, and also of Huber, contains very interesting ideas, based on the conviction – or the hope – that there may be answers at the root of existence. Religion, nature, art and history may be expressions of such answers' (Rosa & Wallenhorst, 2017b, p. 3).

The concept of resonance has certain points in common with Arendt's concept of action: it cannot be harnessed, it emerges, and does not endure beyond what is happening at a given moment in time. Rosa goes so far as to ask whether one of our woes today stems from the Promethean desire for control; in a way, resonance is distanced from the paradigm of project. In Rosa's view, we need to learn how to listen to the world, rather than dominate it. Rosa believes: 'A better world is possible. In order to achieve it, our relationship with the world needs to shift, from one where the aim is to possess and control things to one where the distinguishing characteristic is listening to the world' (Rosa & Wallenhorst, 2017a, p. 28). For Rosa, it is not so much a question of decelerating as of forming a new relationship, based on resonance, with the world – one that allows us to understand the world. Etymologically speaking, the word *resonance* comes from *re-sonare*, meaning to retain, or to echo. This refers to an acoustic phenomenon, where the vibration of one body stimulates another to vibrate, and the two continue to mutually stimulate one another. Resonance is also a musical metaphor: it means being intensely aware of our relationship with the world, and of its power to transform us. The notional universe being constructed in Germany (where the concept of resonance emerged) is marked by two traits that are unusual in traditional sociology: mobilising the existential aspect of the human condition, and sculpting our prospective thinking on the basis of social critique. Rosa exercise aims to lift humanity out of the rut into which it has sunk as a result of unbounded neoliberalism – an immeasurably lofty ambition.

With the concept of resonance, we can identify a necessary shift in how we learn from the world. Could resonance (and the accompanying learning), as opposed to acceleration, provide what we need to find our way in the age of the 'great acceleration' (Steffen et al., 2014, 2015; Waters et al., 2016)? (This is simply another name for the Anthropocene). Indeed, acceleration is apparent even in the geological sediment record, and it is undeniably such accelerated human activity that has led to the uncertain future now facing Earth in the Anthropocene. What lessons can and should be learned from this acceleration? To switch from acceleration to resonance with the world, we need to learn post-Promethean lessons. In post-Promethean education, self-emancipation should not be the ultimate goal, but rather, should be tied in with responsibility toward other people. 'In the modern Western world, being unconnected has come to represent a kind of absolute existence (etymologically, "absolute" means "without links", "untethered")'. In the Ancient world, though, 'being unconnected marked the distinction between slaves and free men: a free man was free precisely because he had a place in society, and could rely on his connections to others' (Flahault, 2008, p. 64). In education, responsibility, with the connections it creates with other people, is a twin to the logic of emancipation, which is a dominant part of many educational paradigms. In the context of resonance, the true beneficiary of education in the Anthropocene is not the student, but other people. We need to educate people on how best to take care of others, and give them the attention they need in order to exist.

The novelty of the concept of resonance, which promotes a paradigmatic shift in educational thinking, is its opposition to alienation. Previously, (self) emancipation was viewed as the counter to alienation. In traditional education, self-affirmation and liberation from alienating social pressures are presented as the counterweight to reification of the subject. Up until now, it has proved difficult to combat this idea in educational thinking, as it was born of the quest for social equality. The limit of the educational contrast between self-emancipation and alienation is the pursuit of the logic of reification. In self-emancipation, the self can sometimes become absolute, and be objectified. It can become the goal of education, which is then depoliticised, referring only to a strictly individual-focused, economic logic. However, the ambition in education must be to combat the alienation of individuals today; such alienation may take various forms (adaptation to market constraints, unequal access to social and financial resources, poor prospects of social integration, and so forth). The concept of resonance offers that alternative, that counterbalance to alienation, without self-reification or excessive focus on the individual. Relationships are the true epicentre of the concept of resonance.

To impart the concept of resonance, in education, our starting point must be not the learner, but rather the relationship between that person and others, things and the world as a whole. Such a shift of perspective is unusual in education. We need to think of learning in a different way, rather than simply the integration of knowledge by the subject (student). By taking intersubjectivity (and therefore relationships) as an anthropological starting point, with emphasis on coexistence, we should be able to repoliticise, and 'de-economise', the act of educating. Rosa draws on the thinking of American sociologist Randall Collins, expounding the idea that humans are marked primarily by interactions of solidarity, rather than by competitive and violent interactions. Rosa's hope is to see a post-growth society take shape. This idea is in line with those of convivialists such as François Flahault (2018), Fred Poché (2019), Corine Pelluchon (2015), Alain Caillé (2019), Pablo Servigne and Gautier Chapelle (2017). The theory does not detract from the importance of opposition and conflict (but the point is that there can be opposition without massacring the opponent, whether figuratively or literally). With resonance, the act of educating is seen as fundamentally post-Promethean: the goal is no longer to gain mastery or dominance over the world, but rather to establish a responsive relationship with it. We need to learn how to heed the world, rather than shackle it. Thus, with this concept, education and educational thinking become part of a post-Promethean existence.

References

Caillé, A. (2019). *Extensions du domaine du don*. Actes sud.

Flahault, F. (2008). *Le crépuscule de Prométhée – Contribution à une histoire de la démesure humaine*. Mille et une nuits.

Flahault, F. (2018). *L'homme, une espèce déboussolée – Anthropologie générale à l'âge de l'écologie*. Fayard.

Pelluchon, C. (2015). *Les nourritures*. Seuil.

Poché, F. (2019). *Le mécontentement*. Chronique Sociale.

Rosa, H. (2013). *Accélération. Une critique sociale du temps*. La découverte, French translation. [Original edition 2010].

Rosa, H. (2018). *Résonance – Une sociologie de la relation au monde*. La découverte, French translation. [Original edition 2016].

Rosa, H., & Wallenhorst, N. (interview with). (2017a). Apprendre à écouter le monde. *Chemins de formation, 21*, 19–31.

Rosa, H., & Wallenhorst, N. (interview with). (2017b). Apprendre ensemble dans la société de l'accélération. *Bildungsforschung, 2017*(1), 1–7.

Servigne, P., & Chapelle, G. (2017). *L'entraide – L'autre loi de la jungle*. Les Liens qui Libèrent.

Steffen, W., et al. (2014). *Global change and the earth system. A planet under pressure* (The IGBP Book Series). Springer.

Steffen, W., et al. (2015). The trajectory of the Anthropocene: The great Acceleration. *The Anthropocene Review, 2*(1), 81–98.

Waters, C., et al. (2016). The Anthropocene is functionally and stratigraphically distinct from the Holocene. *Science, 351*, 137–147.

Nathanaël Wallenhorst is Professor at the Catholic University of the West (UCO). He is Doctor of Educational Sciences and Doktor der Philosophie (first international co-supervision PhD), and Doctor of Environmental Sciences and Doctor in Political Science (second international co-supervision PhD). He is the author of twenty books on politics, education, and anthropology in the Anthropocene. Books (selection): *The Anthropocene decoded for humans* (Le Pommier, 2019, *in French*). *Educate in Anthropocene* (ed. with Pierron, Le Bord de l'eau 2019, *in French*). *The Truth about the Anthropocene* (Le Pommier, 2020, *in French*). *Mutation. The human adventure is just beginning* (Le Pommier, 2021, *in French*). *Who will save the planet?* (Actes Sud, 2022, *in French*). *Vortex. Facing the Anthropocene* (with Testot, Payot, 2023, *in French*). *Political education in the Anthropocene* (ed. with Hétier, Pierron and Wulf, Springer, 2023, *in English*). *A critical theory for the Anthropocene* (Springer, 2023, *in English*).

Sport

Léa Gottsmann and Christophe Schnitzler

Abstract Sport and physical activity are often presented as natural ways to promote sustainability and the development of eco-citizenship. In particular, outdoor sports are associated with many benefits: reconnecting with nature through *Green Exercise*, encouraging physical activity, and improving health and well-being. But the ethics and environmental cost of these activities are typically not questioned. Outdoor activities, which have been booming in the last decades, promote an anthropocentric vision of nature as a playground for human use. In this chapter, we seek to highlight how the anthropocentric values that shape sports practices widen the gap between outdoor education and sustainability literacy. Instead, we argue for the development of a different model for participation in outdoor sports and physical activities, based on the promotion of an eco-centric ethic, and provide examples of such practices across the globe.

Sports participation is imbued with numerous virtues, and involvement projects images of fitness, health and wellbeing, motivation, and social inclusion. In particular, outdoors activities are often associated with practices that contribute to building a better world, also meaning "greener". However, "sport" in its structured form emerged at the turn of the nineteenth century in England from the Industrial Revolution as a means of embodying the cult of performance (Pociello, 1981). Interestingly, this period coincides, at least for some, with the beginning of the Anthropocene epoch (Crutzen, 2002). In this chapter, we seek to discuss how sport is often associated with eco-citizenship, as if it were naturally beneficial for the environment, whereas evidence shows that it accompanies, or even contributes to, the advent of "the era of man", promoting an anthropocentric ethic.

L. Gottsmann (✉)
University Rennes (Laboratoire VIPS2, UR 4636), Rennes, France
e-mail: lea.gottsmann@ens-rennes.fr

C. Schnitzler
Laboratoire E3S (UR 1342), Faculté des Sciences du Sport de Strasbourg, Strasbourg, France
e-mail: cschnitzler@unistra.fr

© The Author(s), under exclusive license to Springer Nature Switzerland AG 2023

N. Wallenhorst, C. Wulf (eds.), *Handbook of the Anthropocene*,
https://doi.org/10.1007/978-3-031-25910-4_226

The International Olympic Committee (IOC) proclaims its engagement for the protection of the environment in its work, in response to rising awareness from the public (Loland, 2006). For example, in the Olympic Charter in Rule 2, Paragraph 13, it is written "that one of the key roles of the IOC is to encourage and support a responsible concern for environmental issues, to promote sustainable development in sport, and require that the Olympic Games are held accordingly". This trend is also present at the educational level. In 2017, an intergovernmental consensus conference on physical education was held in Kazan. It concluded that sport and physical education can help to set sustainable development goals, being directly linked to 7 among 17 goals enacted by the UN 2030 Agenda. However, these highly political declarations did not explain how to achieve any specific goal in relation to the preservation of the biosphere, as they are mostly concerned with health, human protection, and well-being.

Outdoor sports and adventure activities are widely promoted as ways to develop ecological values. There is a growing body of evidence documenting the benefits of outdoor activities and programs for reconnecting children to the natural environment (Merritt et al., 2017; Cho & Lee, 2018; Peden et al., 2016), enhancing physical activity (Mandic et al., 2016) and improving mental, physical and social health (Mygind et al., 2019). Outdoor sports are also called upon to promote active citizenship and reduce anti-social behaviours (Eigenschenk et al., 2019). However, outdoor activity does not necessarily develop environmental awareness, since nature can simply be perceived as a playground for humans (Vignal & Chazaud, 2001). Long et al. (2012) pointed out that professional mountain guides often did not take the opportunity to educate their clients about the beauty of nature, as these clients were instead oriented toward consuming nature as merely another form of entertainment.

The human-centered benefits of outdoor sport also come at a high environmental cost. Outdoor recreation and ecotourism have ecological impacts on aquatic, and terrestrial habitats and wildlife (Liddle, 1997; Xu, 2020). This effect is reinforced by the number of participants in outdoor activities, which has exploded in recent decades. For example, the construction of ski lifts for the purpose of facilitating participation in outdoor winter sports has profoundly modified the alpine landscape through total destruction, degradation, and/or fragmentation of the original natural habitat (Wipf et al., 2005). Other sports lead to the deliberate destruction of wildlife. For example, shark attacks on popular surf spots in Australia have been publicized and have led to programs to reduce the shark population through culling programs. Ironically, various studies have shown that these attacks may be mainly caused by human activities such as overfishing, waste pollution or construction (Lagabrielle et al., 2018). These examples illustrate the gap between the benefits of outdoor activities combining well-being and promotion of eco-citizen values, and the real impact of these sports activities on the environment. Even if sport institutions seem to become aware of environmental issues, their choices of development and lack of action to mediate impact, could be questioned with regard of anthropocentric attitudes (Schnitzler & Saint Martin, 2021).

Outdoor sports federations are becoming more aware of their responsibility to preserve the biosphere. This awareness is reflected in the development of processes and systems for a more sustainable use of the natural environment. Several models and charters are developed to preserve natural areas, manage natural resources, and to develop a more responsible approach toward the environment (Mounet, 2007). These proposals have been exemplified in sport climbing and bouldering (Van der Merwe & Joubert, 2014), in managing shark-human interactions (Lagabrielle et al., 2018) and in the 7 principles for "Leave no trace behind".[1] However, these tools have been developed under the prism of the well-being and safety of the (paying) participant. Most of the time, sport and physical activity programs do not frame how to place their practitioners in a natural environment, and when they do, it is most often to use nature as a simple playground. Most of the time, organized outdoor sports programs conveys anthropocentric values, which generate attitudes where environmental conservation concerns are considered important, but secondary, to participant well-being and pleasure. Therefore, the pedagogical framework that accompanies outdoor and nature sports activities is currently insufficient to develop environmental education in children and adults. It appears essential to complement this anthropocentric position with the promotion of eco-centric attitudes, emphasizing the importance of each component of the ecosystem and promoting its intrinsic value (Cocks & Simpson, 2017; Kopnina, 2015). Some outdoor practices already promote ecocentrism in their outdoor education curriculum. This is exemplified in Scandinavia with the *Frilufstliv*, which seeks "to facilitate a true connectedness to the more-than-human world" (Gelter, 2000). This is not the only example: Germany propose the "Bildung", Brazil promotes the "Buenvivir" and, in Japan "Shinrin-yoku" advocates, literally, the "forest bath" to reconnect to the environment.

Prominent among the major problems we identified in this article is that sport and physical activity remain deeply rooted in the ethics and values of the anthropocentric era, which most administrators and managers embrace. Sustainable activities must be more locally organized, at low environmental cost (food, transportation, equipment, energy costs) and accessible both economically and culturally. Unfortunately, outdoor sport organization remain mainly focused on their financial viability, which lead outdoor sport in the opposite direction. But the recent awareness of the environmental crisis could change this and allow the emergence of a new outdoor sport ethic.

References

Cho, Y., & Lee, D. (2018). 'Love honey, hate honey bees': Reviving biophilia of elementary school students through environmental education program. *Environmental Education Research, 24*(3), 445–460.

[1] https://lnt.org/why/7-principles/.

Cocks, S., & Simpson, S. (2017). Anthropocentric and ecocentric: An application of environmental philosophy to outdoor recreation and environmental education. *The Journal of Experimental Education, 38*(3), 216–227.

Crutzen, P. J. (2002). Geology of mankind—The Anthropocene. *Nature, 415*, 23. https://doi.org/10.1038/415023a

Eigenschenk, B., Thomann, A., McClure, M., Davies, L., Gregory, M., Dettweiler, U., & Inglès, E. (2019). Benefits of outdoor sports for society. A systematic literature review and reflections on evidence. *International Journal of Environmental Research and Public Health, 16*(6), 937. https://doi.org/10.3390/ijerph16060937

Gelter, H. (2000). Friluftsliv as slow experiences in a post-modern experience society. In B. Henderson & N. Vikander (Eds.), *Nature First* (pp. 37–46). Natural heritage books.

Kopnina, H. (2015). Neoliberalism, pluralism and environmental education: The call for radical re-orientation. *Environmental Development, 15*, 120–130.

Lagabrielle, E., Allibert, A., Kiszka, J. J., Loiseau, N., Kilfoil, J. P., & Lemahieu, A. (2018). Environmental and anthropogenic factors affecting the increasing occurrence of shark-human interactions around a fast-developing Indian Ocean Island. *Scientific Reports, 8*, 3676. https://doi.org/10.1038/s41598-018-21553-0

Liddle, M. (1997). *Recreation ecology: The ecological impact of outdoor recreation and ecotourism*. Chapman & Hall Ltd.

Loland, S. (2006). Olympic sport and the ideal of sustainable development. *Journal of the Philosophy of Sport, 28*(2), 217–139. https://doi.org/10.1080/00948705.2001.9714608

Long, T., Bazin, D., & Massiéra, B. (2012). Mountain guides: Between ethics and socioeconomic trends. *Journal of Moral Education, 41*(3), 369–388. https://doi.org/10.1080/03057240.2012.691635

Mandic, S., Williams, J., Moore, A., Hopkins, D., Flaherty, C., Wilson, G., García Bengoechea, E., & Spence, J. C. (2016). Built environment and active transport to school (BEATS) study: Protocol for a cross-sectional study. *BMJ Open, 6*(5), e011196. https://doi.org/10.1136/bmjopen-2016-011196

Merritt, E. G., Rates, C., Greiner, J., Baroody, A., & Rimm-Kaufman, S. (2017). We need trees to line the river to save our little friends. Environmental literacy development through service-learning. *Children, Youth and Environments, 27*(1), 67. https://doi.org/10.7721/chilyoutenvi.27.1.0067

Mounet, J.-P. (2007). Sports de nature, développement durable et controverse environnementale ? *Natures Sciences Sociétés, 15*(2), 162–166.

Mygind, L., Kjedsted, E., Hartmeyer, R., Mygind, E., Bølling, M., & Bentsen, P. (2019). Mental, physical and social health benefits of immersive nature-experience for children and adolescents: A systematic review and quality assessment of the evidence. *Health & Place, 58*, 102–136.

Peden, J. G., Hall, A., Westcost, G., & Police, S. (2016). A walk in the Forest: Elementary students' perceptions and knowledge of forestry principles. *Journal of Park & Recreation Administration., 34*(2), 62–73.

Pociello, C. (1981). *Sports et société. Approche socioculturelle des pratiques*. Vigot.

Schnitzler, C., & Saint Martin, J. (2021). Éduquer aux Activités de Pleine Nature en France: un défi pour l'EPS du XXIe siècle ? *Ejournal de La Recherche Sur L'intervention En Éducation Physique et Sport – eJRIEPS, 49*. https://doi.org/10.4000/ejrieps.5959

Van der Merwe, J. H., & Joubert, U. (2014). Managing environmental impact of bouldering as a niche outdoor-climbing activity. *South African Journal for Research in Sport, Physical Education and Recreation, 36*(1), 229–251.

Vignal, H., & Chazaud, P. (2001). Vers une prise en compte des rapports à la nature dans l'organisation des pratiques de plein air. In *9th international congress of physical activities* (pp. 1–3).

Wipf, S., Rixen, C., Fischer, M., Schmid, B., & Stoeckli, V. (2005). Effects of ski piste preparation on alpine vegetation. *Journal of Applied Ecology, 42*(2), 306–316. https://doi.org/10.1111/j.1365-2664.2005.01.011.x

Xu, S. (2020). Impact of outdoor sports on wildlife and protection. *Revista Científica de la Facultad de Ciencias Veterinarias, 30*(2), 1036–1044.

Léa Gottsmann is an assistant professor at the Ecole Normale Supérieure de Rennes and member of the Laboratory VIPS2. Her research is about physical education practices to promote environmental education in schools.

Christophe Schnitzler is an associate professor at the University of Strasbourg at the Faculty of Sports Sciences (F3S). His research and teaching focus on physical education, and question ways to promote environmental literacy through physical literacy. He also has a strong background in outdoor education, being a mountain guide himself.

Part XIX
Developing Socialization

Citizenship Education

Werner Wintersteiner

Abstract For a long time, citizenship education was understood as educating people to become responsible citizens of a particularly country or state. With the new situation characterized by the Age of Anthropocene, a rethinking of the very idea of education is needed. In particular, it is necessary to focus on the human condition as an elementary component of this education. More specifically, it is a matter of merging the concepts of global citizenship education and education for sustainable development, as well as human rights and peace education and (eco-) feminist education, in such a way that something new emerges, something that looks beyond the individual pedagogies without erasing the specific approaches in the process. This entails a twofold task: striving for global citizenship, i.e. solidarity among all people on Earth, while at the same time realizing planetary citizenship, i.e. solidarity with all beings in the biosphere, our basis of life.

For a long time, citizenship education was understood as educating people to become responsible citizens of a particular country or state, which implied loyalty to the nation, acceptance of democracy and active participation. This concept has since been softened by peace education, education for sustainability, human rights education, education for gender justice and global learning, since all of these pedagogic approaches define their goals not in a national framework, but on a global scale. Thus for some time now, citizenship education has been pushing to assume global responsibility and to be developed into global citizenship education without abandoning the national paradigm. With this recognition of the new situation, defined by the keyword Anthropocene, the old paradigm must finally be overcome in more regards than one. The guiding concepts for this are global citizenship and planetary citizenship.

W. Wintersteiner (✉)
Klagenfurt University, Klagenfurt, Austria
e-mail: werner.wintersteiner@aau.at

© The Author(s), under exclusive license to Springer Nature Switzerland AG 2023

N. Wallenhorst, C. Wulf (eds.), *Handbook of the Anthropocene*,
https://doi.org/10.1007/978-3-031-25910-4_227

The Challenge of the Anthropocene

The decisive characteristic of the Anthropocene is that the ability of the planet to be changed and controlled by humans has increased enormously (Cf. Wulf, 2020b, 33). This refers not only to the direct threat to human life from species extinction, rapid climate change, or self-extinction from nuclear weapons, but also to ambivalent developments such as genetic engineering and Artificial Intelligence. In the words of Donna Haraway, "Anthropocene is about the destruction of places and times of refuge for people and other critters" (Haraway, 100). Anthropocene, however, is a concept that has become common not only in the natural sciences but also in the social sciences, albeit interpreted in different ways, ranging from a glorification of human achievements to a critique of human hubris (cf. Federau, 2017, esp. Chapter IV). Some authors reject the term altogether, accusing it of being blind to capitalism and (neo-)colonialism: "[...] it is not correct to call it Anthropocene, as if all humans have the same degree of responsibility in this planetary catastrophe. It is mainly a fraction of humanity, the richest and the most powerful that are driving our existence into the abyss. It would be more appropriate to use the term Capitalocene or Plutocene or another denomination that highlights the destructive power of the logic of capital and the concentration of power in the hands of a very small minority of rich people" (Pablo, 2017, 187). Anne Fremaux (2019) also criticizes the "ecomodernist" and anthropocentric approach of the Anthropocene discourse and introduces the concept of a "green republican democracy". Donna Haraway (2016) makes a case for the term Chthulucene, not so much as a substitute for Anthropocene, but to describe the collective effort to escape human-made dangers. However, it should be noted that the intentions of critics of the term Anthropocene sometimes coincide with those of authors who do use the term.

This paper shares the critical view of many facets of the term's use, but the term is retained. In short, the fact that humanity has come to leave such a lasting mark on the earth is understood in this context as a warning and an appeal to rethink our relationship to the natural environment. Underlying this is the insight that "the concept of the Anthropocene extends beyond the relationship of humans to nature and has implications for physiological, social, and cultural existence, as well as for human self-understanding and identity" (Wulf, 2020b, 14). This also entails developing a new understanding of what it means to be human.

The struggle to establish this new understanding began long before the term Anthropocene was introduced. For example, Günther Anders (1961) was one of the first to state, in the aftermath of Hiroshima and Nagasaki, that a new age had now begun, while Hans Jonas (1979) pointed to the need to develop a new, expanded and globalized sense of responsibility in view of the power that technology offers humans. This responsibility, he said, must extend not only to human beings but to the entire living world, which, it goes without saying, is also in the interest of human beings themselves. Michel Serres, on the other hand, like quite a few others, has sought to combine this cultural dimension (responsibility) with a structural one (obligation) in the form of his proposal to establish not only all animals but also the earth as a whole as a political subject: "One has forgotten to invite the Earth to the climate conference" (Serres, 2009). Ultimately, what is at stake is what Donna Haraway calls "multispecies ecojustice" (Haraway, 2016, 202).

Rethinking Education

The challenges of the Anthropocene have fundamental implications for education. In this context, it is important to counter the frequently occurring misconception that all the ills of civilization can be eliminated through education, a line of thinking that attempts to instrumentalize education for political goals. Education in the Anthropocene must not be reduced to teaching the Anthropocene or to environmental education. Precisely because our understanding of what it means to be human is ever evolving and because there are numerous competing concepts in this regard – from a reformulation of humanism to transhumanism and posthumanism – it is necessary to focus on teaching the human condition (cf. Wulf, 2020a). Thus, it is not simply a matter of identifying new bodies of knowledge, but instead of a thorough rethinking of the very idea of education. This is also a central thesis in Edgar Morin's essay "Seven complex lessons in education for the Future", written for UNESCO. Education has the task of promoting "cognitive democracy", i.e. providing learners with the knowledge they need to cope with life in an environment that has become complex. This includes, at the very least, the following three aspects: first, it is about dealing with knowledge, about "knowledge about knowledge" and about how prone to error cognitive processes can be, about critical thinking, which must always be holistic thinking: "The predominance of fragmented learning divided up into disciplines often makes us unable to connect parts and wholes; it should be replaced by learning that can grasp subjects within their context, their complexity, their totality" (Morin, 1999, 1). Secondly, it is a matter of teaching the human condition, the threefold definition of homo sapiens-demens as an individual, a social being and a species: the "complex unity of human nature has been so thoroughly disintegrated by education divided into disciplines, that we can no longer learn what human being means. This awareness should be restored so that every person, wherever he might be, can become aware of both his complex identity and his shared identity with all other human beings" (Morin, 1999, 2). And thirdly, it is about "earth identity": "establishment of a relationship of mutual control between society and individuals by way of democracy, fulfillment of Humanity as a planetary community" (Morin, 1999, 4). The notion of Citizenship Education is thus directly addressed.

Rethinking Citizenship Education

At first glance, it would seem that with the concepts of Global Citizenship Education (GCED) and Education for Sustainable Development (ESD) we already have the instruments we need to create a corresponding education policy. Should not all citizenship education take the global dimension into account? And isn't sustainability one of the main goals of any responsible politics? In UNESCO, for example, both concepts are now not only present, but are often represented together (see the biennial Global Forums for GCED and ESD). The UN Sustainable Development Goals (SDGs) as agreed upon by the global community also envisage a bundling of these two pedagogies with additional approaches such as peace education, human rights education and so on according to target 4.7. However, the conceptual foundations so far do not seem to be anywhere near sufficient to successfully face the challenges of the Anthropocene. The following are the main points of criticism:

- Many political pedagogies focus only on their own particular issues, but remain uncritical of traditional knowledge production and indifferent to prevailing educational policies. They fail to note that this often subjects them to the dictates of educational policies that reduce education to measurable competencies. GCED and ESD in particular have struggled with this problem.
- The individual pedagogies are still taught in isolation, and in the worst cases, in competition with each other; ESD too often focuses on purely ecological aspects without examining the global citizenship dimension; GCED does not pick up enough of the impulses provided by ESD and fails to place them at the centre of its cosmopolitan thinking; both ESD and GCED fall short in taking note of the postcolonial criticism of the way the West discusses both ecology and cosmopolitanism (cf. De Souza & Andreotti, 2012; Ferdinand, 2019; Wintersteiner, 2019). And this problem is not exclusive to pedagogic approaches. To give an example: although "Peace Ecology" is by no means a new topic (see, for example, the Earth Charter, 2000), an integrative view is far from being standard in peace research (see, however, Brauch & Oswald Spring, 2021).
- Global citizenship education is generally still classified in a national paradigm, as something supplementary to citizenship education, when it is in fact a matter of "rethinking the whole of citizenship education in the light of the cosmopolitical viewpoint" (cf. Lamarre, 2021, 124/125).
- Even where critical thinking is emphasized, it is not uncommon to ignore the danger of unilaterally emphasizing the threats to humanity and thus sowing the seed of pessimism. In addition to the critical function, however, the function of resistance and of utopian thinking must also be added, as Wallenhorst and Mutabazi (2021, 19) explain conclusively.

Summary: Education for "Homeland Earth"

By now, a wealth of literature already exists on education and the Anthropocene, but there are few publications that actually explore the challenge of the Anthropocene for pedagogy in detail (e.g. education in the anthropocene, 2019; Wallenhorst & Pierron, 2019; Wulf, 2020a; Curnier, 2021; Gough, 2021; Wallenhorst & Mutabazi, 2021).

Thus there is still work to be done, not only in empirical, but also in conceptual terms. Basically, it is a matter of merging the concepts of global citizenship education and education for sustainable development, in addition to human rights and peace education and (eco-) feminist education, in such a way that something new emerges, something that looks beyond the individual pedagogies without erasing the specific approaches, a kind of updated version of what Betty Reardon (1988) has proposed in *Comprehensive Peace Education* ca. 30 years ago. This "something new" applies as far as the understanding of education is concerned, as far as knowledge is concerned, as far as problem awareness is concerned and lastly, as far as action is concerned. The concept of "Homeland Earth", as developed by Edgar Morin and Anne Brigitte Kern (1998), can show us the way. It entails a twofold task: striving for global citizenship, i.e. solidarity among all people on Earth, while at the same time realizing planetary citizenship, i.e. solidarity with all beings in the biosphere, our basis of life. It comes close to what John Barry calls "green republican citizenship" (Barry, 2016), to show another facet of this ongoing discussion.

Note: All translations from German and French by the author.

References

Anders, G. (1961). *Die Antiquiertheit des Menschen. Band I: Über die Seele im Zeitalter der zweiten industriellen Revolution*. C.H. Beck.

Barry, J. (2016). Citizenship and (un)sustainability. In S. M. Gardiner & A. Thomson (Eds.), *The Oxford handbook of environmental ethics* (pp. 333–343). Oxford University Press.

Brauch, H. G., & Oswald Spring, Ú. (Eds.). (2021). *Decolonising conflicts, security, peace, gender, environment and development in the Anthropocene*. Springer Nature.

Curnier, D. (2021). *Vers une école éco-logique*. Le Bord de l'eau.

De Souza, L. M., & Andreotti, V. (Eds.). (2012). *Postcolonial perspectives on global citizenship education*. Routledge.

Education in the anthropocene. (2019). On education. *Journal for Research and Debate, 2*(4).

Federau, A. (2017). *Pour une philosophie de l'Anthropocène*. PUF.

Ferdinand, M. (2019). *Une écologie décoloniale. Penser l'écologie depuis le monde caribéen*. Seuil.

Fremaux, A. (2019). *After the Anthropocene: Green republicanism in a post-capitalist world*. Palgrave Macmillan.

Gough, A. (2021). Education in the Anthropocene. In C. Mayo (Ed.), *Oxford encyclopedia of gender and sexuality in education*. Oxford University Press.

Haraway, D. A. (2016). *Staying with the troubles. Making kin in the Chthulucene*. Duke University Press.

Jonas, H. (1979). *The imperative of responsibility: In search of ethics for the technological age*. University of Chicago Press.
Lamarre, J.-M. (2021). Citoyenneté mondiale et éducation cosmopolitique. In N. Wallenhorst & E. Mutabazi (Eds.), *D'une citoyenneté empêchée à une éducation citoyenne* (pp. 109–128). Le Bord de l'eau.
Morin, E. (1999). *Seven complex lessons in education for the future*. UNESCO.
Morin, E., & Kern, A. B. (1998). Homeland earth. *A Manifesto for the New Millenium*. Hampton Press.
Pablo, S. (Ed.). (2017). Systemic alternatives. Fundación Solón/Attac France/Focus on the Global South 2017.
Reardon, B. A. (1988). *Comprehensive peace education: Educating for global responsibility*. Teachers College Press.
Serres, M. (2009, December 21). On a oublié d'inviter la Terre à la conférence sur le climat. *Le Monde*.
The Earth Charter. (2000). https://earthcharter.org/library/the-earth-charter-text/. Accessed July 13, 2021.
Wallenhorst, N., & Mutabazi, É. (2021). Le temps de la citoyenneté. In N. Wallenhorst & E. Mutabazi (Eds.), *D'une citoyenneté empêchée à une éducation citoyenne* (pp. 7–22). Le Bord de l'eau.
Wallenhorst, N., & Pierron, J.-P. (sous la direction de). (2019). Eduquer en Anthropocène. Le Bord de l'eau.
Wintersteiner, W. (2019). Global citizenship education – Eine pädagogische Antwort auf die ‚große regression'? *Zeitschrift für internationale Bildungsforschung und Entwicklungspädagogik 42* (1), 21–25.
Wulf, C. (2020a). *Bildung als Wissen vom Menschen im Anthropozän*. Weinheim.
Wulf, C. (2020b). Den Menschen neu denken im Anthropozän. Bestandsaufnahme und Perspektiven. *Paragrana, 29*(1), 13–35.

Werner Wintersteiner, former founding director of the "Centre for Peace Research and Peace Education" at Klagenfurt University, Austria. Research fields: peace and global citizenship education, culture and peace, the Alps-Adriatic region, literature and peace, literature education. Recent publications: Die Welt neu denken lernen – Plädoyer für eine planetare Politik (2021). [Learning to Rethink the World – a Plea for a Planetary Politics]; (With Cristina Beretta and Mira Miladinović Zalaznik): Manifestlo Alpe Adria (2020); Herbert C. Kelman: Resolving deep-rooted conflicts. Essays on the Theory and Practice of Interactive Problem-Solving. Eds. Werner Wintersteiner and Wilfried Graf (2017).

Climate Education

Pierre Léna and Lydie Lescarmontier

Abstract This article provides an overview of the progress observed in climate change education in the last two decades. Beginning with a rather slow perception of the issue within international bodies, it grew under the general theme of education for sustainable development. The 2015 Conference of Parties in Paris (COP21) did much to promote it with a specific article in the Paris Agreement on Climate Education. Various international bodies, such as Science Academies and Unesco, took initiatives to shed light on climate change education and to address its absence in most curricula. The mobilisation of youth, observed across the world, demonstrates the demands of the young generation. Curricula are slow to change, and teachers are often unprepared to teach the complex and systemic subjects of climate change, its anthropogenic causes, and adaptation and mitigation strategies. Yet these are identified as the key elements in education systems for contributing to the requests of young people and raising the awareness of populations. To support this evolution, numerous initiatives across the world, both in developed and developing countries, are emerging with the support of climate scientists. They may give hope for future progress.

Climate change and its consequences have been the subject of major scientific work since 1979 (Nichols, 2019), followed by the creation of the IPCC in 1988, whose regular Assessment Reports deal with the evolution of the Earth's climate, as well as with possible mitigation and adaptation scenarios. However, the educational concern itself, specific to climate, knew a slower emergence, analyzed in detail in Borde et al. (2022) and recently well summarized by Alan Reid or Marie-Françoise Gibert

P. Léna (✉)
Observatoire de Paris, LESIA, Meudon, France
e-mail: pierre.lena@obspm.fr

L. Lescarmontier
Office for Climate Education, Sorbonne Université, Paris, France
e-mail: lydie.lescarmontier@oce.global

© The Author(s), under exclusive license to Springer Nature Switzerland AG 2023

N. Wallenhorst, C. Wulf (eds.), *Handbook of the Anthropocene*,
https://doi.org/10.1007/978-3-031-25910-4_228

(Reid, 2019; Gibert, 2020) Mentioned as early as 1992 in the Intergovernmental Panel for Climate Change (IPCC) Convention, it is the subject of the specific international programme UN CC Learn (UNCC Learn, 2009) followed by the IPCC Doha programme. It became more explicit in Art.12 of the Paris Agreement in 2015, which states: "Parties shall cooperate in taking measures, as appropriate, to enhance climate change education, training, public awareness, public participation and public access to information, recognizing the importance of these steps with respect to enhancing actions under this Agreement". Although the Article includes informal as well as formal education, and the general public as well as the schools population, the focus of what follows is limited to primary and secondary schools, with a brief overview of universities.

The world's science academies, gathered within the InterAcademy Partnership for science (IAP), expressed in 2017 the willingness of the scientific community to take up the challenges for schools, posed by Art. 12 (IAP, 2017). In 2015 the publication of the United Nations Agenda 2030, with its 17 sustainable development goals (SDG), included education (Goal #4) and climate (Goal #13). OECD, along the same lines, published in 2020 its Learning Compass (Oecd, 2019). In these, the subject of climate education is included in a broader 'education for sustainable development' (ESD), which can highlight the importance of the climate, but also mask its specificities. After reporting in 2019 on the progress of Art. 12 implementations (Unesco, 2019), Unesco presented in 2021, at the COP26 in Glasgow, a new assessment, showing a slow rise of awareness in many states and their educational systems (Unesco, 2021). Jointly with OECD and the NGO Education International, teachers from across the globe were also invited to share at this COP26 climate action initiatives and projects through short video explainers (OCE, 2021).

In 2020, European Academies undertook a survey of climate education in Europe (Allea, 2020), in phase with decisions of the European Commission to develop further the subject with the creation of a community of actors (EU, 2020).

Youth is doubly concerned by the entering into the Anthropocene. First, children as well as young adults share an anxiety for the future, made explicit in the projection targets of 2050 and 2100 and their major risks, precisely examined by the IPCC reports. The demonstrations worldwide, including the global climate marches, have shown youth's concern and expectations, already well expressed in a White Paper prepared in 2017 at an IPCC Conference in Canada (Goldwaldt & Karsgaard, 2018). From primary school to universities, formal education has to answer these expectations. Moreover, the magnitude of societal changes required by the climate transition, as outlined in the IPCC Assessment Report #6, implies great changes of behaviour from whole populations, a goal out of reach in democratic countries if it were not tenaciously carried by education of the youth.

Since 2020, the profound impact of the pandemic on youth, particularly in developing countries, may make it difficult to obtain a fair assessment of young people's perception of climate change. However, several recent surveys (2021 and 2022), carried out in France or elsewhere, among young people aged 16 to 30, are instructive and reflect an interesting but worrying reality, which can undoubtedly be considered as characteristic of developed countries (La Croix, 2021; Ipsos, 2021;

Institut Montaigne, Hickman et al., 2021), but also of others, such as Nigeria or the Philippines (Hickman et al., 2021). They corroborate and amplify the conclusions of previous ones (Corner et al., 2015). These studies converge to picture a youth who in great majority is aware of a threatening climatic future, often qualified as an expected 'catastrophe'. Climate comes first when they are questioned more generally on 'ecology'. Anxiety for the future, sometimes coined solastalgy, is today present from early ages onward (Hickman et al., 2021). In a call published in 2021, over ten thousand scientists underlined the importance of education, "able to increase awareness and empower learners to action" (Ripple et al., 2021).

Four goals can be identified for undertaking climate education: awareness, knowledge, understanding and action (or agency, which better defines the expectation). Awareness, as just quoted, may not be rational, as often carried in social networks with excessive words, e.g. 'the extinction of humanity'. Awareness has to be based on rationality, hence to be connected to a scientific knowledge, even minimal. Knowledge has to build up on a progressive understanding of the Earth's complex system and its evolution. Finally, action is essential to provide hope in the future, to balance the feeling of being powerless when facing the magnitude of the climate problem.

The long process of schooling, from kindergarten to 12th grade, the place which science holds within it offer an adequate terrain to progressively enter into this difficult and challenging subject. The youngest children deserve to discover the beauty of the Earth's landscapes, its size and location in space, to name and describe the phenomena related to the atmosphere, the soils and the ocean, to experiment with air, water and light: here are the climate basic ingredients. Applications on smart phones, such as 'Windy' developed in Europe, provide a fascinating view in real time of the world's weather and its parameters. At the end of primary school and along middle school, the Earth system can be better understood, with the concepts of energy and its transformations, interdependencies of the subsystems, including the biosphere and the anthropogenic impacts. During these years, a direct contact with nature is important, especially for students living in large cities, where this contact is often lost. Discovering, naming and understanding in situ forests, sea shores, day and night sky represent an unforgettable experience for children and teenagers. It can help bridging the gap with the abstract content of climate issues, reduce the psychological distance to these and instil the desire to contribute by action.

In the final grades, a more advanced knowledge in the natural sciences involved in climate science (physics, chemistry, mathematics, astronomy, Earth science, life sciences) may proceed. The connexion with human and social sciences, caring for adaptation and attenuation issues (sociology, demography, law, etc.), as well as with philosophy presenting ethical issues (e.g. climate justice) may also come at this stage, where the maturity of students allows to better deal with the intricacies of the matter.

Higher education, answering the explicit wishes of students across the world, has to build up on these earlier foundations, but also to account for the specific studies undertaken by the students in view of their professional future. Although beyond the core of the present article, it is worth mentioning, among many, the examples of California campuses, with the course 'Bending the Curve' since 2018, or France, with the studied mobilization of higher education towards the ecological transition, where climate and biodiversity issues play a major part (Ramanathan et al., 2019; Jouzel, 2022).

Climate science was practically born and rapidly developed during the last half-century thanks to space observations collecting terrestrial data and to the power of computers allowing numerical simulations and projections towards the future. This science, dealing with the Earth system, is deeply interdisciplinary, calling upon the whole range of experimental sciences, as underlined above. However, in school programs, these sciences are divided into disciplinary subjects that hide the systemic and essential dimension of the climate problem. The competence of the teachers is itself disciplinary, with the general exception of the elementary school. Neither the content of the programmes they must follow, nor their own competences naturally make a place for climate science. There is no simple solution to this problem, because solving it requires, on the one hand, the production of adapted teaching resources, and on the other hand, the collective work of teachers, each one being a specialist in a different discipline, in order to be able to present to students a systemic vision, without which it becomes impossible to understand the multiple interactions of climate mitigation or adaptation strategies. An interesting set of teaching strategies and resources is proposed by the group 'Trans-disciplinary Research Oriented Pedagogy for Improving Climate Studies and Understanding' or 'Tropicsu', from India (Pune), which develops schemes of implementation of climate science for high schools and bachelor level, while fitting in different course subjects: from Geography to Economics, Humanities to Chemistry, etc. (Tropicsu, 2022).

It is here that it is perhaps useful to emphasize the specificity of climate change education within the broader goal of sustainable development or ecological transition. Students need a framework for structured and clear thinking, as much as the complexity of the subject matter allows. All too often, education for sustainable development takes the form of nice gestures, such as sorting waste or planting shrubs in the schoolyard, which, although symbolic, are far from contributing to the necessary systemic vision. To illustrate this point, let us note that in France in 2019, where there are elementary school with an eco-label which asserts their ecological commitment, 40% of these schools implement waste sorting, but only 3% are interested in climate change. Climate and biodiversity are two subjects whose scientific roots can guarantee a constructed progression of the awareness, knowledge, understanding and agency.

Since 2015, the increasing mobilization of educational systems has highlighted the need for appropriate educational resources for teachers at all school levels. Scientific institutions in developed countries have been the first to provide quality resources, which are now available to teachers in large numbers on the Internet and in various languages. As examples, NASA offers a guide to teach young people about climate change ('Climate Kids') (NASA, 2022) as well as others (NASA Projects 2022). The Jet Propulsion Laboratory in California, which designs space missions to other planets in the Solar system, offers a host of resources to teach about ocean thermal expansion, the role of clouds, and the climatic differences between Earth and Mars. Since the ocean and atmosphere play a central role in the climate machine, the National Ocean and Atmosphere Administration (NOAA), a public agency in the USA, has played a leading role in educational stimulation and offers remarkable resources with its participation to the Climate Literacy and Energy Awareness Network program (CLEAN) (Clean Program, 2022).

In all societies, the successive IPCC Reports play a great role in the perception and understanding of the climate evolution, as well as in the evaluation of potential strategies for mitigation and adaptation. The Summaries for Policy Makers (SPM), which accompany each report in a simplified version, are adequate to communicate widely the main IPCC messages to political, economic and financial circles. These summaries are not fully adequate for teachers, most of them not having the expertise nor the time to read and moreover to conceive for their students the pedagogical translation of the reported knowledge. To answer this and following the IAP recommendation (IAP, 2017), climate scientists and pedagogical experts created in 2018 an Office for Climate Education (OCE) in France, in order to provide 'Summaries and Tools for Teachers', echoing each of the IPCC Reports. Multilingual summaries of the IPCC Reports are published, focusing of the scientific knowledge, accompanied by detailed pedagogical tools aimed at middle and high school teachers, but possibly extended towards primary school. For example, the IPCC Report 'Ocean and Cryosphere' (2018) led to a Summary for Teachers (Lescarmontier, 2020) and a pedagogical guide (Rocha & Wilgenbus 2020). In addition, the Office initiated a regional training program in Latin America (Chile, Colombia, Mexico), is preparing a similar one in Africa and contributes to another in South-East Asia (OCE, 2022). The Office is a Unesco Category 2 Center and an IPCC Observer. During the COP26 in Glasgow (2021), in parallel with the Global Teaching Insights initiative (Climate action, 2021), OCE organised a special event named 'Teachers' COP', prepared with teachers from many developing countries and submitting propositions for implementing climate change education. Their Statement 'Time for Teachers' empowerment' was presented at the COP. It outlines curricula, the professional development of teachers, the educational resources and the place of school in the context of climate change (OCE, 2021).

This short article has presented a very limited sample of thoughts, studies and initiatives related to climate change education in primary and secondary schools. Compared to the situation in 2015, at the time of the unanimous vote leading to the Paris Agreement at the COP21, great progress can be observed and give trust in future developments. Indeed, the climate clock is running fast, time is short, and the role of youth awareness and action is more than ever essential. Top-down decisions from governments, as well as bottom-up initiatives from schools and teachers, are needed. As the issue is global, international cooperations of all kind are needed and already benefit from the contribution of scientists, especially the ones from all over the world, who collaborate to the IPCC reports and appreciate, better than anyone else, the urgency of action, in order to convey to the young generation hope rather than despair.

References

All European Academies (Allea). (2020). *A snapshot of climate change education initiatives in Europe*. An Allea Report' (2020). Retrieved from https://allea.org/publications/

Borde, B., Léna, P., & Lescarmontier, L. (2022). Education as a strategy for climate change mitigation and adaptation. In M. Lackner, B. Sajjadi, & W. Y. Chen (Eds.), *Handbook of climate change mitigation and adaptation*. Springer Nature Reference. Living Edition 2022.

Clean Program (Clean). (2022). *Climate literacy and energy awareness network*. Retrieved from https://cleanet.org/index.html

Climate Action. (2021). *Global teaching insights*. Retrieved from https://www.globalteachinginsights.org/channel/Climate+Action/212779523

Corner A.et al., (2015). 'How do young people engage with climate change? Therole of knowledge, values, message framing, and trusted communication'. WIREs Climate Change, 6:523–534 (2015). doi: https://doi.org/10.1002/wcc.353.

European Union EU. (2020). *Education for climate*. Retrieved from https://education-for- climate.ec.europa.eu/community/home

Gibert A. F. (2020, March).'Éduquer à l'urgence climatique'. Dossier de veille de l'Institut français de l'éducation, #133, École normale supérieure de Lyon.

Goldwaldt, T., & Karsgaard, C. (2018). *Center for global education*. Retrieved from https://tcge.ca

Hickman, C. et al., (2021, December 1). Climate anxiety in children and young people and their beliefs about government responses to climate change: a global survey. *The Lancet* 5, 12, Open access. https://doi.org/10.1016/S2542-5196(21)00278-3.

Institut Montaigne. (2022). *Une jeunesse plurielle. Enquête auprès des jeunes de 18 à 24 ans*. Retrieved from https://www.institutmontaigne.org/publications/une-jeunesse-plurielle-enquete-aupres-des-18-24-ans

InterAcademy Partnership (IAP). (2017). *A statement on climate change and education*. Retrieved from https://www.interacademies.org/statement/statement-climate-change-and-education

IPCC. (2019). Summary for policymakers. In: *IPCC special report on the ocean and cryosphere in a changing climate* H.-O. Pörtner et al. (eds.). Retrieved from https://www.ipcc.ch/srocc/

IPCC. (2021). Summary for policymakers. In: Climate change 2021: The physical science basis. Contribution of working group I to the sixth assessment report of the intergovernmental panel on climate change Masson-Delmotte, V., et al. (eds.). Cambridge University press. Retrieved from https://www.ipcc.ch/report/ar6/wg1/#SPM

IPSOS et Fondation Collège de France (Ipsos). (2021). 'Les jeunes et la science'. IPSOS et Fondation Collège de France (2021). Retrieved from https://www.ipsos.com/fr-fr/les-jeunes-et-la-science

Jouzel, J. (coord.). (2022). 'Enseigner la transition écologique dans le supérieur'. Rapport à la ministre de l'enseignement supérieur. Retrieved from https://www.enseignementsup-recherche.gouv.fr/fr/remise-du-rapport-sensibiliser-et-former-aux-enjeux-de-la-transition-ecologique-et-du-developpement-83903

La Croix & Quantité critique. (2021). *Ecologie: les multiples visages des jeunes de la génération climat*. Retrieved from https://www.la-croix.com/Ecologie-multiples-visages-jeunes-generation-climat-2021-06-14-1201161114

Lescarmontier, L. (coord.) (2020). 'The ocean and cryosphere in a changing climate. Summary for Teachers'. Retrieved from https://www.oceancryosphere.oce.global/en/projects/ocean-and-cryosphere

NASA Climate Kids (NASA). (2022). Retrieved from https://climatekids.nasa.gov/

NASA Projects (NASA). (2022). Retrieved from https://climate.nasa.gov/resources/education/

Nichols, N. (2019, July 23). *The Charney report: 40 years ago, scientists accurately predicted climate change*. Retrieved from https://phys.org/news/2019-07-charney-years-scientists-accurately-climate.html

Oecd. (2019). *OECD Learning compass 2030*. Retrieved from https://www.oecd.org/education/2030-project/teaching-and-learning/learning/learning-compass-2030/

Office for Climate Education (OCE). (2021). *Teacher's COP statement on climate change education: Time for teacher's empowerment*. Retrieved from https://www.interacademies.org/publication/teachers-cop-statement-climate-change-education-time-teachers-empowerment
Office for Climate Education (OCE). (2022). Retrieved from https://www.oce.global
Ramanathan, V., Millar-Ball, A., Niemann, M., & Friese, S. (Eds.). (2019). *Bending the curve: Climate change solutions*. The Regents of University of California (2019).
Reid, A. (2019). Climate change education and research: Possibilities and potentials versus problems and perils? *Environmental Education Research, 25*(6), 767–790.
Ripple, W. J., et al. (2021). World scientists warning of a climate emergency 2021. *Bioscience, XX*, 1–5.
Rocha, M., & Wilgenbus, D. (coord.) (2020). *The climate in our hands. Ocean and cryosphere. teacher's guide book for primary and secondary school*. Retrieved from https://www.ocean-cryosphere.oce.global/en/projects/ocean-and-cryosphere
Tropicsu. (2022). *Climate change education across the curricula, across the globe*. Retrieved from https://tropicsu.org/
UNCCLearn. (2009). Retrieved from https://www.uncclearn.org/
Unesco. (2019). *Country progress on climate change education, training and public awareness: An analysis of country submissions under the united nations framework convention on climate change*. Retrieved from https://unesdoc.unesco.org/ark:/48223/pf0000372164
Unesco. (2021). *Getting every school climate-ready: how countries are integrating climate change issues in education*. Retrieved from https://unesdoc.unesco.org/ark:/48223/pf0000379591

Pierre Léna is an astrophysicist, born 1937, Emeritus Professor at the Observatoire de Paris and the University of Paris, member of the French Académie des sciences. His scientific career dealt with infrared astronomy and the exploration of this domain with instruments on ground-based observatories, aircrafts and satellites. He had a long-lasting interest in the transmission of science to the public, and especially through education. In 1995, he co-founded in France the movement *La main à la pâte* to develop inquiry-based science education in schools, and in 2018 he extended this action to climate issues, with the creation of the Office for Climate education (OCE).

Lydie Lescarmontier is a glaciologist by training, and specialized her work on the impact of climate change on the Antarctic ice sheet. Her interest in science education and then for climate change education started in 2015 when she became vice-president of the French committee of APECS (Association for Polar Early Career Scientists). In 2018 she joined the Office for Climate Education as science officer to coordinate the production of educational resources and to participate to the teachers' professional development.

Collapsonauts

Yves Citton and Jacopo Rasmi

Abstract While collapsologists attempt to elaborate a scientific discourse to anticipate the unescapable collapse of our thermo-industrial civilization, collapsonauts, by contrast, attempt to gather forces and wits in order to navigate as ethically and effectively as possible on the crumbling expectations raised by Western profligacy. This article presents the specificities of the (mostly French) collapsologist movement, before describing the collapsonaut attitude as a pragmatic analysis and anticipation of the structural weaknesses of the social ideology of self-possessive individualism.

The hypothesis that human civilizations as such, or a particular social order, may collapse has haunted mankind at different moments in history. Jared Diamond's book *Collapse* (2005) has widely popularized this theme over the last decades, along with Chris Smith's, 2009 film by the same title, devoted to a long interview with Michael Ruppert. Thomas Moynihan's (2020) rich analysis of the myriad of scholars that have discussed and anticipated the total extinction of mankind, across many centuries, along with anthropologists' studies of the social history of apocalyptic anxieties, feelings of doom and ends of the world (De Martino, 1977), provide a rich background against which to situate collapsology as the dominant form of millenarism in the age of the Anthropocene.

Within this large historical and international context, collapsology itself became surprisingly popular in the French speaking world, during the years 2015–2020, being frequently presented and debated in mainstream media (while it remained virtually absent from the agenda in neighbouring countries like Italy or Spain). According to Yves Cochet, who was Minister of the Environment in the French socialist government in 2001–2002, collapse (*effondrement*) can be defined as "the

Y. Citton (✉)
Université Paris 8, Saint-Denis, France

J. Rasmi
Université Jean Monnet, Saint Etienne, France
e-mail: jacopo.rasmi@univ-st-etienne.fr

© The Author(s), under exclusive license to Springer Nature Switzerland AG 2023

N. Wallenhorst, C. Wulf (eds.), *Handbook of the Anthropocene*,
https://doi.org/10.1007/978-3-031-25910-4_229

process through which basic needs (water, food, housing, clothing, energy, etc.) are no longer provided to a majority of the population by law-abiding collective services" (Cochet, 2011). Édouard Philippe, French Prime Minister between 2017 and 2020, also stated being seriously worried by such a possibility, acknowledging that Jared Diamond's *Collapse* was one of his decisive readings of the previous years.

The topic was made popular by a 2015 best-seller by Pablo Sevigne and Raphaël Stevens (2015), *How Everything Can Collapse: A Manual for our Times*, which was followed by dozens of related in books by different authors in the next 5 years. As to why *effondrement* (briefly?) became a defining feature of French culture (like *baguette*, Camembert and Bordeaux wine), a convincing hypothesis may be that the French people has set in place (and defended against the onslaught of neoliberalism) a most desirable safety net of social services provided by the Welfare State, and feels particularly threatened by the possibility that basic needs could be "no longer provided by law-abiding services".

Collapsologists present their worrying prevision about the doomed future of the thermo-industrial society through a flurry of charts and statistics inspired by the Club of Rome report *The Limits to Growth* (Meadows & Donella, 1972), as well as by Will Stephen's (2004) *Great Acceleration*. Their first publications mostly consisted in presenting the case for the possibility, probability and eventually unavoidability of collapse. They showed their readers that our modes of production were indeed unsustainable and would be more likely to fall apart, under the crush of their own complexity (Tainter, 1988), rather than to transition smoothly towards a less self-destructive mode of production. With time, a growing number of later books and articles—and even a periodical magazine entitled *Yggdrasil*, which was a remarkable editorial success—became devoted to helping readers withstand the psychological trauma of facing the high probability of seeing one's world fall apart. Sometimes spontaneously, more often under the pressing questions of journalists or audiences, collapsologists would on occasion venture into guessing dates for the probable breakdown of the main infrastructures in a country like France—and the predictions hit worryingly close to home, as early as 2023, or 2030, or at the latest 2050.

Apocalyptic and millenarist prophecies have been abundant in previous history. And most reasonable people could agree today that the ways of the Anthropocene cannot be sustained indefinitely, because of climate change as well as because of the current overuse and depletion of essential resources. What makes collapsology unique, and considerably more gripping, is its powerful alloy of scientific data concluding to a calculation of a terribly proximate catastrophe. This time, it is not your usual religious fanatic who announces the end of the world, but the very scientists that used to warn us against religious fanaticism! And their dismal computations do not simply sketch the remote possibility of a major breakdown promised to our grand-grand-sons and daughters: our lifestyle will be crushed within our lifetime! No wonder the carriers of such a terrible message must turn into psychotherapists to make it bearable for their audiences!

The most original (and precious) dimension of French collapsology—especially when compared with the survivalist traditions that have flourished in countries like the USA—is its unassailable commitment to reject a future made of Darwinian struggle for life, ending in a general civil war where everyone would weaponized their bunker and survive at the expense of their neighbours. Pablo Servigne and his collaborators (2017) have been firm in elaborating their worldview as a refutation of the competitive individualistic assumptions that motivate survivalist movements. Their message has always been crystal clear: only through mutual assistance do human populations have a chance to make it through the difficult times ahead, and the best thing they can do is to start growing the networks of solidarity and training the practices of common survival that will soon be urgently needed.

In their third instalment, Servigne et al. (2018) created the neologism 'collapsonauts' to portray those of us in the Western world who are bracing ourselves and each other in order to prepare for our unsustainably comfortable world to fall apart (Citton & Rasmi, 2020). Collapsonauts are less interested in the science of ecology than in the practical wisdom of ecosophy (Næss, 1976). They do not so much attempt to foresee the future as they are eager to learn how collectively to navigate the storms in the present. They reduce their consumptions for the double and convergent goals of diminishing their damaging imprint on the future and of getting used to better with less.

Collapsonauts are not convinced that everything will collapse at once: they realize that many things are *already* falling apart in what supports Western lifestyles. They build the habit of navigating chaotic, repetitive and lasting episodes of present upheavals, more than they hope to survive one future catastrophic disruption. They turn to the 'undercommons' (Harney & Moten, 2013) in the disenfranchised neighbourhoods of Western cities, as well as to Southern societies already collapsed under the onslaught of colonization several centuries ago (Escobar, 2016), in order to learn and share ungoverned forms of sociality that may be our best collective assets in troubled times.

Collapsonauts believe in the need to build and nurture institutions of solidarity in order to provide for our basic and cultural needs, but they train in practices of improvisation in order to anticipate lapses in governance and to face the unprovided-for. They are equally wary of our governments' procrastination in changing the course of our economies, and of the appeal to emergency measures which fail to address the deeper causes of our problems. They work towards developing a tragically under-studied mode of 'collapsonaut attention'—more attuned to backgrounds and textures, environments and surrounds, than focused on affordances, profits and figures (GDP)—that would compensate for the ecocidal shortcomings of our currently hegemonic 'extractivist attention' (Citton, 2021).

Collapsonauts in no way pretend to have elaborated a ready-made one-size-fits-all mindframe or lifestyle for the Anthropocene, to be applied by anyone to any situation. They more humbly call everyone for mutual assistance in complementing our shared incompleteness. And while the gloomy prospect of our thermo-industrial world falling apart is no laughing matter, they strongly believe in the power of comedy and humour to help us take distance towards our self-mutilating fears. When people trip or stumble, when elaborated machines stupidly fail, when high pretence flatly falls down, laughter is often the most spontaneous reaction. Learning to navigate collapse with a vivid sense of its frequent burlesque dimension may be one vital way to face it.

References

Citton, Y. (2021). "Collapsonautattention". Ecodata, Ecomedia, *Ecoaesthetics* Program. Fachhochschule Nordwestschweiz website. https://www.fhnw.ch/de/forschung-und-dienstleistungen/gestaltung-kunst/forschung/forschungsprojekte-des-instituts-kunst-gender-natur-iagn/ecodata-ecomedia-ecoaesthetics/texts/media/citton-collapsonautattention-e-final.pdf
Citton, Y., & Rasmi, J. (2020). *Générations collapsonautes*. Seuil.
Cochet, Y. (2011, May 27). L'effondrement, catabolique ou catastrophique. *Institut Momentum*. https://www.institutmomentum.org/l%E2%80%99effondrement-catabolique-ou-catastrophique/
De Martino, E. (1977). *La fine del mondo. Contributo all'analisi delle apocalissi culturali* (p. 2019). Einaudi.
Diamond, J. (2005). *Collapse: How societies choose to fail or succeed*. Viking Press.
Escobar, A. (2016). Thinking-feeling with the earth: Territorial struggles and the ontological dimension of the epistemologies of the south. *Revista de Antropología Iberoamericana., 11*(1), 11–32.
Harney, S., & Moten, F. (2013). *The Undercommons. Black Study and Fugitive Planning*. Minor Composition.
Meadows, D., & Donella, H. (1972). *The limits to growth*. Potomac Associates.
Moynihan, T. (2020). *X-Risk. How Humanity Discovered Its Own Extinction*. Urbanomic.
Næss, A. (1976). *Ecology, community and lifestyle: Outline of an ecosophy*. Cambridge University Press. 1989.
Servigne, P., & Chapelle, G. (2017). *L'entraide: l'autre loi de la jungle*. Les Liens qui libèrent.
Servigne, P., & Stevens, R. (2015). *How everything can collapse: A manual for our times*. Polity. 2020.
Servigne, P., Stevens, R., & Chapelle, G. (2018). *Une autre fin du monde est possible: vivre l'effondrement (et pas seulement y survivre)*. Seuil.
Smith, C. (2009). *Collapse*. Videograph Films.
Steffen, W. (2004). *Global change and the earth system: A planet under pressure*. IGBP Springer.
Tainter, J. (1988). *The collapse of complex societies*. Cambridge University Press.

Yves Citton is fortunate enough to be paid to study and teach media and literature at the University of Paris 8 Vincennes-Saint Denis and to be co-director of the journal *Multitudes*. He is the author of a dozen book, including *Altermodernités des Lumières* (Seuil, 2022), *Faire avec. Conflits, coalitions, contagions* (Les Liens qui Libèrent, 2021), *Générations collapsonautes* (Seuil, 2020, in collaboration with Jacopo Rasmi), *Mediarchy* (Polity Press, 2019), *Contre-courants politiques* (Fayard, 2018), *The Ecology of Attention* (Polity Press, 2016), *Renverser l'insoutenable* (Seuil, 2012).

Jacopo Rasmi studied in many different places (institutional or not) and in the company of various minds (living and dead), before teaching for the University Jean Monnet – Saint Etienne in the middle of the Covid-19 pandemic. He published *Le Hors-champ est dedans. Michelangelo Frammartino, écologie, cinema* (PUS, 2021) and, with Yves Citton, *Générations Collapsonautes* (Seuil, 2021). Addicted to intellectual activities, he takes part frequently to cine-clubs, programming of screenings, editorial boards of journals (*Multitudes* and *La revue documentaire*) and other artistic initiatives.

Convivialism

Nathanaël Wallenhorst

Abstract The *Convivialist Manifesto* was produced by the MAUSS (*Mouvement Anti-Utilitariste dans les Sciences Sociales*, Anti-Utilitarianism Movement in Social Sciences). The movement's name is a nod to Marcel Mauss, and the idea is an extension of the Frankfurt School's Critical Theory. The manifesto was published in France in 2013, and has been translated into some ten languages. In view of the widespread success of the publication, convivialism has become a school of political thought and movement. Its aim is to radically alter social and political thinking, and in so doing, to transform global society.

Convivialism is the combination of several alternative modes of thinking. What these approaches all share is the idea of coexistence, of living together – of conviviality. To a degree, convivialism is the political arm of the academic MAUSS. In 'convivialism', the MAUSS (spearheaded by French sociologist Alain Caillé), follows in Austro-American political ecological thinker Ivan Illich. In his 1973 book, *Tools for Conviviality*, Illich criticises growth-based ideology. This denunciation of growth, and more broadly, of the hegemony of the economic mentality, became the cornerstone of convivialism (which is embraced mainly by sociologists and economists). The manifesto offers the following definition of convivialism (2013, p. 30): '*Convivialism* is the term used to describe all those elements in existing systems of belief, secular or religious, that help us identify principles for enabling human beings simultaneously to compete and cooperate with one another, with a shared concern to safeguard the world, and in the full knowledge that we form part of that world and that its natural resources are finite'. Management science researcher Pascal Glémain describes convivialism as a 'social movement aiming to bring about a "different" civilisation – a convivialist civilisation that aims to counter the inhumanity of the world' (2017, p. 27). Italian philosopher Francesco Fistetti, for his part, defines it as 'a political philosophy of living together' (2016, p. 247).

N. Wallenhorst (✉)
Catholic University of the West, Angers, France
e-mail: nathanael.wallenhorst@uco.fr

© The Author(s), under exclusive license to Springer Nature Switzerland AG 2023

N. Wallenhorst, C. Wulf (eds.), *Handbook of the Anthropocene*,
https://doi.org/10.1007/978-3-031-25910-4_230

Within convivialism, there are four anthropological principles: (1) togetherness and the possibility of shared freedoms; (2) the rejection of *hubris*, which is understood to be the mother of all threats; (3) transitioning from the satisfaction of needs to the pursuit of desires; and (4) the emergence of a world in which we all coexist harmoniously, on the basis of four principles.

The first of these is *togetherness and the sharing of freedoms.* Etymologically speaking, the word 'conviviality' comprises the particle *con*, meaning with, and *vivere*, living. Conviviality refers to the sharing of living experiences – in much the same way as we do over a good meal! In this concept of existence, emphasis is placed on coexistence. Rather than being an individual affair, existence is shared with other people. *Vivere* also means living or dwelling. It is through sharing experiences (and therefore resources) that we share our existence with others. The idea, with convivialism, is that the way in which we share our existence, in the world, shapes our view of politics. The term 'convivialism' is also semantically related to 'conviviality', which carries the idea of liking, good times, warmth, support, and experiences shared with loved ones. The sharing of sympathies inherent in convivialism may make it more difficult to regulate our freedoms; nevertheless, such regulation is absolutely necessary so that, in exercising those freedoms, we do not do physical or mental harm to others. The Ancient Greeks referred to *hubris* (an extremely powerful form of desire, including destructive folly), as a profound and deep-seated problem in human nature. In opposition to *hubris*, convivialism proposes the art of coexisting and cooperating, together building a new political reality. One of the most crucial political questions is: 'What do we do with our freedoms?'. In convivialism, freedoms are viewed as something that can be shared, and which can come together and even join together. In this context, trust is invested not in an individual, but in the relationships between individuals. The focal point is not the subject (in the Cartesian sense of the word), but relations among subjects. In that sense, the political thinking in convivialism is marked by the contemporary philosophy of intersubjectivity and deconstruction of the subject. Convivialism brings together both individuals and groups. Convivialism teaches us that we can live together (in achieving this goal, a certain amount of courage is needed, which echoes the concept of 'the courage to be as a part', defined by German theologian Paul Tillich, 1999, who emigrated to the United States). Thus, the conflict and tension between individuals, or between groups (which are inevitable when everyone's freedoms are respected and exercised), will not necessarily result in harm.

Here, convivialism is the proposal to move forward from Marxism, rather than a form of neo-Marxism. Similarly, *stricto sensu*, convivialism is not anarchism, despite the trust in relationships 'between us' (which include foreigners, unlike relationships 'between them'); such trust is the crux of politics. Indeed, convivialists do not reject the centralisation of power by the State. In fact, the idea is that re-institutionalisation will make governance easier, by keeping a lid on *hubris*. According to Alain Caillé, convivialism is a contemporary political ideology which aims to allow us to move beyond the four political ideologies of modern times: liberalism, socialism, anarchism and communism, by encapsulating them all in one ideology. Convivialism holds that we should move past these four doctrines – notably with respect to ecological and moral issues – and preserve the democratic ideal. Currently, democracy is subjugated to unsustainable economic activity, and is sure to die out eventually. We need to liberate democracy from the death-grip of economics, and repoliticise it on the basis of interactions 'between us'. The view of politics underlying convivialism is very much in keeping with Hannah Arendt's outlook. Arendt considers politics the art of being together; that art is at the heart of the *Convivialist Manifesto*, and also the *Commons Manifesto* (Bauwens et al., 2018).

The second principle is *the fight against hubris – the worst of all threats*. The *Convivialist Manifesto* analyses a fundamental problem in humans' relationship with the world. It asserts that we need to do two things: firstly, to mutate and evolve; and secondly, to begin thinking about the human adventure in a new way. The *Manifesto* offers an alternative way of dealing with the mother of all threats – that deadly combination of violence and madness: 'Humanity has achieved phenomenal technical and scientific progress, but is powerless as ever to solve its most essential problem: how to deal with rivalry and violence among human beings' (Les convivialistes, 2013, p. 12). The manifesto also asks: 'How do we halt the accumulation of power – hitherto limitless and potentially self-destructive – over men and over nature?'. Faced with this destructive capacity, convivialists choose cooperation. They view humans as being able to give the best of themselves, and to have conflicts without resulting in a massacre.

From this point of view, *hubris* is the anthropological root of many of the threats we face today. The counterweight to *hubris* is cooperation: 'A healthy society acknowledges everyone's desire for recognition, the role of rivalry and the aspiration for self-improvement, and the risk acceptance that this entails. Such sentiments are prevented from becoming excess and *hubris*, and cooperative openness towards other people is encouraged' (Les convivialistes, 2013, p. 14). Hence, conflict must be understood as a force of life, rather than a deadly threat. To control *hubris* is an 'absolute imperative': 'The precondition for rivalry and emulation to serve the common good is that they must be free from the desire for omnipotence, excess and *hubris* (and *a fortiori* from *pleonexia*, the constant desire to possess more). If that condition is met, these feelings of rivalry become tools with which to cooperate more effectively' (Les convivialistes, 2013, p. 14).

The third principle is to *move from the satisfaction of needs to the pursuit of desires*. Convivialism abhors the use of a purely economic frame of reference (with the goal to maximise individual interests predominating), applied to everything in existence. In this situation, *Homo oeconomicus* is the only lens through which people's needs, remedies and aspirations are viewed. Whilst humans do have selfishness and prioritise their own individual interests, it is fallacious to view them solely as *Homo oeconomicus*: they are capable of giving, loving, exchanging and sharing. In an extension to this philosophy, the authors criticise the modern-day madness of the race for growth: 'With an average annual growth rate of 3.5%, for example, the world's GDP would increase 31-fold in a century. Imagine 31 times more oil- or uranium consumption, or CO_2 emissions, in 2100 than today' (Les convivialistes, 2013, p. 18). Similarly: 'Given the ecological state of the planet, we must seek out all possible means of prospering without growth' (Les convivialistes, 2013, p. 33). Are human conflicts truly caused by the scarcity of material resources and difficulty in satisfying material needs? This is the question the *Convivialist Manifesto* asks. Convivialists propose to shift from anthropology based on the satisfaction of needs to one founded on the pursuit of desires. Self-limitation appears to be the key in bringing about the new civilisation that convivialists advocate. However, such self-limitation, or self-control, is learned gradually. Convivialism cannot thrive without an accompanying educational movement, to engineer the anthropological and civilisational shifts needed: 'This is the aim of a new concept of political and human emancipation: to create concrete forms of "living well", sober and convivial, to temper self-interest with altruism, where all goods – including economic goods – are, first and foremost, generators of social links' (Fistetti, 2016, p. 256).

As the *Convivialist Manifesto* points out, the rise of financial individualism was preceded by the circulation of ideas which encouraged individuals to view themselves as separate entities, and focus on optimising their own individual interests. This dominance of mercantile logic has been facilitated by the hegemony of *Homo oeconomicus*, over the course of the twentieth century in the sphere of ideas. Therefore, the convivialist idea of a profound anthropological shift should not be readily discarded. Implementing ideas, however utopian, is never a waste of time (Wallenhorst, 2019; Wallenhorst et al., 2020).

Fourthly, there are *five principles* underlying the reshaping of the world we all share. In addition to denouncing the *hubris* of the hegemonic *Homo oeconomicus* (the authors are notably in favour of establishing a basic living wage – 2013, p. 31), the *Convivialist Manifesto* identifies five principles that are fundamental to coexistence. The first, which was added in the updated edition, *The Second Convivialist Manifesto* (2020), is 'common naturality', which represents the possibility of coexisting harmoniously with nonhumans. The other four are 'common humanity', 'common sociality', 'legitimate individuation' and 'creative opposition'. Thus, convivialism has a direct moral message about politics: 'Politics inspired by convivialism is therefore politics that allows human beings to differentiate themselves by engaging in peaceful and deliberative rivalry for the common good' (Les convivialistes, 2013, p. 27). However, individuals must not 'succumb to excess and the infantile desire for omnipotence (*hubris*, as understood by the Ancient Greeks) – i.e. violate the principle of common humanity and jeopardise common sociality, by claiming to belong to a superior species, or hoarding and monopolising such a quantity of goods or quantum of power that the social existence of those below you is endangered' (Les convivialistes, 2013, p. 29).

References

Bauwens, M., Kostakis, V., & Pazaitis, A. (2018). *The commons Manifesto – Peer to Peer*. http://www.p2plab.gr/en/archives/117

Fistetti, F. (2016). Le convivialisme, "contre-mouvement" du 21ème siècle. *Revue du MAUSS, 48*, 247–258.

Glémain, P. (2017). Penser le convivialisme en économie sociale contemporaine. *RECMA, 4*(346), 27–41.

Illich, I. (1973). *Tools for conviviality*. Harper and Row.

Les convivialistes. (2013). *Manifeste convivialiste*. Le Bord de l'eau.

Les convivialistes. (2020). *Le second manifeste convivialiste*. Acte sud.

Tillich, P. (1999). *Le courage d'être*. Cerf, French translation.

Wallenhorst, N. (2019). Vers un avenir convivialiste. *Revue Projet, 371*, 68–72.

Wallenhorst, N., Mellot, S., & Theviot, A. (Eds.). (2020). *Inter-connectés? Numérique et convivialisme*. Le Bord de l'eau.

Nathanaël Wallenhorst is Professor at the Catholic University of the West (UCO). He is Doctor of Educational Sciences and Doktor der Philosophie (first international co-supervision PhD), and Doctor of Environmental Sciences and Doctor in Political Science (second international co-supervision PhD). He is the author of twenty books on politics, education, and anthropology in the Anthropocene. Books (selection): *The Anthropocene decoded for humans* (Le Pommier, 2019, *in French). Educate in Anthropocene (*ed. with Pierron, Le Bord de l'eau 2019, *in French*). *The Truth about the Anthropocene* (Le Pommier, 2020, *in French*). *Mutation. The human adventure is just beginning* (Le Pommier, 2021, *in French*). *Who will save the planet?* (Actes Sud, 2022, *in French*). *Vortex. Facing the Anthropocene* (with Testot, Payot, 2023, *in French*). *Political education in the Anthropocene* (ed. with Hétier, Pierron and Wulf, Springer, 2023, *in English*). *A critical theory for the Anthropocene* (Springer, 2023, *in English*).

Cooperation

Enzo Pezzini

Abstract This article aims to present cooperation as expressed in its entrepreneurial form, which today covers a wide range of sectors. Cooperatives are the result of a social and ideological process that have permitted the development of a worldwide movement. It is an economic model based on specific anthropological assumptions.

Cooperation, in the etymological sense of a several persons joining together to carry out work or obtain results of common interest, can be considered to have been born with humanity and has developed in parallel with the spirit of association and the sense of solidarity present in human history. Etymologically, the word "cooperation" has a broad meaning and even dissolves into the synonym of collaboration, which manifests itself in people who come together under the pressure of common needs. Forms of association are as old as life in society, across all cultures. They have taken many forms and many different names: *hetairia* in ancient Greece, *sodalitia* in Roman times and, in the Middle Ages, guilds and confraternities, among many others. From the eighteenth century onwards, Friendly Societies multiplied in England, in anticipation of the cooperative movement (Defourny & Nyssens, 2017).

Cooperation in its modern-day form was born under the influence of various inspirers and precursors such as Saint-Simon, Fourier, Owen, King, Proudhon, Buchez, Blanc, Le Play, Raiffeisen and Schulze Delitzsch. These personalities came from different ideological background, among them socialism, Christian solidarism and liberalism (Gueslin, 1987; Pezzini, 2018). As a model of economic business with its own formal structures and regulatory processes, the birth of modern cooperation coincided with the advent of the industrial revolution, as a response to the harmful effects of capitalism and the disruption of social structures, the radical transformation of work and new living conditions marked by precariousness, poverty and insalubrity. By a generally shared consensus, this birth of modern cooperation is dated from the creation of the consumer co-operative of the Rochdale Equitable Pioneer Society near Manchester in 1844 (Desroche, 1976).

E. Pezzini (✉)
Research Centre for Political Sciences (CReSPo), Université Saint-Louis, Brussels, Belgium

© The Author(s), under exclusive license to Springer Nature Switzerland AG 2023

N. Wallenhorst, C. Wulf (eds.), *Handbook of the Anthropocene*,
https://doi.org/10.1007/978-3-031-25910-4_231

Though many cooperative experiences already existed at that time, the cooperative set up by the 28 Rochdale weavers and skilled workers was a unique and exemplary experience through its success in combining social ideals with rigorous management, setting the principles that guaranteed its success, principles from which cooperation, as we know it today, subsequently developed and evolved. These included equality and democratic control (one person, one vote), freedom of membership, economic justice (distribution of profits in proportion to the activities of each member), equity (limited remuneration of capital contributions), political and religious neutrality, and education of members (Bidet, 1997). The principles and rules that these "pioneers" defined as the basis of the consumer cooperative they initiated were quickly taken up to provide the foundations of cooperative movements throughout the world. In the wake of consumer and skilled craftsmen's cooperatives, thanks to the great adaptability of the basic format, cooperatives developed and expanded in banking, agriculture, housing, fishing, transport, education, health and social services and in many other areas until there were no economic sectors excluded (Birchall, 1997).

It is noteworthy that the Rochdale experience, its insights and its rules and regulations became the guidelines that inspired cooperative principles at the international level. These were set out by the International Co-operative Alliance in 1937 and subsequently revised in 1966 and 1995 with the aim of explaining how cooperative practices should be interpreted in a changing world. "A cooperative is an autonomous association of persons united voluntarily to meet their common economic, social and cultural needs and aspirations through a jointly owned and democratically controlled enterprise" (ICA, 1995). This definition was approved in 1995 by the Congress of the International Co-operative Alliance and has since then been accepted and used by all major international bodies such as the UN, ILO and, EU.

In addition to its heritage of values and civic morality, participation and economic democracy, cooperation commends itself for its economic performance, for its contribution to employment, for the development of entire productive sectors, particularly in agriculture. At its best, cooperation is not a phenomenon limited to a few regions of the world, but rather a major global movement, with an estimated billion members at world level. According to the most recent data it provides direct or indirect employment to at least 280 million people, not to mention derivative jobs (Eum, 2017). The United Nations' decision to proclaim 2012 as the International Year of Cooperatives is a recognition of the cooperative model as a factor of economic and social development (Pezzini & Girard, 2018).

It is for this reason that, from a "macro" perspective, the various cooperative movements can be seen as sociologically relevant aggregates within the contemporary economic and political system. As early as 1895, the cooperative movement established an international organisation, the International Cooperative Alliance, which is the guarantor of the cooperative identity and has the ultimate authority for defining the cooperative enterprise and developing the principles on which it should be based (Watkins, 1970; Draperi, 1995). The global progress of the cooperative movement should not, however, detract us from reflecting on the "political" weakness of cooperatives and on the importance for them of organizing in federations, networks and pressure groups, in order to play their role of fully-fledged economic entities and counter the threat of fragmentation and trivialization. In this context one observes the absence of any "systemic" economic theory of the cooperative, possibly explained by the all-too-human "vice" of uncritical acceptance of the idea that all economic agents have individualistic preferences and are guided only by their own interests – as if all actors were *homines economici*. According to Zamagni, empirical evidence shows, on the contrary, that people in the real world become members of cooperatives not only to pursue their own interests, but also out of a genuine concern for values such as democracy, social justice and economic freedom (Zamagni, 2012).

Each economic system is based on anthropological presuppositions. Each system conveys an image of man that has been formed by economic theorists, and this image constitutes the basis upon which these theorists imagine how the system works.

More specifically, in neoclassical political economics, the premises are individualistic, liberalist and utilitarian, with methodological individualism considered an appropriate hypothesis with which to explain human behaviour, that is: self-interest as the predominant pattern (Possenti, 1993).

Reflection on the anthropological dimension of the cooperative enterprise reveals an enterprise model in which the satisfaction of individual and collective human needs is the teleological element of the enterprise's activity, whereas in the capitalist enterprise this element is the making of profit as the result of the investment of financial capital.

A cooperative enterprise unites within itself two dimensions that are distinct but not in competition with each other. On one hand, it is an economic actor that operates in the market by accepting its logic. On the other hand, it has a social dimension, since it pursues meta-economic ends. Indeed, the benefits it generates extend beyond its members to the profit of the wider community as a whole. This is particularly evident in the newly emerging social cooperative-type formats that prioritize social mission and general interest and promote multi-stakeholder participation (Borzaga & Defourny, 2001; Borzaga et al., 2010).

It is this dual nature that makes the cooperative a reality that is difficult to explain and complex to govern. As the almost bicentennial history of the cooperative movement illustrates, cooperativism has gone through alternating phases. During the phases when the "market code" dominated, cooperatives became "common currency" and it became difficult to distinguish them from capitalistic enterprises; at other times, the "code of sociality" was given too much prominence, and cooperatives then went into decline or lost market share. In both cases, the cooperative becomes distorted and loses its specific identity. The real great challenge for the cooperative movement of the twenty-first century is to maintain a dynamic balance between these two codes so that, from their mutual "contamination", strategic complementarities are born (Zamagni & Zamagni, 2008).

Drawing inspiration from the sociologist Zygmunt Bauman, it could be argued that in the global market cooperatives are happy anomalies, "solid bodies" within the dominant "liquid modernity", a society based on circulation, fluidity, flexibility, uprooting and a way of living – and of thinking about the economy – marked by an eternal short-termism (Bauman, 2000).

Solid bodies that have resisted for almost two centuries by remaining faithful to strong principles, whereas, for certain theories they ought to have evolved and been trivialized in a dominant market economy disembedded from the social framework, under pressure from an environment that ought to have brought about their transformation or degeneration.

References

Bauman, Z. (2000). *Liquid modernity*. Blackwell Publishing Ltd..

Bidet, E. (1997). *L'économie sociale*. Le Monde-Editions.

Birchall, J. (1997). *The international co-operative movement*. Manchester University Press.

Borzaga, C., & Defourny, J. (Eds.). (2001). *The emergence of social Enterprise*. Routledge.

Borzaga, C., Depedri, S., & Galera, G. (2010). L'interesse delle cooperative per la comunità. In L. Bagnoli (Ed.), *La funzione sociale delle cooperative* (pp. 125–145). Carocci.

Defourny, J., & Nyssens, M. (2017). *Économie sociale et solidaire*. De Boeck Supérieur.

Desroche, H. (1976). *Le projet coopératif*. Les Éditions Ouvrières.

Draperi, J.-F. (1995). *L'ACI a cent ans: regards sur une histoire mémorable* (Vol. 258, pp. 73–82). Recma.

Eum, H. (2017). *Cooperative and employment, second global report 2017*. CICOPA.

Gueslin, A. (1987). *L'invention de l'économie sociale*. Economica.

ICA – International Cooperatives Alliance. (1995). *Cooperative identity, values & principles*. https://www.ica.coop/en/cooperatives/cooperative-identity. Accessed 22 June 2021.

Pezzini, E. (2018). *Projet coopératif et christianisme social*. Presses de l'Université Saint-Louis.

Pezzini, E., & Girard, J.-P. (2018). *Les coopératives. Une utopie résiliente*. Fides.

Possenti, V. (1993). Presupposti antropologici dei sistemi economici. *La società, 4*, 599–620.

Watkins, W. P. (1970). *The International Co-operative Alliance 1895–1970*. ICA.

Zamagni, S. (2012). *Cooperazione*. AICCON. https://www.aiccon.it/pubblicazione/cooperazione/. Accessed 22 June 2021.

Zamagni, S., & Zamagni, V. (2008). *Le cooperative*. Il Mulino.

Enzo Pezzini has a PhD in Political and Social Sciences from the University Saint-Louis in Brussels, where he is associate researcher at the Research Centre for Political Sciences. He is also a scientific collaborator at the Religions, Spiritualities, Cultures, Societies Research Institute of the Catholic University of Louvain. He is a lecturer at the Lumen Vitae International Centre in Namur and a visiting professor at the Catholic University of Paris. He directed the Brussels office of the Italian Confederation of Cooperatives for 17 years.

He is author of various articles on cooperatives and recent books *Bene commune, partecipazione e democrazia* (Ecra – Edizioni del Credito Cooperativo: Rome, 2019). *Les coopératives. Une utopie résiliente,* [with Jean-Pierre Girard], (éditions Fides, Montreal, 2018). *Projet coopératif et christianisme social*, (Presses de l'Université Saint-Louis, Brussels, 2018).

Didactics

Charlotte Pollet

Abstract This article will introduce some of the important concepts of science didactics before presenting a view on the impact of the concept of the Anthropocene. Within science didactics, there has been a recent paradigm shift from "sustainable development" to "global change". This shift shows that didactics is renewing itself and should not be seen through a lens of catastrophism, but rather as a renewal comparable to the effervescence of a Renaissance.

Didactics is the study of the questions posed by teaching and the acquisition of knowledge in various school subjects. Didactics then serves to transpose knowledge into teachable knowledge, and aims to precisely define each element of knowledge that it wishes to teach, but also to define how this knowledge is taught to learners. The roots of didactics go back to Antiquity with written traces dating back to Greek philosophers and ancient Eastern civilizations like China. From the eighteenth to the twentieth centuries, the theories multiply, with new perspectives opened by the psychologies of education, motivation, as well as the use of computer tools and the internet and lifelong learning. In particular, in the 1970s, didactics developed around mathematics, sciences, languages, life and earth sciences, physical and sports education, history and geography or professional didactics (i.e. Delacôte (1996); Brousseau (1998); Chevallard (1985); Glaeser G (1999)). Originally, didactics was not clearly differentiated from pedagogy which is the science dealing with problems specifically linked to teaching. But gradually, didactics differentiated itself from pedagogy: disciplinary content plays a central role and there is an extra dimension dealing with the nature of the knowledge to be taught.

Taking into account the recent shift towards the Anthropocene requires new ways of producing and disseminating knowledge. The debates around global change and climate change make it urgent to train pupils and students, future citizens and decision-makers, in the social, environmental, political and economic issues of the

C. Pollet (✉)
National Yangming Chiaotung University, Hsinchu, Taiwan

© The Author(s), under exclusive license to Springer Nature Switzerland AG 2023

N. Wallenhorst, C. Wulf (eds.), *Handbook of the Anthropocene*,
https://doi.org/10.1007/978-3-031-25910-4_232

relationship between humanity and the Earth. It is about understanding our times and making them understood, building concrete tools for action at a time when the relationship between human and non-human, between nature and society is radically transforming.

Some Concepts of Science Didactics

Constructivism As the pupils' minds are not a *tabula rasa* and do not simply passively receive knowledge from the teacher, it is necessary to consider their personal conceptions which constitute as many obstacles to the development of new knowledge. This elaboration must go through questioning and construction of these conceptions. Following Jean Piaget (2003), science didactics explored constructivism as the foundation of teaching. But from the 1990s however, the constructivism paradigm was supplemented by other paradigms taking into account the complexity of the act of teaching.

Conceptions The acquisition of knowledge is not the simple memorization of information provided by the outside. This information is filtered, interpreted, put in relation, or competition with prior knowledge. The teaching of certain disciplines then comes up against "spontaneous" conceptions which can stand in the way of learning. Much research in didactics aimed at identifying conceptions in pupils and students by analyzing "errors", their reasoning during problem solving or in practical work situations (L. Viennot).

Situation-Problem These are didactic situations built around a "problem", the term designating a questioning, an enigma, resulting from an object, an observation whose resolution requires student investment with general concrete support. The pupils do not initially have all the means to answer the question. They must first take ownership of the questioning and apply their knowledge and ingenuity to find 'a' solution (R. Douady (1984)).

Didactic Triangle The schematization of a teaching situation is done using the didactic triangle. This triangle – generally represented by equilateral TSC – makes it possible to visualize the possible interactions (sides of the triangle) between 3 poles (the three vertices of the triangle): the T pole of the Teacher, the S pole of the Student and the C pole of Content. When the T-S interaction is privileged by the teacher, it is because he/she favors the relation with students over content of teaching. When the T-C interaction is privileged by the teacher, it means he/she privileges the subject rather than the impact of his teaching on the pupils. In theory, the ideal would therefore be to favour the S-C interaction while in practice the teacher does everything possible to properly measure the mixture of these three interactions. The recent proposal for a "master mediator" goes in this direction: the teacher is no longer the one who delivers the knowledge to the pupil, but the one who helps the pupil to appropriate knowledge.

Didactic Contract The didactic contract is a concept introduced by Brousseau (1998). He defines it as the set of relationships which determine, mainly implicitly, what each partner will be responsible for managing and ways of accountability towards each other. This means that the student interprets the situation presented to him, the questions asked, the information provided to him, the constraints imposed on him, depending on what the teacher reproduces, consciously or not. The effectiveness of the relationship depends on a mutual understanding of each-other's intentions.

Didactic Transposition and Reference Practices Sociologists, in particular M. Verret (1975), have shown that the development of disciplinary content is a complex process, linked to social issues. The play of scholarly references, of a certain image of the discipline and of the associated values, of the purposes attributed to such or such training leads to choices in the content. Even for disciplines such as mathematics or science, the knowledge taught is knowledge reconstructed specifically for teaching. The learned knowledge taken as a reference is a decontextualized knowledge, cut off from its history, transposed to be taught at a given level.

Prescribed, Real, Hidden Curriculum A curriculum, in the educational field, represents a training course. Ph. Perrenoud (2016) suggests distinguishing between two levels: (1) the "programming" of an educational path, particularly in the mind of the educator; this is the level of the prescribed curriculum: a set of texts and representations. (2) The experiences that the learner lives and that transform him; this is the level of the real curriculum. Even when the prescribed curriculum is fully respected, the expected learning only takes place for a fraction of the students. But in this real curriculum two parts are to be distinguished: a manifest and a hidden part. The manifest part is the more or less faithful translation of an intention to instruct, the implementation of a prescribed curriculum. The hidden part regularly generates formative experiences without the knowledge of those concerned or at least without such learning having been voluntarily favoured: the hidden curriculum.

Educating in the Anthropocene

The changes in school curricula seem to respond to scientific developments: in the early 2010s, sustainable development was the common thread of certain programs, as in geography in France. To this was added a strong institutional injunction to practice education for sustainable development in all subjects and at all levels. While this injunction has not disappeared, the expression "sustainable development" is now less used and the expression "global change" has appeared. Taking into account the Anthropocene seems to have involved a shift in the approach to environmental degradation linked to the actions of companies. Sustainable development insists, in its consideration of ecology, on the imperative need to continue to ensure economic and social development for all of humanity, while global change and the Anthropocene are putting the emphasis on the irreversibility and severity of the global ecological crisis. Projects are emerging to think about environmental education differently and in particular in a fairer way (Martusewics et al., 2011).

Environmental education is now taught in several countries. But the results remain largely insufficient to face the challenge. This may be due to the form of environmental education which ignores the nature of ecological issues and so often the subjects are taught separately, suggesting that each discipline is autonomous. There is also a specifically productivist logic that has shaped the school since the first half of the nineteenth century (standardized organization of lessons and timetables, layout of the classroom, competition between pupils, schools and countries, etc.) and the primacy of economic goals weighs heavily on the decisions of political institutions. However, the school is the subject of "new social demands" (Audigier, 2015), that is to say pressures to take charge of social issues traditionally entrusted to the private sphere. These political injunctions, formulated in terms of knowledge and know-how, are part of a desire for institutional innovation.

This paradigm shift represents an unparalleled opportunity to rethink our pedagogies. Several interdisciplinary experiments are underway. These include the Anthropocene curriculum initiated in 2013 at the House of World Cultures in Berlin in collaboration with the Max Planck Institute where scientific, human and social dimensions are deeply integrated into the world of the arts. The Anthropocene questions the meaning of didactics between a past fed with guilt and a future fed with uncertainty. Far from the media catastrophism or environmental education reduced to the teaching of eco-gestures, it is possible to conceive educational projects. By its richness and its intrinsic interdisciplinarity, the notion of Anthropocene could link human, social and exact sciences in their complexity and complementarity. An essential value of the concept of the Anthropocene is therefore its plural definition, which makes it an interdisciplinary well in which no one is an expert, thus offering new didactic freedom. Here we are entering an era which may resemble the Renaissance. New questions are thus appearing, at the crossroads of exact and experimental sciences, human and social sciences, and educational sciences.

References

Audigier, F. (2015). Domaine généraux de formation, compétence, éducation à… : Les curriculums et les disciplines chahutées. D'un cadrage historique et pédagogique a l'éducation en vue du développement durable comme exemple emblématique. *Revue suisse des sciences de l'éducation, 37*, 427–460.

Brousseau, G. (1998). *Théorie des situations didactiques*. Grenoble.

Chevallard, Y. (1985) *La transposition didactique*. Grenoble : La Pensée Sauvage. (nouvelle édition augmentée de « *Un exemple de la transposition didactique* » avec M.-A. Johsua).

Delacôte, G. (1996). *Savoir apprendre: les nouvelles méthodes*. Odile Jacob.

Douady, R. (1984). Jeux de cadres et dialectique outil-objet, *Cahier de Didactique* n°3. IREM PARIS 7.

Glaeser, G. (1999). *Une introduction à la didactique expérimentale des mathématiques*. Grenoble.

Martusewics, R. A., et al. (2011). *EcoJustice Education. Towards diverse, democartic and sustainable communities. Sociocultural, political and historical studies in education*. Routledge.
Perrenoud, P. H. (2016). *La pédagogie différenciée des intentions à l'action*. ESF éditeur.
Piaget, J. (2003). *La représentation du monde chez l'enfant*, Quadrige, PUF. (1re édition : PUF, 1926).
Verret, M. (1975). *Le temps des études*. Thèse. Université de Lille III.

Dr. Charlotte Pollet is an associate professor in Yangming Chiaotung University (Taiwan). She received her dual Ph.D from Paris 7 University and National Taiwan Normal University in didactics and history of mathematics. She works on the development and promotion of philosophy for children and new philosophical practices in Taiwan. Her research interests concern the transmission and elaboration of mathematical objects and procedures in Asia. She is the author of "The Empty and the Full: Li Ye and the Way of Mathematics" (World Scientific, Singapore) and "Guangong dit oui" (L'asiathèque. Paris).

Gesture

Christoph Wulf

Abstract Gestures play an important role in the Anthropocene. Many lead to the negative effects with which humans are destroying nature. However, they can also help to stop negative anthropogenic developments and transform them into sustainable actions and behaviours. Destructive gestures include those which regard nature only as an object or a thing. In this case, nature serves primarily to satisfy human needs. The mining of coal and other natural resources as well as their excessive use is a good example of this. We can make a distinction between four different angles in the study of gestures in the Anthropocene: (1) gestures as movements of the body; (2) gestures as expression and representation; (3) gestures as forms of meaning-making; (4) gestures as forms of education. To read and to decode gestures they must be perceived mimetically. Someone perceiving a gesture understands it by recreating it internally in their imagination and thereby grasps its specific physical expression and representation.

Gestures play an important role in the Anthropocene. Many lead to the negative effects with which humans are destroying nature. However, they can also help to stop negative anthropogenic developments and transform them into sustainable actions and behaviours. Destructive gestures include those which regard nature only as an object or a thing. In this case, nature serves primarily to satisfy human needs. The mining of coal and other natural resources as well as their excessive use is a good example of this. There are also many gestures in other areas of life through which humans violate nature (Wallenhorst & Wulf, 2023). Many of these gestures are performed globally in different places not only once but again and again and therefore have a profound effect on nature. There are hardly any areas left that are not influenced by people's actions and behaviours, and thus by their gestures (Wulf, 2022b). Examples of such gestures are driving, flying and electronic communication, all of which contribute to the mobility and acceleration of life. In addition,

C. Wulf (✉)
Freie Universität Berlin, Berlin, Germany
e-mail: christoph.wulf@fu-berlin.de

© The Author(s), under exclusive license to Springer Nature Switzerland AG 2023

N. Wallenhorst, C. Wulf (eds.), *Handbook of the Anthropocene*,
https://doi.org/10.1007/978-3-031-25910-4_233

there are the gestures of exchange of energy, commodity and capital across our planet, the effects of which make a clear distinction between nature and culture problematic. In antiquity humans were part of the *physis* (nature) and in the Middle Ages part of God's creation, but with the arrival of industrialization and modernity we have become detached from nature. As subjects we experience nature as something which stands opposed to us – it has become the object of our actions. The idea of nature as being an integral part of our world has given way to a subject-object relationship with many negative effects characteristic of the Anthropocene.

Gestures are influenced first by *collective ideas and practices*, second by *institutional conditions and traditions*, and third by *individual circumstances*. They are significant movements of the body that can be understood as actions (Gebauer & Wulf, 1995, 1998). The etymology makes this clear. The term "gesture" goes back to the Latin word *gestus*, which in a general sense denotes a movement or posture of the body and in a special sense a movement of a body part and especially of the hand. There are smooth transitions between the idea of the deed that the word gesture denotes and the action itself. In etymological terms, the word refers to the body moving in the world, to activities of the hand, to human actions, to movements of individual body parts expressing and representing emotions, and to the performativity of the body and social actions as a whole. Gestures are performed not only by individuals, but also by groups and societies. In a figurative sense, however, the term gesture is also used when several partial actions are combined in one activity, as is the case in the gesture of coal mining, in which several partial actions are combined in the image of a gesture. It is irrelevant whether it is the gestural action of a person or the gesture of a clearly defined community or a collective gesture.

We can make a distinction between four different angles in the study of gestures in the Anthropocene: (1) gestures as movements of the body; (2) gestures as expression and representation; (3) gestures as forms of meaning-making; (4) gestures as forms of education (Wulf, 2010, 285). Gestures are bodily representations. This can be seen, for example, in interactions, where gestures express facts and emotions due to their performative character. They are associated with linguistic utterances. Some of these are often only produced with the help of the corresponding gestures. By articulating something that the speaker themself is not or only barely aware of, gestures reinforce or complement linguistic utterances. Gestures can express something that is at odds with what the speaker intends to say. Gestures contribute to creating, expressing and representing meaning. They make it possible to understand and respond to the meaning that is formed in human interactions and thus create social relationships. Gestures, and especially the gesture of pointing, play an important role in education. With their help, things are highlighted, contexts are clarified, explained, and mimetically processed. However, the significance of gestures goes further in the Anthropocene. They express an existential relationship to the world. Gestures can be understood as expressions in a compressed form of human behaviour towards nature, towards other humans, and towards ourselves (Jousse, 2008; Flusser, 1991). The gesture of coal mining is a good example of this enhanced use of the word. Here we find a situation where human beings access nature for their own advantage and benefit.

Gestures must be read in their historical and cultural context (Bremmer & Roodenburg, 1992). Historical and ethnological studies show how differently gestures are interpreted in different cultures. The more iconic the character of a gesture is, the more it is understood by people in a variety of cultures. The more symbolic the character is, the more its understanding requires a particular cultural background. Gestures occur in a historical and cultural context which has its own power structures, and it is only by taking account of this context that we can grasp their meaning. They give us information about core values of a society and insight into "mentality structures". Gestures are significant movements of the body, which are directed by an intention. Their forms of expression and representation cannot, however, be reduced to their underlying intention. The difference between gestures as expression and representation of the body and the semantic meaning of gestures cannot be ignored. In the structure of gestures, there is a meaning which cannot be interpreted linguistically, and which can only be understood in a mimetic process, in which the gesture is performed. This can be either physically or internally. One of the gestures that characterizes the industrial age is the extraction of lignite in open-cast mines, with the deployment of earth moving machinery and the loading of the coal onto trains with the numerous machine-driven activities this entails. It is a gesture that enables human beings to unlock the resources of nature for their own use. Such a gesture encompasses a goal, a design and a consequence of actions that are structured by time and language.

The language of gesture is older than spoken or written language (Tomasello, 2008). From a historical point of view there is no doubt gestures contributed considerably to the creation of spoken language. Even today gestures help us to articulate emotions and thoughts. They help us to understand the meaning of the spoken language and of social and cultural behaviour and are essential elements in human communication. Gestures are developed, learnt, and shaped through practice. Whereas with facial expressions it is very difficult to separate expression from emotion, or form from content, gestures highlight the differences between these aspects. Thus, gestures can be shaped intentionally. Perfect gestures often express a high degree of "artificial naturalness" (Wulf & Fischer-Lichte, 2010; Wulf et al., 2010).

Since gestures are forms of cultural expression that are rooted in the body, social subjects can come to know themselves by understanding their own gestures. Insofar as subjects can perceive themselves in gestures. In a mimetic relationship to their gestures, social subjects experience themselves in these presentations. In the use of gestures, corporeal being can be transferred into the active mode of performing a gesture and having one's body in this performance. Social subjects express themselves in facial expressions and gestures and experience the reactions of others to these forms of expression. The iconic and corporeal language of gestures is the result of a historical and cultural process; this language shapes the individual, who at the same time continuously modifies the language. As social subjects acquire gestures through a process of mimesis, they participate in corporeal and iconic traditions, which they can apply creatively to new social situations. A corporeal figuration is expressed in gestures, whilst in the performance of gestures an internal relation to oneself and a mediated relation to the world come into existence. In the corporeal character of a gesture, it is hard to separate the emotions from the physicality of the gesture (McNeill, 1992, 2005; Kendon, 2004; Kraus & Wulf, 2022). If we take the coalmining that takes place all over the world to be a gesture, the physical, action-related aspect of this activity is emphasized by the fact that it is performative.

Gestures play an important role in the process of civilization and human self-domestication. Most social institutions develop specific sets of gestures. The demands social institutions make on the use of specific gestures in an institutional context force an individual to behave in particular ways. The performance of new gestures is not a superficial change. Since these new gestures express the incorporation of new values, attitudes, and feelings, they can be understood as signs of profound changes. By learning a new gesture language, the individual is made ready for a new life. Gestures are one way in which institutions can express their social power – in this case by making people perform institutionally acceptable gestures (Wulf & Fischer-Lichte, 2010; Wulf et al., 2011a, b). Where people fail to live up to institutional expectations, this is seen by the representatives of that institution as a criticism of the institution, which often results in sanctions being taken. Therefore, the acquirement of gesture competence and of practical gesture knowledge is an important goal to help human beings to cope with institutions successfully.

To read and to decode gestures they must be perceived mimetically (Wulf, 2013). Someone perceiving a gesture understands it by recreating it internally in their imagination and thereby grasps its specific physical expression and representation. Although gestures are meaningful and can be analyzed in terms of their meaning, it is mimetic re-creation which permits a full understanding of their sensual, emotional, and symbolic content. This is also true of the mise-en-scène of gestures. Reproducing them mimetically helps us to assimilate the specific corporeal expression of another person and it is the mimetic recreation of their gestures that allows what is specific to their body to be perceived. In the mimesis of another person's gestures, what takes place is an encounter with the world of the other person, in the way they use their body to express and present themself. Thus, it becomes possible to relive the sensual experience of the Other. Getting to know another person's world is often experienced as enriching and pleasurable, since it provides the individual with the experience of the outside world. The mimetic process is thus a way of enlarging the world of individuals, leading to wide-ranging experiences outside those that they have within them (Wulf, 2022a).

References

Bremmer, J., & Roodenburg, H. (Eds.). (1992). *A cultural history of Gesture*. Cornell University Press.

Flusser, V. (1991). *Gesten. Versuch einer Phänomenologie*. Bollman.

Gebauer, G., & Wulf, C. (1995). *Mimesis. Culture – Art – Society*. University of California Press.

Gebauer, G., & Wulf, C. (1998). *Spiel, Ritual, Geste. Mimetisches Handeln in der sozialen Welt*. Rowohlt.

Jousse, M. (2008). *L'anthropologie du geste*, 3 vols. Gallimard.
Kendon, A. (2004). *Gesture: Visible action as Utterance*. Cambridge University Press.
Kraus, A., & Wulf, C. (Eds.). (2022). *Handbook of embodiment and learning*. Palgrave Macmillan.
McNeill, D. (1992). *Hand and mind. What gestures reveal about thought*. University of Chicago Press.
McNeill, D. (2005). *Gesture and thought*. Chicago University Press.
Tomasello, M. (2008). *Origins of human communication*. MIT Press.
Wallenhorst, N., & Wulf, C. (Eds.). (2023). *Handbook of the Anthropocene*. Springer.
Wulf, C. (2010). Der mimetische und performative Charakter von Gesten. In C. Wulf & E. Fischer-Lichte (Eds.), *Gesten. Inszenierung, Aufführung, Praxis* (pp. 283–297). Wilhelm Fink.
Wulf, C. (2013). *Anthropology. A continental perspective*. University of Chicago Press.
Wulf, C. (2022a). *Humans and their images. Imagination, mimesis, performativity*. Bloomington.
Wulf, C. (2022b). *Education as human knowledge in the Anthropocene. An Anthropological perspective*. Routledge.
Wulf, C., & Fischer-Lichte, E. (Eds.). (2010). *Gesten: Inszenierung, Aufführung, Praxis*. Wilhelm Fink.
Wulf, C., Althans, B., Audehm, K., Bausch, C., Göhlich, M., Sting, S., Tervooren, A., Wagner-Willi, M., & Zirfas, J. (2010). *Ritual and identity. The staging and performing of rituals in the lives of young people*. The Tufnell Press.
Wulf, C., Althans, B., Audehm, K., Blaschke, G., Ferrin, N., Kellermann, I., Mattig, R., & Schinkel, S. (2011a). *Die Geste in Erziehung, Bildung und Sozialisation: Ethnographische Fallstudien*. Verlag für Sozialwissenschaften.
Wulf, C., Suzuki, S., Zirfas, J., Kellermann, I., Inoue, Y., Ono, F., & Takenaka, N. (2011b). *Das Glück der Familie: Ethnographische Studien in Deutschland und Japan*. Springer.

Christoph Wulf is Professor of Anthropology and Education and a member of the Interdisciplinary Centre for Historical Anthropology, the Collaborative Research Centre (SFB, 1999–2012) "Cultures of Performance," the Cluster of Excellence (2007–2012) "Languages of Emotion," and the Graduate School "InterArts" (2006–2015) at the Freie Universität Berlin. His books have been translated into 20 languages. For his research in anthropology and anthropology of education, he received the title "*professor honoris causa*" from the University of Bucharest and the honorary membership of the German Society of Educational Science. He is Vice-President of the German Commission for UNESCO. *Major research areas:* historical and cultural anthropology, educational anthropology, imagination, intercultural communication, mimesis, aesthetics, epistemology, Anthropocene. Research stays and invited professorships have included the following locations, among others: Stanford, Tokyo, Kyoto, Beijing, Shanghai, Mysore, Delhi, Paris, Lille, Strasbourg, Modena, Amsterdam, Stockholm, Copenhagen, London, Vienna, Rome, Basel, Saint Petersburg, Moscow, Kazan, Sao Paulo.

Intercultural Education

Frédérique Brossard Børhaug

Abstract Enhancing cultural and religious diversity has been well-explored within intercultural educational discourse, but to what extent does it also raise crucial awareness about and provide tools for enhancing biodiversity? This contribution proposes that intercultural education may have a powerful role in intensifying the preservation of multiple diversities in the Anthropocene by interweaving the desire for social justice and justice for the climate with political intercultural engagement.

Towards Equity for All in Multicultural Societies

Intercultural education pursues the aim of social justice and cultural/religious diversity although it is implemented diversely in different multicultural countries. The field also includes varied and sometimes contradictory terms, e.g. Anti-racist Education (Lee, 2009), Ethnic Studies (Sleeter & Zavala, 2020), Intercultural Communicative Competences (Byram, 2021), Reference Framework of Competences for Democratic Culture (Barrett, 2020), Intercultural Education (Portera, 2008) and Multicultural Education (Banks, 2009).

In sum, the field of research aims to give way for equitable education based on the student's particular needs, and diverse cultural competencies; it counteracts in democracies a frequent superficial understanding of tolerance where one tolerates cultural/religious diversity from minorities "up to a certain point", determined by powerful majority leaders. Intercultural education on the contrary seeks to change a discriminatory social order by promoting systemic social change and inclusive cultural practices based on 'deep equality', fair-minded dialogue with all community members and intercultural didactics (Beaman, 2014; Buchardt & Fabrin, 2011). However, such critical work necessitates extensive efforts and a steady political

F. B. Børhaug (✉)
VID Specialized University, Stavanger, Norway
e-mail: frederique.borhaug@vid.no

© The Author(s), under exclusive license to Springer Nature Switzerland AG 2023

N. Wallenhorst, C. Wulf (eds.), *Handbook of the Anthropocene*,
https://doi.org/10.1007/978-3-031-25910-4_234

engagement, which are threatened by neoliberalist output-oriented educational ideology (Sleeter, 2014), and undemocratic forces such as assimilative policies and nationalist-populist thinking (Taguieff, 2012).

Strengthening the Contribution of Intercultural Education in Anthropogenic Climate Change

A current challenge is to link more explicitly biodiversity and cultural/religious diversity. For instance, despite its continuous focus on equity and social justice, it appears that intercultural education still does not sufficiently relate these fundamental dimensions of diversity in academic discourse, and it represents a missed valuable opportunity (Brossard Børhaug, 2021). However, the call of first nation natives for a stronger recognition of their cultural ties in pact with nature is familiar (e.g., Escobar, 2016). There is also a growing focus on decolonial ecology, bridging environmentalism and decolonization by disclosing long-lasting exploitation of racialized bodies and lands (see for example the Caribbean case, Ferdinand, 2019). Furthermore, young people globally are getting increasingly engaged in nature preservation and often embrace cultural/religious diversity as well. Linking even more explicitly the two claims in youth activism would strengthen an overall interest in protecting diversity in its multiple dimensions and make intercultural education more mainstream (Brossard Børhaug, 2021, p. 396). Thus, a great potential of intercultural education exists in the Anthropocene despite it being currently under communicated.

Transformative Intercultural Education for Biocultural Diversity

Education must prepare agents of climate and social justice in multicultural states. In the intercultural field, one advocates for a systemic approach where educators not only examine the subject's positionalities and responsibilities in present discriminatory practices in and out of schools, but also reflect on the remediation of the system's injustice through new critical cultural and social practices (Gorski & Dalton, 2019). Becoming an agent of change thus raises existential questions about subjectivity. It entails that educators aim to promote what is called 'subjectification' going beyond the common objectives of socialisation and qualification. Not only should educators help learners to become highly knowledgeable and socialised in particular cultural frames, they must also give to upcoming generations the ability to say NO to an unjust climate and social order and become independent from that order when necessary. Promoting subjectivity which goes beyond socialization and qualification implies a politicization of learning and a dialogical emancipation. It brings to the fore the critical reflection about 'the desirability of desires' and one's own irreplaceable responsibility, i.e. 'how' I am, not foremost 'who' I am in the world (Biesta, 2014). In our current climate urgency, youth does not wait for adults to decide; global school-strikes initiated by Greta Thunberg are a typical example of subjectification. As the Anthropocene raises unprecedented existential questions about the perpetuation of human living, subjectification must be strengthened by all citizens, reflecting over how to take concrete actions, and developing the '*courage d'être*' (Hétier & Wallenhorst, 2021). Intercultural education has a long tradition of developing resilient transformative learning in oppressive social order and efficient didactical tools, and it also claims that education must be decolonized (Gorski, 2008); it thus may be a beneficial contribution for changing destructive climate, social and cultural practices.

Furthermore, innovative concepts are needed. The emergent concept of 'biocultural diversity' is crucial to put forward (Maffi, 2018; UNESCO/Terralingua/WWF, 2003). This notion argues that all manifestations of life on earth are interrelated where local environments and cultures have coevolved in complex interdependent systems generating rich linguistic, cultural, and biological diversity. In other words, through close interaction with the environment, people adapted to and modified their local biological context, they also developed a unique and valuable knowledge of it (Maffi, 2018). Nevertheless, globalised social and climate injustice put the multiple diversities on Earth at great risk; languages, cultures and species are dying (e.g., UNESCO/Terralingua/WWF, 2003). Without strong intercultural practices, humans are on the verge of losing the vivacity of local cultures which are always supported by vivid local languages. Without robust local languages and cultures, humans respond less adequately to their local environment's needs and eventually mistreat their ethical responsibility of letting each existence be worthy for itself, not only for the sole utility of humans needs.

In times of climate change the need to root education back into the vitality of earth is therefore fundamental (Hétier & Wallenhorst, 2021), and it supposes enhancing local cultures and languages that have been developed in close interaction with nature. Curricula in schools thus must reverse a common assimilative trend where the majority language becomes the priority language for transmitting knowledge and values at school. For instance, the concept of sustainable development is included in all syllabuses of the current Norwegian curriculum for primary up to upper secondary school (*The Knowledge Promotion Curriculum* 2020). However, this key notion is not included in mother tongue education for minority pupils, and it can be considered as a typical example of assimilative policies. On the contrary, state education must give visibility and genuine recognition for all minority languages in place; a call that was made many years ago by intercultural educators (e.g., Cummins, 1981). Because local cultures and languages also shed light on varied and rich ways of understanding and interacting with nature, a fruitful dialogue may take place. Seeds were the first travellers; languages, cultures and local environments are also sown through migration, and intercultural education may play a decisive role in protecting our inter-reliant diversity.

References

Banks, J. (2009). Multicultural education: Dimensions and paradigms. In J. Banks (Ed.), *The Routledge international companion to multicultural education* (pp. 9–32). Routledge.

Barrett, M. (2020). The Council of Europe's reference framework of competences for democratic culture: Policy context, content and impact. *London Review of Education, 18*(1), 1–17. https://doi.org/10.18546/LRE.18.1.01

Beaman, L. G. (2014). Deep equality as an alternative to accommodation and tolerance. *Nordic Journal of Religion and Society, 27*(2), 89–111.

Biesta, J. J. G. (2014). *The beautiful risk of education*. Paradigm Publishers.

Brossard Børhaug, F. (2021). Missing links between intercultural education and anthropogenic climate change? *Intercultural Education, 32*(4), 386–400. https://doi.org/10.1080/14675986.2021.1889984

Buchardt, M., & Fabrin, L. (2011). *Interkulturel didaktik: Introduktion til teorier og tilgange [intercultural didatics: Introduction to theories and approaches]*. Hans Reitzels Forlag.

Byram, M. (2021). *Teaching and assessing intercultural communicative competence: Revisited* (2nd ed.). Multilingual Matters.

Cummins, J. (1981). The role of primary language development in promoting educational success for language minority students. In Office of Bilingual Bicultural Education (Ed.), *Schooling and language minority students: A theoretical framework* (p. 3–49). Evaluation, Dissemination and Assessment Center. Accessed from: Cummins1981paperinCaliforniaTheoreticalFramework(2).pdf

Escobar, A. (2016). Thinking-feeling with the earth: Territorial struggles and the ontological dimension of the epistemologies of the south. *Revista de Antropologica Iberoamericana, 11*(1), 11–32. 10_VOL11_1_ART_1_INGLES.indd (aibr.org)

Ferdinand, M. (2019). *Une écologie décoloniale: Penser l'écologie depuis le monde caribéen*. Anthropocène Seuil.

Gorski, P. (2008). Good intentions are not enough: A decolonizing intercultural education. *Intercultural Education, 19*(6), 515–525. https://doi.org/10.1080/14675980802568319

Gorski, C. P., & Dalton, K. (2019). Striving for critical reflection in multicultural and social justice teacher education: Introducing a typology of reflection approaches. *Journal of Teacher Education, 71*(3), 357–368. https://doi.org/10.1177/0022487119883545

Hétier, R., & Wallenhorst, N. (2021). *Enraciner nos enfants dans la vitalité de la Terre pour les élever _ quelques éléments pour penser l'éducation en* Anthropocène. Enraciner nos enfants dans la vitalité de la Terre pour les élever – Quelques éléments pour penser l'éducation en Anthropocène – La pensée écologique (lapenseeecologique.com).

Lee, E. (2009). Taking multicultural, anti-racist education seriously: An interview with Enid Lee. In W. Au (Ed.), *Rethinking multicultural education: Teaching for racial and cultural justice* (pp. 9–16). Rethinking Schools.

Maffi, L. (2018). Sustaining biocultural diversity. In K. L. Rehg & L. Campbell (Eds.), *The Oxford handbook of endangered languages*. Oxford University Press. https://doi.org/10.1093/oxfordhb/9780190610029.013.32

Portera, A. (2008). Intercultural education in Europe: Epistemological and semantic aspects. *Intercultural Education, 19*(6), 481–491. https://doi.org/10.1080/14675980802568277

Sleeter, E. C. (2014). Multiculturalism and education for citizenship in a context of neoliberalism. *Intercultural Education, 25*, 85–94. https://doi.org/10.1080/14675986.2014.886357

Sleeter, E. C., & Zavala, M. (2020). *Transformative ethnic studies in schools: Curriculum, pedagogy, and research*. Teachers College press.

Taguieff, P.-A. (2012). *Le nouveau national-populisme*. CNRS Éditions.

UNESCO/Terralingua/WWF. (2003). *Sharing a world of difference: The earth's linguistic, cultural and biological diversity*. UNESCO. Accessed from: Sharing-English (terralingua.org)

Learning for Climate Action

Robert B. Stevenson and Hilary Whitehouse

Abstract In 2021 the Federal Court of Australia handed down a judgment, in response to a class action brought against the Australian Environment Minister, by a group of young people, acting on behalf of their peers, that the Minister owes a duty of care to young people not to cause them harm from climate change. This court concluded that quality of life, including opportunities to grow and prosper and enjoy good health and nature's treasures, *"will be greatly diminished,"* while

> *Lives will be cut short. Trauma will be far more common*
>
> *None of this will be the fault of nature itself. It will largely be inflicted by the inaction of this generation of adults, in what might fairly be described as the greatest inter-generational injustice ever inflicted by one generation of humans upon the next.*
>
> *To say that the children are vulnerable is to understate their predicament.*

The global community is awakening to the fact that we now live in a completely different world to the one we inhabited previously, or thought we did. The urgency and kinds of problems and crises individuals, communities and societies confront today have escalated dramatically. Global warming, pandemics, species extinctions, deforestation, desertification, plastic oceans and threatened loss of coral reefs, and diminishing sources of food, and safe drinking water are major socio-ecological crises resulting from human impacts on the natural processes of the planet, and which now define the current epoch of the Anthropocene.

That young people understand the growing uncertainties as to their future quality of life, the stabilities of their cultures and societies, and of the habitability of the planet, is evidenced by global protest movements, such as the school climate strikes and Extinction Rebellion. Although there are some recent signs of progressive change, activism is a clear signal of the failure of educational institutions, politicians, policy makers, corporations and governments to directly address young peoples' heightened concerns. Young people are making it clear that the lack of urgent

R. B. Stevenson (✉) · H. Whitehouse
James Cook University, Cairns, Australia
e-mail: bob.stevenson@jcu.edu.au; hilary.whitehouse@jcu.edu.au

© The Author(s), under exclusive license to Springer Nature Switzerland AG 2023

N. Wallenhorst, C. Wulf (eds.), *Handbook of the Anthropocene*,
https://doi.org/10.1007/978-3-031-25910-4_235

global and national action in the face of climate disruption, has pushed them to collective activism to address such inertia.

The complexity, uncertainty, and contentious characteristics of the known (and escalating) threats demands individual, group and community resilience and flexibility, as well as capacities for collaborative community-based research, learning and citizen action skills for responding relatively rapidly to emerging perils.

Today's children and youth will have to develop, along with resilience, resourcefulness and flexibility, capacities for collaborative community-based knowledge building and action taking for responding relatively rapidly to known and emerging crises. A community-based action dimension in schools enables young people to prepare for responding to a global issue at their local community level by discussing alternative actions and testing the appropriateness and efficacy of actions in accordance with their understandings and values. Unless students can act on their understandings and value judgments, they will not consider themselves capable of rectifying an issue of concern (Stevenson, 1987/2007).

Our approach to not only economic growth and development, but also to education needs to radically change. Historically, schools have been essentially concerned only with passing on knowledge and practices of the Holocene, a stable period of climate during which human societies expanded across the globe. Unfortunately, the Holocene is no more, and we now live in a period of great climate and environmental uncertainty entirely of our own making.

Learning about established knowledge and practices needs to be balanced with more attention to current and emerging realities facing children, youth and their communities. This is recognised by the United Nations Educational, Scientific and Cultural Organization (UNESCO) ESD Roadmap (UNESCO, 2020) that prescribes the responsibilities of all signatory nations to implement education for sustainable development including climate change education. For all signatory nations, the ESD Roadmap requires that education for sustainable development (ESD) be integrated into all national and state educational policies in recognition that it is appropriate for nations to acknowledge the Anthropocene in contemporary learning. Scaling up ESD within all education institutions, communities and informal learning settings means that established and emerging ecological and socio-environmental learnings will occupy their place alongside established knowledge and practices.

The ESD Roadmap positions children and young people as capable and active agents of change, a responsibility many are willing to carry despite personal risk (see Laing, 2021). To quote the ESD Roadmap; "this whole-institution approach to ESD calls for learning environments where learners learn what they live and live what they learn" (UNESCO, 2020, p. 28). The previous division between formal learning and taking positive environmental actions is challenged by this revisioning of learning.

In line with United Nations Sustainable Development Goals (UNSDGs) and Agenda 2030, educators are asked to be better prepared in terms of enacting the knowledge, skills, values and actions of ESD, for which there is large body of published evidence on what is effective and what practices result in material change at different scales of engagement. The ESD Roadmap recognises young people "continue to envision the most creative and ingenious solutions to sustainability challenges" (p. 32), and a stated focus on intergenerational justice acknowledges the capabilities of children and young people to make decisions in their own interests.

The ESD Roadmap identifies the importance of community-scale actions, especially for enabling learning partnerships for change, promoting partnerships for learning and active cooperation between learning institutions, communities, and business enterprises. Learning is recognised as one of the means by which a more creative and sustainable future can be built for the largest generation of young people in history.

Individuals, groups and society make progress by learning not just from the past but also acknowledging the limits of current understanding and the need for new ways of thinking and acting. Schools need to shift some of their focus from individual academic achievement to developing many of the individual and collective capacities needed for dealing with the emerging threats to human survival.

This demands collaborative and participatory forms of education that prepare students to collectively address the urgent, uncertain and contentious world of planetary and pandemic crises. New student understandings and action capabilities can emerge in the process of investigating and taking actions to test solutions at the local community level, the scale at which sustainability education is likely to be most effective and meaningful.

Current school curriculum and organizational structures and policies, however, especially in secondary education, do not make it easy for educators to provide opportunities to engage their students in learning, thinking and acting in and with their local communities. Research on climate change education at James Cook University in Australia has investigated teachers' perspectives on and classroom practices in climate change education (Nicholls, 2017) and examined young people's informal environmental learning and activism within social media (Field, 2016). In the first case, the limited teachers who reported addressing climate change indicated it was relegated to short informal discussions outside the formal curriculum. In the second, youth from around the globe reported participating in environmental-interest social media networks because of the lack of inclusion of climate change in their school curriculum.

Most nations do not have sufficient educational policy structures in place to promote the goals of the ESD Roadmap nor meet the expectations of the young that their futures are not going to be in complete turmoil. Robust policy tends to promote robust action and its long past time for policymakers to adjust their thinking to address current socio-ecological and thermodynamic realities. Without such adjustments, educators are facing a crisis of trust. Children and young people are using the peer learning opportunities afforded by social media to inform themselves of how little action their governments, corporations and educational systems have taken over the last decades to protect their futures. They are using these opportunities to connect with a wide audience, work with peers and other community members and develop their leadership skills to promote change (Field, 2016).

The youth climate justice movement has gained rapid momentum – even during a pandemic – as young people draw on their own capabilities for learning for active citizenship. Now that young people better understand the root causes of climate turmoil, they can feel abandoned and disheartened by historic and contemporary economic, political and educational structures and this learning motivates their activism. The emerging trust gap between young people and educational systems will widen unless formal education can pivot sufficiently to recognise the needs of their students and act effectively to address their concerns.

The extent of change needed appears transformational rather than reformational. What is required now is significant time and curriculum space that explicitly recognises that students' present and becoming future is indeed worthy and valued within the strictures and "deep grammars" of formal education (Fraser, 2009). For interdisciplinary and action-oriented community studies, the problem-based curriculum approach addresses "problems that matter" (Ziplin, 2017) in such a way as to both recognise and materially address the inherent problems of intergenerational injustice and the multiplying injustices experienced by the natural world.

Instead of comprising pre-determined content, sections of school curriculum might be viewed as a "space of emergence" (Osberg & Biesta, 2007) in recognition that Anthropocene learning is emergent and creative: where problems that matter to students are posed with a strong focus of developing materially relevant actions for change. Enabling school students to engage in critical and creative thinking and building knowledge for co-creating solutions is a pro-active learning approach. Creating more time for problem-based learning actively brings student concerns into the formal educational space. Strategies include identifying what really matters to young people; explicitly developing student capacity for collaborative inquiry; interacting with a diverse range of community members, community groups and specialists; conducting investigations to collect and examine evidence; identifying underlying values and conflicts; determining appropriate responses and testing them in action.

For the last 20 years, Student Action Teams (SATs) have been a feature of many Victorian (Australian) schools. Students in these teams identify and investigate "an issue of concern within their school and/or (hopefully) wider community and then plan and take action to improve" the situation (Mayes & Holdsworth, 2020). These examples of Student Action Teams are set in the "*present realities and shared concerns of young people, their teachers and their communities, rather than deferred 'until' students achieve 'future citizenship' in the world*" (Mayes & Holdsworth, 2020). This approach can be an important means for young people to construct visions of and possibilities for alternative, positive futures.

Educators are coming to realise that transformational learning in the Anthropocene by necessity means learning for responsibility and response, that is, learning for action and learning for change. Business as usual is not going to lead us anywhere pleasant. Children and young people are not future but current citizens. They are asserting their rights to a more sustainable future now. They demonstrate responsibility for their learning, they are aware, active and engaged. Given their only real choices at this point in human history are nihilism and despair, or active protest and learning for action, the majority will likely choose action. Whether or not our educational learning systems pivot or remain in stasis in response is a question yet to be answered.

References

Fraser, N. (2009). *Scales of justice: Reimagining political space in a globalizing world.* Columbia University Press.

Field, E. (2016). *Constellations of environmentalism: An exploration of learning and activism within youth-created social media interest groups.* Unpublished Ph.D. thesis, James Cook University, Australia.

Laing, O. (2021, October 17). Gen Z on how to save the world: Young climate activists speak out. *The Guardian.* Download from https://www.theguardian.com/environment/2021/oct/17/gen-z-on-how-to-save-the-world-young-climate-activists-speak-out?

Mayes, E., & Holdsworth, R. (2020). Learning from contemporary student activism: Towards a curriculum of fervent concern and critical hope. *Curriculum Perspectives, 40*, 99–103. https://doi.org/10.1007/s41297-019-00094-0

Nicholls, J. (2017). *Understanding how Queensland teachers' views on climate change and climate change education shape their reported practices.* Unpublished Ph.D. thesis, James Cook University, Australia.

Osberg, D., & Biesta, G. J. J. (2007). Beyond presence: Epistemological and pedagogical implications of 'strong' emergence. *Interchange, 38*(1), 31–51. https://doi.org/10.1007/s10780-007-9014-3

Stevenson, R. B. (1987/2007). Schooling and environmental education: Contradictions in purpose and practice. In I. Robottom (Ed.), *Environmental education: Practice and possibility.* Deakin University Press. (Reprinted in *Environmental Education Research, 13*(2): 139–153).

UNESCO. (2020). *Education for sustainable development: A roadmap.* United Nations Educational, Scientific and Cultural Organization. Download from: https://www.gcedclearinghouse.org/sites/default/files/resources/200782eng.pdf

Ziplin, L. (2017). Pursuing a problem-based curriculum approach for the sake of social justice. *Journal of Education, 69*, 67–92. http://www.scielo.org.za/pdf/jed/n69/04.pdf

Robert (Bob) Stevenson is Adjunct Professor, formerly Professor and Tropical Research Leader (Education for Environmental Sustainability), The Cairns Institute, James Cook University and Honorary Professor of Education, University of Queensland, Australia. His scholarship focuses on theory-policy-practice relationships in environmental/sustainability education and its history and marginalized status as an educational reform in schools. He was lead editor of the International Handbook of Research in Environmental Education (AERA/Routledge, 2013) and Editor-in-Chief of the Journal of Environmental Education, 2010–18.

Hilary Whitehouse is a Fellow of the Cairns Institute and Deputy Dean of the Graduate School at James Cook University in Cairns, Queensland. She is known for her scholarship on climate change education, and education for sustainability and transformation. She is an editor for both the Journal of Environmental Education and the Australian Journal of Environmental Education. She volunteers as a researcher and educator with the small, conservation group, the Bats and Trees Society of Cairns. https://research.jcu.edu.au/portfolio/hilary.whitehouse

Mimesis

Christoph Wulf

Abstract From early childhood on, mimetic processes are extremely important both for retaining and passing on the negative effects of the Anthropocene as well as for changing them and making fundamental reforms in the relationship between human beings and nature. Plato and Aristotle understood people depend on mimetic processes for their individual and collective, cultural, and social development. This insight has been confirmed by research in Historical Anthropology, in Evolutionary Anthropology and in Neuroscience. People learn to a very large extent in mimetic processes, that is, through imitating, making themselves similar, representing. These processes are not simple copying processes like making photocopies. They are productive processes in which people behaving mimetically take an "impression" of the behaviour of other people which they then integrate into their imaginary.

The way we live in today's world is marked by the fact that we human beings have a powerful influence on the fate of the planet. This has given rise to the term "Anthropocene" where the traditional distinction between nature and nature no longer applies. For there is hardly any nature left upon which humans have not left their mark (Wulf, 2022b; Wallenhorst & Wulf, 2022, 2023). Among the things causing this situation are the numerous destructive effects of industrialization and modernization. These include global warming, destruction of biodiversity, atomic and hydrogen energy, the cyborgization of humanity, genetics, pollution, the production of plastic and the eradication of non-renewable energy sources.

From early childhood on, mimetic processes are extremely important both for retaining and passing on the negative effects of the Anthropocene as well as for changing them and making fundamental reforms in the relationship between human beings and nature. Plato and Aristotle understood people depend on mimetic processes for their individual and collective, cultural, and social development (Plato, 2012; Aristotle, 2013). This insight has been confirmed by research in Historical

C. Wulf (✉)
Freie Universität Berlin, Berlin, Germany
e-mail: christoph.wulf@fu-berlin.de

© The Author(s), under exclusive license to Springer Nature Switzerland AG 2023

N. Wallenhorst, C. Wulf (eds.), *Handbook of the Anthropocene*,
https://doi.org/10.1007/978-3-031-25910-4_236

Anthropology (Wulf, 2013; Gebauer & Wulf, 1995, 1998; Resina & Wulf, 2019), in Evolutionary Anthropology (Tomasello, 1999, 2008) and also in Neuroscience (Rizzollati & Sinigalia, 2008; Jacoboni, 2008). People learn to a very large extent in mimetic processes, that is, through imitating, making themselves similar, representing. These processes are not simple copying processes like making photocopies. They are productive processes in which people behaving mimetically take an "impression" of the behaviour of other people which they then integrate into their imaginary.

The concept of "mimesis" contains suggestions of mimicry, representation, imitation, reproduction, simulation, autopoiesis. As an anthropological concept mimesis helps us to understand and explain processes of socialization and education and also social action and aesthetic experience. Here "mimesis" is a useful tool in describing and analyzing interdisciplinary and transdisciplinary processes (Wulf, 2022a). The spectrum of meanings of the concept is ambivalent. On the one hand, as processes of mimicry, mimetic processes can result in an adaptation to something that is given, set in stone and lifeless, and thus also result in the negative conditions of the Anthropocene (Horkheimer & Adorno, 1972). On the other hand, such processes also inspire many hopes for the processing and incorporation of better ways of life, practices, attitudes and values. They can lead to "living experiences" (Adorno) of the outside world, of the Other and one's own self. Mimetic processes can initiate movements of "broken intention", space for what is non-identical, possibilities for a non-instrumental approach to nature and the world, in which the particular is protected as opposed to the universal, and people and things are conserved. Given the present situation of social and cultural development, the ambivalence in mimetic processes is inevitable.

While modern, rational thinking is directed towards the individual, isolated subject of perception, mimetic processes are often embedded in the complex relationships between people. The mimetic creation of a symbolic world refers to other worlds and their creators and draws other people into one's own world. It recognizes the exchange between world and human beings and the aspect of power that this contains. The history of mimesis is a history of the power struggle over the creation of symbolic worlds, over the freedom to portray others and to interpret the world according to one's own ideas. In this respect, particularly in education and socialization, mimesis is part of the history of power relationships.

Mimetic processes are not merely processes of reproducing or taking imprints. On the contrary, they need to be shaped individually by children, young people, or adults. Here the degree of individual difference will vary on the basis of differing conditions. Many mimetic processes cannot be distinguished from the processes of desiring or wishing for something, and of sensual perception and experience (Wulf, 2022a, b).

Although, in view of the negative effects of the Anthropocene on the planet, it is necessary for people's behaviour to fundamentally change, there is the danger that the destructive developments endangering the future of the planet will continue without being diminished at all or only slightly changed. The younger generation take over the attitudes and behaviour patterns that prevail in everyday life in society through mimetic processes. Since it is extremely difficult to resist the attraction and the power of these behaviours, they continue to spread. One way of removing oneself from the influence of ways of life that are perceived as negative is to prohibit them and banish them from everyday life. This prevents them spreading, allowing them to be replaced with new behaviours that are regarded as positive. Another possibility, that also recognizes the mimetic pull of negative ways of life, aims to challenge the effects of negative ways of life and to use our critical insight to protect ourselves from their power and influence. Here it is not prohibitions but insight and recognition that are seen as driving the change in destructive social practices and ways of life.

The example of consumer behaviour serves as a good example of this. If we accept that the excessive consumption indulged in by many people is a central characteristic of the Anthropocene, then we consider how this behaviour can be changed to reduce the destructive nature of human consumption. People buy things and replace them just for fun. They enjoy the luxury which makes them feel good. They enjoy purchasing things, even if they do not necessarily need them. People do not care if the things they replace are still in good condition or not. In mimetic processes children, young people and adults learn to behave the same way. To build a sustainable society, people must change their behaviour of consumption. Otherwise, due to the power of mimetic processes it will not be possible to develop a sustainable society. Despite a rhetoric of sustainable development, the behaviour of extensive consumption will be a powerful model encouraging people to continue the practices of non-sustainable behaviour.

In order to prevent this, we can ban excessive consumption or support people in challenging the destructive sides of their consumer behaviour and thus help them change their behaviour through insight. There are different preferences for dealing with the problem depending on the social system or democratic tradition in question. Both alternatives' ways of acting recognize that current practices of consumer behaviour are based on intergenerational ways of life that continue to be preserved and spread through mimetic processes. These must, however, be changed, if we are to develop a society that values sustainability. New practices and ways of life are necessary if such a society is to come into being. These practices must also become the starting point for mimetic processes which will help to disseminate them. The first alternative involves preventing mimetic processes that relate to negative practices by means of prohibition and exclusion. The second involves developing new ways of life that are viewed as positive on the basis of a critical view of negative practices.

By assimilating previously experienced sustainable behaviour, people acquire the ability to behave in a sustainable way. By mimetically participating in sustainable practices human beings expand their behaviour and create new possibilities for sustainable actions. People re-create sustainable situations or behaviors or practices they have experienced in the past and by repeating them make them their own. It is in the confrontation with sustainable situations or practices that they acquire the possibility of behaving sustainably. Outer and inner world continuously resemble each other and can only be experienced in their interrelation. Similarities and correspondences between the sustainable inner and the sustainable outer arise. Subjects make themselves similar to the sustainable outside world and change their behavior in this process (Wulf, 2022a, b; Gebauer & Wulf, 1995, 1998).

The mimetic capacities of children, young people and adults are closely tied up with physical and social processes and counteract tendencies towards abstraction in society (Kress et al., 2021). They form a bridge between them and the outside, the world and the other person. They attempt to reduce the sharp divide between subject and object and the clear difference between what is and what should be. There is an understanding of what is "between", that is experienced in a subject "becoming similar" to an outside world or another person. Mimetic processes contain rational processes but go far beyond this (Kraus & Wulf, 2022). In these processes people step out of themselves, become similar to the world around them and are able to integrate the outside world into their inner world. Mimetic processes result in the embodying of attitudes and the preservation of sustainable practices. They help us to become close to objects and the Other and are therefore necessary for our understanding.

In mimetic processes an imitative change and shaping of preceding worlds takes place. Herein lies the innovative moment of mimetic acts and the chance to contribute to sustainable development. Social practices are mimetic if they refer to other actions and can themselves be understood as social arrangements that represent independent social practices as well as having a connection to other actions. Social acts become possible through the emergence of practical knowledge in the course of mimetic processes. The practical knowledge relevant to social acts has a physical, playful, historical, and cultural side; it is formed in face-to-face situations and is semantically ambiguous; it has imaginary components, cannot be reduced to intentionality, contains an excess of meaning and is manifested in the social productions and performances of religion, politics. and everyday life (Kress et al., 2021; Kraus et al., 2021).

Conclusion

Most social and cultural processes have mimetic components that are often not easily distinguishable from their innovative aspects. From an early age human beings develop through relating mimetically to others, above all to those they are close to. Through relating mimetically to them they take an imprint of how they behave which they then use to develop their own ways of behaving. Similarly, nature and the surrounding world as well as their cultural products, objects and artifacts are also appropriated through mimesis. Mimetic processes enable people to interact with their common heritage of nature and culture in a responsive way. In our social lives and day to day actions such processes are also an essential part of innovation and creative action. In art, literature, and music they expand our horizons of experience and play an important role in the creation of individual and collective imaginaries. These mimetic powers remain at play as post-literary generations enter deeper into the age of the Anthropocene. In addition to new human and nonhuman challenges, the "broken intentionality" of mimetic processes opens up new possibilities for a caring approach, leading to a less violent sustainable world. If today's societies are to undergo a transformation and become less violent and more oriented towards sustainability, then many different examples of "best practice" are needed. These models can become the starting point for mimetic processes of processing and spreading sustainable practices and can play an important role in creating a world that promotes sustainability.

References

Aristotle. (2013). *Poetics*, translated by Anthony Kenny. Oxford University Press.

Gebauer, G., & Wulf, C. (1995). *Mimesis. Culture, art, society*. University of California Press.

Gebauer, G., & Wulf, C. (1998). *Spiel, Ritual, Geste. Mimetisches Handeln in der sozialen Welt*. Rowohlt.

Horkheimer, M., & Adorno, T. W. (1972). *The dialectic of enlightenment: Continuum*. Seabury Press.

Jacoboni, M. (2008). *Mirroring people*. Farar, Straus, and Giroux.

Kraus, A., & Wulf, C. (Eds.). (2022). *Palgrave handbook of embodiment and learning*. Palgrave Macmillan.

Kraus, A., Budde, J., Hietzge, M., & Wulf, C. (Eds.). (2021). *Handbuch Schweigendes Wissen*. Beltz Juventa.

Kress, G., Selander, S., Säljö, R., & Wulf, C. (Eds.). (2021). *Learning as social practice. Beyond education as an individual enterprise*. Routledge.

Platon. (2012). *The republic*, trans. Christopher Rowe. Penguin.

Resina, J. R., & Wulf, C. (Eds.). (2019). *Repetition, recurrence, returns. How cultural renewal works*. Lexington Books/Roman & Littlefield.

Rizzollati, G., & Sinigalia, C. (2008). Mirrors in the brain. In *How our minds share actions and emotions*. Oxford University Press.

Tomasello, M. (1999). *The cultural origins of human cognition*. Harvard University Press.

Tomasello, M. (2008). *Origins of human communication*. MIT Press.

Wallenhorst, N., & Wulf, C. (Eds.). (2022). *Dictionnaire d'anthropologie prospective*. Vrin.

Wallenhorst, N., & Wulf, C. (Eds.). (2023). *Handbook of the Anthropocene*. Springer Nature.

Wulf, C. (2013). *Anthropology. A continental perspective*. The University of Chicago Press.

Wulf, C. (2022a). Humans and their images. In *Imagination, mimesis, performativity*. Bloomington.

Wulf, C. (2022b). *Education as human knowledge in the Anthropocene. An anthropological perspective*. Routledge.

Wulf, C. (2022c). Embodiment through mimetic learning. In A. Kraus, & C. Wulf (Eds.). *The Palgrave Handbook of Embodiment and Learning*. Palgrave Macmillan.

Christoph Wulf is Professor of Anthropology and Education and a member of the Interdisciplinary Centre for Historical Anthropology, the Collaborative Research Centre (SFB, 1999–2012) "Cultures of Performance," the Cluster of Excellence (2007–2012) "Languages of Emotion," and the Graduate School "InterArts" (2006–2015) at the Freie Universität Berlin. His books have been translated into 20 languages. For his research in anthropology and anthropology of education, he received the title "*professor honoris causa*" from the University of Bucharest and the honorary membership of the German Society of Educational Research. He is Vice-President of the German

Commission for UNESCO. *Major research areas:* historical and cultural anthropology, educational anthropology, imagination, intercultural communication, mimesis, aesthetics, epistemology, Anthropocene. Research stays and invited professorships have included the following locations, among others: Stanford, Tokyo, Kyoto, Beijing, Shanghai, Mysore, Delhi, Paris, Lille, Strasbourg, Modena, Amsterdam, Stockholm, Copenhagen, London, Vienna, Rome, Basel, Saint Petersburg, Moscow, Kazan, Sao Paulo.

Rituals

Christoph Wulf

Abstract The negative effects of the Anthropocene are repeated, spread, and intensified through ritualizations, ritual arrangements and rituals. The fact that these are performative results in many destructive behavior patterns becoming rooted in institutions and the practices of everyday life. This can be seen in almost all areas of everyday life, e.g. in the consumer world, in the transportation of people and goods, in our approach to the climate, energy and natural resources. For the development of sustainable behavior that corrects the negative developments of the Anthropocene new ritual arrangements and rituals are necessary, in which subjects stage, perform and repeat sustainable actions and behavior. An analysis of many reforms shows the possibilities that exist. In this context it is important first to explain the social and individual significance of rituals and to show the potential that rituals have to bring about ecological, economical, social and cultural reforms.

The negative effects of the Anthropocene are repeated, spread, and intensified through ritualizations, ritual arrangements and rituals (Wulf, 2013a, 2022b; Resina & Wulf, 2019). The fact that these are performative results in many destructive behavior patterns becoming rooted in institutions and the practices of everyday life. This can be seen in almost all areas of everyday life, e.g. in the consumer world, in the transportation of people and goods, in our approach to the climate, energy and natural resources. If we want to make humans treat nature, the world, and other people in a less destructive way, what is needed first and foremost is a critical examination of the violence inherent in the way we have behaved up to now. We must also analyze the origins and effects of this behavior (Wulf, 2022b; Wallenhorst, 2021; Federau, 2017; Gil & Wulf, 2015). Then it is a question of developing plans and examples of other cultural and social practices which value sustainability and the reduction of violence. Finally, these must become rooted in the way people live their daily lives through ritualizations, ritual arrangements and rituals. If we are to

C. Wulf (✉)
Freie Universität Berlin, Berlin, Germany
e-mail: christoph.wulf@fu-berlin.de

© The Author(s), under exclusive license to Springer Nature Switzerland AG 2023

N. Wallenhorst, C. Wulf (eds.), *Handbook of the Anthropocene*,
https://doi.org/10.1007/978-3-031-25910-4_237

achieve a reduction in the destructive conditions of the Anthropocene there must be a complete change in many of our current ritual practices.

For the development of sustainable behavior that corrects the negative developments of the Anthropocene new ritual arrangements and rituals are necessary, in which subjects stage, perform and repeat sustainable actions and behavior. An analysis of many reforms shows the possibilities that exist (Wallenhorst & Wulf, 2022, 2023; Federau, 2017; Wulf, 2022a, b). In this context it is important first of all to explain the social and individual significance of rituals and to show the potential that rituals have to bring about social reforms.

What Is a Ritual?

In international ritual studies there are different definitions and an agreement that different concepts of rituals are possible. To begin with I would like to make a distinction between four types of rituals: firstly, religious rituals which one may also call liturgy – the holy mass, for example; secondly, ceremonies: e.g. the visit of a state president to another country; thirdly, festivals such as weddings or Christmas parties; fourth, everyday rituals. All approaches to classifying rituals are faced with the fact that rituals are the product of repetitive multidimensional processes of symbolization and construction. The phenomena studied are more complex than the concepts and theories used to describe them. This also applies to the attempt to organize the field of ritual studies by types of occasion and to distinguish, for instance, the following kinds of rituals:

- Rituals of transition (birth and childhood, initiation and adolescence, marriage, death)
- Seasonal rituals (birthdays, days of remembrance, national holidays)
- Rituals of intensification (eating, celebrating, love, sexuality)
- Rituals of rebellion (peace and ecological movements, rituals of youth)
- Rituals of interaction (greetings, taking leave, conflicts) (Gebauer & Wulf, 1998, 130).

Rituals are tied to time and space, and their cultural and historical conditions are embodied in these terms. Different spaces have differing effects on the structure, quality, and style of the rituals that take place within them. Ritual spaces differ from physical spaces. Rituals create ritual stagings, performances, and ritual spaces, using body movements, settings, and symbolic and indexical frames. Rituals and space are not related in terms of subject/object or cause and effect, but interactively.

Family rituals transform everyday spaces such as living rooms into festival rooms. This happens, for instance, at Christmas, when living rooms are decorated and made into festive Christmas spaces. This reorganization of space and time is carried out every year when the living room is transformed into a space for celebrating Christmas. In most families, this is done by performing appropriate rituals, the staging of which remains the same and the performance of which varies with each repetition (Wulf, 2006; Wulf & Zirfas, 2007; Kress et al., 2021; Kraus & Wulf, 2022). In an ethnographic study carried out at the homes of three German families and three Japanese families, we showed how family members in both countries create contentment and happiness by repeating their Christmas and New Year's rituals every year (Wulf et al., 2011b).

The performativity of ritual repetition brings the body into play, which leads to the development of bodily, sensory experiences (Michaels & Wulf, 2012, 2014). Their performativity can be interpreted in different ways. However, even if its interpretation varies, the performativity of a ritual repetition can contribute to the integration of a community. Mimetic processes also play an important role in ritual repetitions. They relate to previously performed ritual actions, the current repetition of which is the result of a creative, mimetic act of reference in which not sameness, but similarity and difference are engendered. Here we have the diachronic dimension, which is oriented towards the past.

In addition to this, there is also a synchronic dimension of mimesis in repetition which is also important and in which the participants relate to each other in their ritual activities. This mimetic reference to each other is necessary for the staging of the ritual arrangement to be successful in a functional and aesthetic sense (Michaels & Wulf, 2012, 2014; Wulf, 2016).

During the performance of rituals, the participants orient themselves simultaneously and directly towards the actions of other participants. They do so largely by means of mimetic processes, using the senses, the movements of the body, and a joint orientation towards words, sounds, language, and music. A ritual can only take place as a structured whole if all actions are successfully coordinated, precisely orchestrated, and adequately embodied. If the interaction is to be harmonious, the ritual activities must be mimetically coordinated with each other. If this is achieved, energies can "flow" between the ritual participants, and this is experienced as intense, pleasant, and bonding.

In this process, the images, schemes, and meanings that are produced become part of the participants' imaginaries. At the same time, the movements of the ritual are incorporated into the participants' bodies, resulting in the development of practical knowledge. Practical sustainable knowledge is implicit or tacit knowledge and as such difficult to investigate in research (Kraus et al., 2021). It is a specific form of knowledge, which Gilbert Ryle has called "knowing how," as opposed to "knowing that." With his distinction between "knowing how and knowing that," Gilbert Ryle drew attention to the fact that there are different forms of knowledge, the practical implementations of which that are referred to as "knowing how" are difficult to research (Ryle, 1990). With these methods, the focus is not on the acquisition of factual knowledge which can be expressed linguistically. On the contrary, "knowing how" refers to a skill which enables the person to act and is learned in mimetic processes by relating to the practices of other people. Rituals are an example of this. Rituals and ritual repetitions are not statements, reasons, or explanations. They must be staged and performed. The knowledge required for rituals is a performative, practical kind of knowledge. This differs from the knowledge which is needed to describe, interpret, and analyze rituals. "Knowing how" is thus a practical form of knowledge—a skill which is incorporated and visible in a person's repetitive performances. Other examples of this knowledge which are expressed as skills include games and actions in sports (e.g., football), dance, music, painting, drama, and performance. Knowing how, i.e. "skills," is also required as a pivotal form of knowledge that is acquired through repetition and used in everyday activities such as driving a car, cooking, using a mobile phone or navigation system. In mimetic processes today, mobile phones, smartphones, and tablets merge with the body and with their assistance our immediate bodily boundaries are expanded.

Practical sustainable knowledge is acquired through repetition of the ritual of sustainable development. The significance of ritual actions for the embodiment of the values, attitudes and emotions of sustainability and the development of practical knowledge lies in the role of repetition, in the creation of this form of knowledge, which is so important for sustainable behavior. It is learned mimetically in body-oriented, sensory processes which enable us to act sustainably in institutions and organizations. This kind of knowledge is an important aspect of practical sustainable knowledge, and it is how sustainability becomes rooted in the human body, enabling us to orient ourselves accordingly (Kraus et al., 2021). Images, schemes and movements are learned in mimetic processes, and these render the individual capable of action. Since these repetitive processes also involve sustainable products of history and culture, scenes, arrangements and performances, these processes are among the most important ways of handing down a culture of sustainability from one generation to the next (Wulf, 2022a; Huppauf & Wulf, 2009).

Central Functions of Rituals

The following section will examine the most important findings of the performative approach to ritual research in education, where the focus is on the performative arrangement and the practical and bodily side of rituals and ritualized educational practices. It touches on many forms of theoretical and empirical research and demonstrates the complexity of ritual structures and activities and their great potential for education for sustainable development.

1. Rituals create social relationships and social communities. As the social movement "Fridays For Future" shows, the ritual of "demonstration" is of central importance for the establishment of the movement. Without the ritualization, the social community of the demonstrators would not exist. The symbolic and performative content of the ritual practice creates and stabilizes the identity of the demonstrators. The ritual of "demonstration" creates a structure in which all demonstrators participate. The ritual structure of the demonstration is both real and rooted in the imaginary of the participants. It gives the participants a feeling of belonging together. It relates the ritual practices of the demonstrators to each other in such a way that they respond to each other. A community is formed in

this ritual activity as a performative community (Wulf et al. 2001a, b, 2004, 2007, 2010, 2011a, b; Wulf et al., 2011b).

2. During the performative arrangement of rituals in education for sustainable development, a new social reality is created. Taking earlier rituals as a basis, every performative arrangement in education creates a new ritual reality and a new ritual community. This ritual community can develop among the children or people who carry out the ritual practices for the first time, but it can also involve a repetition, whereby the community confirms its status as such. The actual performance of ritual practices is essential for the forming of social and educational communities and the development of sustainable behavior. The community expresses itself in the performative style of the performance. The ritual presentation enables the expression of something that cannot be expressed otherwise.
3. Performativity yields its full effect in the staging and performative arrangement of rituals in education. (In other words, staging and performative arrangement matter.) The term staging in this case refers to the way in which the ritual scene is set in education for sustainable development. Spontaneous demonstrations are examples of rituals in education in which staging and performance largely coincide (Butler, 1990; Wulf & Zirfas, 2007; Wulf, 2006). Especially in such cases, the question arises as to who is staging the educational ritual—who is the agent and who is the agency of its performance? Is it a tradition, a group, a person, or a collective imaginary and practical knowledge which emerges from the ritual?
4. The bodies of the participants are implicitly involved in the staging and performance of rituals in education for sustainable development (Lakoff & Johnson, 1999; Wulf, 2013a, b). How do the bodies appear in a ritual? How do they take their place in the scene? What does their arrangement in the ritual tell us about the community, the individuals, and their culture? The movements and practices of bodies require our attention. How is the ritual space measured in terms of bodies and what rhythm do they follow? The distance between bodies and the way they approach each other and keep their distance is significant. What positions do they take? Are they standing or sitting? The figurations of bodies are symbolically coded and are used to communicate messages. The "logic" of the body, its presentation and expression play an important role in the performance of rituals in education. This is especially true for the preconscious perception of bodily expressions, which forms the basis on which the atmosphere of ritual arrangements is sensed. The bodies of other people look at us before we become consciously aware of them, and they determine our perception of them in this way. For the performance of rituals to result in community-forming processes, children need to experience the flow of energies and force between people—a physical and psychological process which takes place at the outer reaches of our consciousness (Wulf & Zirfas, 2007; Wulf & Fischer-Lichte, 2010).
5. Social hierarchies and power structures are staged and placed in context in ritual performances. Ritual power structures are not always easy to recognize. Judith Butler (1990) has illustrated in several works that ritual repetition is one of the most effective social strategies for establishing and securing power structures in education. Even belonging to a gender is tied to ritual repetitions, which are

required to create our initial identities in this respect. Power issues between the genders and generations are also dealt with in everyday rituals at the family breakfast table; this occurs in a seemingly casual manner that is more effective for its relaxed appearance. Ritual staging and performance allow several matters to be handled simultaneously in education. The coherence of educational settings and communities depends on the distribution of power and therefore the control of this distribution is one of the central tasks of rituals. A stable balance of power is maintained, regardless of whether issues of authority are addressed directly, dealt with in passing, or analyzed in detail. To achieve gender equality and empower all women and girls, Sustainable Development Goal 5 must be considered. This aims to eliminate all forms of discrimination and violence against women.

6. Rituals play an important role in the treatment and handling of difference and alterity in education (Wulf, 2016). In the multicultural context of inner-city schools, they are important for interaction between children of different ethnicities. They support children to approach others with different cultural backgrounds, overcome differences, and live together in harmony. School communities offer examples of both success and failures in this area; the imaginary, symbolic, and performative elements are equally important here (Huppauf & Wulf, 2009).
7. The synchronous and diachronic aspects of mimetic processes are vital for the success of ritual practices in education. During the performance of rituals and ritual practices, the participants relate immediately and directly to the actions of other participants. This takes place in a largely mimetic manner, using the senses, the movements of the body, and the common understanding of words, sounds, language, and music. A complete arrangement and complete occurrence of a ritual only takes place when all ritual actions are successfully coordinated and precisely orchestrated. A prerequisite for this is the staging, but the performance itself is the decisive factor, as the ritual actions must be in exact relation to each other. Otherwise, the results are farcical, and the ritual is deemed as having failed. Harmonious interaction in education requires that the ritual practices relate to each other mimetically. If this happens, energies can "flow" between the ritual participants and they can be experienced as intensive, pleasant, and bonding. Just as in dance or wooing, the rational control of actions also has its limits in rituals. The feeling that a ritual has succeeded only occurs if a mimetically created harmony that is beyond rational control occurs in bodies, movements, and gestures. This mimetic occurrence is the basis for the feeling of belonging and community as well as the experience of the sacred.
8. Mimetic processes are also essential because they enable the learning of the practical sustainable knowledge necessary for the ritual practices (Wulf, 2006; Gebauer & Wulf, 1995, 1998). Ritual knowledge, which allows children to develop the skills required for rituals, evolves from real or imaginary participation in ritual activities. Children take part in ritual practices by means of mimetic processes; these processes are corporeal and are independent actions as well as actions which relate to other ritual ceremonies or arrangements. These processes

incorporate ritual figurations, scenes, consequences, images, and behavior patterns which are all composite parts of the correct execution of a ritual practice.

If rituals and ritualizations do not change it will not be possible to correct the negative effects of the Anthropocene. A transformation of the reality of society so that we begin to treat nature, other people and ourselves in a less violent way requires a fundamental change in rituals and ritual arrangements (Paragrana, 2018).

References

Butler, J. (1990). *Gender Trouble. Feminism and the subversion of identity*. Routledge.

Federau, A. (2017). *Pour une philosophie de l'Anthropocène*. Presses Universitaires de France.

Gebauer, G., & Wulf, C. (1995). *Mimesis. Culture, art, society*. University of California Press.

Gebauer, G., & Wulf, C. (1998). *Spiel, Ritual, Geste. Mimetisches Handeln in der sozialen Welt*. Rowohlt.

Gil, C. I., & Wulf, C. (Eds.). (2015). *Hazardous future: Disaster, representation and the assessment of risk*. de Gruyter.

Huppauf, B., & Wulf, C. (Eds.). (2009). *Dynamics and performativity of imagination. Images between the visible and the invisible*. Routledge.

Kraus, A., & Wulf, C. (Eds.). (2022). *The Palgrave handbook of embodiment and learning*. Palgrave Macmillan.

Kraus, A., Budde, J., Hietzge, M., & Wulf, C. (Eds.). (2021). *Handbuch Schweigendes Wissen. Erziehung, Bildung Sozialisation und Lernen* (2nd ed.). Beltz Juventa.

Kress, G., Selander, S., Säljö, R., & Wulf, C. (Eds.). (2021). *Learning as social practice. Beyond education as an individual enterprise*. Routledge.

Lakoff, G., & Johnson, M. (1999). *Philosophy in the flesh. The embodied mind and its challenge to Western thought*. Basic Books.

Michaels, A., & Wulf, C. (Eds.). (2012). *Emotions in rituals and performances*. Routledge.

Michaels, A., & Wulf, C. (Eds.). (2014). *Exploring the senses: Emotions, performativity, and ritual*. Routledge.

Paragrana. (2018, January 1). Rhythmus, Balance, Resonanz. In G. Brandstetter, M. Buchholz, A. Hamburger, & C. Wulf (Eds.), *Internationale Zeitschrift für Historische Anthropologie*. De Gruyter.

Resina, J. R., & Wulf, C. (Eds.). (2019). *Repetition, recurrence, returns. How cultural renewal works*. Lexington Books/Roman & Littlefield.

Ryle, G. (1990). Knowing how and knowing that. In *Collected papers* (Vol. 2, pp. 212–225). Thoemmes.

Wallenhorst, N. (2021). *Mutation. L'aventure humaine ne fait que commencer*. Le Pommier.

Wallenhorst, N., & Wulf, C. (Eds.). (2022). *Dictionnaire d'anthropologie prospective*. Vrin.

Wallenhorst, N., & Wulf, C. (Eds.). (2023). *Handbook of the Anthropocene*. Springer Nature.

Wulf, C. (2006). Praxis. In J. Kreinath, J. Snoek, & M. Stausberg (Eds.), *Theorizing rituals: Issues, topics, approaches, concepts* (pp. 395–411). Brill.

Wulf, C. (2013a). *Anthropology. A continental perspective*. The Chicago University Press.

Wulf, C. (2013b). *Das Rätsel des Humanen*. Wilhelm Fink.

Wulf, C. (Ed.). (2016). *Exploring alterity in a globalized world*. Routledge.

Wulf, C. (2022a). *Human beings and their images. Mimesis, imagination, performativity*. Bloomsbury.

Wulf, C. (2022b). *Education as human knowledge in the Anthropocene. An anthropological perspective*. Routledge.

Wulf, C., & Fischer-Lichte, E. (Eds.). (2010). *Gesten: Inszenierung, Aufführung, Praxis*. Wilhelm Fink.
Wulf, C., & Zirfas, J. (Eds.). (2007). *Pädagogik des Performativen*. Beltz.
Wulf, C., Althans, B., Audehm, K., Bausch, C., Göhlich, M., Sting, S., Tervooren, A., Wagner-Willi, M., and Zirfas, J. (2001a), Das Soziale als ritual. Zur performativen Bedeutung von Gemeinschaft. , Leske und Budrich.
Wulf, C., Göhlich, M., & Zirfas, J. (Eds.). (2001b). *Grundlagen des Performativen. Eine Einführung in die Zusammenhänge von Sprache, Macht und Handeln*. Juventa.
Wulf, C., Althans, B., Audehm, K., Bausch, C., Jörissen, B., Göhlich, M., Mattig, R., Tervooren, A., Wagner-Willi, M., & Zirfas, J. (2004). *Bildung im Ritual. Schule, Familie, Jugend, Medien*. Springer, Verlag für Sozialwissenschaften.
Wulf, C., Althans, B., Blaschke, G., Ferrin, N., Göhlich, M., Jörissen, B., Mattig, R., Nentwig-Gesemann, I., Schinkel, S., Tervooren, A., Wagner-Willi, M., & Zirfas, J. (2007). *Lernkulturen im Umbruch. Rituelle Praktiken in Schule, Medien, Familie und Jugend*. Springer, Verlag für Sozialwissenschaften.
Wulf, C., Althans, B., Audehm, K., Bausch, C., Göhlich, M., Sting, S., Tervooren, A., Wagner-Willi, M., & Zirfas, J. (2010). *Ritual and identity: The staging and performing of rituals in the lives of young people*. Tufnell Press.
Wulf, C., Althans, B., Audehm, K., Blaschke, G., Ferrin, N., Kellermann, I., Mattig, R., & Schinkel, S. (2011a). *Die Geste in Erziehung, Bildung und Sozialisation: Ethnographische Feldstudien*. Springer, Verlag für Sozialwissenschaften.
Wulf, C., Suzuki, S., Zirfas, J., Kellermann, I., Inoue, Y., Ono, F., & Takenaka, N. (2011b). *Das Glück der Familie: Ethnografische Studien in Deutschland und Japan* (p. 2013). Springer VS (Japan).

Christoph Wulf is Professor of Anthropology and Education and a member of the Interdisciplinary Centre for Historical Anthropology, the Collaborative Research Centre (SFB, 1999–2012) "Cultures of Performance," the Cluster of Excellence (2007–2012) "Languages of Emotion," and the Graduate School "InterArts" (2006–2015) at the Freie Universität Berlin. His books have been translated into 20 languages. For his research in anthropology and anthropology of education, he received the title "*professor honoris causa*" from the University of Bucharest and the honorary membership of the German Society of Educational Science. He is Vice-President of the German Commission for UNESCO. *Major research areas:* historical and cultural anthropology, educational anthropology, imagination, intercultural communication, mimesis, aesthetics, epistemology, Anthropocene. Research stays and invited professorships have included the following locations, among others: Stanford, Tokyo, Kyoto, Beijing, Shanghai, Mysore, Delhi, Paris, Lille, Strasbourg, Modena, Amsterdam, Stockholm, Copenhagen, London, Vienna, Rome, Lisabon, Basel, Saint Petersburg, Moscow, Kazan, Sao Paulo.

School

Jean-Marc Lange

Abstract Accepting the idea of the Anthropocene is, for the institution of school, a fundamental challenge. Effectively, its aims and systems, even its content, need to be re-examined with a view to transforming the institution completely. Tension and potential risks emerge that need to be dealt with. Nevertheless, the research that has already taken place in the field of education on the environment, and also that of education in sustainable development, provide us with a certain number of marker. School can be seen as an efficient lever in the construction of a new relationship with the world, as well as in developing a collective narrative that mobilises people without being a source of anxiety.

Contemporary school is an institution that is given the mission to instruct, educate or train the new generations by local, national, or international groups. This mission is entrusted to professionals who have themselves been trained for this purpose. The academic institution historically finds its source in religious congregations before becoming institutionalised as part of the philosophical project of the Enlightenment to train enlightened citizens through the appropriation of elementary scientific and technical knowledge with a view to encouraging social Progress. This initial worldwide project was enhanced on the one hand by a political project to make it possible to emancipate subjects to encourage the rise in democracy (Dewey, 1916) and, on the other, in many countries, to construct unity and national unity through the construction of a unifying and mobilising narrative.

On the sociological level, School is both a place for appropriating majority standards and a place for emancipating the upcoming generations (Durkheim, 1922). Providing these new generations with the means of bringing about evolution in standards, or even the emergence of new standards, is thus a source of tension that research is aiming to highlight and render understandable.

J.-M. Lange (✉)
University of Montpellier, Montpellier, France
e-mail: jean-marc.lange@umontpellier.fr

© The Author(s), under exclusive license to Springer Nature Switzerland AG 2023

N. Wallenhorst, C. Wulf (eds.), *Handbook of the Anthropocene*,
https://doi.org/10.1007/978-3-031-25910-4_238

In terms of content, the academic institution produces standards in the form of prescriptions. These then take the form of either study plans, or national programmes, or local or national curricula, that is, educational careers considered for their cultural, political, and pedagogical coherence. The format of study plans or programmes is the dominant force in countries under French-speaking influence. The curriculum format is dominant in English-speaking countries. Modes of coexistence are possible, as is currently the case in France (Lebeaume, 2019). In all cases, it is a question of both transmitting the cultural inheritance judged necessary by the generation responsible for the group, but also of providing the keys for understanding the contemporary world (Forquin, 2008). The result is tension between two legitimate points of view, one academic and the other social. Academic institutions draw up prescriptive texts that aim to be normative and which attempt to hold together in a coherent whole these sometimes contradictory aims. Alistair Ross, a researcher in the field of education (2000), established that there are three main modes of piloting curricula that differ in terms of their political aims and centration: piloting through content organised into subjects, piloting through objectives expressed as skills, and piloting by means of an educational process centred on experiences to be had. Once again, each country favours one particular piloting mode.

However, very quickly large geopolitical groups tried first to share and then to unify or even influence the content and/or teaching systems and methods in the form of recommendations or curriculum transfers. This internationalisation has accelerated and intensified with globalisation. Examples of such groups include UNESCO, UNICEF, OECD, WHO, or the EU. With regard to the field that we are looking at here, increasingly insistent pressure has been exerted since the Second World War in terms of public health or environmental matters, gradually replacing pre-existing themes such as health and hygiene and discovering nature in local environments. For example, WHO has drawn up successive recommendations on the forms of health education, such as the current "health promotion" programme with its humanist aims. For environmental questions, the pressure is less stable, more recent, and with more diverse sources. Thus Environment-Related-Education (ERE) was established in the 1970s on the initiative of researchers and practicians. Education in Sustainable Development (ESD) progressively rivalled it before supplanting it in the 2000s through international roadmaps, such as that of UNESCO, whose 2015 version is called "Education 2030" and concerns Sustainable Development Goals (Barthes et al., 2021). These international recommendations are taken up at the local level in a variety of formats. Nevertheless, they all promote transversal education focusing on systemic questions that aim to develop "soft skills" such as psychosocial skills from WHO, as well as societal values, but no longer social ones such as commitment, solidarity, responsibility or even creativity. Finally, the recommendations cover education models such as global education and inclusion. These successive recommendations fit more or less easily into the local modes instituted.

What, then, becomes of the "School" project in the Anthropocene? We shall now examine this question in terms of the purposes, pedagogical strategies, and content.

The purpose of School in the twentieth century was, as we have seen, more often than not that of a modernist project used as a vector for social progress, based on declared universal values and targeting both an emancipated subject and one with a cultural background that was the source of a feeling of belonging to a community, most often a national one. The idea of the Anthropocene completely overturns this goal. It is specifically the idea of social progress itself resulting from scientific and technical progress developed on behalf of economic growth that is effectively brought into question. The metanarrative(s) for the Anthropocene cover the fundamental concept of planet limitations, and thus the failure of the Promethean world that developed from the Enlightenment (Wallenhorst, 2021). We are collectively becoming aware that we live in an unstable world that is changing at great speed, in which all dimensions, both geosystemic and planetary, are impacted by our development choices and the weight that our species has taken up within the biosphere. To repeat the phrase of Bruno Latour (2021), we need to learn how to live in this new world. We can clearly see, then, just how the explicitly or implicitly progressive narratives, as well as the national ones of which School was until recently the vector, find themselves at odds in the Anthropocene. This change in era has come about precisely because of an inadequate relationship with the world (Lussault et al., 2017), for which School has its own share of the responsibility. Nevertheless, although today's world is the result of a constructed academic project, another School project can make it possible to make a transformation of it. The means for doing so are indeed those of understanding the global stakes of this uncertain and limited world, and collectively finding possibilities for solutions that will allow us to avoid fear, rejection of the other, and thus barbarity (Curnier, 2021). Critical, systemic, and complex thought; reliance, uncertainty, interculturality; and collective, participative, and territorialised actions are the key words for education in policy, itself at the service of the development of our abilities to collectively put together a new relationship with the world. The universal must thus be sought in otherness, and *Enlivenment* replaces *Aufklärung* for a post-modern and post-Promethean School (Wallenhorst, 2021). School in the Anthropocene can thus be seen no longer only as an institution designed to teach how to distinguish what is true from what is false, but also, because of the uncertainty with which we are confronted, as a place for debate and problem-oriented life experiences, accepting and integrating a heterotopy of statements and milieus, in Foucault's sense (Lange & Kebaïli, 2019). It thus becomes the place in which the new generations must learn to become Earthlings, in the sense of the term used by Bruno Latour (2021). It is a question of learning to adapt, to live in a context of uncertainty, as well as transforming the world of humans into significance that is compatible with our survival as a species. To achieve this, it is necessary to train author pupils who are proactive and reflexive, familiarised with complex, and neither simplifying nor dogmatic, thought, capable of understanding the implications of their actions and decisions, and the point of view of others (both human and non-human).

In terms of strategic choices, several curricular options are possible without necessarily excluding any of them. The toolbox obtained from research work in the ERE and ESD periods is both rich and diverse. It remains valid in the Anthropocene:

immersive experiences of nature; a detour via the history of countryside; territorial prospective; experience of anti-values… They advocate for, and most often take the pedagogical form of, both the project and the exploration. For our part, we retain the strategy of assembling experiential, effective, and territorial actions, completed with surveys focusing on the local stakes that need to be linked to the global stakes (Lange & Martinand, 2014). Questions of Environment and Development (QED) are unparalleled opportunities for learning about policy because of the political potential that they have in place of the customary neutralisation, a source of inhibition and renouncement of a transformative desire (Slimani et al., 2020; Slimani et al., 2021). These questions are thus central for education in the Anthropocene.

In terms of content, curricular centration on experience-based activities nevertheless runs the risk of *anomie*, in Durkheim's sense, that is, the absence of regulatory standards in the social relationship, which is a source of instability, and that of heteronomy, an inability to regulate one's thoughts, which is a source of cognitive confusion. For all that, understanding the stakes and challenges of the Anthropocene, and their political problematisation, in theory leaves considerable room for the concepts and methods found in geography, life sciences, geosciences, and environmental sciences because there is a need for knowledge before we can act. However, although the input of science in general, and these sciences in particular, is essential for thinking about the world, QED cannot be reduced to these fields alone at the risk of naturalising and technologizing them. The curricular choices to be implemented in this case must effectively take into account the diversity found in cultural affiliations and references, including ethical choices, values, and epistemology. The choice of contents, modes of organisation, and teaching methods must refer culturally to the contexts on which they are based, to the teachers who must teach them, but also to the public at whom they are directed. This approach makes it possible to break free of ethnocentrism, which is generally expressed in an implicit manner. This also refers to the notion of the hidden curriculum, which historically underpins all domination policies. How, then, can we reflect on these apparent contradictions? One promising avenue is that developed by Jickling and Wals (2008). For these authors, it is a question of reflecting on environmental matters, not through academic subjects but on the contrary by seeing academic subjects as contributing to understanding environmental matters. More recently, one French author has gone into detail on this idea (Lebeaume, 2019). We have also adopted it for a curriculum relating to QED. The key to organising this content is to question the function: in this way, Lebeaume suggests distinguishing constitutive practices from contributory and/or constitutive knowledge. By thus respecting a hierarchical principle that consists in interrogating the role and function of teaching content with a view to an explicit endpoint, the perspective of curricular reconfigurations becomes possible. These will make it possible to collectively construct a new metanarrative that mobilises rather than being a source of anxiety (Lange, 2017).

We can thus see that the entry into the Anthropocene era shakes the foundations of the institution of School in terms of its purposes, systems, and contents if, as an institution, it hopes to take up the challenges of the Anthropocene with more or less vigour depending on the academic traditions in place. It may thus be a question of metamorphosis: curricular choices involving rupture will need to be made in certain cases. Nevertheless, the idea of "one health" – the idea that human health, social health, and ecosystem health form an indissociable whole –, of "climate changes", "biodiversity" and geosystemic "limits" will, we can be sure, take pride of place.

References

Barthes, A., Sauvé, L., & Torterat, F. (2021). Questions environnementales et éducation au politique ? *Numéro 63, Éducation & Socialisation*. (A paraitre).

Curnier, D. (2021). *Vers une école éco-logique*. Le bord de l'eau, critiques éducatives.

Dewey, J. (1916). *Democracy and education: An introduction to the philosophy of education*. Macmillan.

Durkheim, E. (1922). *Éducation et sociologie*. Félix Alcan, coll. Bibliothèque de philosophie contemporaine.

Forquin, J.-C. (2008). Organisation des savoirs. In A. van Zanten (Ed.), (dir.) *Dictionnaire de l'éducation*. PUF.

Jickling, B., & Wals, A. (2008). Globalization and environmental education: Looking beyond sustainable development. *Journal of Curriculum Studies, 40*(1), 1–21.

Lange, J.-M. (2017). Éducations a-disciplinaires, entre récits et pratiques: un paradoxe didactique? Le cas de l'éducation au développement durable. *Éducations, [En ligne], 17(1)*.

Lange, J.-M., & Kebaïli, S. (2019). Penser l'éducation au temps de l'anthropocène: Conditions de possibilités d'une culture de l'engagement. *Éducation et Socialisation, [En ligne], 51*(51).

Lange, J.-M., Martinand, J.-L. (2014). Principes d'élaboration et de structuration d'une éducation au développement durable scolaire. In Jacques Brégeon et Fabrice Mauléon (coord.) *Développement durable, compétences 21, comprendre et développer les compétences collectives* (pp. 129–145). ESKA.

Latour, B. (2021). *Où suis-je? Leçons du confinement à l'usage des terrestres*. Les empêcheurs de penser en rond.

Lebeaume, J. (2019). Précisions sur la forme curriculaire et distinction entre pratiques constitutives et savoirs contributifs. *Éducation et didactique [En ligne], 13*(1).

Lussault, M., Fort, F., Jacques, M., Brugère, F., Blanc, L., & Guillaume. (2017). *Constellation.s. Nouvelles manières d'habiter le Monde*. Actes Sud.

Ross, A. (2000). *Curriculum: construction and Critique*. Routledge Falmer.

Slimani, M., Barthes, A., & Lange, J.-M. (2020). Les questions environnementales au miroir de l'évènement Anthropocène: tendance politique et hétérotopie éducative. *Le Télémaque., 58*(2), 75–88.

Slimani, M., Lange, J.-M., & Håkansson, M. (2021). The political dimension in environmental education curricula: Towards an integrative conceptual and analytical framework. *Environmental Education Research, 27*(3), 354–365.

Wallenhorst, N. (2021). *Mutation, l'aventure humaine ne fait que commencer*. Le Pommier.

Jean-Marc Lange is a professor in Education and Training Sciences at the University of Montpellier. His research, carried out from a curricular point of view, aims at developing coherent curricula that are acceptable to the actors of formal education in the field of transversal education. His work also integrates issues of educational policy orientation and their historical roots. He is the author of numerous publications and co-author with Angela Barthes and Nicole Tutiaux-Guillon of the Dictionnaire critique des enjeux et concepts des " éducations à " published by L'Harmattan in 2017.

Section VI
Violence and Peace

Nathanaël Wallenhorst and Christoph Wulf

Many articles in this handbook examine the tension between violence and peace – peace in the sense of the absence of or reduction of violence. Where it is not explicitly defined as being the subject, "violence" forms the background, overwritten as in a palimpsest, against which issues in the different articles are discussed. The articles in this next section of the handbook are about violence or the reduction of violence, although they do not claim to deal with the subject in an exhaustive way. In other sections of the handbook too, there are articles which tackle the question of violence. The same goes for the articles which investigate ways and opportunities to reduce violence. The explicit or implicit ideas and concepts of violence and peace differ depending on the historical and cultural context.

The amount of violence used by human beings against the world, other people and themselves is a characteristic of the Anthropocene. Actions and behaviour patterns that affect people, animals, plants and objects and cause harm to them can be termed violent. The World Health Organisation defines violence as "the intentional use of physical force or power, threatened or actual, against oneself, another person, or against a group or community, which either results in or has a high likelihood of resulting in injury, death, psychological harm, maldevelopment, or deprivation" (Krug et al., 2002). This definition omits nature as an object of violence.

Looking at violence differently we could say that there are three broad categories of violence, (1) self-directed violence, (2) interpersonal violence, (3) collective violence. The first category includes the forms of violence that people inflict upon themselves, the second the forms of violence that arise between people, and the

N. Wallenhorst (✉)
Catholic University of the West, Angers, France
e-mail: nathanael.wallenhorst@uco.fr

C. Wulf
Freie Universität Berlin, Berlin, Germany
e-mail: christoph.wulf@fu-berlin.de

third the forms of collective violence that are enacted by groups, communities, or societies. Of particular significance here is war, where two or several societies engage in actions of collective violence. The objective is mostly asserting one's own interests in finding a solution to territorial or other conflicts. Ideological, religious, ethnic, economic, or revolutionary reasons usually play a central role in these clashes.

Violence can be seen in the illegitimate use of force and is implemented by power, that is ultimately the power over the lives or deaths of other people.

A characteristic of human beings is that they can injure other people and are themselves also vulnerable. Violence is a central dynamic in our socialization. It is a possibility of human action that can always be used. "The power to kill and the powerlessness of the victim are latent or manifest determinants of the structure of social coexistence" (Popitz, 1986, 82 f.). If we accept this view, then we must accept the terrifying realisation that violence has universal characteristics that are part of our human make-up. This destructive potential is thus part of the *conditio humana*. It can be seen in the fact that humans are the only primates "who engage in killing others of their species on a large scale and in a planned and enthusiastic way. War is one of their greatest inventions" (Enzensberger, 1996, 9). Many disciplines research the various forms of violence and their specific conditions in the Anthropocene (Porchon, 2021; Wallenhorst, 2020, 2021, 2022; Wulf, 2013, 2022a, b). In peace and conflict research a distinction is drawn, for example, between manifest structural violence (Galtung, 1969, 1982, 1990; Scheper-Hughes & Bourgois, 2005) and symbolic violence (Bourdieu, 1998). In ethnology a differentiation is made between an 'etic' and an 'emic' form of violence research. In the case of 'etic' research, the norms, and values of the western system of science are the reference points of ethnography. In 'emic' approaches, violence is studied and interpreted in the context of the culture in which it takes place, using its own values and concepts to do so (Elwert, 2002). Looking at the concept of violence presents many difficulties, and this is seen in many research studies in the Anthropocene, where it is difficult to determine the normative frame of reference.

In the Anthropocene the collective, violent actions of humans towards nature and towards other people have such negative effects that they endanger the basis of human existence and the whole of life on this planet. For a long time, few people were aware of this. Although there is underlying violence in many actions and modes of behaviour, it is only when it spreads and becomes intensified collectively that it becomes dangerous. Well-known examples are climate change, biodiversity, the biogeochemical cycles, the acidification of the oceans, environmental pollution, and the consumption of non-renewable energies. Negative effects are also brought about by modern ways of working the soil, the huge number of materials that are produced through chemical syntheses, breeding and genetic engineering, dumping in lakes, rivers, seas, etc. Violence towards other people in the form of slavery, colonialism and racism have a negative effect on human co-existence. Wars and atomic disasters as well as environmental destruction and man-made catastrophes are a threat to the future of life on our planet (Gil & Wulf, 2015; Kamper & Wulf, 1989, 1994). Many cultures have knowledge of apocalyptic situations. Are these references to avoidable dangers or to the unavoidable fate of humanity? There have been many different attempts to identify these menacing developments, for example by the Club of Rome and the Brundtland Commission. Intensive efforts have been made to try and initiate global developments to improve the situation. The reduction of violence is the goal in many of these efforts. It is also hoped that this goal can be achieved by successfully developing actions and behaviour patterns that will significantly reduce violence. In this process education plays an important role (Hétier & Wallenhorst, 2021, 2022a, b, c; Heitmeyer & Hagan, 2002; Wulf, 1973a, b, 1974; Obrillant et al., 2017).

Peace denotes people living together calmly and safely within a nation or with people of other nations (Gießmann & Rinke, 2019), and also the state of harmony and calm through security in God (Kueng, 2001). There is no doubt that a central condition of peace is the absence of violence. When societies are not caught up in conflicts of war, then quite rightly that is called peace. From a historical perspective, war and violence always present a danger to living in peace. Our hope that there can be a "nuclear peace", in other words peace achieved through human beings arming themselves with nuclear weapons, has become less likely since the war in Ukraine. The assumption that atomic war can be ruled out because it endangers the survival of humans has become questionable. When we consider the destruction of peaceful ways of life through uprisings, civil wars, and wars, we realise that it is a huge and almost impossible task to preserve even those ways of life that are not impacted by manifest violence.

Within a state, peace is protected by the state monopoly on the use of violence, which means that only the state and state institutions are permitted to threaten violence or to use it to maintain non-violent situations. Admittedly the state monopoly on the use of violence is only legitimised in constitutional states, where there is a separation of powers, and laws that are founded in a constitution. One of the state's most important tasks is to see that social conflicts within different levels of society and groups are resolved in a non-violent way, thus ensuring "social peace". If this is the goal then the negative concept of peace will be broadened, becoming a positive concept of peace, containing the idea of social justice.

To create and maintain peace, territorial, economic and social conflicts must be resolved without the use of violence. This is also the case where there are cultural and ideological differences, i.e. differences in conceptions of nature, the world, human beings and society (Senghaas, 1995, 1997, 2007). Differences in religious beliefs also result in conflicts being resolved by violence. Hans Kueng saw this and proposed four theses to deal with this situation: "No peace among the nations without peace among the religions. No peace among the religions without dialogue between the religions. No dialogue between the religions without global ethical standards. No survival of our globe without a global ethic, a world ethic" (Kueng, 2001).

There is no doubt that peace is a utopia that can only be achieved gradually. Since humans are dependent on creating themselves by their own actions, they need visions and goals to help them focus their actions. The reduction of violence against nature, against other people and oneself is a goal that, from an ethical point of view, can be justified and is essential. Therefore, peace is a necessary utopia, an aspiration that needs to be reflected upon critically, so that the much desired peace does not turn into the exact opposite, as was warned by Horkheimer and Adorno with regard to the Enlightenment (Horkheimer & Adorno, 1972).

References

Bourdieu, P. (1998). *Practical reason: On the theory of action*. Stanford University Press.

Elwert, G. (2002). Sozianthropologisch erklärte Gewalt. In W. Heitmeyer & G. Albrecht (Eds.), *Internationales Handbuch der Gewaltforschung*. Westdeutscher Verlag.

Enzensberger, H. M. (1996). *Aussichten auf den Bürgerkrieg* (2nd ed.). Suhrkamp.

Galtung, J. (1969). Violence, peace and peace research. *Journal of Peace Research, 6*(3), 167–191.

Galtung, J. (1982). *Strukturelle Gewalt. Beiträge zur Friedens-und Konfliktforschung*. Rowohlt.

Galtung, J. (1990). Cultural Violence. *Journal of Peace Research, 27*(3), 291–230.

Gießmann, H., & Rinke, B. (Eds.). (2019). *Handbuch Frieden* (second ed.). Springer VS.

Gil, I. C., & Wulf, C. (Eds.). (2015). *Hazardous future: Disaster, representation and the assessment of risk*. De Gruyter.

Heitmeyer, W., & Hagan, J. (Eds.). (2002). *Internationales Handbuch der Gewaltforschung*. Westdeutscher Verlag.

Hétier R., & Wallenhorst N. (Eds.). (2021). L'éducation au politique en Anthropocène, *Le Télémaque, 58*.

Hétier, R., & Wallenhorst, N. (2022a). The COVID-19 pandemic: A reflection of the human adventure in the Anthropocene. *Paragrana, 30*, 41–52.

Hétier, R., & Wallenhorst, N. (2022b). *Enseigner à l'époque de l'Anthropocène*. Le Bord de l'eau.

Hétier, R., & Wallenhorst, N. (2022c). Promoting embodiment through education in the Anthropocene. In D. A. Kraus & C. Wulf (Eds.), *Palgrave handbook of embodiment and learning* (p. xxx). Palgrave Macmillan.

Horkheimer, M., & Adorno, T. W. (1972). *The dialectic of Enlightenment: Continuum*. Seabury Press.

Kamper, D., & Wulf, C. (Eds.). (1989). *Looking Back on the end of the world*. The MIT Press.

Kamper, D., & Wulf, C. (Eds.). (1994). *Anthropologie nach dem Tode des Menschen. Vervollkommnung und Unverbesserlichkeit*. Suhrkamp.

Krug, E. G., et al. (Eds.). (2002). *World report on violence and health*. World Health Organization.

Kueng, H. (2001). Kein Frieden ohne Frieden der Religionen. Über die Rolle der Religionen nach den Anschlägen in den USA vom 11. September. *Reader's digest*. 11/2001, 12 ff.

Obrillant, Damus/Wulf, Christoph/Saint-Fleur, Joseph, and Jeffrey, Denis. eds. 2017. Pour une éducation à la paix dans un monde violent. : L'Harmattan.

Popitz, H. (1986). *Phänomene der Macht*. Mohr Siebeck.

Porchon, D. (2021). L'auteur de la fin du monde: quelques aspects de la catastrophe anthropologique en Anthropocène. *Recherches et éducation, 23*.

Scheper-Hughes, N., & Bourgois, P. (Eds.). (2005). *Violence in war and peace: An anthology*. Blackwell.

Senghaas, D. (Ed.). (1995). *Den Frieden denken. Si vis pacem, para pacem*. Suhrkamp.

Senghaas, D. (Ed.). (1997). *Frieden machen*. Suhrkamp.

Senghaas, D. (2007). *On perpetual peace: A timely assessment*. Berghahn.

Wallenhorst, N. (2020). *La vérité sur l'Anthropocène*. Le Pommier.

Wallenhorst, N. (2021). *Mutation. L'aventure humaine ne fait que commencer*. Le Pommier.

Wallenhorst, N. (2022). *Qui sauvera la planète ?* Actes Sud.

Wulf, C. (Ed.). (1973a). *Kritische Friedenserziehung*. Suhrkamp.

Wulf, C. (Ed.). (1973b). *Friedenserziehung in der Diskussion*. Piper.

Wulf, C. (Ed.). (1974). *Handbook on peace education*. International Peace Research Association.

Wulf, C. (2013). *Anthropology. A continental perspective*. University of Chicago Press.
Wulf, C. (2022a). *Education as human knowledge in the Anthropocene. An anthropological perspective*. Routledge.
Wulf, C. (2022b). *Human beings and their images. Imagination, mimesis, and performativity*. Bloomsbury.
Wulf, C., & Kamper, D. (Eds.). (2002). *Logik und Leidenschaft. Erträge Historischer Anthropologie*. Reimer.
Wulf, C., et al. (2010). *Ritual and identity: The staging and performing of rituals in the lives of young people*. Tufnell Press.

Nathanaël Wallenhorst is Professor at the Catholic University of the West (UCO). He is Doctor of Educational Sciences and Doktor der Philosophie (first international co-supervision PhD), and Doctor of Environmental Sciences and Doctor in Political Science (second international co-supervision PhD). He is the author of twenty books on politics, education, and anthropology in the Anthropocene. Books (selection): *The Anthropocene decoded for humans* (Le Pommier, 2019, *in French). Educate in Anthropocene (*ed. with Pierron, Le Bord de l'eau 2019, *in French*). *The Truth about the Anthropocene* (Le Pommier, 2020, *in French*). *Mutation. The human adventure is just beginning* (Le Pommier, 2021, *in French*). *Who will save the planet?* (Actes Sud, 2022, *in French*). *Vortex. Facing the Anthropocene* (with Testot, Payot, 2023, *in French*). *Political education in the Anthropocene* (ed. with Hétier, Pierron and Wulf, Springer, 2023, *in English*). *A critical theory for the Anthropocene* (Springer, 2023, *in English*).

Christoph Wulf is Professor of Anthropology and Education and a member of the Interdisciplinary Centre for Historical Anthropology, the Collaborative Research Centre (SFB, 1999–2012) "Cultures of Performance," the Cluster of Excellence (2007–2012) "Languages of Emotion," and the Graduate School "InterArts" (2006–2015) at the Freie Universität Berlin. His books have been translated into 20 languages. For his research in anthropology and anthropology of education, he received the title "*professor honoris causa*" from the University of Bucharest. He is Vice-President of the German Commission for UNESCO. *Major research areas:* historical and cultural anthropology, educational anthropology, imagination, intercultural communication, mimesis, aesthetics, epistemology, Anthropocene. Research stays and invited professorships have included the following locations, among others: Stanford, Tokyo, Kyoto, Beijing, Shanghai, Mysore, Delhi, Paris, Lille, Strasbourg, Modena, Amsterdam, Stockholm, Copenhagen, London, Vienna, Rome, Lisbon, Basel, Saint Petersburg, Moscow, Kazan, Sao Paulo.

Part XX
The Risk of Violence

Apocalypse

Christoph Antweiler

Abstract Global and, from a geological point of view, abrupt man-made change has given rise to a particular form of personal anxiety and also of societal perception of crisis. Many people feel helpless, especially in the face of climate change and species extinction. Texts on the "Age of Man" often work with an alarmist vocabulary and convey negative images of the future. Catastrophe narratives and dystopias are often accompanied by misanthropic ideas about humanity. This entry shows that the Anthropocene is qualitatively different from earlier apocalyptic views and can also motivate positive responses.

With the twenty-first century, uncertainty about the future of the world creates a specific sense of threat to our freedom (Stoner & Melathopoulos, 2015). This sense of constriction is not only an individual one, but also manifests itself in the image that societies make of themselves. Here, the Anthropocene forms just one of several occasions for catastrophic imagery, albeit a particularly monstrous image (Guiliani, 2021: 83, 140–193, Sepkoski, 2020: 85-86100-101). Regarding the Anthropocene, we read drastic dystopian metaphors, such as about a "mega-crisis" or even of "the demise of humanity."

Historically, the late modern awareness that the planet is ecologically existentially threatened goes back to a phase before the recognition of anthropogenic climate change. It can be traced to nuclear science and the perception of the impact of atomic bombs (Masco, 2010). Atomic bombs revived the always latent catastrophism or eschatological attitudes toward human history. What was crucially new was that they made it empirically clear for the first time that humans had the potential to plunge the geosphere into catastrophe. This awareness of a possible global human-triggered catastrophe was then almost simultaneously spurred by species extinctions and the realization of a possible global "sixth extinction event," challenging any assumption of uniformity, even geologically.

C. Antweiler (✉)
Institute of Oriental and Asian Studies, University of Bonn, Bonn, Germany
e-mail: christoph.antweiler@uni-bonn.de

© The Author(s), under exclusive license to Springer Nature Switzerland AG 2023

N. Wallenhorst, C. Wulf (eds.), *Handbook of the Anthropocene*,
https://doi.org/10.1007/978-3-031-25910-4_239

In the discussion about the Anthropocene, scientific findings come together with a universal narrative and primarily negative language images, dystopian metaphors, and alarmist and melodramatic narratives. Many representations of the planet's future take the form of pessimistic narratives, such as those about the "end of nature," the end of Western or modern civilization, or even the "end of the world" (Danowski & Viveiros de Castro, 2017). Texts on the Anthropocene often appear as narratives, meaning-making narratives that are potentially political or socially mobilizing (Dürbeck, 2018). Narratives on the Anthropocene share with other historiographies that they are present-oriented narratives. What makes Anthropocene narratives special, however, is that they are not only related to present interests, problems, and concerns, but are also future-oriented.

Some scholars think that modern civilization is "already dead" because we already live in a new world. This civilization also includes all the achievements and hopes of modernity, such as freedom, self-determination, mobility, knowledge and the vision of equal access for all. In the face of total rupture, it is a matter of "learning to die" (Scranton, 2015: 23). Some anthropologists believe that the prophecy of the end of the world must be announced performatively so that it does not actually materialize (Danowski & Viveiros de Castro, 2017: 108). Such extreme positions, however, harbor a problem: they paralyze thought and social action. They make the problem seem barely approachable in earnest and thus dissolve the central motive of generating responsibility in an ultimately de-politicizing way (Purdy, 2015: 4–5).

In an oft-cited turn of phrase, Anna Lowenhaupt Tsing speaks of us living in the "ruins of capitalism," defining ruins as spaces abandoned for the pro-duction of assets. However, this phrase of ruins is then used in other publications in a much broader sense (Tsing et al., 2020: 193). In the context of the "more-than-human" concept, it is argued to overcome the separation between humans and other living beings. Not only humans, but also other living beings are "on the run." In the earth-historical present, characterized by capitalist "rui-nes," refuges for refugees of all kinds would become a necessity for survival. In view of the comprehensive and profound environmental change, it often comes to exaggerated formulations. This already starts when talking about man-made "change of the planet", when in fact it "only" affects the geosphere. Statements are exaggerated, such as that the insight into the worldwide environmental crisis of apocalyptic proportions is new only for Western societies. Indigenous communities in the Americas, for example, had already realized this for 500 years and processed it in their dystopias or their science (fiction) (Hoelle & Kawa, 2021: 660, Whyte, 2018: 228–234).

Apocalyptic images of the future are always to be seen in the context of positive counter-images of the future, whether these are religious or secular in character. Much rarer than catastrophe narratives are positive-minded Anthropocene narratives. They cast the representation of the Anthropocene in a value-based and, if necessary, encouraging narrative or utopia. Here, humans in the Anthropocene appear as "gardeners," as "shapers," or as nurturing humans, as guardians, "steward", or as advocates, e.g., for the living world or nature. Positively coloured or optimistic metaphors and words claiming power conceive the role of humans in a trusteeship. They are found primarily in such writings from the earth sciences that highly value humanity's ability to find answers to the Anthropocene. Representatives of so-called "ecomodernism" optimistically derive from the Anthropocene catastrophe a hope for a purposeful "salvation" of the Earth and therefore speak of the "good Anthropocene" (Asafu-Adjaje et al., 2015: 1, 4). Here, hopeful solutions are sought that should make a nature-compatible civilization possible. To this end, technical measures are often envisaged, prototypically a deliberate influencing of the atmosphere (climate engineering, geo-engineering).

Apocalyptic views of a coming catastrophe thus have a function to trigger the search for positive solutions. Political, economic and social innovations are needed. However, it is precisely the social and human dimensions that are still missing from most models of the future that rely heavily on technology and infrastructure. Apocalyptic views of a coming catastrophe trigger the search for positive solutions. Critics of ecomodernists complain their naïve trust in technological solutions: the proponents of the good Anthropocene would pay homage to a technicistic perspective ("techno-fix") and assume that they can solve everything (Sklair, 2019; Asayama et al., 2019: "solutionist science"). A deeper critique states that proponents of the good Anthropocene continue to think in terms of human and technological systems, whereas it is important to realize that nature is now active in a de facto socio-ecological system (Dryzek & Pickering, 2019: 10).

There are also non-modernist approaches that seek to counter the widespread dystopias with positive utopias. One call is to develop positive voices beyond technology (techné), such as a poetics that emphasizes the principle of creative aliveness. This can build equally on insights into conditions for commons from economics as on knowledge of forms of symbiosis from biology (Weber, 2019: 10–14). Another example of positive images of the future triggered by apocalyptic fears is that of the so-called "pluriverse." Borrowing from Zapatista conceptions, this umbrella term is used to refer to a "world of many worlds," a compilation of concepts, worldviews, and practices from around the world that emphasizes diversity rather than modernist universalism Humanity, like nature, is inherently multiverse and thus in no way reducible to any unity (Escobar, 2018; De la Cadena & Blaser, 2020; Canaparo, 2021).

In view of the impending apocalypse the now globalized activist project of the pluriverse or "multiple (one) world" aims to transform human civilization. The aims are an entangled diversity and a conception of development oriented toward socio-ecological environmental justice, promoted through experimental alternatives (Kothari et al., 2019: xvii). However, these diversity-focused concepts tend to leave open the question of humanity as a community of destiny. Nevertheless, an apocalyptic narrative, in addition to emphasizing a threat, can and often does result in concrete instructions for action. Since the pessimistic narratives dominate so strongly, it should not be forgotten that we need good reasons and positive goals to enable cultures, societies, and politics worldwide to work against the Anthropocene.

Apocalyptic views of man-made global environmental change may at first seem premature. Catastrophic views of the world or of humanity as well as apocalyptic self-descriptions of societies come and go. In contrast, however, the Anthropocene forms a new quality. It is precisely the notion of the earth as a system combined with a deep-time perspective of the Anthropocene that can empirically support catastrophic thinking. What appears to be a slow change or a creeping environmental crisis from a human perspective actually is a sudden geospheric catastrophe from a geological and paleontological perspective.

References

Asafu-Adjaye, J., Blomquist, L., Brand, S., Brook, B., DeFries, R., Ellis, E., Foreman, C., Keith, D., Lewis, M., Lynas, M., Nordhaus, T., Pielke, R., Pritzker, R., Roy, J., Sagoff, M., Shellenberger, M., Stone, R., & Taege, P. (2015). *An Ecomodernist Manifesto*. www.ecomodernism.org/manifesto-english/. Accessed 21 Sept 2021.

Asayama, S., Sugiyama, M., Ishii, A., & Kosugi, T. (2019). Beyond solutionist science for the Anthropocene: To navigate the contentious atmosphere of solar geoengineering. *The Anthropocene Review, 6*(1–2), 19–37.

Canaparo, C. (2021). *El mundo de atrás. Efecto antropoceno y especulación en los ámbitos periféricos*. Peter Lang.

Danowski, D., & Viveiros de Castro, E. B. (2017). *The ends of the world*. Polity Press.

De La Cadena, M., & Blaser, M. (Eds.). (2020). *A world of many worlds*. Duke University Press.

Dryzek, J. S., & Pickering, A. (2019). *The politics of the Anthropocene*. Oxford University Press.

Dürbeck, G. (2018). Narrative des Anthropozän – Systematisierung eines interdisziplinären Diskurses. *Kulturwissenschaftliche Zeitschrift, 2*(1), 1–20.

Escobar, A. (2018). *Designs for the Pluriverse. Radical interdependence, autonomy, and the making of worlds*. Duke University Press (New Ecologies for the Twenty-first Century).

Guiliani, G. (Ed.). (2021). *Monsters, Catastrophes and the Anthropocene. A Postcolonial Critique*. Routledge (Routledge Environmental Humanities).

Hoelle, J., & Kawa, N. C. (2021). Placing the Anthropos in the Anthropocene. *Annals of the American Association of Geographers, 111*(3), 655–662.

Kothari, A., Salleh, A., Escobar, A., & Demaria, F. (Eds.). (2019). *Pluriverse. A post-development dictionary*. Tulika Books & Authorsupfront.

Masco, J. (2010). Bad weather: On planetary crisis. *Social Studies of Science, 40*(1), 7–40.

Purdy, J. (2015). *After nature. A politics for the Anthropocene*. Harvard University Press.

Scranton, R. (2015). *Learning to die in the Anthropocene. Reflections on the End of Civilzation*. City Lights.

Sepkoski, D. (2020). *Catastrophic thinking. Extinction and the value of Diversity from Darwin to the Anthropocene*. The University of Chicago Press (Science Culture).

Sklair, L. (2019). The corporate capture of sustainable development and its transformation into a 'Good Anthropocene' historical Bloc. *Civitas, 19*(2), 296–314.

Stoner, A., & Melathopoulos, A. (2015). *Freedom in the Anthropocene. Twentieth-century helplessness in the face of climate change*. Palgrave Macmillan (Palgrave Pivot).

Tsing, A. L., Deger, J., Keleman, A. S., & Zhou, F. (curs., and eds.). (2020). *Feral Atlas. The more-than-human anthropocene*. Stanford University Press. https://feralatlas.org/

Weber, A. (2019). Enlivenment. *Toward a Poetics for the Anthropocene*. Cambridge, Mass. & London: The MIT Press (Untimely Meditations, 16).

Whyte, K. P. (2018). Indigenous science (fiction) for the Anthropocene: Ancestral dystopias and fantasies of climate change crises. *Environment & Planning E: Nature and Space, 1*(1–2), 224–242.

Christoph Antweiler is an anthropologist with a background in geology and paleontology. His main interests are socio-cultural evolution, pan-cultural patterns (human universals) and vernacular cosmopolitanism. His main research region is Southeast Asia. Among related book publications are *Inclusive Humanism. Anthropological Basics for a Realistic Cosmopolitanism* (Taipei: National Taiwan University Press, 2012), *Our Common Denominator. Human Universals Revisited* (New York and Oxford: Berghahn, 2018) and a new comprehensive treatise *Anthropologie im Anthropozän* (Darmstadt: WBG, 2022).

Atomic Destruction

Anna Weichselbraun

Abstract Atomic destruction—the central modality of the Cold War nuclear arms race—is integral to understanding the Anthropocene. Geologists have identified plutonium 230 as an appropriate golden spike to mark the beginning of the new geological epoch. Atomic destruction has not only scattered plutonium 239 across the globe but has also shaped the world into what it is today and what it might be tomorrow.

Atomic destruction—the central modality of the Cold War nuclear arms race—is integral to understanding the Anthropocene. To mark the beginning of the Anthropocene, the International Stratigraphic Commission has proposed the year 1952 because it coincides with a vast increase in nuclear weapons testing, the fallout from which deposited radioactive particles across the globe. One of these is the isotope plutonium 239. Stratigraphers have recommended this isotope as the "golden spike"—a physical reference point "that marks the lower boundary of a succession of rock layers" (Waters et al., 2015, 48)—to mark the beginning of the Anthropocene.

While it is a geological technicality which yokes the beginning of the Anthropocene to the mid-century nuclear weapons testing, the relation is less arbitrary than it seems. Anthropologist Joseph Masco argues that the Cold War arms race at once mobilized technoscience to understand "the planet as an ecosystem" while it simultaneously instrumentalized "the power of the bomb to block collective thought and action" about planetary risk (2010, 9). Further, "the Cold War nuclear arms race was central to the production of the Anthropocene as both a geological epoch and as an object of knowledge, intervention and reflection" (van Munster, 2021, 70). Building on these insights, I propose that the project of atomic destruction has been constitutive of the political, social, cultural, affective, and epistemic (infra)structures of the Cold War and into the post-Cold War period, a period

A. Weichselbraun (✉)
University of Vienna, Vienna, Austria
e-mail: anna.weichselbraun@univie.ac.at

© The Author(s), under exclusive license to Springer Nature Switzerland AG 2023
N. Wallenhorst, C. Wulf (eds.), *Handbook of the Anthropocene*, https://doi.org/10.1007/978-3-031-25910-4_240

coterminous with the Anthropocene. The project of atomic destruction has not only scattered plutonium 239 across the globe but has also shaped the world into what it is today, and what it might be tomorrow. In other words, atomic destruction entails construction and imagination.

Radioactive particles from fallout did not only spread across the globe, floating to various surfaces, and coming to rest at the bottom of earth's bodies of water. Such particles are also found in us. "Every human being alive carries in his or her body radioisotopes from nuclear test explosions, the largest collective source of radiation exposure by human hands" (Ruff, 2015, 813). Notwithstanding this global distribution of radionuclides, nuclear testing impacted populations differentially. This differential impact on marginalized and oppressed populations—a form of nuclear colonialism—has been amply documented by scholars and activists (Petryna, 2002; Hecht, 2012; Stawkowski, 2016), and as such mirrors the differential impact which anthropogenic climate change has on these populations.

During the period of atmospheric nuclear testing, it was thought prudent to select sites that were remote and apparently isolated from relatively denser human settlement, which coincided with places where Indigenous or colonized populations lived. Deserts and islands proved attractive. The first American test, Trinity, was carried out in the New Mexico desert at a site 210 miles south of Los Alamos, where the Manhattan Project itself was located. One hundred atmospheric tests were carried out at the Nevada Test Site located 65 miles northwest of Las Vegas from whose downtown hotels tourists could glimpse mushroom clouds. St. George, Utah experienced the brunt of the fallout, evidenced by rising cancer rates. The US also conducted tests of its biggest thermonuclear weapons in the Pacific. The Bikini islanders were displaced from their homelands and relocated to other places, their eventual return prohibited by persistently high levels of radiation (Johnston, 2009). France also conducted between 175 and 181 nuclear tests on the Moruroa and Fangataufa Atolls in Polynesia from 1966–1996 (the bulk of these underground) after having carried out 22 detonations at two sites in Algeria from 1960–1966. The UK conducted 45 tests in Australia, on Christmas and Malden Islands, as well as at the Nevada Test Site. The Soviet Union conducted its 715 tests predominantly at the test site in the Kazakh steppe as well as in Novaya Zemlya, an archipelago in the Arctic Ocean, with other tests conducted at other sites in Kazakhstan, Uzbekistan, Ukraine, and Turkmenistan. Chinese nuclear testing took place predominantly in Lop Nur, a dried out salt lake in a desert in a region where Uyghurs live. The tendency to test on land belonging to or near Indigenous or otherwise minoritized populations is thus abundantly clear. These populations suffer disproportionately of radiogenic health effects as well as social forms of stigma and isolation.

In order to build and test nuclear weapons one needs to find uranium deposits in sufficient concentration for mining. The bulk of uranium suitable for nuclear fuel is produced in mines in Kazakhstan, Canada, Australia, Namibia, Niger, and Russia. Uranium mining tells a common Anthropocene story of exploited labour and ignored health risks. Historian Gabrielle Hecht chronicles these aspects for Africa (specifically Gabon, Madagascar, Namibia, and South Africa). The health impacts of uranium mining were ignored in a newly spectacular way which required categorizing uranium ore as less nuclear than other forms of processed nuclear material which would be "made nuclear" and then become subject to stricter forms of occupational health measures (Hecht, 2012). "Toxic infrastructures" describe homes built with tailings from uranium mills not just in Gabon, but also Grand Junction, CO, and Navajo territory (Hecht, 2018, 130). Atomic destruction powered atomic construction.

Since July 1945, 2121 nuclear tests have been conducted involving 2476 nuclear devices. Only the two detonations in Japan were part of combat. These "tests" are full scale nuclear detonations. Nuclear testing was carried out under conditions where effects from the weapon could be measured. War planning and nuclear weapons development were based on calculations of the weapons' supposed impact that did not include estimated destruction from the fire storms produced by the weapon (Eden, 2006). These effects, arguably more immediately destructive than the radiation itself, were not included because effects from fire were too difficult to reliably predict. As a result, nuclear weapons yields grew larger and larger to accommodate ever more destructive strikes.

As this example illustrates, atomic destruction has also been carried out virtually, in the war planning exercises of the US defence department but also in other nation-states. The construction of civil defensive infrastructure and concomitant propaganda readied the American population for its own decimation, while at the same time acclimating them to the idea of a post-nuclear war existence (Masco, 2008).

While the project of atomic destruction generated the technoscientific systems to understand the planet as an ecosystem, the imperative to understand weather patterns for nuclear war planning also produced specific "knowledge infrastructures" which eventually produced a climate science that has observed and measured anthropogenic climate change (Edwards, 2010). The US effort to trace the flows of fallout in the atmosphere led to the establishment of a sophisticated system for detecting nuclear activity which subsequently served as the model for the Comprehensive Test Ban Treaty's global monitoring system (Higuchi, 2020). Knowledge infrastructures that conceptualize the planet as a whole also permitted modelling the climatic effects of nuclear war. In the 1980s the concept of "nuclear winter" (Turco et al., 1983) along with other global atmospheric problems such as acid rain and the ozone hole, "paves the way for widespread awareness of the greenhouse problem" as the cause for global warming (Edwards, 2010, 380–88).

The practice of atomic destruction led some to express a desire to limit it. Global actors and activists' have attempted to curtail various forms of atomic destruction with treaties (such as the Partial Nuclear Test Ban Treaty prohibiting atmospheric nuclear tests, and the Nuclear Nonproliferation Treaty limiting the number of legal nuclear weapons states) and organizations (most prominently, the International Atomic Energy Agency, founded to promote and monitor civilian nuclear technologies to ensure no illegal nuclear weapons are built). These infrastructures reflect and perpetuate the geopolitical hierarchies of the Cold War nuclear order in which states with nuclear weapons could dictate what kinds of atomic destruction to permit. To counter these orders, regional nuclear weapon free zones (NWFZ) were agreed. Today, there are seven NWFZ in Africa, Antarctica, Latin America and the

Caribbean, South Pacific, South East Asia, Central Asia, as well as the "single State" NWFZ in Mongolia.

While the end of the Cold War brought a significant reduction in the sizes of the US and Russia's nuclear arsenals, nuclear weapons persist, and the deterrence logics by which they are legitimated on "security" grounds permit no (peaceful) future without nuclear weapons. Against the continued hegemony of the nuclear order, the Treaty on the Prohibition of Nuclear Weapons (TPNW) agreed in 2017 and entered into force in 2021 is a veritably Anthropocene treaty. Its objective is to eliminate nuclear weapons and the risks that emanate from their continued existence, while acknowledging the atomic destruction that nuclear weapons have produced on the planet so far. Its preamble recognizes the profound damage to humans and the planet wrought by nuclear weapons, and also specifically notices the differential, multigenerational impact of atomic destruction.

As an icon of atomic destruction, the mushroom cloud is a particularly salient image. While in the US, the mushroom cloud became domesticated as "atomic kitsch," in Western Europe it became symbol for nuclear disarmament or as object of nuclear envy as when Swiss and GDR armies used mushroom cloud dummies in military exercises (Marti, 2019, 236–37). If in the 1950s, Las Vegas tourists could observe mushroom clouds in the desert from downtown hotels (Latson, 2015), a ban on atmospheric testing led nuclear weapons to be tested underground and virtually (Masco, 2004). Atomic destruction grew increasingly abstract. Simulations like historian Alex Wellerstein's popular "Nukemap" allows users to select different kinds of bombs, place them anywhere on earth and then see the concentric circles of fire, blast, and radiation visited upon the location. Billed as an educational tool, Nukemap also provides a perverse pleasure of being the author of atomic destruction. It is this tension between horror and the sublime which makes the tool popular. Nevertheless, not all virtual representations of atomic destruction are welcome or popular. In January 2021, to mark the entry into force of the Treaty on the Prohibition of Nuclear Weapons, the Austrian foreign minister held a press conference which included a video simulation of the destructive effects a nuclear weapon on the city of Vienna. The mayor of Vienna tweeted in outrage that such fear-mongering was unacceptable and inappropriate, especially in times of Covid-related public stress. While some of this can be explained by party political strategy, being offended by a virtual representation of nuclear risk seems indicative of a resistance to being reminded of the existence of nuclear weapons. A journalist aptly asked why nuclear weapons have completely disappeared from (Austrian) public discourse (Liechtenstein, 2021).

The project of atomic destruction has material, but also imaginative and infrastructural effects that shape how we understand the Anthropocene. In destroying the world, we realize we are destroying the world. Yet, if the imaginative practices of atomic destruction have acclimated us to the end of civilization, have rendered our self-extinction banal, or have even led us to forget about its possibility, is this also preventing us from dealing with the planetary risk emanating from our destruction of the planet? What lessons might we draw from understanding the nuclear aspects of the Anthropocene for the human condition and our agentive possibilities in this new geological epoch?

References

Eden, L. (2006). *Whole world on fire: Organizations, knowledge, and nuclear weapons devastation*. Cornell University Press.

Edwards, P. N. (2010). *A vast machine: Computer models, climate data, and the politics of global warming* (1st ed.). The MIT Press.

Hecht, G. (2012). *Being nuclear: Africans and the global uranium trade*. The MIT Press.

Hecht, G. (2018). Interscalar vehicles for an African Anthropocene: On waste, temporality, and violence. *Cultural Anthropology, 33*(1), 109–141. https://doi.org/10.14506/ca33.1.05

Higuchi, T. (2020). *Political fallout: Nuclear weapons testing and the making of a global environmental crisis*. Stanford University Press.

Johnston, B. R. (2009). Atomic times in the Pacific. *Anthropology Now, 1*(2), 1–9.

Latson, J. (2015, January). When mushroom clouds were all the rage. *Time*.

Marti, S. (2019). "Atombombe." In *Europa und Erinnerung* (pp. 227–242). transcript-Verlag.

Masco, J. (2004). Nuclear Technoaesthetics: Sensory politics from trinity to the virtual bomb in Los Alamos. *American Ethnologist, 31*(3), 349–373.

Masco, J. (2008). 'Survival is your business': Engineering ruins and affect in nuclear America. *Cultural Anthropology, 23*(2), 361–398.

Masco, J. (2010). Bad weather: On planetary crisis. *Social Studies of Science, 40*(1), 7–40. https://doi.org/10.1177/0306312709341598

Petryna, A. (2002). *Life exposed: Biological citizens after chernobyl*. Princeton University Press.

Ruff, T. A. (2015). The humanitarian impact and implications of nuclear test explosions in the Pacific region. *International Review of the Red Cross, 97*(899), 775–814. https://doi.org/gksb98

Stawkowski, M. E. (2016). 'I am a radioactive mutant': Emergent biological subjectivities at Kazakhstan's Semipalatinsk nuclear test site. *American Ethnologist, 43*(1), 144–157.

Liechtenstein, S. (2021). *@StLiechtenstein*. https://twitter.com/StLiechtenstein/status/1353385996319150081

Turco, R. P., Toon, O. B., Ackerman, T. P., Pollack, J. B., & Sagan, C. (1983). Nuclear winter: Global consequences of multiple nuclear explosions. *Science, 222*(4630), 1283–1292. https://doi.org/dd75fm

van Munster, R. (2021). The nuclear origins of the Anthropocene. In D. Chandler, F. Müller, & D. Rothe (Eds.), *International relations in the Anthropocene: New agendas, new agencies and new approaches* (pp. 59–75). Springer International Publishing. https://doi.org/10.1007/978-3-030-53014-3_4

Waters, C. N., Syvitski, J. P. M., Gałuszka, A., Hancock, G. J., Zalasiewicz, J., Cearreta, A., Grinevald, J., et al. (2015). Can nuclear weapons fallout mark the beginning of the Anthropocene epoch? *Bulletin of the Atomic Scientists, 71*(3), 46–57. https://doi.org/10.1177/0096340215581357

Anna Weichselbraun is a postdoctoral research and teaching associate at the University of Vienna. She holds a doctorate in linguistic and sociocultural anthropology from the University of Chicago. Her previous publications include "From Accountants to Detectives: How Nuclear Safeguards Inspectors Make Knowledge at the International Atomic Energy Agency" (2020) and "Of Broken Seals and Broken Promises: Attributing Intention at the IAEA" (2019).

COVID-19

Renaud Hétier

Abstract It is tricky to strike the right balance in an entry dedicated to COVID-19. On the one hand, the worldwide impact that it has had, and the economic and social fallout it has caused, mean that we cannot fail to address it (lest it become an 'elephant in the room'). On the other, a great deal of uncertainty still remains as to its origin, its side effects, and ultimately how long the phenomenon will last, and therefore how important it will prove to be in the grand scheme of things. In particular, it is apparent that, though COVID-19 vaccines have been developed with unprecedented swiftness, they cannot solve all the problems that the virus has caused. Nevertheless, we believe that, beyond scientific knowledge of the pandemic such as it is at the time of writing, this crisis is indicative of a certain state of the world, and what might become of us, and this gives us ample food for thought. In fact, we can glimpse the 'mirage' of the possible fate of the world – a shaky, nebulous image on the horizon. Against the backdrop of the Anthropocene, of which COVID-19 can be seen as clearly symptomatic, which is curtailing our freedoms, certain changes appear to be coming about, while others are having trouble getting off the ground. The experience of lockdown has certainly opened the door to new ways of living, but the pressing need for an economic 'recovery' would seem to confirm the dominance of the neoliberal paradigm. As so often happens after an 'accident', our resilient societies are keen to return to 'business as usual', getting back in the saddle, so to speak. In this context, we must pay close attention to what may actually change in the long term.

With the global spread of the coronavirus (COVID-19) pandemic, humankind is presented with a new type of challenge. The astonishingly rapid spread of the virus is a very fitting expression of a globalised society, in which anyone, and anything, can travel anywhere (human beings, and the illnesses they carry). Notably, this virus

R. Hétier (✉)
UCO, Angers, France
e-mail: Renaud.hetier@uco.fr

© The Author(s), under exclusive license to Springer Nature Switzerland AG 2023
N. Wallenhorst, C. Wulf (eds.), *Handbook of the Anthropocene*,
https://doi.org/10.1007/978-3-031-25910-4_241

is revealing: it has highlighted the state of the health systems in the various countries (like a crash test), and the state of health of the people affected. Massive disparities have come to light, between well-funded systems (such as that in Germany), which have been capable of responding swiftly, and have a large number of resuscitation beds, and a range of others (those in France, Italy, Spain and also Brazil), which were initially caught short. It will take time to produce an effective remedy, testing the power of science, technology and medicine, financial resources (funding for healthcare, prevention, handling of unemployment, etc.), and the acceptance of State authority (restrictions, containment measures, compulsory vaccination for certain categories of workers, testing, and so forth). More generally, we cannot fail to be struck by the global response (everywhere, the scientific approach to the disease has prevailed, containment measures have been taken, including in countries which are poverty stricken, and/or are wrestling with other diseases, etc.). Yet despite this seeming 'alignment' over COVID-19, political differences are glaringly obvious. For example, in the majority of European countries, the welfare State machine was put into high gear, to support those affected; however, authorities in Brazil, and for a time, those in the USA, cynically refused to intervene, preferring to sacrifice many lives than to allow the economy to slow down.

COVID-19 tends, in particular, to kill people over the age of 65 (18% in France) – particularly the over-75s (71% in France). Being overweight is also an aggravating factor. Thus, the deleterious effects of this epidemic also reveal the state of the population (their average age and state of health). At the same time, the pandemic, notably at the height of its spread, is the tree which hides the forest: if we focus essentially on certain figures (the numbers of deaths and resuscitation patients, firstly, and then the daily numbers of positive cases), we see that we have tended to lose sight of the fact that the 'rest' of life goes on (other diseases, other accidents, other emergencies, and a significant portion of economic activity that continued, unbowed). In addition, the measures taken to save lives could have very harmful 'secondary' effects: disengagement from school, social withdrawal, domestic violence, etc.

Now that facemasks have become available in the majority of countries, and measures to slow the spread of the virus have been identified, a number of questions remain: To what extent are people capable of exercising the necessary discipline? What do they do with activities where social distancing is not possible? What price are they willing to pay to protect the vulnerable? General lockdown measures, which have largely been adhered to, have had a very significant effect on the propagation of the virus, and a major impact on the economy. Logically, they needed to be backed up by self-discipline on the part of all. However, over time, as restrictions were imposed anew, with so-called 'emergency' measures being implemented (in France, for example), we began to see a certain refusal to comply with the restrictions being imposed. Undeniably, it is difficult to adhere to certain deprivations over the long term – all the more so given that these restrictions weigh most heavily on a particular sector of the population, with high exposure (socially active young people), while statistically, that same group is least at risk from serious forms of the disease. Whatever social group one belongs to, though, it is difficult to go for a long time without essential activities and contacts, without kissing or embracing loved ones. People undergo a sort of 'bereavement', but cannot actually grieve. In addition, a number of activities – sporting, cultural, educational activities, and so forth – require physical proximity to others, or indeed, a *corps-à-corps* interaction. These activities are not all, and not exclusively, economic. They are what binds the fabric of society together. In particular, they offer children entertainment and occupations. Most of these activities have resumed with differing degrees of protection, in the belief that children were at less of a risk, and that they were not virulent carriers of the disease. This leads us to look at groups who are vulnerable and need protection (the elderly, sufferers of chronic illness, the overweight). Selective shielding, though, raises the problem of social exclusion – it is an issue which certainly exists already, and would be worsened by such measures. For example, elderly people in care homes have been deprived of contact with what the few family relations they had left.

Of course, this is not the first pandemic to afflict humankind (Ruffié & Sournia, 1993; Delaporte, 2004). There are two major features of the COVID-19 pandemic that are comparable to ones our ancestors suffered, bringing with them a climate of fear – particularly before the age of modern scientific medicine (Bretelle-Establet & Keck, 2014). The global proportions of it are reminiscent of the ravages of the Black Death in Europe – Genoese sailors brought the Plague to Marseille in 1347, and it spread rapidly across the continent through the trade routes (Dedet, 2010). It is also not dissimilar to the decimation of the indigenous South American populations by diseases brought by the European invaders. More recently, HIV became a global pandemic as a result of travel. Thus, the globalisation that is characteristic of a pandemic is not something never before seen. However, COVID-19 has spread more rapidly than anything that has gone before, and paradoxically, seems to be impossible to stop, although countries did have some warning and therefore the opportunity to try to protect themselves (by controlling the flow of travellers). It has spilled over the relatively firm north/south divide (the global south suffers recurrent, poorly controlled epidemics; in the north, epidemics tend to be brought under control and defeated) (Bourdelais, 2003). The mobilisation of State authorities, healthcare systems and research laboratories has been most impressive, engaging in a frantic race against time. By comparison to other diseases that have killed more people (such as hepatitis and tuberculosis), what is striking about COVID-19 surely relates to its impact on social and economic activities, given how easily it spreads (Le Monde, 2020). In that sense, it is symptomatic of globalisation itself, the harmful (social and environmental) effects of which have become clear, but which appears to be irreversible in today's society. The second characteristic of the current pandemic is that it originated in animals. Again, this is not actually new. The Plague, carried by rats and their fleas, is the obvious example, but there are many other diseases that human beings can contract from contact with domesticated or tamed animals. It can therefore be held that the state of balance achieved by a population is always the product of history, which has led to the development of natural defences and/or the use of vaccines and medicines (Brossolet et al., n.d.). COVID-19 most likely originated in animals, and as is the case with HIV, stemmed from contact with a wild animal.

Although the living world has been depleted in a spectacular fashion (in terms of loss of biodiversity), human beings are continuing to encroach further upon it (into the last uncultivated lands, as is attested by deforestation), and to take new risks. In other words, on a planet that is overpopulated with humans, urbanised and colonised mainly by agriculture, the interactions with what remains of the living world look to be difficult: invasive species (transported by humans) and deadly organisms (liberated from their natural by human activities, like viruses suspended in permafrost). Thus, it has come to light that one of the most rudimentary of lifeforms could wipe out the most complex (human society).

The COVID-19 pandemic, the measures taken as a result of it, the social and political upheaval it has caused, the lasting effects that are becoming clear, can all be interpreted as manifesting – and shedding light upon – the general state of humanity in the Anthropocene. From this standpoint, we shall speak of 'falling by the wayside' – the paradox relating to distance and proximity, the manifestation of the phenomenon of collapse, and finally, the question of the balance between freedom and protection of life.

We well know that the lockdown measures taken in different countries have had a major impact on the economy, slowing it greatly (resulting in less production, and far lower consumption). A second wave of effects is now being felt, with businesses going bankrupt and dismissing staff, leaving a long-lasting impression on society. However, there is also a third wave: changes in practices and behaviours which may become permanent. In the process of widespread homeworking, and particularly remote working, remotely held meetings have been found to be viable. We might legitimately wonder whether lockdown could possibly have worked in the long term without that digital 'solution'. That is, digital technology made lockdown possible, and tools which had previously not been widely used have suddenly become absolutely essential. Remote work and exchanges have become the new norm – this approach is more economical (in terms both of time and of money), and indirectly, more environmentally friendly.

This naturally leads us to wonder about distance and proximity. Digital devices have been produced as a result of recent technological progress, and have reached a given point. Indeed, they are at a stage, now, where they can be used to carry on a significant number of activities which dominate in western countries (activities in the tertiary sector). However, this use of digital technology during the pandemic merely reinforces a trend that was already ongoing: the computerisation and digitisation of work. Distances are erased by the immediacy offered by digital… which allows people to 'meet' but remain apart from one another. As everyone can simply remain at home, the choice may be made to put down stronger roots there. The time spent at home may mean that more stringent requirements come into play about the quality of our housing, but above all, the constraints of lockdown have led to reflection on our lifestyles, interest in having an outdoor space (particularly for children), and in being able to access goods that can be accessed by short circuits. Thus, this pandemic, which is the result of an ecological problem (the overspill of the human race and encroachment on other animal species) is accelerating the emergence of certain ecological solutions – in particular, the desire for stronger roots.

Behind this crisis and the changes it has brought about in how we live and work, we can see the shadow of a social collapse, the prospect of which might, up until now, have seemed only theoretical. In particular, we can point to the collapse of the living world (the sixth mass extinction in Earth's history), and the collapse of civilisations (linked to the rarefaction of resources, entire regions becoming uninhabitable, and mass migration), as consequences of climate change (which is ongoing, with worse sure to come). However, the 'accident' that is COVID-19 has suddenly catapulted us into the future. Certainly, States may continue to support pollution-heavy sectors such as aviation, but in the meantime, a number of individuals have become aware of the harmful effects of certain habits, choosing to draw on different solutions instead (e.g. a more grounded life, and local activities instead of far-away ones).

Ultimately, this pandemic pulls into sharp focus the question of how freedom can be reconciled with life. We have seen vigorous protests against the general or partial lockdowns imposed, and the reduction of activity in certain sectors (culture, leisure, hospitality, etc.). Stringent measures have been introduced, and certain people have complained loudly about the fact that all are having their freedom curtailed – especially young people, to whom freedom is most important in concrete terms, or the fact that vaccination is becoming practically compulsory. The worry by which certain people are seized may seem ridiculous to others, who do not feel themselves exposed. However, the feeling of worry is a life preserver – it is an instinctual way of safeguarding not only our own lives, but those of others as well. However, can the preservation of life be regarded as 'sacred', and if so, does it justify the imposition of a unified approach for the whole of society, with the sacrifices that this entails, and potentially authoritarianism on the part of the State? On the other hand, should absolute priority be given to individual freedom and individual responsibility, irrespective of the way in which anyone and everyone can act as a carrier for the virus? Does the sense of solidarity still have a place in a society where freedom is understood as *individual* (rather than communal) freedom?

References

Bretelle-Establet, F., & Keck, F. (2014). 'Les épidémies entre « Occident » et « Orient »', *Extrême-Orient Extrême-Occident* [Online], 37 | 2014, uploaded on 1 September 2014, consulted on 20 September 2020. http://journals.openedition.org/extremeorient/327

Brossolet, J., Duby, G., Gachelin, G., & Miège, J.-L. (n.d.). 'ÉPIDÉMIES', *Encyclopædia Universalis* [Online], consulted on 20 September 2020. https://www.universalis.fr/encyclopedie/epidemies/

Dedet, J.-P. (2010). *Les épidémies, de la peste noire à la grippe A/H1N1*. Dunod.

Delaporte, F. (2004). Épidémie. In D. Lecourt (Ed.), *Dictionnaire de la pensée médicale* (pp. 418–425). PUF.

Le Monde. (2020). https://www.lemonde.fr/les-decodeurs/article/2020/05/14/non-le-covid-19-n-est-pas-seulement-au-17e-rang-mondial-en-nombre-de-morts_6039679_4355770.html

Ruffié, J., & Sournia, J. C. (1993). *Les épidémies dans l'histoire de l'homme: Essai d'anthropologie médicale*. Flammarion.

Renaud Hétier is a Professor at the Catholic University of the West (UCO). He holds a Doctorate in Educational Sciences. He has authored books on education, and anthropology in the Anthropocene. Books (selection): *Humanity versus the Anthropocene* (PUF, 2021, *in French*); *Cultivating attention and care in education* (PURH, 2020, in French); *Creating an educational space with fairy tales* (Chronique sociale, 2017, *in French*); Education between presence and mediation (L'Harmattan, 2017, *in French*).

Disasters

Yoann Moreau

Abstract When the question of a prospective anthropology of disasters (PCA) arises, two fields of use should be first distinguished. On the one hand, a descriptive (or taxonomic) use, in which the word disaster is understood as a relatively homogeneous general category, grouping together sudden and rare phenomena that profoundly affect human societies and their environments. On the other hand, a normative usage, in which the use of the word disaster is the expression of a judgment that varies according to (a) the values advocated by individuals, (b) the ideals defended by an era and (c) the instituted standards of existence.

In the first case (descriptive), the use of the word disaster as a universal phenomenon is intended to reflect a reality that has no reason to be modified later on: the recurrence of disasters throughout the ages. From this perspective, it is recognized that future generations will also be confronted – sooner or later and on a large scale – with the trial of loss, irreversible destruction and death. In the second case (normative), on the contrary, uncertainty can and *has to* be mitigated because the occurrence of disasters is considered to be intimately correlated with the various ways of inhabiting space and the imagination. In this case, it becomes necessary to put in place tools and methods capable of anticipating the way in which mentalities, technical systems, ecological conditions and modes of governance will evolve.

Descriptive Use

Taking into account the inevitability of disasters deploys a small variety of anthropological postures with regard to the future, oscillating between a "prophetic" and a "spiritual" (or metaphysical) trope. The first consists in thinking and acting as if some events in the future could be considered certain; the second leads to fleshing out organic life with elements capable of withstanding disasters, for example, by assuming that every individual is endowed with a soul. In both cases, the absurd and

Y. Moreau (✉)
Mines ParisTech, Paris, France

© The Author(s), under exclusive license to Springer Nature Switzerland AG 2023

N. Wallenhorst, C. Wulf (eds.), *Handbook of the Anthropocene*,
https://doi.org/10.1007/978-3-031-25910-4_242

inevitable nature of disasters induces counterfactual rhetorical postures (i.e. *act as if x*, without being able to prove x) that escape the dominant rationality of modernity. As such, some of them can become the opposite and complement the classic prospective approaches based on accounting aspects (risk calculation, probabilities) and legislative aspects (prevention, precautionary principle). By choosing the certainty of the future disaster as the central element of decision making, "a fixed future we do not want", Dupuy (2009) has developed a philosophical version of prophetism, a rhetorical sleight of hand that takes seriously – and undertakes to counteract – the anthropological tendency to force disasters out of the realm of possible outcomes. His "enlightened disasterism" (*catastrophisme éclairé*) is a strong proposal to circumvent the tendency to deny the arrival of a disaster that "we do not believe will happen even though we have every reason to know it will happen" (Dupuy, 2009, p. 84). Others may provide a basis for a PCA that is off the beaten track of modern rationality and the motivations of possessive individualism (Macpherson, 1971). Through his critique of the Heideggerian catastrophist ontology of a 'being towards death' (sein zum Tode), the Japanese philosopher Watsuji Tetsurô proposed an ontological horizon where the certainty of future catastrophes leads to a surpassing of individual existence. If human existence is conceived as being, in essence, a relationship, then its primary motivation becomes 'being towards life' (sei e no sonzai) (Watsuji, 1996). Thus, according to Watsuji, human existence is accomplished in its relationship to others, places and things, in short, through coexistence within an environment (fûdo) (Watsuji, 2011). With this "mesological" perspective (i.e., considered according to a logic of milieu, not individuals (Berque, 2000)), the certainty of that disaster that threatens everyone (individual death) leads to its overcoming in the accomplishment of a non-individualistic existence (the life of environments). To "be towards life" is to exist on the level of a collective, in the sense of maintaining ecological and social vitality. The human subject is therefore not necessarily reduced to an ego confronted with the anguish of an inevitable death and the enjoyment of all the morbid fantasies that are correlated with it (Jeudy, 1990); he is not condemned to be "a tragic species" (Descola & Prochiantz, 2011) because he is just as much an epic species: Parallel to individual tragedies, marked by the certainty of an ineluctable and irreversible death, the epic of generations and environments conducive to life is played out. This perspective initiated by Watsuji leads us to consider the possibility of a PCA based on the awareness of an "ontological leaflet" alternating tragic and epic horizons: a "social epic" traverses and overcomes the tragedies of individual deaths (as also shown by Durkheim (Durkheim, 1967)) and, similarly, a "physiological epic" traverses and overcomes daily cellular tragedies, an "ecological epic" traverses the tragedies of species, an "atomic epic" traverses molecular tragedies, and an "energetic epic" traverses all the upheavals of matter (on this scale, to paraphrase, nothing is lost, nothing is created, everything is transformed). This nesting of scales of existence reflects the "anomalous" nature of disasters (Canguilhem, 2013), inextricably linked to the choice of scales of impact (Moreau, 2017, pp. 119–160), and leads us to pay close attention to the "transcendental damage" (Bourg, 2018) and to the metaphysical rustle that the increase in the certainty of an imminent ecological disaster provokes – or even resurrects.

Normative Use

From a normative point of view, what is qualified as a disaster varies considerably according to cultural worlds (Douglas & Wildavsky, 1983), social spheres (Revet, 2007), eras (Corbin, 1982; Mercier-Faivre & Thomas, 2008; Citton, 2009) and disciplines (Moreau, 2017). This relativism in evaluating what impacts a collective leads PCA to focus its attention on what is likely to be valued in the future. This is an eminently delicate operation, just as it was unimaginable for a person at the end of the nineteenth century to envisage the air disasters that would begin only a few decades later or, a fortiori, the upheaval in the earth's ecology due to human activities detected less than a century later. Fundamental ignorance of the evolution of future lifestyles and environments leads the PCA to deal with uncertainty by imagining and simulating scenarios, making plans and testing models. Preparedness is likely to be the dominant option in modern institutions (Keck, 2016; Sassa et al., 2012). Noting this, one should consider, on the one hand, raising awareness of the dramaturgical models and imaginary patterns underlying these productions, comparing them with literary productions (SciFi, novels of anticipation, etc.) and, on the other hand, taking into account the disasters generated by the very application of what constitutes – in effect – a formatting of the imaginary and the governing of disasters (Langumier & Revet, 2013).

These incentive and anticipation tools contain the possibility of becoming propaganda tools leading to the institution of new standards of influence and reflexivity that change the very nature of the risks incurred (Beck, 1992). This is already the case of rating agencies which choose and impose their prescriptions (Ouroussoff, 2013), the generalisation of hazard management in terms of risk (Giddens, 2013) the biopolitics of disasters (Neyrat, 2008) and the globalization of words, norms and structures of disasters management (Revet, 2019). In other cases, disasters may be linked to civilizational transformations initially perceived as positive, such as the notions of technological progress, economic growth and environmental control (Morin, 2002; Virilio, 2005). In this case, "business as usual is disaster" (Benjamin, 1979, p. 342; cited in Fressoz, 2011). In still other cases, the very possibility of which is terrifying, a disaster situation can continue over time, become commonplace and lead to a new standard of existence (Levi, 1987), new health standards and annual radiation doses (Daubas-Letourneux et al., 2014), and a renormalization of acceptable cancer rates (Boudia & Jas, 2014). In this context, it is understandable that a PCA should also address the issue of a lack of use of the word catastrophe or equivalent concepts.

This last aspect exacerbates a final category of disasters, which is systemic in nature. Any generalization of a social norm, whether symbolic, technical, economic or otherwise, tends to generate systemic hazards, and the larger and more sustained the social structures over time, the more "structural disasters" they generate (Matsumoto, 2012). In this case, one way to avoid the systematization of disasters and the inherent risk of their trivialization would be to continue to promote and defend the diversity of human ways of making the world (Lévi-Strauss, 1987) in order to "fragment the world" (Rafanell i Orra, 2017) to keep future systemic effects within a sustainable order of magnitude.

References

Beck, P. U. (1992). *Risk society: Towards a new modernity*. SAGE.

Benjamin, W. (1979). *Charles Baudelaire*. Payot.

Berque, A. (2000). *Ecoumène: introduction à l'étude des milieux humains*. Belin.

Boudia, S., & Jas, N. (2014). *Powerless science?: Science and politics in a toxic world*. Berghahn Books.

Bourg, D. (2018). *Une nouvelle Terre*. Desclée De Brouwer.

Canguilhem, G. (2013). *Le normal et le pathologique*. Presses universitaires de France.

Citton, Y. (2009). *La passion des catastrophes N° 9,* Janvier-Févrie XVIe-XXIe siècles, La revue internationale des livres & des idées, 9.

Corbin, A. (1982). *Le miasme et la jonquille. L'odorat et l'imaginaire social, XVIIe-XIXe siècles*. Aubier.

Daubas-Letourneux, V., Frigul, N., Jobin, P., & Thébaud-Mony, A. (2014). *Santé au travail: Approches critiques*. La Découverte.

Descola, P., & Prochiantz, A. (2011, juillet 31). *L'homme est-il un animal?* In France Culture. Consulté à l'adresse https://www.franceculture.fr/emissions/croisements/lhomme-est-il-un-animal

Douglas, M., & Wildavsky, A. (1983). *Risk and culture*. University of California Press.

Dupuy, J.-P. (2009). *Pour un catastrophisme éclairé. Quand l'impossible est certain: Quand l'impossible est certain*. Le Seuil.

Durkheim, E. (1967). *Le suicide*. PUF.

Fressoz, J.-B. (2011). *Les leçons de la catastrophe*. La Vie des idées. Consulté à l'adresse http://www.laviedesidees.fr/Les-lecons-de-la-catastrophe.html

Giddens, A. (2013). *The Consequences of Modernity*. John Wiley & Sons.

Jeudy, H.-P. (1990). *Le désir de catastrophe*. Aubier.

Keck, F. (2016). *Preparedness*. Cultural Anthropology Website. Consulté à l'adresse https://culanth.org/fieldsights/961-preparedness

Langumier, J., & Revet, S. (2013). *Le gouvernement des catastrophes*. KARTHALA Editions.

Levi, P. (1987). *Si c'est un homme*. Pocket.

Lévi-Strauss, C. (1987). *Race et Histoire*. Gallimard.

Macpherson, C. B. (1971). *La théorie politique de l'individualisme possessif. De Hobbes à Locke (M. Fuchs, Trad.)*. Gallimard.

Matsumoto, M. (2012). *Kôzô-sai. Kagaku gijutsu shakai ni hisomu kiki* 構造災—科学技術社会に潜む危機.

Mercier-Faivre, A.-M., & Thomas, C. (2008). *L'invention de la catastrophe au XVIIIe siècle: Du châtiment divin au désastre naturel*. Librairie Droz.

Moreau, Y. (2017). *Vivre avec les catastrophes*. PUF.
Morin, E. (2002). *Pour une politique de civilisation*. Arléa.
Neyrat, F. (2008). *Biopolitique des catastrophes*. Éditions MF.
Ouroussoff, A. (2013). *Triple A: une anthropologue dans les agences de notation*. Editions Belin.
Rafanell i Orra, J. (2017). *Fragmenter le monde*. Éditions Divergences.
Revet, S. (2007). *Anthropologie d'une catastrophe: les coulées de boue de 1999 au Venezuela*. Presses Sorbonne Nouvelle.
Revet, S. (2019). *Les coulisses du monde des catastrophes naturelles*. MSH éditions.
Sassa, K., Rouhban, B., Briceño, S., McSaveney, M., & He, B. (2012). *Landslides: Global risk preparedness*. Springer Science & Business Media.
Virilio, P. (2005). *L'accident originel*. Galilée.
Watsuji, T. (1996). *Watsuji Tetsuro's Rinrigaku: Ethics in Japan*. SUNY Press.
Watsuji, T. (2011). *Fûdo, le milieu humain*. CNRS Éditions.

Yoann Moreau is an anthropologist working at the Centre de recherche sur les Risques et les Crises (CRC Mines ParisTech / PSL Research University). Visiting professor at the Research Institute for Humanity and Nature (Kyoto) and associate researcher at the Institut interdisciplinaire d'anthropologie du contemporain (CNRS-EHESS), Yoann Moreau is currently living in Japan where he is developing a research programme on ways of living in extreme situations and the consequences of Fukushima. He is notably the author of Vivre avec les catastrophes (PUF, 2017). With the Cie Jours tranquilles (Lausanne), he has directed a dozen shows dealing with the issue of disasters that do not correspond to the canons of the dominant media and dramaturgy (demographic explosion of overweight people, generalized depression and loss of vitality, etc.).

Emergency

Agnès Sinaï

Abstract In the context of the Great Acceleration of the Anthropocene, emergency becomes a permanent state. However, there are two main reasons why the urgency is not necessarily perceived: the fact that we have become habituated to portrayals of the destruction of nature and also the fact that the catastrophe is beyond anything we can imagine. An emergency requires a political approach to time. A permanent time in which politicians can act.

In the sixties of the last century, the comedian Pierre Dac said: "It is still too early to know if it is already too late". Today, a form of temporal contraction is expressed by a set of indicators disturbing enough to estimate that we are in a countdown. In the context of the Great Acceleration of the Anthropocene, emergency becomes a permanent state. It is to be thought of as a critical threshold that must be avoided. The accelerated transformation of nature by human activities is of an unprecedented scale and pace. The idea of a planet under pressure is inherent in the anthropocene condition. All components of the global environment are influenced by human activities. Industrial societies operate on timescales much faster than the rates of natural variability. Taken together in terms of extent, magnitude, speed and simultaneity, their activities have produced an unparalleled state in the dynamics and functioning of the Earth system. Measurements of ecological processes, past and present, have led to the conclusion that in the past 60 years the planet has moved well outside the range of natural variability over the past half a million years. Adapting timely to such an accelerated change may be impossible for industrial societies.

In addition, the effects can interact in cascade. Human impacts on the Earth system do not operate in single and separate effect loops. Particularly significant risks are associated with the abrupt threshold change behavior that occurs when the system is forced beyond a certain limit. The temporality of the Anthropocene is that of a Great Acceleration, inherent in a form of modernity and mechanized life subjected

A. Sinaï (✉)
Sciences Po, Paris, France
e-mail: Agnes.sinai@sciencespo.fr

© The Author(s), under exclusive license to Springer Nature Switzerland AG 2023

N. Wallenhorst, C. Wulf (eds.), *Handbook of the Anthropocene*,
https://doi.org/10.1007/978-3-031-25910-4_243

to the regime of shock. The thermo-industrial revolution instills a new temporal order in the modes of existence. The systemic interdependence of human settlements and environmental cycles suggests that other seemingly independent crises, from the psychological to geopolitical, combine to mark a historic tipping point.

However, the urgency is not necessarily perceived, for two main reasons: the drift of representations, and the supraliminal dimension of the disaster. The shock regime induced by industrial modernity induces an erosion of sensitive experience and a drift of values that the oceanographer Daniel Pauly (1995) has called shifting baselines. According to this theory, our representational frames drift as the world changes around us. Destruction of nature becomes the norm because of this cognitive addiction. In addition, the trajectory of collapse in which we are engaged is supraliminal. In the sense understood by the philosopher and journalist Günther Anders in 1956, the supraliminal designates the crossing of a threshold of perception. This threshold is characteristic of contemporary overshoot, which concerns both the saturation of the faculty of feeling and the colonization of nature by human overactivity. There is a gap between the values of an Enlightenment modernity and the speed of the ongoing transformations of the Earth system. The volume of what we can produce, do or think exceeds the capacity of our imagination and our feelings (Anders, 1960).

Emergency is a matter of politics. Emergency suggests an order of priority, a hierarchy in the decisions to be made and implemented. If the emergency is a reality documented by the figures attesting to the disruption of the Earth system, it is also a perception, giving rise to one or more strategies. Emergency arises from a social construction of the representation of risk. Thus, the gigatonne gap attested each year by the United Nations Environment Program (UNEP, 2019) reports does not trigger proportionate policies. To avoid warming above 1.5 ° C, we must reduce emissions by 7.6% each year by 2030. Each year that we do not act, the level of difficulty and the cost of reducing emissions are increasing. According to the IPCC (2021), CO_2 emissions into the atmosphere in 2019 were the highest on record for two million years with a rate of 410 ppm. Same for methane (CH_4) and nitrous oxide (N_2O), which reached a concentration in the atmosphere never equalled for 800,000 years. According to the report, this increase is very sudden: in 271 years, the concentrations of CO_2 and CH_4 have increased more than during the transition from an ice age to an interglacial area.

However, these acceleration phenomena are not the basis for an emergency policy. Faced with the imminent danger and the associated uncertainties about the future of the Earth system, there is no escape. Emergency calls for a political construction of time. A permanent time on which the politician can act. The experience of modern time refers to the notion of duration as perpetual time. Duration represents this belief in continuity in the capacity of political actors to negotiate the framework and the goals of the production of politics. The end of the twentieth century imposes a series of questions on the distortions between time of nature and time of society (Semal & Villalba, 2013). There is indeed a gap between the duration of the policy and the reality of ecological phenomena. The duration must face up to the age of catastrophes which is now ours. The delay is presented as a time measurement instrument more suited to these constraints (Anders, 1960). The delay makes it possible to highlight one of the temporal paradoxes of our current societies. While urgency becomes a political leitmotif, duration continues to impose itself as a framework for understanding and acting on public decision-making. The delay assumes that the urgency will eventually impose situations on us in a non-negotiable manner. Conversely, the duration hypothesis assumes that we can always postpone the emergency. The delay mobilizes a catastrophist – or alarmist – vision of politics. Such a prospect forces us to renegotiate the order of priorities of the political project.

The current gap between emergency and political action is all the more surprising since a characteristic feature of mature, robust and sustainable civilizations is the ability to see things in the long term, to aim for desirable but realistic futures, and to deploy strategies and insurance mechanisms to manage surprise and uncertainty. Our industrial civilization should have been able to allocate resources to the development of strategies capable of facing, at the technical, infrastructural, organizational, cultural and personal levels, the consequences of the conjunction of climate change and the depletion of fossil fuels. The shortsightedness of the information and decision-making processes of our larger organizations can be interpreted as one of the many signs of cultural decline, and as a clue that our stock of human capital may well be declining at the same time (Holmgren & Scenarios, 2009). Without having properly anticipated current ecological, climatic and energy changes, humanity will have to face without preparation the abrupt natural upheavals, now inevitable, and social upheavals potentially destructive of all civilization (Cochet, 2009).

In the emergency scenario (Holmgren & Scenarios, 2009), the supply of high-quality fossil fuels declines rapidly, the economy fails, and if human contributions to global warming fall, the lagging effects on the system of past contributions continue to accelerate global warming. Local wars, including the use of nuclear weapons, accelerate the collapse in some areas, but the ruin of national systems of power means that a world war is avoided. Successive waves of famine and disease ruin the economy on a larger scale than the Black Plague did in medieval Europe and, within decades, halved the world's population.

For the Deep Adaptation movement, we live in a non-linear world. According to collapsologists, it is too late to avert an environmental catastrophe and a runaway climate change, nine tipping points being already active, including near-term massive release of methane from the Arctic Ocean. They appeal to a massive campaign and policy agenda to transform agriculture and restore ecosystems globally is needed right now. Deep adaptation would mean to uphold the norms of the current society so as to maintain sustainability even in troubled situation. The sense of emergency would lead to face the risk of societal collapse and endorse a new range of politics. Once attained a post-denial state, we should aim at the valued norms and behaviours that human societies will wish to maintain as they seek to survive such as relinquishment to some harmful forms of consumption and production, restoration such as re- wilding landscapes, so they provide more ecological benefits and require less management, changing diets back to match the seasons, rediscovering non-electronically powered forms of play, and increased community-level productivity and support. Another area for Deep Adaptation could be termed "reconciliation." How we reconcile with each other and with the predicament we must now live with will be key to how we avoid creating more harm by acting from suppressed panic (Bendell, 2020).

In France, the "Ecological emergency" list presented on the occasion of the European elections of May 26, 2019 estimates that making the choice of ecology to preserve a habitable planet is now a matter of life and death. In addition, we are witnessing municipal initiatives all over the world around the theme of the climate emergency. The mayor of Strasbourg (France) Jeanne Barseghian during the installation of the City Council in Strasbourg on July 4, 2020 calls for the climate emergency as a founding act of his mandate as mayor, which must infuse for 6 years in all decisions and through all the policies carried out. Preceded in this area by Paris, Barcelona, New York and the European Parliament, the Alsatian capital intends to meet the expectations formulated by the youth and all the inhabitants who have realized that it is necessary to act quickly.

References

Anders, G. (1960). *Le temps de la fin* (Vol. 2007). L'Herne.

Bendell, J. (2020). Deep adaptation: *A map for navigating climate tragedy* (IFLAS occasional paper 2, Revised 2nd ed. released July 27th 2020).

Cochet, Y. (2009). *Antimanuel d'écologie*. Bréal.

Holmgren, D., & Scenarios, F. (2009). *How communities can adapt to peak oil and climate change*. Greenbooks.

IPCC. (2021). *Climate change 2021: The physical science basis. Contribution of working group I to the sixth assessment report of the intergovernmental panel on climate change* (Masson-Delmotte, V., P. Zhai, A. Pirani, S. L. Connors, C. Péan, S. Berger, N. Caud, Y. Chen, L. Goldfarb, M. I. Gomis, M. Huang, K. Leitzell, E. Lonnoy, J. B. R. Matthews, T. K. Maycock, T. Waterfield, O. Yelekçi, R. Yu, & B. Zhou (eds.)). Cambridge University Press. In Press.

Pauly, D. (1995). Anecdotes and the shifting baselines syndrom of fisheries. *Trends in Ecology and Evolution, 10*(10), 430.

Semal, L., et Villalba, B., (2013). Chapitre 4. Obsolescence de la durée. La politique peut-elle continuer à disqualifier le délai ?, Franck-Dominique Vivien éd., *L'évaluation de la durabilité.* Éditions Quæ, pp. 81–100.

United Nations Environment Programme. (2019). *Emissions gap report 2019*. UNEP.

Agnès Sinaï is a professor at Sciences Po, environmental journalist and author, founder of the Momentum Institute, a network for reflection on Anthropocene policies. Doctor in spatial planning and town planning (PhD, Université de Paris Est). Co-author of *Le Grand Paris après l'effondrement. Pistes pour une Ile-de-France bioregionale* (Wildproject, 2020) and of various books, including *Small treatise on local resilience* (with Pablo Servigne, Raphaël Stevens, Hugo Carton, ECLM éditions, 2015), *Walter Benjamin facing the storm of progress* (Le Passager clandestin, 2016) (all these publications are in French).

Fake News

Sharon Rider and Mats Hyvönen

Abstract Fake news, in the sense of lying (the circulation of disinformation) is, of course, nothing new. While it is today typically associated with political propaganda or conspiracy theories such as climate change denial, it is as old as human communication itself, and a seemingly inevitable aspect of public discourse. Falsehoods can be used for numerous ends, from myth to satire to outright deception; the effects of fake news can range from laughter to death (Burkhart JM. Libr Technol Rep 53(8):1–36, 2017, p. 5). One might doubt the utility of a buzzword such as "fake news", but the phenomena indicated by the term nonetheless constitute a serious challenge to modern notions of the "public sphere" as the space in which commonality is established and maintained. The very idea of the Anthropocene, and the contexts in which it appears, is perhaps one of the best illustrations of the risks involved, as well as of the problematic nature of "fake news" as a concept.

In this paper, we briefly review the literature on fake news in order to locate common themes. One of these in particular stands out as central, namely, the perceived threat to liberal democratic institutions and the viability of the very idea of a public sphere on which they rely. We argue that while empirical studies and categorization of actors, motives, effects, and techniques of dissimulation and distortion are useful for understanding the nature of fake news, they tend not to pay careful attention to what is perhaps the greatest challenge, namely, the expansion and consequent reconstitution of the public sphere as a genuine plurality, with all the risks of disintegration that is entailed. We conclude with a proposal, inspired by Steve Fuller, that we acknowledge the post-truth condition as a kind of "fact" of contemporary life in which the most important function of the public sphere is, as Richard Rorty would say, to keep the conversation going. This means that the notion of "fake news"

S. Rider (✉)
Department of Philosophy, Uppsala University, Uppsala, Sweden
e-mail: sharon.rider@filosofi.uu.se

M. Hyvönen
Department of Cultural Anthropology and Ethnology, Uppsala University, Uppsala, Sweden
e-mail: mats.hyvonen@antro.uu.se

© The Author(s), under exclusive license to Springer Nature Switzerland AG 2023

N. Wallenhorst, C. Wulf (eds.), *Handbook of the Anthropocene*,
https://doi.org/10.1007/978-3-031-25910-4_244

should not be taken at face value, but seen as itself a trope in that discourse. The article ends with a consideration of the fate of the Anthropocene as a scientific term to exemplify this point.

Fake news, in the sense of lying (the circulation of disinformation) is, of course, nothing new. While it is today typically associated with political propaganda or conspiracy theories such as climate change denial, it is as old as human communication itself, and a seemingly inevitable aspect of public discourse. Falsehoods can take the form of myth, satire and outright deception; the effects can range from laughter to death (Burkhart, 2017, p. 5). One might doubt the utility of a buzzword such as "fake news", but the phenomena indicated by the term nonetheless constitute a serious challenge to modern notions of the "public sphere" as the space in which commonality is established and maintained. The very idea of the Anthropocene, and the contexts in which it appears, is perhaps one of the best illustrations of the risks involved, as well as of the problematic nature of "fake news" as a concept.

Fake News: The Who, What, When, How and Why

Ever since the publication of the first newspapers in the seventeenth century, journalistic reporting has been an unabated source of inaccurate, exaggerated, politically skewed and sensationalistic content. The line of demarcation between fake and real news is not so distinct as some might be inclined to think (Lippmann, 1922). Lazer et al. (2018, p. 1094) define fake news as "fabricated information that mimics news media content in form but not in organizational process or intent". On this account, fake news looks like real news, but is produced without adherence to the editorial norms and procedures necessary to ensure that the information published is accurate and reliable. This conception is fairly standard in the literature, where the term "fake news" refers to reports aimed at deceiving the reader (Farmer, 2021, p. 1). The phenomena so described are, however, heterogeneous. Who produces fake news, for what reasons and for whom, the degree to which the intent is to mislead or do harm, and perhaps most importantly, to what extent that which is reported can be described as simply false, varies significantly depending on what kind of "fake news" is at issue.

Several typologies have recently been proposed to delineate different types of fake news and the motivations behind them. Edson et al. (2018, p. 148) identify two dimensions – facticity and intention – which they use to map the various types of fake news discussed in previous research. While satire and parody are not as such designed to deceive, news fabrication, media content manipulation, advertising, and propaganda all have the objective of promoting a belief that is or may be inaccurate, slanted or false. In contrast to fabrication, however, satire and advertising most often presuppose some degree of facticity. Wardle (2017) distinguishes between seven types of misinformation across a continuum. In addition to the types discussed above, she highlights false connection (e.g., headlines that do not reflect the content), misleading framing of information, deceptive contextualization of accurate reports, and bluff sourcing. "Fake news" thus covers everything from subterfuge to gossip, memes and bad journalism.

All news, fake or real, requires the dissemination of information. Celliers and Hattingh (2020, pp. 226–231) name five factors involved in the sharing and spreading of fake news on social media: the desire for acceptance and popularity within a group; the capacity for sifting information; the ambition to influence others; monetary incentives to steer or exploit social media users; malicious intent to spread rumour. Thus while intent and content are important, their effects cannot be properly understood in abstraction from the behaviour of the audience; in particular, the degree to which it is given or seizes the opportunity to assess the veracity of claims. But if all news is to a greater or lesser degree co-constructed by its reception and use, and an audience does not treat a piece of information as if it were an accurate report, when is it fake news, irony, evangelism or simply fiction? (Nagle, 2017) This difficulty is one of the thorniest problems in understanding the post-truth condition.

Much of the current research on fake news links it to populist political movements. In many of these accounts, fake news is presented as posing a threat to objectivity and, ultimately, to liberal democracy. The standard approach is to examine who is misleading whom about what, in which way and for what reason. But the deepest issue with regard to our political institutions and the public sphere is not that there are actors bending or breaking implicit or explicit rules of good publicistic conduct. Rather, it is that the rules themselves are being contested and renegotiated. Fuller (2018, p. 185) defines fake news as "a calling card of the post-truth condition, whereby the contesting parties accuse each other of imposing the wrong conceptual framework for telling what is true and false". Instead of focusing on the problem of fake news on the internet, Fuller urges us to see post-truth and fake news as the contemporary manifestation of the age-old struggle over who decides the conceptual framework for making and adjudicating between truth claims.

Fake News, the Where: The Public Sphere

Arendt argued that a fundamental precondition for political action is commonality, the appearance and sharing of a public world, the objectivity of which "relates and separates men at the same time" (1958, p. 52). Plurality (a genuine diversity of perspectives on the same phenomenon), a prerequisite for all political action, presumes a minimum of agreement in judgments, practices, and facts. Or, as Horsthemke (2017, p. 275) puts it: "A democratic debate requires provable facts about which there exists agreement. The debate then concerns what actually follows from these facts."

Not long ago, most media content "came from a limited number of synchronized central sources", and the opportunities for media production "were very limited and tied to the operation and gatekeeping power of those central sources" (Couldry & Hepp, 2017, p. 52). Modern electronic mass media made possible simultaneous transmission of media content across vast spaces; through their cyclical production, they generated "shared rhythms of simultaneous experience and new narratives of commonality" (Couldry & Hepp, 2017, p. 47). From a historical perspective, the material limits of the mass media (e.g., the costs of running a newspaper) played an important part in the emergence of the nation-state with a mediated national "public sphere" (Habermas, 1989) that was essential to the establishment of the nation as an "imagined community" (Anderson, 1991). But the era of a monopoly on news seems to have reached its terminus. Anyone with a cell phone and a modicum of technical know-how can gather, formulate, and disseminate information in a media landscape teeming with facts and factoids competing for attention. As expertise and authority have been fragmented and dispersed through a multiplicity of actors and agents, so too have the communicative fora for the dissemination of knowledge; there is "no main venue in which a trusted authority can definitively debunk truths." (Harsin, 2015, p. 332).

The internet provides access to an unprecedented amount of content; we now carry a universe of diverse perspectives in our pockets. This ceaseless flow of news, fake or not, is likely to expand and accelerate. Insofar as more people than ever can engage in creating and disseminating information and ideas that can reach large audiences, the surge in fake news has to be seen as an effect of the ongoing democratization of the media landscape. In this digitized, mediatized epistemic environment, it is increasingly difficult to institute commonality. The resultant situation is paradoxical. On the one hand, the algorithms of search engines and social media (as well as individualized news feeds from "old" media) make it possible for us to navigate the unfathomable amount of information out there. But they also induce us to tailor our reception to reflect our unreflecting inclinations and interests, and thus reinforce contingent preconceptions and existing habits of mind. By "suggesting contents similar to the ones we are usually exposed to", exposure to a plurality of views is limited, which clears the way for "the formation of groups of like-minded users framing and reinforcing a shared narrative, that is, echo chambers" (cf. Nagle, 2017).

The problem, it has been argued, is not the democratization of the right to give, demand and hear arguments from all quarters, but the capitalization on the techniques and processes that made it possible to begin with (Barbrook & Cameron, 1996). It remains to be seen whether the ongoing technological, social and political developments will allow for a re-invention of something that resembles the most important facet of the ideal-type of a shared "public sphere". At present, the outlook looks rather grim. The media (old and new) seem to lack not only the requisite institutional capacity and authority, but also the ambition to shoulder the responsibility involved in being truly public in this respect. A further and perhaps even more menacing prospect is that, for similar reasons, there may be no other candidates in a position to take on the mantle (Hyvönen, 2018; Rider, 2020).

Looking Forward: Post-truth Reflections on the Anthropocene

In late March 2020, less than three weeks after the WHO declared COVID-19 a global pandemic, the secretary-general of the United Nations, António Guterres, tweeted his concerns about an ongoing infodemic in the wake of the pandemic: "Our common enemy is #COVID19, but our enemy is also an 'infodemic' of misinformation" (Guterres, 2020). In the first pandemic of the digital age, the internet and social media are used on a massive scale to inform, but also to confuse, frighten and enrage. The same might be said of the Anthropocene, about which the philosopher Peter Sloterdijk remarked, "The proliferation of this concept can mainly be traced back to the fact that, under the guise of scientific neutrality, it conveys a message of almost unparalleled moral-political urgency" (Cited in Davison, 2019). On the one hand, the notion derives from a context in which a group of scientists met to describe dramatic planetary changes. On the other, the institutionalization of a "fact" such as the designation of a new epoch on the stratigraphical timescale is an arduous, slow collective process of extraction, analysis, measurement and calculation, requiring enormous intellectual effort and substantial economic and technical resources. Even if the International Commission on Stratigraphy (ICS) should determine that the evidence justifies a clear demarcation from the Holocene epoch, that decision will have to be ratified by the International Union of Geological Sciences (IUGS). The "fact" so determined will thus be a judgment based on the findings of an enormous collective enterprise, examined by experts invested with the authority to further the scientific agenda by coming to agreement as to how to proceed. In this respect, the scientific fact of the matter is as much decision as discovery.

The word "fact" derives from the Latin factum (from *facere*, to do), and, as a matter of "fact", was first used in English with precisely that meaning: something done or performed. That usage is now obsolete, having been replaced with the common usage of "something that has really occurred or is the case" in the middle of the sixteenth century. So originally, one could say, the difference between fact and fiction wasn't between truth and falsity, but between action and invention, or doing and making. This distinction fits with the one named earlier between manipulation and fabrication, where the first is tied to some kind of acknowledged shared practices, while the latter is not.

A way forward out of the entrenched and polarized positions that make democratic debate about the right course of action given what we know is to acknowledge that one might be sceptical of the adequacy of a certain action or decision, such as bringing together radically diverse phenomena under one name, without doubting the scientific consensus that the human impact on the environment has been global and perhaps irreversible. A further step would be to compile and construe existent data and the established interpretation thereof in such a way as to negate the severity of the crisis, in order to effect political opinion regarding what would be necessary or desirable action. Neither of these responses in themselves constitute a considered effort to misrepresent the state of affairs. A clear-cut case of "fake news" in this instance would be rather to "make stuff up" or distort the findings by changing the form or content of the studies cited or cast doubt on the reliability of the sources through innuendo, hype or rumour.

If researchers find representative markers in the rock record that identify the point at which human activity "exploded to such a massive scale that it left an indelible signature on the globe" (Subramanian, 2019), that insight can be a shared point of reference for further debate in which the plurality of possible perspectives is acknowledged. The danger of using "facts" as if they were simple observations rather than hard-won results of ongoing investigation in order to coerce the polity into silence and submission, pulls science into the sphere of the political, delegitimizing it as a common resource in a shared world. As such, it belongs to the netherworld of "fake news", an attitude of conviction, sanctimonious or sincere, rather than of open and reasoned discourse in a liberal, democratic spirit in which we encourage the expansion of the "community of inquirers", consider truth as simply the endpoint of the process of inquiry and embrace the thought that "the solution to poor opinions is more opinions" (Peirce, cited in Legg, 2018).

References

Anderson, B. R. O. (1991). *Imagined communities: Reflections on the origin and spread of nationalism*. Verso.

Arendt, H. (1958). *The human condition*. University of Chicago Press.

Barbrook, R., & Cameron, A. (1996). The Californian ideology. *Science as Culture, 6*(1), 44–72.

Burkhart, J. M. (2017). Combating fake news in the digital age. *Library Technology Reports, 53*(8), 1–36.

Celliers, M., & Hattingh, M. (2020). A systematic review on fake news themes reported in literature. In M. Hattingh et al. (Eds.), *Responsible design, implementation and use of information and communication technology* (I3E 2020. Lecture notes in computer science, vol 12067). Springer. https://doi.org/10.1007/978-3-030-45002-1_19

Couldry, N., & Hepp, A. (2017). *The mediated construction of reality*. Polity Press.

Davison, N. (2019, May 20). The Anthropocene epoch: Have we entered a new phase of planetary history? *The Guardian*.

Edson, C., et al. (2018). Defining "Fake News": A typology of scholarly definitions. *Digital Journalism, 6*(2), 137–153.

Farmer, L. S. J. (2021). *Fake news in context*. Routledge.

Fuller, S. (2018). *Post-truth: Knowledge as a power game*. Anthem Press.

Guterres, A [@antonioguterres]. (2020, March 28). *Our common enemy is #COVID19, but our enemy is also an "infodemic" of misinformation* [Tweet]. Twitter. https://twitter.com/antonioguterres/status/1243748397019992065

Habermas, J. (1989). *The structural transformation of the public sphere: An inquiry into a category of bourgeois society*. Polity Press.

Harsin, J. (2015). Regimes of post-truth, postpolitics, and attention economies. *Communication, Culture & Critique, 8*(2), 327–333.

Horsthemke, K. (2017). '#FactsMustFall'? – Education in a post-truth, post truthful world. *Ethics and Education, 12*(3), 273–288.

Hyvönen, M. (2018). As a matter of fact. Journalism and scholarship in the post-truth era. In M. A. Peters et al. (Eds.), *Post truth, fake news: Viral modernity and higher education*. Springer.

Lazer, M. J., et al. (2018). The science of fake news. Addressing fake news requires a multidisciplinary effort. *Science, 359*(6380), 1094–1096. https://doi.org/10.1126/science.aao2998

Legg, C. (2018). 'The solution to poor opinions is more opinions': Peircean pragmatist tactics for the epistemic long game, post truth. In M. A. Peters et al. (Eds.), *Post truth, fake news: Viral modernity and higher education*. Springer.
Lippmann, W. (1922). *Public opinion*. Harcourt, Brace.
Nagle, A. (2017). *Kill all Normies. Online culture wars from 4Chan and Tumblr to trump and the Alt-Right*. Zero Books.
Rider, S. (2020). Going public: Higher education and the democratization of knowledge. In M. A. Peters et al. (Eds.), *Knowledge socialism – The rise of peer production: Collegiality, collaboration and collective intelligence*. Springer.
Subramanian, M. (2019). Humans versus Earth: The quest to define the Anthropocene. *Nature, 572*, 168–170. https://doi.org/10.1038/d41586-019-02381-2
Wardle, C. (2017, February 16). Fake news. It's complicated. *First Draft Footnotes*. https://medium.com/1st-draft/fake-news-its-complicated-d0f773766c79. Accessed 15 June 2021.

Sharon Rider is Professor of Philosophy at Uppsala University, specializing on the cultural conditions for knowledge and rational agency. She was Vice Dean of the Faculty of Arts 2008–2014, and is currently Deputy Director of the Swedish Research Council-funded Engaging Vulnerability Program. Rider is a government appointee to the Scientific Advisory Board for the Swedish International Cooperation Agency, and an elected member of the Royal Society of Humanities. Recent publications include (as co-editor and contributor) *Post-Truth, Fake News: Viral Modernity & Higher Education* (Springer, 2018).

Mats Hyvönen is Media Scholar at Uppsala University and coordinator for the Engaging Vulnerability Research Program. Hyvönen's research interests are mainly in media history, especially the study of the public sphere as a vulnerable space, and how the media both resist and facilitate that vulnerability. Recent publications include (as co-editor and contributor) *Post-Truth, Fake News: Viral Modernity & Higher Education* (Springer, 2018).

Feral

Julie Beauté and Salomé Dehaut

Abstract There is semantic ambiguity in the term "feral" since it can designate species, landscapes, dynamics or ethics. Challenging dualisms, it accounts for out-of-control historical and spontaneous trajectories from the domestic to the wild. Feral entanglements refer to the latent uncertainty and instability of matter at play in the more-than-human Anthropocene. They are subtended by imaginaries (colonial, ruins and liberal imaginaries), but open the way to counter-centric and more-than-human narratives.

Przewalski's horses, long deemed the last wild horses, are actually descendants of horses herded by the Botai around 5500 years ago (Gaunitz et al., 2018). They are therefore not considered fully wild any more, yet due to their build, behaviour and modern history of life in the wild, they are not seen as domestic either: these horses come to challenge the wild/domestic dualism and fall into the delineation of what we call feral.

The Oxford Dictionary of English defines "feral" as meaning "in a wild state, especially after escape from captivity or domestication". Mainly used to describe animals, it derives from the Latin fera, wild animal. Contemporaneous uses of "feral" stem from evolutionary ecology: in this field, it describes a species or an individual considered wild but descending from domestic or captive ancestors. The term thus infers a trajectory from the domestic to the wild. It often takes on a

J. Beauté (✉)
Aix Marseille Univ, CNRS, EFS, ADES, Marseille, France

ENS Ulm, PSL, Pays Germaniques, Paris, France
e-mail: julie.beaute@univ-amu.fr

S. Dehaut
Univ. Grenoble Alpes, CNRS, Sciences Po Grenoble, Pacte, Grenoble, France

Univ. de Lausanne, Lausanne, Switzerland
e-mail: salome.dehaut@umrpacte.fr

© The Author(s), under exclusive license to Springer Nature Switzerland AG 2023

N. Wallenhorst, C. Wulf (eds.), *Handbook of the Anthropocene*,
https://doi.org/10.1007/978-3-031-25910-4_245

pejorative meaning when used to qualify species or entities that transgress the norm and call for management. For this reason, scientists sometimes prefer to use the term "wild" interchangeably with "feral", and there exists discrepancies between its technical definitions (Clancy et al., 2021).

Today, the adjective also qualifies spaces, landscapes or ecosystems that were cultivated but that now develop beyond human control: "feral" is used to describe abandoned agricultural lands and industrial wastelands turning into fallow lands and forests (Schnitzler et al., 2011). In this sense, feral landscapes are rewilding themselves: trees and other plants are reclaiming spaces from which infrastructure projects had excluded them. Applied to landscapes and ecosystems, "feral" tends to be more positively perceived.

Ferality more generally designates entanglements, assemblages, processes that particularly characterize the Anthropocene. Contemporary social, political and environmental crises have indeed led to a complexification of temporalities and spatialities. Systems of capitalist production have escaped individual scales, challenges and needs have accelerated, unexpected deviations of all types have proliferated, ecologies have been turned upside down. Capital and geology come together, creating unintentional and proliferating effects of more-than-human negotiations (Bubandt & Tsing, 2018). Therefore, anthropocene dynamics themselves are deemed feral.

"Feral" is finally used as a socio-political concept urging Western societies to ethically rethink their relation to wildlife and the environment. Monbiot for instance calls for a rewilding of not only other-than-human ecosystems but also of human lives, through a "re-involvement" in nature (2014). In that sense societies that come to find a more relational way of becoming with other-than-human beings can also be deemed feral. This proposal goes hand-in-hand with rewilding approaches aiming at releasing natural processes from human control (Prior & Ward, 2016). In that respect, rewilding calls for a feralization of natural processes.

Ferality challenges the very idea of a boundary between the wild and the domestic but also the native and the non-native (Clancy et al., 2021). It calls into question the naturalist ontology grounded in the nature/culture divide (Descola, 2013). Feral donkeys are considered wild in the United States, but do not belong to the "wildlife" category in the same way as wolves or bald eagles (Clancy et al., 2021). Feral entities are categorized relationally, and challenge the fixity of pre-existing and stable categories. They also temper the dichotomy between the natural and the cultural realms by manifesting hybrid spaces, and question the direction of its trajectory as going from the domestic and towards the wild: ferality doesn't arise from the most non-domestic places but from the most human-impacted ones.

Ferality refers to complex spatio-temporal processes: it depicts the historical trajectories of living beings and environments. Feral dynamics open up spaces for new ecosystems (Hobbs et al., 2013) and therefore look more to the future than to the past. It is not a matter here of a return to nature nor of a return of nature: such a romantic vision tends to ignore the irreversible dynamics that shape landscapes, species and societies. The term "feral" describes what is no longer domesticated, but what has not suddenly become simply wild: a feral dog is indeed not a wolf. The feral embodies a strange wildness that is not pristine but matches the hybrid realities of the Anthropocene present and futures. Ferality thus accounts for dynamic trajectories and processes that escape the human domestication.

It is indeed characterized by a spontaneity that challenges and exceeds human control. While the term feral often expresses frustration with the lack of control over species and ecosystems, it also highlights the fact that organisms and landscapes cannot fit into preconceived patterns. Ferality thus opens up a reflection on more-than-human autonomy (Collard et al., 2015), as it insists on the independent yet relational agency of living existences (Bubandt & Tsing, 2018). Feral entanglements can evolve in an unpredictable way. They go against the logic of the One and politics of Identity, and thwart the valorization of the determined: "the feral is the one that does not fit, that refuses to conform to dominant standards of propriety, order and culture" (van Dooren, 2015).

The term "feral" is especially appropriate to capture the contemporary world, which is deeply affected by social, political and environmental crises: it paves the way to a more-than-human Anthropocene. This idea is explored in the multimedia project called Feral Atlas. The More-Than-Human Anthropocene (Tsing et al., 2021). It combines sciences, humanities, and arts to understand the feral dynamics of the more-than-human modernity. It specifies the scope of the notion, by linking urbanism and architecture: ferality occurs when an entity, transformed by a human project of infrastructure, pursues a trajectory beyond human control. The atlas thus explores feral effects occurring in ecological worlds where more-than-human entities become entangled with human projects, radically transforming them and even putting them in check.

The notion of "feral" is decisively opposed to projective thinking, that is to say, planning and deterministic thinking. It describes uncertain becomings and unplanned material effects. Feral doesn't describe a fixed state, but a process of becoming less domestic or wilder. This involves being considered in relation to other entities and in particular to certain forms of human power that vary over time and space. Considering feral realities therefore implies taking into account the instability of anthropocene environments. In particular, Tsing et al. (2019) describe feral proliferations as enabled by and deeply entangled with modular patches of simplifications that characterize colonial and capitalist ecologies. Anthropocene milieux are both the scene and the result of unstable materialities and feral dynamics emerging from modular landscape structures.

More-than-human entanglements reflect then a latent uncertainty that characterizes contemporary ecologies. Four aspects of the feral can be identified (Tsing, 2015): feral entanglements are the result of mutualistic partnerships within a confused game; they are not good for all beings: far from the idea of a naïve harmony of nature, they make space for some but leave others out; they are hardly institutionalizable: they are initiated only through infringement or infection; and they are immersed in trouble, where humans never fully hold control. The feral approach thus makes it possible to account for the unstable, the uncertain, the possible and the latent.

These feral entanglements are subtended by imaginaries: colonial, ruins and liberal imaginaries. Firstly, feral dynamics come as counter-points to human dominion over nature, and as such also question the desirability of this dominion, in line with contributions from anarchist anthropology. Ferality emerges where domestication shows its limits. Feral dynamics indeed characterize not only what escapes from but also what transgresses the domesticated sphere. In French, the term marron (maroon) works as a synonym for "feral" when describing an animal species. Like in English, it was also used to qualify former slaves who had freed themselves and established communities outside of the colonial rule. Thinking with ferality therefore allows for consideration beyond colonial imaginaries.

Feral entanglements also induce an imaginary of instability and vulnerability, often articulated around the image of ruins. Numerous research works on ferality seem to postulate a generalized precarity, by inviting to study life forms in capitalist and liberalist ruins (Tsing, 2015). Ruins tend to come at the centre of discussions on multi-species cohabitations. But then don't we indirectly draw from them an aesthetic and an ethics of precarity? If ruins are a particularly operative motif in feral ecology, they can obscure the fact that some people are more vulnerable to ecological devastation than others. The issue here is not to encourage such a worldview, but to think about the concrete, material effects of feral entanglements.

Ferality indeed appear dangerously compatible with imperialist and neoliberalist values, insofar as they promote an image of fluidity and malleability (Lorimer, 2020). Feral dynamics could work as a tool to promote liberal conservation practices such as laissez-faire, without considering the inequities they might bring between entities that thrive in feral environments and entities that wither. A political ecology of feral entanglements is therefore needed to better understand the more-than-human power relations at play within these environments.

Although the notion of the feral can be both descriptive and axiological, it definitely has the advantage of challenging the dualistic categories of thought and the tendencies towards anthropocentrism and occidentalocentrism. It highlights the importance of multiple entangled autonomies and thus opens the way for counter-centric and more-than-human narratives.

References

Bubandt, N., & Tsing, A. L. (2018). Feral dynamics of post-industrial ruin: An introduction. *Journal of Ethnobiology, 38*(1), 1–7.

Clancy, C., Cooke, F., & Row, Z. (2021). Entanglement, autonomy and the co-production of landscapes: Relational geographies for free-roaming 'Feral' donkeys (*Equus Asinus*) in a rapidly changing world. *Geoforum, 123*(July), 66–77.

Collard, R.-C., Dempsey, J., & Sundberg, J. (2015). A manifesto for abundant futures. *Annals of the Association of American Geographers, 105*(2), 322–330.

Descola, P. (2013). *Beyond nature and culture* (J. Lloyd, Trans.). University of Chicago Press.

Gaunitz, C., et al. (2018). Ancient genomes revisit the ancestry of domestic and Przewalski's Horses. *Science, 360*(6384), 111–114.

Hobbs, R. J., Higgs, E. S., & Hall, C. (2013). *Novel ecosystems: Intervening in the new ecological world order*. Wiley.
Lorimer, J. (2020). *The probiotic planet: Using life to manage life*. University of Minnesota Press.
Monbiot, G. (2014). *Feral: Rewilding the land, sea and human life*. Penguin.
Prior, J., & Ward, K. J. (2016). Rethinking rewilding: A response to Jørgensen. *Geoforum, 69*(february), 132–135.
Schnitzler, A., Aumaitre, D., & Schnitzler, C. (2011). From rurality to ferality: A case study from the upper Moselle, France. *Revue d'Écologie, 66*(2), 117–134.
Tsing, A. L. (2015). *The mushroom at the end of the world: On the possibility of life in capitalist ruins*. Princeton University Press.
Tsing, A. L., Mathews, A. S., & Bubandt, N. (2019). Patchy Anthropocene: Landscape structure, multispecies history, and the retooling of anthropology: An introduction to supplement 20. *Current Anthropology, 60*(S20), S186–S197.
Tsing, A. L., Deger, J., Saxena, A. K., & Zhou, F. (2021). *Feral atlas: The more-than-human Anthropocene*. Stanford University Press.
Van Dooren, T. (2015). On Ferals. *Thom van Dooren* (blog). http://www.thomvandooren.org/2015/12/05/on-ferals/. Accessed 21 Jan 2022.

Julie Beauté is a PhD candidate in philosophy at the ENS Ulm and at Aix-Marseille Université. In her research, she proposes to renew the relation between ecology and architecture. She is particularly interested in the role of material, human and non-human agencies in architecture, and draws on environmental humanities, new materialisms, feminist epistemologies and field philosophy.

Salomé Dehaut is a geographer interested in the more-than-human politics of rewilding and more broadly in human/other-than-human relations in the context of nature conservation. She is a PhD candidate in Geography at the Universities of Grenoble (France) and Lausanne (Switzerland).

Nuclear War

Jun Yamana

Abstract War is defined differently depending on the historical and cultural context of those describing it. From the perspective of the Anthropocene, however, the emergence of nuclear weapons is of decisive importance because it enables planetary destruction and the extermination of human beings from the entire ecosystem. Since the "nuclear revolution," we have been living with the risk of an unimaginably huge man-made catastrophe. Many who have examined this issue refer to the importance of education for peace. Hiroshima and Nagasaki are considered symbolic starting points when thinking about the hard-to-imagine devastation caused by nuclear weapons. The attempt of Hiroshima City to provide peace education is a concrete example of an effort to consider the possibilities and challenges that we face.

Questioning the nature of war is almost synonymous with inquiring about what a human being is and is similar to questioning what culture is. Even if the problematic perspectives regarding war are limited historically (McNeill, 1982) or philosophically (Kimmnich, 1976), no clear answers can be expected to emerge. War is defined differently depending on the historical and cultural context of those who question it. "War remains, as it always has been, one of the chief human mysteries" (Svetlana Alexievich, *The Unwomanly Face of War*, 1984). Generally speaking, war is an organized conflict waged with great violence and involving the use of weapons. Groups or individuals considered enemies are wounded or killed, and the infrastructure in which they live destroyed. War inflicts great damage on the social and economic systems of "enemies." Both those who reject war, and those who approve it agree that war causes immense violence that cannot occur in peacetime.

Despite the difficulty of defining of war, there is no doubt that nuclear war is one of the main starting points for a discussion on the theme of "war" in relation to the key word "Anthropocene." Nuclear energy is a technology that harnesses the power

J. Yamana (✉)
The University of Tokyo, Tokyo, Japan
e-mail: jyamana6s@p.u-tokyo.ac.jp

© The Author(s), under exclusive license to Springer Nature Switzerland AG 2023

N. Wallenhorst, C. Wulf (eds.), *Handbook of the Anthropocene*,
https://doi.org/10.1007/978-3-031-25910-4_246

of nature on a scale that far surpasses anything in human history (Pálson, 2020). It has been pointed out that the birth of nuclear energy technology is one of the most important characteristics of the Anthropocene (see the article "Atomic Destruction" in this handbook). If nuclear energy is used as a weapon, ecosystems consisting of plants and animals including humans could be instantly destroyed on a planetary scale. Nuclear weapons would immediately annihilate the distinctions between winners and losers, allies and enemies, past and future, as well as happiness and unhappiness. Günther Anders (1902–1992), a philosopher concerned with "the nuclear threat," describes the hopeless situation of a potential future catastrophe, where "there will be no difference … between those weeping and those wept over, because mourners will drift in the waters beside the dead" (Anders, 1972: 8).

In the first half of the nineteenth century, Carl von Clausewitz (1780–1831) had already described the logic of "absolute war" as an endless mutual increase of force (Von Clausewitz, 1989: 76). He presented it as a fiction distinct from "real war." In the twentieth century, however, the principle of "absolute war" revealed its ultimate development in the form of nuclear war. The impact of absolute war was manifested in the atomic bombing of Hiroshima and Nagasaki, two cities with a large number of civilian inhabitants. And "over two decades following those bombings, innovations in weapons design, decisions about arsenals, and geopolitical crises had cohered into a rethinking not only of the status of nuclear weapons but of the entire geopolitical order" (Gordin & Ikenberry, 2020: 3). "Nuclear revolution" (Mandelbaum, 1981; Jervis, 1990) is the name given to such radical global changes in society brought about by the emergence of a culture that seeks to harness immense nuclear power at great risk.

Roger Caillois (1913–1978) was a thinker who considered war in the atomic era and provided deep insights into humanity and culture. He published his book "*BELLONE ou la pente de la guerre* (*BELLONE or the slope of the war*)" (Caillois, 1963 in French). At the time he had been working for UNESCO since 1948. UNESCO had been established 2 years previously as a central organization whose aim was to help to achieve world peace through the promotion of education and culture. According to Caillois, modernity promoted a transformation from "aristocratic warfare" to "national warfare," and it ultimately brought "total war" as a result, in which the maximum amount of human and material resources have to be invested. Moreover, Caillois emphasizes the irony of the situation, in that nuclear war also destroys the framework of "national war" with its unlimited violence.

Another philosopher who has discussed the issue of nuclear war intensively is the aforementioned German philosopher Anders. He says that we are living in an era in which any place at any time, or even the entire world, could become a new Hiroshima, since an atomic bomb was dropped there. Despite the "apocalyptic" situation of the "era of Hiroshima", people today are "unable to imagine the real extent and real effects of what they effectively produced. They believe that being able to cope with today's products by using yesterday's categories and treatment modes is "antiquated"." (Anders, 1972: 13). Anders names this situation the "Promethean fall" after the Greek god who gave fire to mankind.

Anders imagines a new kind of planetary community where the world is facing nuclear risks, in which all boundaries should become meaningless and future mankind should also be included as community members. "Since the horizon of our technical achievements is global, the horizon of our moral solidarity must become global as well" (Anders, 1972: 52). However, if we are to approach the ideal global community, we must inevitably overcome the "Promethean fall" and comprehend the "apocalyptic" situation that humanity is facing. How is this possible?

Many who have dealt with the issue of the enormous risk that followed the "nuclear revolution" have anticipated the effectiveness of education (Caillois, 1963; Jervis, 1990; Gordin & Ikenberry, 2020; Wigger et al., 2017; Wigger, 2020; Wulf, 2020, 2021). In the context of the problem of Hiroshima, Anders also recommends enlightenment through stories that magnify people's anxieties about the nuclear threat, although he does not directly use "education" as a key word. His thought influenced the French philosopher Jean-Pierre Dupuy (1941-) who has developed the theory of the "enlightened catastrophe". According to him, it is crucial in the present to design a better future situation through imagining the worst-case scenario by recalling past catastrophes (Dupuy, 2005). This way of thinking is also supported by recent collective memory studies. "The discussions about climate change, the Anthropocene, and posthumanism also give Memory Studies [= interdisciplinary and international studies dealing with corrective memories and remembering cultures] the task to include the future in its models and to think actively about forms of remembering in the sense of "future memory" which have lasting effects on the upcoming world" (Erll, 2017: 121). Peace education that addresses the theme of war and violence can be situated within this context.

It is argued that contemporary peace education should be redefined as a subject that addresses global issues. "Peace education today assumes that constructive engagement with the planetary problems of violence [...] is part of a lifelong learning process that begins in childhood and should continue in later life" (Wulf, 2021: 471). As Christoph Wulf suggests, new current concepts such as "global citizenship education" (UNESCO, 2015) can and should include education for a world without war. Education about crises and violence after the "nuclear revolution" can form the central part of such education for peace in a broader sense. In this context, it is noteworthy that peace education has been conducted in Hiroshima and Nagasaki to remember the events of August 1945 and to pass them on to the next generation in order to prevent same tragic situations from happening again in the future. As an example, the basic characteristics and contemporary trends of Hiroshima's peace education reveal its possibilities and challenges.

Hiroshima was the first city to suffer damage from atomic bombing during the war, and since then it has been continuously engaged in peace education activities to remember its devastation. On the basis of various attempts at peace education undertaken by several schools and teachers, the City of Hiroshima began to organize peace education more comprehensively at the educational administrative level in the 1970s. The goal for peace education set by the City of Hiroshima Board of Education was "to foster an understanding of the sanctity of lives and the dignity of each individual human being, and to develop motivations and attitudes to contribute to the realization of lasting worldwide peace and to become a member of this cultural city for international peace based on the experiences of the atomic bombing" (1970). Since then, the city has published "*Peace Education Guide* (*heiwa kyōiku no tebiki*)" and "*Collection of Peace Education Instructional Materials for Peace Education* (*shidō shiryō*)" for elementary, middle, and high schools to "make efforts to provide guidance so that each school can independently promote initiatives in accordance with their actual conditions" (Hiroshima City Board of Education, 2014: 2).

Self-reflection regarding the peace education in Hiroshima established that the overall systematization was insufficient. As a result, a cross-curricular integrated program for peace education was formulated for all levels of school (see Fig. 1). In addition, the city created a series original teaching materials "*Peace Notebook* (*heiwa note*)" (2015) to accompany the program. Furthermore, to fit into the contemporary context that emphasizes the importance of global sustainability, the Hiroshima Peace Education Program also states that "the content should enable students to grasp the reality of the atomic bombing and other facts, to be future-oriented through those facts, and to acquire the following knowledge and abilities necessary for shaping a peaceful and sustainable society" (*Hiroshima Municipal Peace Education Program Outline*, from the website of The City of Hiroshima Board of Education).

Peace education in Hiroshima is not completed within the realm of school education, but is supported by the public staging of history in memory through broader "remembering cultures (*Erinnerungskulturen*)" (Assmann, 2007; Erll, 2017). It is

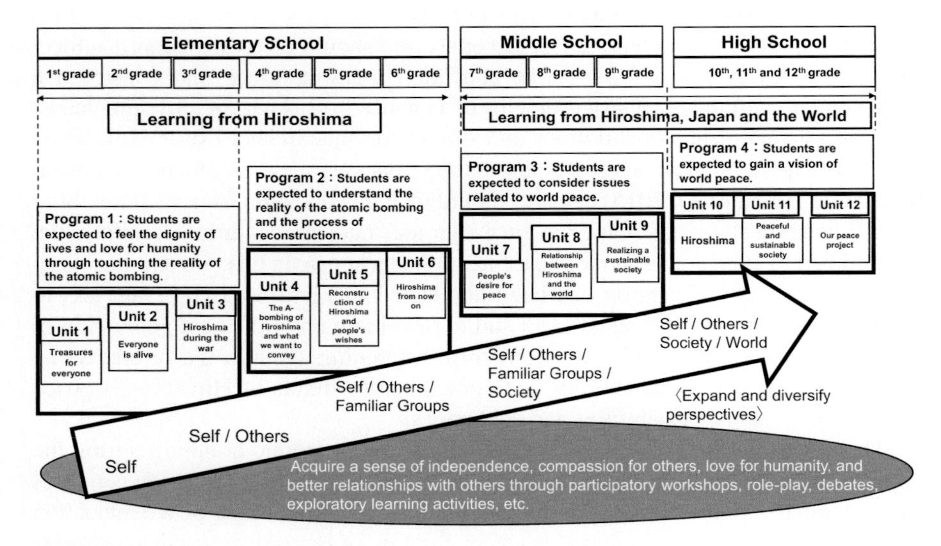

Fig. 1 Image of learning in the Hiroshima City Peace Education Program. (Created by simplifying the figure posted on the Hiroshima City Board of Education website "Peace Learning Promotion Project for the next generation and the Succession and Dissemination of Hiroshima by Elementary, Middle and High School Students.") https://www.city.hiroshima.lg.jp/uploaded/attachment/52824.pdf (URL: 3 May 2022)

considered to constitute a composite structure made up of the physical and electronic spaces of museums and urban structures as well as other institutions outside schools. As a concrete place and medium for remembering practices, as its resources it uses literature, animations, films, music, paintings, and sculptures related to the atomic bombing. One of the most significant features of peace education in Hiroshima is its connection to world cultural heritage education. The architecture of the Hiroshima Peace Memorial Park and its surrounding area comprise a complex site dedicated to the remembrance of the atomic bombing. The area includes the Atomic Bomb Dome (registered as a World Cultural Heritage Site in 1996), the Peace Memorial Museum (Schäfer, 2018), and many other memorial monuments. At the same time, it is a place to learn about the significance of the system for preserving World Cultural Heritage Sites (Yamana, 2020).

While Hiroshima's architecture of remembrance plays an important role as regards "cultural memory (*Kulturelles Gedächtnis*)" (Assmann, 1988), there is another central practical form for peace education as regards "communicative memory (*Kommunikatives Gedächtnis*)." Namely, it is the testimonial activity undertaken by A-bomb survivors who meet regularly, mainly in the Hiroshima Peace Memorial Museum. Many people, including students on school excursions, listen to their words. One unique example is the project called "Picture of Atomic Bomb (*genbaku no e*)", in which high school students at *Municipal Motomachi High School* draw pictures of what they had learned from A-bomb survivors at many interactive meetings with them (Yamana, 2017).

According to Assmann, the effectiveness of communicative memory is usually limited to approximately 70 to 90 years. More than 75 years have already passed since the atomic bombing. Hiroshima is facing the problem of how to pass on A-bomb testimonies to the future when no living survivors remain. The City of Hiroshima has launched a "training program for secondary storytellers of A-bomb experiences" 2012 (Hiroshima City website). Also, the "HIROSIMA ARCHIVE" can be considered a unique example of attempts at peace education that looks also toward the post-survivor era. It was launched by students at *Hiroshima Jogakuin High School* with the help of media scientists (Hiroshima Archive Production Committee, 2015-). This city map website shows where survivors were, and basic information about them and their testimony can be accessed by clicking on their photographic portraits.

The remembering culture of Hiroshima has been critically discussed from various angles, for instance, regarding the atmosphere of "self-pity" through a pseudo-religious culture of staging remembrance (Buruma, 1994), the deformation of memory politics through tourism (Yoneyama, 1999), and insufficient analysis of the meaning of the symbol "Hiroshima" from a global perspective (Zwigenberg, 2014) etc. In terms of education regarding war during the Anthropocene, we must be aware of other questions that can be linked to these but that are also different. What kind of narrative forms should the next generation employ in the post-A-bomb survivor era? How should we think about issues related to the secondary trauma of "postmemory" generations (Hirsch, 2012)? How should peace education about risks after the "nuclear revolution" be situated within the broader peace education program directed at the many issues related to the Anthropocene? In other words, how can we handle the limited capacity of the memory economy, in which information regarding global catastrophes must inevitably be ranked and chosen for a suitable human scale memory capacity? How should we confront the merits and demerits of the increasing manipulability of memory through the development of media technology? What continues to be necessary as regards peace education outside authentic locations such as Hiroshima and Nagasaki? If biological weapons, chemical weapons, new weapons such as unmanned aerial vehicles, and cyberattacks using new scientific technology can be considered components of Anthropocene warfare in addition to nuclear weapons, how can peace education in relation to the "nuclear revolution" be linked to them?

Education is only a part of the social system as a whole. Obviously, it is not an all-powerful means of governing society. "It cannot be overlooked that war and violence are often caused by problems in the macrostructural system. It is difficult to reduce war and violence through education" (Wulf, 2021: 471). While keeping in mind the limitations of education, it is also important to consider how to link the educational system with other social systems, such as those of politics, economics, law, and medicine. It is said that "our understanding of the meaning of Hiroshima - how it has been lived and understood across generations and places- is always incomplete, even as we respond to a moral obligation to search for understanding" (Gordin & Ikenberry, 2020: 13). As difficult as it is to fully understand warfare in the Anthropocene, we cannot turn away from the fact that we are constantly in the midst of this crisis brought on by human beings themselves. Educational outreach to the next generations is then essential, because the "new community" includes people who have not yet been born (Anders, 1982). Since there is no correct answer to Anthropocene warfare, the outreach must take the form of intergenerational dialogue, not simply one-way instruction from one generation to the next.

References

Alexievich, S. A. (2017). *The unwomanly face of war: An oral history of women in world war II*. Random House. (Original: 1984).

Anders, G. (1972). *Endzeit und Zeitende. Gedanken über die atomare Situation*. C.H. Beck.

Anders, G. (1982). Der Mann auf der Brücke. Tagebuch aus Hiroshima und Nagasaki (1959). Ders. *Hiroshima ist überall*. C.H. Beck: 1–189.

Assmann, J. (1988). Kollektives Gedächtnis und kulturelle Identität. In J. Assmann & T. Hölscher (Hrsg.), *Kultur und Gedächtnis* (pp. 9–19). Suhrkamp.

Assmann, A. (2007). *Geschichte im Gedächtnis: Von der individuellen Erfahrung zur öffentlischen Inszenierung*. C. H. Beck.

Buruma, I. (1994). *Wages of guilt: Memories of war in Germany and Japan*. Farrar, Straus and Giroux.

Caillois, R. (1963). *Bellone ou la pente de la guerre*. Renaissance du livre.

Dupuy, J.-P. (2005). *Pitite métaphysique des tsunamis*. Seuil.

Erll, A. (2017). *Kollektives Gedächtnis und Erinnerungskulturen. Eine Einführung*, 3. Aktualisierte und erweiterte Aufl. J.B. Mezler Verlag.

Gordin, M. D., & Ikenberry, G. J. (Eds.). (2020). *The age of Hiroshima*. Prince University Press.
Hirsch, M. (2012). *The generation of Postmemory: Writing and visual culture after the holocaust*. Columbia University Press.
Jervis, R. (1990). *The meaning of the nuclear revolution*. Cornell University Press.
Kimmnich, O. u.a. (1976). Krieg. In J. Ritter & K. Gründer (Hrsg.), *Historisches Wörterbuch der Philosophie. Bd., 4*, 1230–1235.
Mandelbaum, M. (1981). *The nuclear revolution: International politics before and after Hiroshima*. Cambridge University Press.
McNeill, W. H. (1982). *The pursuit of power: Technology, armed force, and society since A.D. 1000*. University of Chicago Press.
Pálson, G. (2020). *The human age*. Welbeck Publishing Gorup Limited.
Schäfer, S. (2018). *Das Atombombenmuseum Hiroshima: Erinnern jenseits der Nation (1945–1975)*. Bielefeld: transcript Verlag.
UNESCO. 2015. Global citizenship education.
Von Clausewitz, C. (1989). On war. Princeton University Press. (Original: 1832).
Wigger, L. (2020). The legacy of 'Hiroshima' in Germany: History and current problems. *E-Journal of Philosophy of Education: International Yearbook of the Philosophy of Education Society of Japan., 5*, 21–41.
Wigger, L., et al. (Eds.). (2017). *Nach Fukushima? Zur erziehungs- und bildungstheoretischen Reflexion atomarer Katastrophen*. Julius Klinkhardt.
Wulf, C. (2020). *Bildung als Wissen vom Menschen im Anthropozän*. Belz Juventa.
Wulf, C. (2021). Global Citizenship Education: Bildung zu einer planetarischen Weltgemeinschaft im Anthropozän. *Vierteljahrsschrift für Wissenschaftliche Pädagogik, 97*, 464–480.
Yamana. (2017). Hiroshima als architektonischer Raum der Erinnerung: Zur Problematik der Pädagogisierung eines geschichtlichen Ortes. *Jahrbuch für Historische Bildungsforschung, 22*, 61–79.
Yamana. (2020). Catastrophe, commemoration and education: On the concept of memory pedagogy. *Educational Philosophy and Theory, 52*(13), 1375–1387.
Yoneyama, L. (1999). *Hiroshima traces: Time, space, and the dialectics of memory*. University of California Press.
Zwigenberg, R. (2014). *Hiroshima: The origin of global memory culture*. Cambridge University Press.

Website

Hiroshima Archive Production Committee (2015-). Hirosima Archive.
https://hiroshima.mapping.jp/index_en.html (URL: 3 May 2022).

Jun Yamana is a Professor at the University of Tokyo. He has taught numerous courses on the Philosophy of Education. His recent special area of interest and research is "Memory Pedagogy," which analyzes educational theories and practices from the perspective of cultural memory (collective memory) and commemoration developed in the field of Memory Studies. He has collaborated with German educational researchers on various philosophical and historical research projects since he studied at the Humboldt University in Berlin in the 1990s. He has been published widely in Japanese and German.

Nuclear Waste

Christine Eriksen and Stephen Herzog

Abstract Nuclear waste epitomizes the Anthropocene. Scientific discovery of nuclear fission in the 1930s ushered in the atomic age. The onset of nuclear weapons and nuclear energy production in the 1940s and 1950s then created a uniquely human problem with planetary implications. Today, 33 countries operate 442 nuclear power reactors, and nine countries possess nearly 13,000 nuclear arms. The result is high-level waste that is dangerously radioactive for millennia to come. Yet, there has *never* been a permanent waste solution in place. Technically feasible long-term nuclear waste storage options exist, but nearly all governments prefer riskier interim plans hidden from public view and debate. This chapter considers the likelihood of societies addressing the contentious environmental and economic politics of deep geological repositories; and it asks, how long will obfuscation of the risks of this unique Anthropocene challenge continue?

This chapter is not about technical aspects of nuclear waste, but rather, positioning nuclear waste as a problem embodying the very essence of the Anthropocene. People create nuclear energy for commercial (electricity, medical technology) and political–military (weapons) applications. In turn, societies must deal with nuclear waste like irradiated spent uranium fuel from nuclear power plants (NPP). The half-lives of certain radioisotopes in such high-level waste (HLW) mean this problem will not disappear in the lifetimes of all living generations, and possibly not in the lifetime of humanity. Just because fossil fuels present a more immediate threat to life on Earth does not mean the long-term consequences of HLW can be wilfully ignored.

The scope of this enduring challenge is global. A total of 33 countries produce electricity from 442 nuclear power reactors, with 57 reactors under construction and a growing interest among developing countries (International Atomic Energy Agency, 2023). The waste problem is hardly confined to civilian nuclear power,

C. Eriksen (✉) · S. Herzog
Center for Security Studies, ETH Zürich, Zürich, Switzerland
e-mail: christine.eriksen@sipo.gess.ethz.ch; stephen.herzog@sipo.gess.ethz.ch

© The Author(s), under exclusive license to Springer Nature Switzerland AG 2023

N. Wallenhorst, C. Wulf (eds.), *Handbook of the Anthropocene*,
https://doi.org/10.1007/978-3-031-25910-4_247

as plutonium reprocessing for nuclear weapons development also generates HLW (Garwin & Charpak, 2001). Nine countries maintain nearly 13,000 nuclear weapons, many of which are undergoing modernization (Kristensen & Korda, 2022). So long as their leaders rely on nuclear arms for security and military power projection, others will be similarly incentivized to pursue the bomb, as is human nature. Even if all NPP ceased operations and nuclear weapons disarmament was achieved, the problem of nuclear waste storage will remain for millennia (World Nuclear Association, 2021).

The magnitude of the nuclear waste problem is perhaps unsurprising. Human entanglement with the atom arguably created the Anthropocene via the release of atomic particles into the atmosphere from the mid-twentieth century (AWG, 2020). Nuclear waste embodies the dramatic environmental transformation of this epoch. That is, impacts of the human on the planetary system are felt at all levels from the geological, to the arboreal, to the atmospheric (Eriksen & Turnbull, 2022).

Although production of nuclear weapons and nuclear energy has occurred since the 1940s and 1950s, respectively, there has *never* been a permanent HLW solution (Alley & Alley, 2013). Checks and balances on nuclear waste were missing from the start, largely due to the World War II and Cold War security environments. Nuclear capabilities were first developed for military power, subject to relentless classification and lacking public visibility (*e.g.*, Rhodes, 1986). They were then transformed into a commercial energy source shrouded in secrecy and power struggles. The rhetoric of security concerns regarding nuclear proliferation—and later, terrorist attacks—has long contributed to a transparency deficit in storage of nuclear materials and waste products.

But despite dangers, countries have flocked to the atom. The nuclear industry presents an appealing façade, a futuristic power source with transformative properties for societies. As Anna Tsing (2015, p.3) argues: 'Grasping the atom was the culmination of human dreams of controlling nature. It was also the beginning of those dreams' undoing.'

Public debates about atomic energy focus on visible manifestations of nuclear technology. Policymakers and their constituents usually engage with NPP operational safety, accident risks, and nuclear weapons development (Baron & Herzog, 2020). Meanwhile, polling firms contracted by the nuclear industry consistently attempt to convince publics that nuclear power is clean energy (*e.g.*, Bisconti, 2016). Missing from this picture is the handling of nuclear waste that has accumulated around the world for over 75 years. Yet, for the nuclear industry, 'it is issues surrounding nuclear waste disposal that is its Achilles heel' (Sanders & Sanders, 2020, p.23).

These debates often centre on whether to increase nuclear energy production. There is an inherent tension between catastrophic impacts of nuclear accidents—and use/testing of nuclear arms—and low carbon dioxide emissions from nuclear power. On the one hand, nuclear energy's proponents draw on legitimate fears of insufficient electricity supply and growing clean energy needs. On the other hand, critics highlight the dangers of the atom in the hands of unstable nations as well as radioactive pollution from NPP meltdowns, leakages, and ageing infrastructure.

That more reactors will intensify the HLW problem, however, is generally ignored. Arguments about emissions also tend to overlook the high kilowatt-hour costs of nuclear electricity and the environmental consequences of uranium mining and NPP construction/decommissioning. Moreover, notable experts maintain there is insufficient time for nuclear energy innovation to be the focal point of global decarbonization given the escalating climate crisis (*e.g.*, Macfarlane, 2021).

Nuclear waste should be a much more significant part of these conversations, as expanded nuclear power production will exacerbate the problem. After all, nuclear waste is the quintessence of the "high-tech risks" described in Ulrich Beck's seminal work *Risk Society: Towards a New Modernity* (first published in 1986 after the Chernobyl disaster). Beck (1992, p.22) was unequivocal in describing their harms: 'In the affliction they produce they are no longer tied to their place of origin—the industrial plant. By their nature they endanger *all* forms of life on this planet.'

Current HLW storage has been problematic at best; it risks lasting environmental degradation. It is telling that storage mostly takes place far from population centres, often on ancestral land of Indigenous peoples, and is almost always "interim." Consider one widely used method, dry casks. Spent reactor fuel rods are typically submerged in deep water pools at reactor sites, but some pools have reached capacity. Consequently, spent nuclear fuel may be placed in a metal cask, surrounded by inert gas, and transported offsite from the NPP. Cask numbers continue to grow while risks from cracks, natural hazards, and terrorism remain. Likewise, as more NPP reach decommissioning age, strategies like entombment and so-called safe enclosure have emerged for managing materials and equipment. The aforementioned risks as well as soil and groundwater contamination count among their challenges.

Given the far-reaching temporal and spatial scales of nuclear waste's impacts, is there a feasible and safe solution for this problem in the Anthropocene? There is strong international scientific agreement that deep geological repositories are the only viable long-term solution. Yet, progress has stalled despite over 30 years of research supporting this endeavour (*e.g.*, Andra, 2021; Nagra, 2021; Posiva Oy, 2021; SKB, 2021). Two main sets of constraints are to blame. First, 'institutional obstacles remain formidable, and the public's aversion to having a repository in its 'backyard' continues' (Adamantiades & Kessides, 2009, p.5161). Second, leaks, mismanagement, and engineering failures throughout the history of nuclear waste storage have shaken confidence in authorities' ability to operate repositories (Ramana, 2018).

No option will be foolproof or without setback, but unwillingness to confront the politics of nuclear waste has arguably been the tallest hurdle of all. To self-interested politicians and government officials, the timescale offers few incentives and many liabilities. It is easier, of course, to "kick the can down the road" and leave nuclear waste far from public view. Numerous countries have researched deep geologic disposal, operated underground test facilities, and even built interim repositories. The only one, so far, to have overcome the environmental, social, and economic politics of the matter is Finland. Its Onkalo facility, a *permanent* deep geological repository, is scheduled to begin storage in 2025 (Posiva Oy, 2021). HLW canisters will be placed in the repository until around 2120 when the facility access tunnel will be sealed.

In many respects, 'An argument can be made that nuclear waste management concerns are primarily a social/political dilemma and not so much of a technical/scientific argument' (Sanders & Sanders, 2020, p.4). The crux of the matter is that the best solution to a problem created by humans is now subject to the constraints of human emotion and impulse. Issues of (in)stability, irreversibility, and fear prevent its resolution. This is ironic given that the best interim storage facilities are temporary, and the worst are akin to ticking time bombs (Alley & Alley, 2013). This begs the questions: What is being buried in the ongoing nuclear waste dilemma, the risk itself (radioactive waste) or knowledge of the risk? What are the acceptable risks, to whom, and why?

References

Adamantiades, A., & Kessides, I. (2009). Nuclear power for sustainable development: Current status and future prospects. *Energy Policy, 37*, 5149–5166.

Alley, W. M., & Alley, R. (2013). *Too hot to touch: The problem of high-level nuclear waste*. Cambridge University Press.

Andra. (2021). French National Radioactive Waste Management Agency. https://international.andra.fr/. Accessed 20 July 2021.

Anthropocene Working Group (AWG). (2020). Results of binding vote by AWG. Subcommission on Quaternary Stratigraphy. http://quaternary.stratigraphy.org/working-groups/anthropocene/. Accessed 20 July 2021.

Baron, J., & Herzog, S. (2020). Public opinion on nuclear energy and nuclear weapons: The attitudinal nexus in the United States. *Energy Research & Social Science, 68*, 101567.

Beck, U. (1992). *Risk society: Towards a new modernity* (M. Ritter, Trans.). Sage.

Bisconti, A. S. (2016). Public opinion on nuclear energy: What influences it. *Bulletin of the Atomic Scientists*. April 27. https://thebulletin.org/2016/04/public-opinion-on-nuclear-energy-what-influences-it/. Accessed 20 July 2021.

Eriksen, C., & J. Turnbull. (2022). Insure the volume? Sensing air, atmospheres and radiation in the Chernobyl exclusion zone. In *Climate, society, and elemental insurance,* ed. K. Booth, et al.: Routledge.

Garwin, R. L., & Charpak, G. (2001). *Megawatts and megatons: A turning point in the nuclear age?* Knopf.

International Atomic Energy Agency. (2023). Power reactor information system. https://pris.iaea.org/PRIS/home.aspx. Accessed 24 March 2023.

Kristensen, H. M., & Korda, M. (2022). Status of world nuclear forces. Federation of American Scientists. Updated February 23. https://fas.org/issues/nuclear-weapons/status-world-nuclear-forces/. Accessed 24 March 2023.

Macfarlane, A. (2021). Nuclear energy will not be the solution to climate change. *Foreign Affairs*. July 8. https://www.foreignaffairs.com/articles/2021-07-08/nuclear-energy-will-not-be-solution-climate-change. Accessed 20 July 2021.

Nagra. (2021). Swiss National Cooperative for the Disposal of Radioactive Waste. https://www.nagra.ch/en. Accessed 20 July 2021.

Posiva Oy. (2021). We have a solution which is an example for the whole world. https://posiva.fi/en/. Accessed 20 July 2021.

Ramana, M. V. (2018). Technical and social problems of nuclear waste. *Wiley interdisciplinary reviews: Energy and environment, 7*, e289.
Rhodes, R. (1986). *The making of the atomic bomb*. Simon & Schuster.
Sanders, M. C., & Sanders, C. E. (2020). *Nuclear waste management strategies: An international perspective*. Academic Press.
SKB. (2021). Managing the Swedish nuclear waste. Swedish Nuclear Fuel and Waste Management Company. https://www.skb.com/. Accessed 20 July 2021.
Tsing, A. (2015). *The mushroom at the end of the world: On the possibility of life in capitalist ruins*. Princeton University Press.
World Nuclear Association. (2021). Storage and disposal of radioactive waste. https://world-nuclear.org/information-library/nuclear-fuel-cycle/nuclear-waste/storage-and-disposal-of-radioactive-waste.aspx. Accessed 20 July 2021.

Dr. Christine Eriksen is a Senior Researcher in the Center for Security Studies at ETH Zürich. She gained international research recognition by bringing human geography, social justice, and environmental hazards into dialogue. With a particular interest in social dimensions of disasters, her widely published work examines social vulnerability and risk adaptation in the context of environmental history, climate change, cultural norms, and political agendas. She is the author of 80+ articles and two books: *Alliances in the Anthropocene: Fire, Plants, and People* (2020) and *Gender and Wildfire: Landscapes of Uncertainty* (2014).

Dr. Stephen Herzog is a Senior Researcher in the Center for Security Studies at ETH Zürich and an Associate of the Project on Managing the Atom at Harvard University. He was formerly a U.S. Department of Energy nuclear arms control official and a researcher at the Federation of American Scientists. His research focuses on government strategies and public opinion on military and civilian nuclear risk reduction. He has authored studies on these topics in journals such as Science, *Arms Control Today*, the *Bulletin of the Atomic Scientists*, *Energy Research & Social Science*, the *Nonproliferation Review*, and the *Journal for Peace and Nuclear Disarmament*.

Pollution

Neil L. Rose, Sarah L. Roberts, and Agnieszka Gałuszka

Abstract A wide array of inorganic and organic chemicals, particulates and gases may be defined as pollutants and, for many, their release to the environment has increased markedly since the start of the Anthropocene. Although national and international policies have resulted in significant emissions reductions for some pollutants, over 80,000 chemicals are currently in commercial use, and new chemicals are emerging each year. Due to the marked increase in emissions starting in the mid-twentieth century, a number of pollutants including PCBs, artificial radionuclides, black carbon and microplastics have been proposed as stratigraphic markers for the start of the epoch. However, the vast number of chemicals in circulation through Earth ecosystems means that the Anthropocene may also be defined as an epoch of increasingly complex exposure and accumulation of a multi-pollutant contamination burden to all organisms, including humans. This burden may be remarkably inequitable between communities.

Pollutants may be defined as substances that cause a detrimental change when released into the environment. They include a wide array of chemicals from excessive nutrients, such as nitrogen and phosphorus causing eutrophication in surface

N. L. Rose (✉)
Environmental Change Research Centre, Department of Geography, University College London, London, UK
e-mail: n.rose@ucl.ac.uk

S. L. Roberts
Environmental Change Research Centre, Department of Geography, University College London, London, UK

Ecotoxicology and Chemical Risk, Pollution Science Division, UK Centre for Ecology and Hydrology, Wallingford, UK
e-mail: SarRob@ceh.ac.uk

A. Gałuszka
Institute of Chemistry, Jan Kochanowski University in Kielce, Kielce, Poland
e-mail: agnieszka.galuszka@ujk.edu.pl

© The Author(s), under exclusive license to Springer Nature Switzerland AG 2023

N. Wallenhorst, C. Wulf (eds.), *Handbook of the Anthropocene*,
https://doi.org/10.1007/978-3-031-25910-4_248

waters, and gaseous emissions of CO_2 and acidifying compounds to the atmosphere, to non-essential trace metals such as mercury (Hg), lead (Pb) and cadmium (Cd), radionuclides released from atomic weapons testing and reactor accidents, as well as a vast number of organic pollutants including organochlorine pesticides, brominated flame retardants, per- and polyfluoroalkyl substances (PFAS), organophosphates, dioxins and furans (PCDD/Fs), personal care products (many of which contain antibacterial and antifungal agents) and a variety of pharmaceutical compounds. To this list may be added particulates such as black carbon from fossil-fuel combustion sources, microplastics and, by some definitions, excessive light and noise.

All organisms on Earth therefore inhabit a multiple stressor world whereby they accumulate a multi-pollutant contaminant burden over the duration of their lives. However, while anthropogenic pollution of the environment is recognised to have occurred over millennia, releases of many pollutants have increased dramatically over the last 70 years – since the start of the Anthropocene. Therefore, while pollutants are important in the Anthropocene as potentially useful stratigraphic markers for the start of the new epoch, they are also indicative of an increase in pollutant exposure to biota on a global scale since the mid-twentieth century. Hence, there is a need to recognise that the Anthropocene also marks the start of an increase in the scale and complexity of contaminant burden to all ecosystem compartments and all biological organisms, including humans.

Fate and Stratigraphy of Anthropogenic Pollutants

Since the mid-twentieth century, a rapid expansion in human population with concomitant increases in many socio-economic indicators including energy consumption, resource exploitation and industrial-scale agriculture has occurred (Syvitski et al., 2020). This upsurge of activity, which has become known as the 'Great Acceleration', also resulted in the major deliberate and accidental release of many pollutants to the atmosphere and to surface waters. For example, global anthropogenic emissions of black carbon were estimated to be around 0.6 Tg/year in the late-nineteenth century rising to 6.6–7.2 Tg/year between 2000 and 2010 (Klimont et al., 2017). By 2015, this combustion, especially that of coal, and along with other anthropogenic sources such as artisanal gold and silver extraction, had also increased atmospheric Hg concentrations by an estimated 450% over pre-industrial levels. Similarly, following the discovery of its insecticidal properties in 1939, 5-year total emissions of DDT to the atmosphere had increased from around 30,000 tonnes (t) in the early-1940s to an estimated 750,000 t in 1970–75 (Schenker et al., 2008), while emissions of the polybrominated diphenyl ether BDE209, widely used for its fire-retardant properties until production ceased in 2018, increased from less than 1 t/year in 1970 to a peak of around 145 t/yr in 2008 (Abbasi et al., 2019). The annual global production of plastic also increased from around 0.5 megatonnes (Mt) in 1950 to almost 370 Mt in 2019 and the abundance of microplastics in the environment is thought to closely follow this trend.

Once released, anthropogenic pollutants may travel long distances at hemispheric and global scales via atmospheric and ocean current transport. Many are now considered ubiquitous, and even remote and formerly pristine regions of the globe, including the Arctic and Antarctic, are sinks for them (Vecchiato et al., 2020). PFAS have been found in surface waters across the globe since the early-2000s (Muir & Miaz, 2021) while a range of organochlorine pesticides as well as elevated concentrations of trace metals are reported from ice cores and recent snows in Antarctica. Single-use plastic has been recorded as ubiquitous in the world's deepest oceans.

Naturally accumulating archives such as lake and marine sediments, peat and coral sequences, ice cores, tree rings and speleothems store historical records of depositing contamination. Due to the increase in emissions of many pollutants since the mid-twentieth century, a number of these records have been considered as markers for the start of the Anthropocene, including polychlorinated biphenyls (PCBs) (Gałuszka et al., 2020), microplastics (Bancone et al., 2020), black carbon (Han et al., 2018) and industrial fly-ash (Rose, 2015). However, it is the record of radionuclide fallout, and, in particular, that of the radioisotopes of plutonium (^{239}Pu, ^{240}Pu), that are currently considered most likely to provide a globally synchronous signal for the start of the Anthropocene as they show a marked and consistent increase in 1952 across most accumulating environments (Waters et al., 2018).

Pollution Exposure in the Anthropocene

The real-world situation for all biota in the Anthropocene is one of chronic exposure to multiple pollutants over their whole lives, with intake principally via feeding, drinking and breathing. Due to transgenerational transfer, exposure often starts prior to birth and, while growth, reproduction, cognitive development and immunodeficiency are all adversely impacted, effects may be greater in early life-stages during periods of rapid development. For example, a broad range of persistent organic pollutants (POPs) can act as endocrine disruptors at very low concentrations whereby they mimic or block hormonal activity, although the full effects of these disruptions may not be observed until adulthood. Humans are no exception to pollutant exposure or to the detrimental effects of an accumulating contaminant burden. A 2012 study identified over 3200 chemicals in the blood plasma of seven mammalian species, including humans, and highlighted that with over 80,000 chemicals registered for commercial use and with additional transformations of these within the biosphere, around half a million chemicals may pose a level of risk (Park et al., 2012).

Over 90% of the global human population is now estimated to live in areas with unsafe air quality according to the World Health Organisation (defined as exceeding an annual average $PM_{2.5}$ concentration of 10 $\mu g/m^3$; where $PM_{2.5}$ is particulate matter with an aerodynamic diameter of less than 2.5 μm). Furthermore, the fossil-fuel derived component of $PM_{2.5}$, largely derived from vehicular emissions and stationary combustion of fossil-fuels, was estimated to be responsible for the premature deaths of 10.2 million people in 2012 (Vohra et al., 2021) mostly in China (3.9 million) and India (2.5 million). At a smaller geographical scale, communities living close to oil production and refining centres on the Niger Delta in Nigeria may be exposed to average dry season atmospheric polycyclic aromatic hydrocarbon (PAH) concentrations 4 orders of magnitude higher than communities with a "low industrial presence" just 60 km away (Ana et al., 2012).

Such inequity of exposure is true for all intake pathways. Indigenous Arctic communities who traditionally consume a fat-rich diet of high trophic-level animals receive a high intake of methylmercury (MeHg; the toxic bioaccumulative form of Hg) and organic pollutants which have bioaccumulated and biomagnified within the Arctic food-web (de Wit et al., 2006). In Nunavik Canada, Hg concentrations in the blood of mothers and women of child-bearing age fall above the upper safe limit (5 μg/L) while legacy PCBs and brominated flame retardants are similarly elevated. Tropical riverine communities in the Amazon are also exposed to high levels of MeHg as a result of artisanal gold-extraction and other activities contaminating their fish-based diet. Hair-Hg concentrations exceeding 15 μg/g in these riparian communities indicate a scale of contamination amongst the highest in the world and there is some evidence for neurobehavioural deficit and cardiovascular toxicity (Passos & Mergler, 2008). Clearly, the Anthropocene is an epoch of striking disparity where some communities are exposed to far greater levels of contaminant burden.

Pollution Controls and Legacy Release

Although the Great Acceleration signifies a major increase in the release of many pollutants to the environment starting in the 1950s, national and international policies have led to more recent, but significant, emissions reductions of some chemicals. The Stockholm Convention, a globally ratified agreement on POPs which came into force in 2004, led to the cessation of production of penta- and octaBDEs as well as commercial decaBDE mixtures (Abbasi et al., 2019), while a number of other regulated chemicals are now decreasing globally. Similarly, the Montreal Protocol which came into effect in 1989 resulted in a two-thirds reduction in the emissions of the chlorofluorocarbon refrigerant CFC-12 to the atmosphere by 2000 and some recovery from stratospheric ozone depletion over Antarctica is now being recognised. Other regulatory efforts have resulted in a dramatic decrease in anthropogenic metals emissions. The United States Environment Protection Agency shows that Pb emissions declined by ~98% between 1980 and 2014 and the European Environment Agency reports a similar reduction across Europe (~93% between 1990–2017) primarily due to the phase out of leaded vehicle fuel. The UNECE Heavy Metals Protocol (adopted in 1998) also contributed to emissions reductions of Cd, Pb and Hg from industrial sources, combustion processes and waste incineration, and anthropogenic emissions of Hg and Cd have decreased (~72% and ~ 64% respectively, 1990–2017) across Europe.

However, while these represent significant improvements, emissions of pollutants, like the Anthropocene itself, need to be viewed at the global scale. So, while Hg emissions in Europe and North America have declined, global anthropogenic Hg emissions continue to grow (1.8% per year between 2010 and 2015) mainly due to emissions in Asia and other industrialising regions (Streets et al., 2019). Levels of emerging contaminants of concern such as some perfluorinated acids are also increasing and new chemicals are developed every year for a wide range of applications. Furthermore, for some pollutants there is a significant 'lag' within the environment following their release. The degradation of macro- to microplastics, for example, can take place in locations and environments remote from where they are finally deposited. This can result in anachronistic inputs and complications in interpreting archival records (Bancone et al., 2020). However, the scale and persistence of plastic debris already present in the environment, means that inputs of microplastic contamination will continue for many decades, and possibly centuries, regardless of current or forthcoming regulation.

For other pollutants, the effects of a changing climate may cause their re-emission or release from where they have been stored in terrestrial and aquatic ecosystems following, in some instances, centuries of emissions and deposition. Sediment concentrations of PCBs, DDT and PCDD/Fs have been seen to increase in Lake Oberaar, Switzerland following meltwater inputs from the retreating glacier that feeds it (Bogdal et al., 2010), while climate-enhanced catchment soil erosion has also been shown to elevate transfer of stored trace metals into surface freshwaters (Rose et al., 2012). Arctic permafrost soils and peatlands in circumpolar regions are global reservoirs of atmospherically deposited contaminants, and their thawing may release these back into the environment. Releases of Hg to the atmosphere have been reported from thawing permafrost in Canada and these are predicted to increase across the Arctic through the twenty-first century under all climate scenarios (Schaefer et al., 2020).

References

Abbasi, G., Li, L., & Breivik, K. (2019). Global historical stocks and emissions of PBDE. *Environmental Science & Technology, 53*, 6330–6340.

Ana, G. R. E. E., Sridhar, M. K. C., & Emerole, G. O. (2012). Polycyclic aromatic hydrocarbon burden in ambient air in selected Niger Delta communities in Nigeria. *Journal of the Air & Waste Management Association, 62*, 18–25.

Bancone, C. E. P., Turner, S. D., Ivar do Sul, J. A., & Rose, N. L. (2020). The paleoecology of microplastic contamination. *Frontiers of Environmental Science, 8*, 574008.

Bogdal, C., Nikolic, D., Lüthi, M. P., Schenker, U., Scheringer, M., & Hungerbühler, K. (2010). Release of legacy pollutants from melting glaciers: Model evidence and conceptual understanding. *Environmental Science & Technology, 44*, 4063–4069.

de Wit, C. A., Alaee, M., & Muir, D. C. (2006). Levels and trends of brominated flame retardants in the Arctic. *Chemosphere, 64*, 209–233.

Gałuszka, A., Migaszewski, Z. M., & Rose, N. L. (2020). A consideration of polychlorinated biphenyls as a chemostratigraphic marker of the Anthropocene. *The Anthropocene Review, 7*, 138–158.

Han, Y. M., An, Z. S., & Cao, J. J. (2018). The Anthropocene – A potential stratigraphic definition based on black carbon, char, and soot records. In D. A. DellaSala & M. I. Goldstein (Eds.), *The encyclopedia of the Anthropocene* (Vol. 1, pp. 171–178). Elsevier.

Klimont, Z., Kupiainen, K., Heyes, C., Purohit, P., Cofala, J., Rafaj, P., Borken-Kleefeld, J., & Schöpp, W. (2017). Global anthropogenic emissions of particulate matter including black carbon. *Atmospheric Chemistry and Physics, 17*, 8681–8723.

Muir, D., & Miaz, L. T. (2021). Spatial and temporal trends of perfluoroalkyl substances in global ocean and coastal waters. *Environmental Science & Technology*. doi.org/10.1021/acs.est.0c08035

Park, Y. H., Lee, K., Soltow, Q. A., Strobel, F. H., Brigham, K. L., Parker, R. E., Wilson, M. E., Sutliff, R. L., Mansfield, K. G., Wachtman, L. M., & Ziegler, T. R. (2012). High-performance metabolic profiling of plasma from seven mammalian species for simultaneous environmental chemical surveillance and bioeffect monitoring. *Toxicology, 295*, 47–55.

Passos, C. J., & Mergler, D. (2008). Human mercury exposure and adverse health effects in the Amazon: A review. *Cadernos de Saúde Pública, 24*, 503–520.

Rose, N. L. (2015). Spheroidal carbonaceous fly-ash particles provide a globally synchronous stratigraphic marker for the Anthropocene. *Environmental Science and Technology, 49*, 4155–4162.

Rose, N. L., Yang, H., Turner, S. D., & Simpson, G. L. (2012). An assessment of the mechanisms for the transfer of lead and mercury from atmospherically contaminated organic soils to lake sediments with particular reference to Scotland, UK. *Geochimica et Cosmochimica Acta, 82*, 113–135.

Schaefer, K., Elshorbany, Y., Jafarov, E., Schuster, P. F., Striegl, R. G., Wickland, K. P., & Sunderland, E. M. (2020). Potential impacts of mercury released from thawing permafrost. *Nature Communications, 11*, 1–6.

Schenker, U., Scheringer, M., & Hungerbühler, K. (2008). Investigating the global fate of DDT: Model evaluation and estimation of future trends. *Environmental Science & Technology, 42*, 1178–1184.

Streets, D. G., Horowitz, H. M., Lu, Z., Levin, L., Thackray, C. P., & Sunderland, E. M. (2019). Global and regional trends in mercury emissions and concentrations, 2010–2015. *Atmospheric Environment, 201*, 417–427.

Syvitski, J., Waters, C. N., Day, J., Milliman, J. D., Summerhayes, C., Steffen, W., Zalasiewicz, J., Cearreta, A., Gałuszka, A., Hajdas, I., Head, M. J., Leinfelder, R., McNeill, J. R., Poirier, C., Rose, N. L., Shotyk, W., Wagreich, M., & Williams, M. (2020). Extraordinary human energy consumption and resultant geological impacts beginning around 1950 CE initiated the proposed Anthropocene epoch. *Communications Earth & Environment, 1*, 32.

Vecchiato, M., Gambaro, A., Kehrwald, N. M., Ginot, P., Kutuzov, S., Mikhalenko, V., & Barbante, C. (2020). The great acceleration of fragrances and PAHs archived in an ice core from Elbrus, Caucasus. *Scientific Reports, 10*, 1–10.

Vohra, K., Vodonos, A., Schwartz, J., Marais, E. A., Sulprizio, M. P., & Mickley, L. J. (2021). Global mortality from outdoor fine particle pollution generated by fossil fuel combustion: Results from GEOS-Chem. *Environmental Research, 195*, 110754.

Waters, C. N., Zalasiewicz, J., Summerhayes, C., Fairchild, I. J., Rose, N. L., Loader, N. J., Shotyk, W., Cearreta, A., Head, M. J., Syvitski, J. P. M., Williams, M., Wagreich, M., Barnosky, A. D., Zhisheng, A., Leinfelder, R., Jeandel, C., Gałuszka, A., Ivar do Sul, J. A., Gradstein, F., Steffen, W., McNeill, J. R., Wing, S., Poirier, C., & Edgeworth, M. (2018). Global boundary Stratotype section and point (GSSP) for the Anthropocene series: Where and how to look for potential candidates. *Earth-Science Reviews, 178*, 379–429.

Neil L. Rose is Professor of Environmental Pollution and Palaeolimnology at University College London (UCL) and Visiting Professor at the University of Johannesburg. His research uses natural archives, especially lake sediments, to assess the spatial and temporal distributions of pollutants including fly-ash particles, trace metals, persistent organic compounds and microplastics especially in remote and upland regions. He has authored and co-authored 200 scientific publications and a further 25 book chapters. He has been a member of the Anthropocene Working Group (AWG) since 2018.

Sarah L. Roberts is a Researcher within the Pollution Science Division at UK Centre for Ecology and Hydrology. Her research examines ecotoxicological data and assesses spatial and temporal trends in pollutants. Previously she completed fellowships at University College London and the Canada Centre for Inland Waters laboratories at Environment and Climate Change Canada.

Agnieszka Gałuszka is Professor of Earth and Environmental Sciences at the Jan Kochanowski University. Her research covers a broad range of topics including trace element geochemistry and biogeochemistry, acid mine drainage, geochemical background, geochemical anomalies, geogenic and anthropogenic sources of pollution, chemostratigraphy of the Anthropocene. She is an author and co-author of one hundred scientific publications, three books on environmental geochemistry and nine chapters in edited books. She joined the Anthropocene Working Group as a member in 2013.

Racism

Ruprecht Mattig

Abstract Recent literature on the Anthropocene calls for abandoning the generalized view of humankind and addresses lines of difference within humanity, including racism. This article presents the thesis, advanced in the literature, that the Anthropocene has a fundamental racial character, which, however, remains largely invisible and can best be discerned from a historical perspective. The literature points to the emergence of global capitalism after the arrival of Europeans in the Americas, which has a pattern that has remained stable despite historical changes. It implies the double exploitation of nature and racialized people and is described in terms of dispossession, pollution, migration, degradation, and extinction. The literature also criticizes the concept of the Anthropocene itself as racist because it obscures this pattern and thus reproduces asymmetrical power relations and social inequalities. Finally, the article summarizes suggestions on how to constructively address the problems mentioned.

Anthropocene is usually used to describe humanity's relationship with the earth: It is 'humankind' that is causing Anthropocene changes such as climate change, and it is 'humankind' that is suffering from these changes and will suffer even more in the future. Recently, however, there has been an increasing call in the social sciences and the humanities to shift perspectives and examine the Anthropocene in terms of human differences such as gender, social status, and race. In the following, an overview is given of how race and racism are discussed in the literature on the Anthropocene. The thesis advanced in the literature is that the Anthropocene has a fundamental racial character, but it is mainly invisible. A historical approach is necessary to make it visible.

The terms race and racism are often confusingly broad and applied to various phenomena. Race proposes differences of social groups based on inborn characteristics, the most common of them being skin colour. Race is often inextricably linked

R. Mattig (✉)
TU Dortmund University, Dortmund, Germany
e-mail: ruprecht.mattig@tu-dortmund.de

© The Author(s), under exclusive license to Springer Nature Switzerland AG 2023

N. Wallenhorst, C. Wulf (eds.), *Handbook of the Anthropocene*,
https://doi.org/10.1007/978-3-031-25910-4_249

with other social discriminators such as culture, nationality, gender, social status, or geographic origin, making the concept somewhat fuzzy. Although attempts have failed to define race biologically, from a social science perspective, racism must be viewed as an enduring social fact that permeates thought and action at the explicit and implicit levels (cp. Moody-Adams, 2005). As a social fact, racism is bound to power asymmetries and social inequalities. Different forms of racism can be distinguished, which have emerged historically.

From a historical perspective, the literature refers to the time of the arrival of Europeans in the Americas and the developments that took place subsequently. In particular, it looks at the emergence of the capitalist economic system, which spread throughout the world over time. The European conquest of the Americas resulted in the deaths of an estimated 50 million individuals of indigenous peoples from war, enslavement, famine, and European diseases between 1492 and 1650. According to Lewis and Maslin (2015), this near extinction of indigenous populations can be seen directly in geological data. They refer to records from Antarctic ice cores showing that CO_2 emissions declined significantly between 1570 and 1620, which they attribute to the decline of agriculture and fire use by indigenous peoples and subsequent regeneration of forests, savannas, and grasslands (for a discussion of this study in terms of race see Pulido, 2018: 124f.; Saldanha, 2020: 25–27).

Yusoff (2018: 33–39) highlights the year 1452, when the first Africans were forced to work on plantations in Madeira. After the European's 'discovery' and settlement of the Americas, this practice evolved into the transatlantic slave trade, which involved massive forcible relocation of people to exploit nature and racialized people.

The genocide of the indigenous population, the appropriation of their land, and slavery, ideologically supported by the notion of white supremacy, created the conditions for primitive accumulation, which prepared the ground for capitalism. Capitalism, therefore, had a racist background from the beginning (Pulido, 2018: 125f.). From the perspective of the Anthropocene, it should be noted that the capitalist system also set matter and energy in motion around the globe. Through the exploitation of racialized labour, raw materials were extracted and shipped to the rich countries of the global North. For Saldanha, a crucial point in this development was reached around 1750, when an economy based on the growth imperative emerged in northern England, which enabled "a radically different way of appropriating minerals and labour-power" (2020: 28f.). And Padovan and Alietti (2019: 10) point to the significant acceleration of colonial territorial acquisition based on the ideology of white supremacy by Europeans and U. S. Americans in the late nineteenth century. These historical developments involved the exploitation of land, plants, animals, and racialized people. At the same time, the white countries were able to accumulate capital and secure their economic and political power.

After the Second World War, a new postcolonial world order emerged. Slavery was abolished, but global inequalities remained and perpetuated colonial racism in other forms (Saldanha, 2020: 28). Meanwhile, the traditional banking system was replaced by high finance, which continues to accelerate the extraction and consumption of the earth's resources and, at the same time, perpetuates inequalities and, thereby, racism (Padovan & Alietti, 2019: 15).

Padovan and Alietti point out that although the distinction between whites and non-whites may have been accurate for the time of European colonialism, contemporary global racial capitalism is more differentiated. The former difference between (white) core and (non-white) periphery has been replaced by a model of (white) core, (non-white) semi-periphery (with countries like China, India, Brazil, Indonesia, or Nigeria), and (non-white) periphery. The semi-periphery represents formerly poor and exploited countries that have improved their position and now, in turn, exploit the periphery countries. The semi-periphery appears as exploiter and exploited, while the periphery suffers from exploitation by the core and the semi-periphery. Accordingly, "Global racism is engendered by a myriad of processes that produce racialization of people in different stages and places" (Padovan & Alietti, 2019: 11). Moreover, racism is observed not only between but also within countries (e.g., Laird, 2017).

The historical perspective points to the successive establishment of specific political and economic structures and their ideological underpinnings. In the capital-driven economy, capital is invested to make profits. Since capital tends to be invested where the highest profits can be expected, this economy is subject to a growth imperative (cf. Saldanha, 2020). At the same time, costs must be kept as low as possible. The accumulation strategy of global capitalism is based on "cheap nature": "Nature is 'cheap' in a double sense: to make Nature's elements 'cheap' in price; and also to cheapen, to degrade or to render inferior in an ethico-political sense, the better to make Nature cheap in price" (Moore, 2016: 2–3)—with "nature" including racialized human labour-power. This economic system involves the extraction, processing, and consumption of raw materials such as petroleum, coal, metals, etc., and therefore causes the very changes associated with the Anthropocene: climate change, degradation and pollution of land and sea, etc.

There are dividing lines between the people who can invest capital and make profits and those who mine the raw materials but receive the smallest share (if they receive anything at all); between those who consume most of the products made from raw materials and those who suffer the most from the environmental consequences of consumption. And these dividing lines have a racial character. It is repeatedly pointed out that most of the CO_2 emissions responsible for climate change are caused by a relatively small proportion of the world's population, namely in the wealthy countries of the global North. At the same time, the consequences of climate change mainly affect the global South (e.g., Padovan & Alietti, 2019: 13; Saldanha, 2020: 13; Pulido, 2018: 117ff.). To be sure, all of humanity will be affected by the Anthropocene changes. Yet, the evidence clearly hints that the wealthy and white few will be mainly "inconvenienced", while "it is overwhelmingly places occupied primarily by non-white peoples that will pay the highest price for global warming: death" (Pulido, 2018: 118). In combination with other social and political problems, such consequences of the Anthropocene lead to massive migration movements: "Millions leave their country, and they are rarely white" (Padovan & Alietti, 2019: 14).

The political borders, however, are difficult to cross. Wealthy countries are eager to let no one in, like the "Fortress Europe and its racialized policies of 'letting die' in the Mediterranean" (Yusoff, 2017: 261). As a result, racialized communities and peoples are doomed to live on depleted soils, suffer from pollution, dry up in droughts, or drown in floods.

Anthropocene theorists reject the progress-oriented view (often called 'liberal' or 'neoliberal') that this global capitalist system, even if it widens the gap between rich and poor, will ultimately lead to better living conditions for all, including the poor. Instead, the Anthropocenic world order is seen as intrinsically racist and violent. For Yusoff (2017: 259), for example, neoliberal discourse omits "what properly constitutes Anthropocene politics—capitalism, inequality, racism, colonialism, and uneven energy geographies".

More than that, it is argued that liberal policies and worldviews rather mask the racism upon which they are based. Pulido holds that evasion and indifference must be considered a contemporary form of racism. The decision-making politicians fail to assist the developing countries most affected by the consequences of the Anthropocene: "Indifference is a form of racism, because not only does it serve to reproduce racial inequality, but also this inequality enables the well-being of those destined to live" (Pulido, 2018: 121). Therefore, the inherent racism of global capitalism is obscured: "It is understandably difficult to accept the fact that we value lives differently—as this is contrary to contemporary human rights values, as well as the idea of racial progress. As a result, one of the primary responses to this contradiction is evasion, as we consistently seek to avoid addressing race" (ibid.: 122).

Along these lines, it is often argued that even the concept of the Anthropocene, with its focus on a generalized 'humankind', obscures inherent Anthropocene violence and thus must be seen as racist (e.g., Yusoff, 2018; Hoelle & Kawa, 2021).

In the Anthropocene literature, different approaches of suggestions to overcoming "the deeply immoral nature of the Anthropocene" (Pulido, 2018: 120) can be found. From an epistemological or ideological perspective, the racist character of the Anthropocene has been addressed. A critical reflection of the notion of "anthropos" is demanded (e.g., Wulf, 2020). Cultural anthropology, which examines practices and ideologies other than Western racist, capitalist, and sexist ones, can make an important contribution in this context (e.g., Hoelle & Kawa, 2021). Perhaps it is also necessary to find another name for the Anthropocene—'Chthulucene' is an example that is deliberately meant to irritate (Haraway, 2016).

From an educational perspective it has been proposed to induce change via a new education for the coming generations. Wulf (2021) considers 'global citizenship education' as an inclusive approach comprising education for sustainable development, heritage education, human rights education, and peace education. According to Laird (2017), education needs to focus on 'Learning to live in the Anthropocene', which includes, among others, acquiring new habits, skills, and ways of thinking as well as ecological literacy. Jones (2019), again, proposes critical and embodied food pedagogy for racialized youth to cultivate resilience.

From the perspective of politics and economics, finally, it has been suggested to consider forms of socio-ecological redistribution and re-appropriation, which would compensate for both past wrongs and future damages to racialized people (cf. Folkers, 2020: 600). Also, the abolition of the described relations of power and exploitation is demanded (cf. ibid.: 599). Saldanha (2020: 27) considers it essential to "shun the thirst for profit".

In summary, a new field of research on racism and the Anthropocene is emerging. Overall, the accounts appear relatively abstract; from an anthropological viewpoint, empirical case studies are needed, in order to develop the claims further. Although there are stimulating suggestions for dealing with the problems presented, ultimately there are no suggestions in the literature as to how the powerful mechanisms of global racial capitalism might be overridden.

References

Folkers, A. (2020). Was ist das Anthropozän und was wird es gewesen sein? Ein kritischer Überblick über neue Literatur zum kontemporären Erdzeitalter (review essay). *Zeitschrift für Geschichte der Wissenschaften, Technik und Medizin, 28*, 589–604.

Haraway, D. (2016). Staying with the Trouble. Anthropocene, Capitalocene, Chthulucene. In J. Moore (Ed.), *Anthropocene or Capitalocene? Nature, history, and the crisis of capitalism* (pp. 34–76). Kairos books.

Hoelle, J., & Kawa, N. C. (2021). Placing the Anthropos in Anthropocene. *Annals of the American Association of Geographers, 111*(3), 655–662. https://doi.org/10.1080/24694452.2020.1842171

Jones, N. (2019). 'It tastes like heaven': Critical and embodied food pedagogy with black youth in the Anthropocene. *Policy Futures in Education, 17*(7), 905–923. https://doi.org/10.1177/1478210318810614

Laird, S. (2017). Learning to live in the Anthropocene: Our children and ourselves. *Studies in Philosophy and Education, 36*, 265–282. https://doi.org/10.1007/s11217-017-9571-6

Lewis, S., & Maslin, M. A. (2015). Defining the Anthropocene. *Nature, 519*, 171–180. https://doi.org/10.1038/nature14258

Moody-Adams, M. (2005). Racism. In R. G. Frey & C. H. Wellman (Eds.), *A companion to applied ethics, Blackwell Companions to Philosophy* (pp. 89–101). Blackwell Publishing.

Moore, J. W. (2016). *Anthropocene or Capitalocene? Nature, history, and the crisis of capitalism*. Kairos books.

Padovan, D., & Alietti, A. (2019). Geo-capitalism and global racialization in the frame of Anthropocene. *International Review of Sociology, 29*(2), 172–196. https://doi.org/10.1080/03906701.2019.1641263

Pulido, L. (2018). Racism and the Anthropocene. In G. Mitman, M. Armiero, & R. Emmett (Eds.), *Future remains: A cabinet of Curiosities for the Anthropocene* (pp. 116–128). University of Chicago Press. https://doi.org/10.7208/9780226508825-014

Saldanha, A. (2020). A date with Destiny: Racial capitalism and the beginnings of the Anthropocene. *Environment and Planning D: Society and Space, 38*(1), 12–34. https://doi.org/10.1177/0263775819871964

Wulf, C. (2020). Den Menschen neu denken im Anthropozän. Bestandsaufnahme und Perspektiven. *Paragrana, 29*(1), 13–35.

Wulf, C. (2021). Global citizenship education. Bildung zu einer planetarischen Weltgemeinschaft im Anthropozän. *Vierteljahresschrift für wissenschaftliche Pädagogik, 97*, 464–480.

Yusoff, K. (2017). Politics of the Anthropocene: Formation of the commons as a geologic process. *Antipode, 50*(1), 255–276.

Yusoff, K. (2018). *A billion black Anthropocenes or none*. University of Minnesota Press.

Ruprecht Mattig is a professor of systematic educational studies and methodology of educational research at TU Dortmund University. His research and academic teaching comprises of educational anthropology, qualitative educational research, research on rituals and gestures, language and education, Wilhelm von Humboldt studies, global citizenship education, and youth research. His publications include *Wilhelm von Humboldt als Ethnograph. Bildungsforschung im Zeitalter der Aufklärung* (2019) and *Rock und Pop als Ritual. Über das Erwachsenwerden in der Mediengesellschaft* (2009).

Slavery

Christoph Antweiler

Abstract Along with wet rice cultivation, commodity extraction and plantation economies have historically been the socioeconomic formations that permanently transformed landscapes. Plantation economies and extractive economies were based in part on slave labour. These economies profoundly transformed local and, in some cases, regional landscapes and their living environments. In contrast, the causal contribution of slavery to global change is difficult to determine empirically. In the context of Anthropocene scholarship, the emphasis on slavery contributes to a "provincialization" of the Anthropocene concept.

Slavery is an extreme form of unfree labour combined with social dependency. Within the Anthropocene literature, slavery is important first in terms of the chronology of the Anthropocene and second for a critical reflection on Western dominance in scholarship. When it comes to questions of the historical beginning of the Anthropocene, the debate is whether Atlantic slavery emblematically, but also quite specifically, marks the beginning of the very economic mode that produced early Anthropocene effects on the biosphere.

Given the almost global colonial outreach of colonial economies and similar dependent economies, it is debated whether the Anthropocene did not begin in the twentieth century or with the advent of industry and social modernity in the late eighteenth century, but already with the outreach of colonialism in the early modern period, at the latest from the early seventeenth century. Central to this was the intensive exchange of people and biota across the Atlantic ("Columbian Exchange"). Climatologically, this corresponds with a brief low point in atmospheric CO_2 levels due to reforestation caused by colonial-induced population collapse, the so-called "orbis spike" in 1610 (Lewis & Maslin, 2018:147–148). However, the quantitative

C. Antweiler (✉)
Institute of Oriental and Asian Studies, University of Bonn, Bonn, Germany
e-mail: christoph.antweiler@uni-bonn.de

© The Author(s), under exclusive license to Springer Nature Switzerland AG 2023

N. Wallenhorst, C. Wulf (eds.), *Handbook of the Anthropocene*,
https://doi.org/10.1007/978-3-031-25910-4_250

contribution of slavery is hard to distil from the variety of colonial and other socio-economic and natural effects, such as epidemics and depopulation and resulting abandonment of agricultural land (Koch et al., 2019).

However, Atlantic slavery and other forms of extreme dependence are thus at least potentially significant for historically informed alternatives to the currently dominant assumption on the part of Earth system sciences of the onset of the Anthropocene in the twentieth century. The bulk of empirical studies of slavery focus on transatlantic forms of slavery and, second, on forms of slavery in ancient Mediterranean societies, particularly in the Greco-Roman world. The history of Atlantic slavery is closely intertwined with the emergence of the modern West as an economic formation, leading to Anthropocene relations and effects. The concept makes clear that humans had to pay for the development of the Anthropocene even before the environmental catastrophes of the present. Such a periodization thus fits with the thesis of authors who seek to decolonize the concept of the Anthropocene (Todd, 2015; Davis & Todd, 2017; Whyte, 2017; Simpson, 2020; Driscoll, 2020). Placing the beginning of the Anthropocene with the colonization of the two Americas emphasizes the forcible dispossession and marginalization of cultures (up to and including genocides) and their anthropogenic impacts on ecosystems (Joralemon, 1982). This also has implications for anthropology, as this approach would mean doing more work on the cumulative effects of capitalist economic activity, making much more use of indigenous voices and sources on this than has been done in the past (e.g., Whyte, 2017).

Humans make natural habitats and their ecosystems (biomes) significantly shaped by humans, so-called anthromes. A specific link may exist between slave labour and the Anthropocene in the form of tropical plantation agriculture. As a form of industrial agriculture, tropical plantations combine drastic and permanent environmental change with certain forms of labour, such as forced labour and slavery, to produce world market products. Here, human economic activity permanently reshapes landscapes. This means geomorphological changes, but also a strong change in the living conditions of organisms.

Plantations as a tropical export-oriented form of operation reshape diverse local ecosystems into monocultures. Furthermore, plantations are externally determined and supply external spaces. Slave labour formed a labour base of plantation economies. In fundamental terms, slavery can be considered emblematic of an extractive form of economy and imperial way of life that brought into the world a special social relationship with nature in which natural resources are extracted as much as human bodies are exploited.

The development of capitalism was closely linked to colonialism, which led to area-wide changes in socioecological processes. The colonial economy, especially in the Americas, was largely based on plantation agriculture. Research in environmental history and colonial history emphasizes the global standardization of the landscape, which was triggered by geographically distant powers and organized by means of monocultures and slave labour in the plantation operating system. The plantation system is based on the long-distance transportation of plants, animals, and people, implies an abstract relationship between owner and investment area. In plantations, plants become investment resources and the system involves an alienation of people from the local ecosystem. At the heart of this form of enterprise is the exploitation of workers and slaves in monocultural export-oriented agriculture, which, after all, still exists today.

Insights into colonial early forms or precursors of the Anthropocene have prompted the concept of the Plantationocene as an alternative to the Anthropocene. The Anthropocene concept is thereby seen as monolithic and Eurocentric. Against this background, and in exchange with the eco-Marxist critique, Donna Haraway coined an alternative term to the Anthropocene similar in content to the Capitalocene idea: "Plantationocene" (Haraway, 2016: 99–103). From a Plantationocene perspective, the standardization of nature through rationalized agricultural production, which is equally an exploitation of the environment and human labour including racism, constitutes the plantation. As an early example of causes of Anthropocene change, plantations historically precede capitalism and thus place the beginning of this era earlier than industrial capitalism. Moreover, the term appears more descriptive and less ideologically or normatively loaded than that of the "Capitalocene" (Tsing, in Haraway et al., 2016: 556–557).

As a form of industrial agriculture, tropical plantations combine drastic and permanent environmental change with certain forms of labour, such as forced labour and slavery, to produce world market products. According to Tsing, the logic of the plantation is different from that of the commons (Tsing, in Haraway et al., 2016). It centres on displacements of living matter, that is, displacements of plants, animals, and microbes as well as humans. Plantations combine regimes of human existence with the destination of non-human life ("regimentation"). Transfers often over long distances formed the basis of these extractive economies (Lewis & Maslin, 2018).

Recent studies of sugarcane plantations emphasize colonial relations between humans and nonhuman beings (Moore et al., 2021), and ethnohistorical studies have documented translocal, regional, and even global effects of these new relations of nature as early as the mid-1980s using the sugar economy as a case study (Mintz, 1985).

The subsistence of workers in the plantation period deserves more interest in Anthropocene scholarship. Within the worldwide expansion of plantation monocultures today we find subaltern smallholder farming systems and bio-cultural refugia. Some have a historical continuity to slavery times. Within the In New World tropical landscapes today some Afrodescendant smallholder farming systems that prioritized agrobiodiversity and agroecological practices survived. These "slave gardens" as subaltern food systems emerged during the plantation era, when slaves leveraged subsistence precarity for the right to independent production and a limited autonomy over their labour (Carney, 2021: 1075, 1090–1094). Another example of studies inspired by the Plantationocene concept is soil archaeological investigations and combined art actions to "remember the soil." Soil finds on a Maryland plantation reveal the significant role of slavery in plantation economies. Such information could not be found in the historical archives accessible to date. This also has significance for historical memory, as archaeologically documented traces of slavery can serve as a memory anchor that illuminates aspects that remain hidden on postcards, for example (Martens & Robertson, 2021).

Drawing on approaches to coloniality (Bendik-Keymer, 2020; Baldwin & Erickson, 2021), slavery represents only part of the colonial changes in the pre-industrial Anthropocene. Plantations form only one aspect of comprehensive transformations that go beyond landscape change. On the surface, it is only about capitalist-oriented labour relations and exploitation. In fact, however, not only material goods and economy changed, but power-relevant categories emerged in the wake of the transformation of colonized landscapes. Examples relevant to anthropocene environmental change include cartography, surveying, descriptions, and externally dominated forms of travel in the tropics ("worlding," Spivak, 1985). This is also evident in recent empirical studies of slavery (e.g., Hauser, 2021).

Critics here also place geology as the leading science of the Anthropocene within the framework of slavery and extraction. In eloquent writings, geographer and professor of inhumanity (!) Kathryn Yusoff notes that the "grammar of geology" was fundamental to the establishment of extractive economies. She emphasizes that this significantly shaped both the personal physical and psychological lives of people as well as geological circumstances (Yusoff, 2018). This is partly true, but there is a considerable narrowing of geology to a colonialist extractive enterprise associated with this analysis. This is not covered in the history of science, given the multifaceted role of geology in thematizing the Anthropocene (Davies, 2016).

In conclusion, slavery is an element of such human economies that produce an anthropogenic nature. Just as commodity exploitation is an extraction of natural resources, slavery is an extraction of labour and life energy. Commodity mines and plantation systems brought both forms of extraction together in the form of an operating system that generated local asymmetrical relationships. In the world framework, slavery was part of an emergent environmentally transformative world system that tied the Global South to the Global North in asymmetrical ways. Nevertheless, it should not be forgotten that the causal contribution of slavery to Anthropocene environmental change is difficult to assess in a geological sense. Given the nevertheless limited spatial scale of plantations, the contribution of slavery to global environmental change, and isochronous to it, is certainly more limited than, say, wet rice cultivation (Fuller et al., 2011).

Forms of agricultural slavery also existed outside of plantations. Given the significant environmental role of forms of inequality similar to slavery to this day (Antweiler, 2022), the role of unfree labour and strong forms of dependency is a desideratum of Anthropocene research. The study of slavery as a factor in Anthropocene economies forms an important aspect of research on histories of extraction and general efforts to "decolonize" and "indigenize" the Anthropocene.

References

Antweiler, C. (2022). *On dependence, dependency, and a dependency turn. An essay with systematic intent. Bonn: Excellence cluster beyond slavery and freedom (concept paper 2).* https://www.dependency.uni-bonn.de/en/publications/bcdsss-publishing-series/bcdssconcept-papers. Accessed 19 Mar 2022.

Baldwin, A., & Erickson, B. (2021). Introduction: Whiteness, coloniality, and the Anthropocene. *Environment and Planning D: Society and Space, 38*(1), 3–11.

Bendik-Keymer, J. D. (2020). *Involving anthroponomy in the Anthropocene: On decoloniality*. Routledge (Routledge Research in the Anthropocene).

Carney, J. A. (2021). Subsistence in the Plantationocene: Dooryard gardens, agrobiodiversity, and the subaltern economies of slavery. *The Journal of Peasant Studies, 48*(5), 1075–1099.

Davies, J. (2016). *The birth of the Anthropocene*. The University of California Press.

Davis, H., & Todd, Z. (2017). On the importance of a date, or decolonizing the Anthropocene. *ACME: An International E-Journal for Critical Geographies, 16*(4), 761–780.

Driscoll, M. W. (2020). *The whites are enemies of heaven. Climate Caucasianism and Asian ecological protection*. Duke University Press.

Fuller, D. Q., van Etten, J., Manning, K., Castillo, C. K., Kingwell-Banham, E., Weisskopf, A., Qin, L., Sato, Y.-I., & Hijmans, R. J. (2011). The contribution of Rice agriculture and livestock pastoralism to prehistoric methane levels: An archaeological assessment. *The Holocene, 21*, 743–759.

Haraway, D. J. (2016). Staying with the trouble. In *Making kin in the Chthulucene*. Duke University Press.

Haraway, D., Ishikawa, N., Gilbert, S. F., Olwig, K., Tsing, A. L., & Bubandt, N. O. (2016). Anthropologists are talking – About the Anthropocene. *Ethnos. Journal of Anthropology, 81*(3), 535–564.

Hauser, M. W. (2021). *Mapping Water in Dominica. Enslavement and Environment under Colonialism*. University of Washington Press. (Culture, Place, and Nature: Studies in Anthropology and Environment).

Joralemon, D. (1982). New world depopulation and the case of disease. *Journal of Anthropological Research, 38*, 108e127.

Koch, A., Brierley, C., Maslin, M. A., & Lewis, S. L. (2019). Earth system impacts of the European arrival and great dying in the Americas after 1492. *Quarternary Science Reviews, 207*, 13–36.

Lewis, S. L., & Maslin, M. A. (2018). *The human planet. How we created the Anthropocene*. Yale University Press (also London: Pelican Books).

Martens, R. L., & Robertson, B. (2021). *How the soil remembers plantation slavery.* https://edgeeffects.net/soil-memory-plantationocene/

Mintz, S. (1985). *The sweetness of power. The place of sugar in modern history*. Penguin Books.

Moore, S. S., Alewaert, M., Gómez, P. F., & Mittman, G. (2021). *Plantation legacies*. https://edgeeffects.net/plantation-legacies-plantationocene/

Simpson, M. (2020). The Anthropocene as colonial discourse. *Environment & Planning D: Society and Space, 38*(1), 53–71.

Spivak, G. C. (1985). The Rani of Sirmur: An essay in reading the archives. *History and Theory, 24*(3), 247–272.

Todd, Z. (2015). Indigenizing the Anthropocene. In H. Davis & E. Turpin (Eds.), *Art in the Anthropocene. Encounters mong aesthetics, politics, environments and epistemologies* (Critical climate change) (pp. 241–254). Open Humanities Press.

Whyte, K. P. (2017). *Indigenous climate change studies: Indigenizing futures, decolonizing the Anthropocene*. https://kylewhyte.marcom.cal.msu.edu/wp-content/uploads/sites/12/2018/07/IndigenousClimateChangeStudies.pdf

Yusoff, K. (2018). *A billion black Anthropocenes or none*. University of Minnesota Press (Forerunners: Ideas First).

Christoph Antweiler is an anthropologist with a background in geology and paleontology. His main interests are socio-cultural evolution, pan-cultural patterns (human universals) and vernacular cosmopolitanism. His main research region is Southeast Asia. Among related book publications are *Inclusive Humanism. Anthropological Basics for a Realistic Cosmopolitanism* (Taipei: National Taiwan University Press, 2012), *Our Common Denominator. Human Universals Revisited* (New York and Oxford: Berghahn, 2018) and a new comprehensive treatise *Anthropologie im Anthropozän* (Darmstadt: WBG, 2022).

Solastalgia

Tracey Skillington

Abstract We do not simply know the natural world in which we live. We also feel it and engage with it intimately through sight, touch, sound and smell. We may note the significance of our sense relations with this world to understanding how climate-related loss (loss of landscape, species, stable weather patterns, traditional ways of life, etc.) elicits sadness, grief and symptoms of 'solastalgia'. This chapter assesses research documenting how the Anthropocene provokes a wounding of the mind, as much of the body of the living subject and considers the significance of these conditions in triggering moments of resonance (Rosa (2019) Resonance: a sociology of our relationship to the world. Polity, Cambridge) when the emotional energies opened up by the adverse conditions of an overheating planet provoke a critical 'thinking beyond' (Adorno (2001) Metaphysics: concepts and problems (trans: Jephcott E F N). Stanford University Press, Stanford.) damaged life and a desire to create better possibilities.

We may concur with the pragmatist John Dewey that all rational understandings of the challenges we face are bound up with 'holistic forms' of experiencing the world around us. Ongoing deteriorations in ecological conditions, for instance, are qualitatively disclosed to us from a perspective of engaged involvement. We do not simply know the world in which we live. We also feel and engage with it intimately through touch, sight, sound and smell. We may note the significance of these engagements in creating an attitude of openness to an endangered natural order (Skillington, 2019). As environmental conditions continue to decline, they evoke concern, even grief in some instances. 'Ecological grief' (Cunsolo et al., 2018, p. 275) is expressed as a combination of sadness, anger, despair, fear and hopelessness, arising in response to real or anticipated losses of landscape, species, traditional way of life, etc. No longer disputed today is the capacity of such destruction to provoke a wounding of the mind, as much as the body of the Anthropocene

T. Skillington (✉)
University College Cork, Cork, Ireland
e-mail: t.skillington@ucc.ie

© The Author(s), under exclusive license to Springer Nature Switzerland AG 2023

N. Wallenhorst, C. Wulf (eds.), *Handbook of the Anthropocene*,
https://doi.org/10.1007/978-3-031-25910-4_251

subject. Unlike other forms of grief, ecological grief is directed at the very fabric of our relationship with a dying natural order. It is a grief that is difficult to appease without some major transformation in our relationship with Anthropocene worlds. Addressing ecological grief means addressing the type of problems that arise in our relational ties with damaged nature and the slow violence of deep climate change. For Cunsolo et al. (2018), ecological grief is a 'disenfranchised grief' in the sense that is one that has not yet been accorded sufficient recognition within policy discourse, in spite of the fact that it affects growing numbers. Craps describes ecological grief as equivalent to the feelings felt when a good friend passes away (Craps, 2020, p. 1–7). However, for Menning (2017, p. 39–40), ecological grief differs in important ways from grief inspired by human loss in that we are nearly always complicit in ecological loss, 'if only by virtue of living in the Anthropocene… … Guilt entwines with sorrow, complicating the grieving process'. What we grieve, therefore, is not only what we have lost in ecological terms but, also, that we have knowingly destroyed (Skillington, 2021). Guilt sentiments thereby intensify ecological grief and add further to sentiments of alienation.

While coverage of the physical repercussions of deteriorating climate conditions on communities has been extensive, it is only in more recent times that the emotional impacts of these changes have received greater attention. The research of Ryan (2017), for instance, offers insights on how the loss of a plant species from the landscape of South-west Australia (i.e., boronia, a traditional plant of the Perth region) and its accompanying perfume were interpreted by local residents as a loss of long-established ecological intimacy with the flora of their homeland. Ryan recounts how locals expressed real sorrow over the disappearance of this plant from the landscape of Perth and the consequent impoverishment of their sensory worlds. Ryan notes a complex ecology of grief and mourning at work here, with mourning centring on the physical loss of the plant species and an associated milieu for whom memories of its blooming season, perfume, as well as the cultural practices that evolved around its presence were entangled. As Ryan and, indeed, Albrecht (2005, p. 46) note, some of the most poignant moments of ecological grief occur when individuals experience the loss of familiar components of their environment. Albrecht introduces the term 'solastalgia' to describe the melancholia associated with such loss, noting how it can give rise to feelings of 'intense desolation' or grief (Albrecht, 2005, p.44). Witnessing the clearing of a local forest area or a greenfield site, for instance, can provoke a degree of 'mental anguish' (Albrecht, 2005, p. 46) for the subject.

In this moment, the individual is said to be deprived of the 'solace' normally derived from one's relationship to 'home'. 'Solastalgia' combines the concept of 'solace' (with meanings linked to alleviation of distress or to the provision of comfort) with that of 'desolation' (abandonment and loneliness) and 'algia', which translates as pain or suffering, 'Solastalgia' thus conveys the pain provoked by a loss of solace and a sense of isolation created by some disturbance in one's relationship to home (Albrecht, 2005: 45–6). Albrecht further proposes the concept of 'environment' be replaced with that of 'symbiocene' to better convey the realization that all planetary bodies and ecological fates, human and non-human, are intertwined. 'Home' in this instance is Earth itself rather than some specific segment of it. Whether or not such a change in perspective occurs, Albrecht is correct in his assumption that one of the biggest challenges we face today moving forward is our ability to navigate the turbulent emotional terrain of largescale environmental breakdown. Already, there is a growing body of empirical research that supports of this view. Ellis and Albrecht (2017), for instance, examine the effects of prolonged drought on farming communities in the Australian wheatbelt and parts of India and show how exposure to chronic weather conditions elicits experiences of ecological grief, depression and even suicide amongst some farmers. Unable to maintain healthy farming conditions due to worsening seasonal dryness and wind erosion, farmers report feelings of shame, especially when animal stock start to die of malnutrition and thirst, or when dams run empty. Being a 'good farmer' is seen as dependent on one's ability to cope with the challenges extreme weather conditions bring (Ellis & Albrecht, 2017). In their research on subarctic Inuit communities, Cunsolo et al. (2014) note important differences in emotional responses to fluctuating weather conditions, with female respondents reporting higher levels of sadness and distress when encountering changes in environmental conditions compared to their male counterparts who were more likely to admit to substance abuse, especially those in climate sensitive employment (e.g., farming, fishing). This research offers many important insights into the local implications of fluctuating weather conditions and their effects on mental wellbeing (see, also, Berry et al., 2018).

We should not assume, however, that the type of grief and losses recounted here are exclusive to humans. In *On the Origin of Species* (1859) and *The Expression of the Emotions in Man and Animals* (1872), Darwin defends the idea of a continuity between humans and other animals in their emotional and cognitive lives. Darwin observes important differences between species but such differences, he adds, are more a matter of degree (Bekoff, 2000, p. 864). Comparative evolutionary and interdisciplinary research reveals the complex nature of grief, joy and sadness as strongly animal emotions. The first event to emerge into consciousness for all is the ability to experience the sensation of pleasure and displeasure. In her study of the behaviour of chimpanzees, Jane Goodall (1990) recounts the grieving behaviour of Flint, an eight-and-a-half-year old chimp who withdrew from his group, stopped feeding and eventually died after his mother, Flo, passed away. Similarly, Poole's (1998) research depicts the mourning rituals of elephants who stand guard over a still born baby for days with their head and ears hanging low. If animals can display grief in response to the loss of a significant other in ways similar to humans, they are also capable of experiencing grief when familiar features of their natural surroundings are destroyed by wildfires or storm surges, for instance. Following devastating bushfires in Australia in January 2020, the image of a grieving koala bear sitting near the dead body of his companion in charred surroundings was published by global media press. The photo in question was taken by volunteers at the Humane Society International (HIS) searching for surviving animals after wildfires burned through 40,000 square miles of bush lands, rainforests, and national parks throughout Australia, killing an estimated 1 billion animals (Ng, 2020). The image of the suffering koala bear touched the hearts of millions and drew attention to the multiple losses climate change imposes upon non-human, as much as human communities. The debates which followed highlighted how experiences of ecological grief and solastalgia, whilst painful for the subjects involved, can also trigger moments of resonance (Rosa, 2019) when raw emotional energies opened up by the very real conditions of an overheating planet provoke a critical 'thinking beyond' (Adorno, 2001) damaged life and a desire to create better possibilities.

References

Adorno, T. (2001). *Metaphysics: Concepts and problems* (E.F.N. Jephcott. Trans.). Stanford University Press.

Albrecht, G. (2005). Solastalgia: A new concept in health and identity. *PAN: Philosophy, Activism, Nature, 3*, 45–46.

Albrecht, G. (2007). Solastalgia: The distress caused by environmental. *Change' Australasian Psychiatry, 15*(September), S95–S98. https://doi.org/10.1080/10398560701701288

Bekoff, M. (2000). Animal emotions: Exploring passionate natures. *Bioscience, 50*(10), 861–870.

Berry, H. L., Waite, T. D., Dear, K. B. G., et al. (2018). The case for systems thinking about climate change and mental health. *Nature Climate Change, 8*(4), 282–290. https://doi.org/10.1038/s41558-018-0102-4

Craps, S. (2020). Introduction: Ecological grief. *American Imago, 77*(1), 1–7.

Cunsolo, A. W., Stephenson, E., Allen, J., Bourque, F., Drossos, A., Elgarøy, S., Kral, M. J., Mauro, I., Moses, J., Pearce, T., Petrasek MacDonald, J., & Wexler, L. (2014). Examining relationships between climate change and mental health in the circumpolar north. *Family Medicine and Biobehavioral Health, 15*, 169–182. https://doi.org/10.1007/s10113-014-0630-z

Cunsolo, A., Ellis, N., & R. (2018). Ecological grief as a mental health response to climate change - related loss. *Nature Climate Change, 8*, 275–281.

Darwin, C. (1859). *On the origin of species by means of natural selection*. Murray.

Darwin, C. (1998[1872]). *The expression of the emotions in man and animal.* 3rd ed. Reprint. Oxford University Press.

Ellis, N. R., & Albrecht, G. A. (2017). Climate change threats to family farmers' sense of place and mental wellbeing: A case study from the Western Australian Wheatbelt. *Social Science & Medicine, 175*, 161–168. https://doi.org/10.1016/j.socscimed.2017.01.009

Goodall, J. (1990). *Through a window*. Houghton-Mifflin.

Menning, N. (2017). Environmental mourning and the religious imagination. In A. Cunsolo & K. Landman (Eds.), *Mourning nature: Hope at the heart of ecological loss* (pp. 39–63). McGill-Queens University Press.

Ng, K. (2020, January 18). Australia wildfires: koala pictured 'grieving' over dead friend. *Independent.* https://www.independent.co.uk/news/world/australasia/australia-wildfires-koala-picture-grieving-friend-a9289581.html. Accessed 21 June 2021.

Poole, J. H. (1998). An exploration of a commonality between ourselves and elephants. *Etica & Animali, 9/98*, 85–110.

Rosa, H. (2019). *Resonance: A sociology of our relationship to the world*. Polity.

Ryan, J. C. (2017). Where have all the Boronia Gone? A posthumanist model of environmental mourning. In A. Cunsolo & K. Landman (Eds.), *Mourning Nature: Hope at the Heart of Ecological Loss & Grief* (pp. 117–143). McGill-Queen's University Press.

Skillington, T. (2019). *Climate change and intergenerational justice*. Routledge.

Skillington, T. (2021). Natural resource inequities, domination and the rise of youth communicative power. *Distinktion, 22*(1), 23–43.

Tracey Skillington is Director of the BA (Sociology), Department of Sociology, University College Cork. Recent monographs include *Climate Justice & Human Rights* (Palgrave), *Climate Change & Intergenerational Justice* (Routledge) and forthcoming, *A Critical Theory of Climate Trauma* (Routledge). Her publications have appeared in many journals, including the *European Journal of Social Theory*, the *British Journal of Sociology*, *The International Journal of Human Rights, Distinktion: Journal of Social Theory*, *Sociology*, and *Sustainable Development*.

Tragedy

Asmus Trautsch

Abstract The Anthropocene and its core dynamics, particularly climate change, are increasingly described as tragic. This article examines the conceptual framework for interpreting the Anthropocene or its processes as a planetary tragedy. It compares the structural elements and philosophical discourse of Greek tragic drama to the Anthropocene condition and argues that in the face of catastrophic developments the ancient form of theatre does indeed provide a lens through which the connection between agency and suffering, progress and deterioration, power and helplessness can be better understood. It argues that learning from suffering will be vital for a new human self-understanding in the Anthropocene.

The Anthropocene, and in particular climate change, have been called tragedies in recent years (Gardiner, 2011, Buck, 2019, Toly, 2019, Malkmus, 2020, Wallace, 2020). Obviously, the Anthropocene is not a festive theatre performance that presents a mostly mythic plot situated in a specific location with a few protagonists and a chorus for a regional audience. Rather, it is a non-experienceable historical-physical process without projected end that intertwines geological and historical time and changes the entire planet. It abrogates the architecturally manifested separation of actors and spectators and makes the stage itself a part of the play as the technosphere is constantly transforming the earth system which is increasingly retroacting on it. Nonetheless, it makes sense to think of the Anthropocene as an "*existential* tragedy" (Malkmus, 2020, p. 101) as a comparison of six key features of Greek tragedy and the Anthropocene may illuminate.

(i) Irreversible catastrophic turns: Ancient tragic dramas and many of their successors since the Renaissance are characterized by a moment of a radical turn in the course of action, which Aristotle calls "reversal", *peripeteia* (*Poetics*, 1452a). The reversal as key feature of tragedies with a complex plot provokes an unexpected sudden change, *metabolē* or *metabasis* (*Poetics*, 1452a), from,

A. Trautsch (✉)
University of Greifswald, Greifswald, Germany
e-mail: astrautsch@gmail.com

© The Author(s), under exclusive license to Springer Nature Switzerland AG 2023

N. Wallenhorst, C. Wulf (eds.), *Handbook of the Anthropocene*,
https://doi.org/10.1007/978-3-031-25910-4_252

generally, fortune into misfortune. With the reversal, this turn happens to the agents not despite but precisely because of their actions, although they did not aim for it (like Phaedra) or deliberately intended to avoid it (like Oedipus). The consequences irreversibly destroy the conditions of their and their kins' good lives and cause existential suffering, despair and death. This non-linear momentum of shift to an irreversible catastrophe has played a key role as "tragic dialectic" (Szondi, 2002) in the modern philosophy of tragedy from Schelling and Hegel to Critical Theory. The Anthropocene can be understood as such a tragic dialectic on a planetary scale. It is characterized by the fact that the accumulation of actions, especially those of powerful agents in the industrialized nations, have been changing the earth system over decades so radically that the conditions for healthy, peaceful and prosperous life on earth are in a process of deterioration, while abrupt, non-linear and irreversible changes in the earth system – and, according to some, in technological developments of AI, too – are becoming more likely. Progress-oriented modernity in both West and East, with its prospect of sustained power and continuous improvement of living conditions, has tragically produced a turn (*strophē*) downwards (*kata*) which puts the ecological conditions of its reproduction and expansion at existential risk. As in tragedy, this hasn't merely happened due to contingent external accidents but precisely because of the striving for amelioration and growth by means of fossil fuel based economies with an accelerating and increasing requirement of natural resources. This is expressed in the widely shared – but not undisputed – diagnosis of climate change as a tragedy of the global commons (McLachlan, 2019). By repeatedly using and exploiting unmanaged commons without allocated costs, agents – corporations, nations, individuals – have collectively contributed to a dramatic degradation of local, regional and planetary goods that make their future use as resources or dumping grounds for emissions highly problematic. It is this kind of human agency on various scales within modern, particularly capitalist civilization that has irreversibly affected its ecological conditions.

(ii) Complex causality: In most tragedies, several causes of different kinds interfere to induce sudden turns to maximum disaster. The protagonists' responsibility lies in their mistake (*hamartia*), which appears as a necessary but hardly sufficient condition for the change into irreversible suffering. In addition to the individual actions and the language charged with tragic irony, contingent factors contribute to tragic causation, such as the behavior of other human, divine or creatural agents, the structure, position and movement of things or the environmental situation. In a modern discourse on tragedy a form of hindsight bias as retroactive pessimism identifies the whole causation as unforeseeable fate rather than a contingent interplay of factors shaped by fallible knowledge and decisions that could have been different, e.g., more cautious. In the Anthropocene, causation is also complex, much more than on stage, because agency and its consequences unfold intergenerationally on manifold scales. Its dynamics like land deterioration are not defined by a single sudden radical turning point due to a protagonist's action and its side effects but a series of sometimes abrupt, sometimes more steady, more or less irreversible changes caused by "an endless number of factors and agents contributing to it." (Horn, 2018, p. 80). A fundamental experience of tragedy and the Anthropocene alike is the impossibility to ensure control of complex causation which agency is a part of.

(iii) The entanglement of power and powerlessness: As in tragedy, the power to act in the Anthropocene is linked to experiences of powerlessness in the face of massive environmental feedback mechanisms. This dialectical relationship lies at the heart of its concept. For the realization that the *anthropos* – abstracted from concrete (historically mainly white and male) agents with power in petro-modernity – is competing with the great powers of nature emerges at the very moment when the fragile conditions of Holocene civilization become recognizable. The integration of the autopoietic technosphere into the earth system with finite resilience is itself irreversible and generates a cybernetic momentum with feedback mechanisms against which the selective activity of individuals, institutions or even states seems rather impotent. In tragedy as in the Anthropocene, the sublime, which in Kant's view strengthens a person's relationship to him- or herself as a rational and moral being in the face of natural superiority, has turned into the uncanny: a destructive physical event such as a daunting flood can no longer be distanced or othered as "nature" since rationality itself contributed to its production without being able to undo it. Oedipus cannot distance himself from the crisis he and the city find themselves in in order to overcome it, because it turns out to be a crisis of his intelligence and his making.

(iv) Existential consequences: Many effects of agency in the Anthropocene are destructive to life on Earth. The losses, e.g., of human lives and biodiversity, are like those of tragedy "irremediable" (*Poetics*, 1453b). Large-scale destruction of ecosystems and cultures is already happening and some planetary boundaries have been dramatically exceeded. The consequences of climate change that are still to come from an increase in extreme weather events to the strain on societies to the point of collapse are existential as they affect the survival of countless organisms and the conditions of survival and good life for billions of people. The existential dimension of massive intergenerational suffering is prefigured in tragedy by the characters and choruses lamenting great losses of their loved ones and a grim outlook on the rest of life. The turn to disaster affects them deeply and thwarts any prospect of a good life in the future.

(v) Emotional reactions: In tragedies, the characters and choruses experience and reflect strong emotions such as pain, guilt, shame, despair, grief and anxiety as suffering spectators of their own tragic failures. These emotions are taken up by the audience and responded to with aesthetically generated affects such as fear, anger, or compassion. In the Anthropocene, most people find themselves in the role of more or less helpless spectators of planetary processes with typically tragic sentiments spreading and affecting the mental health especially of vulnerable people. Shame, guilt, grief, anxiety, melancholy, despair as well as compassion are prevalent in the awareness and communication of global environmental damage. In some cases a specific terminology of reflexive emotions has been developed such as ecological grief, solastalgia, eco-paralysis or environmental despair (Panu, 2020). Moreover, emotions that strongly motivate to act are aroused by activists to mobilize for the avoidance of future damage, such as prohibitive dread (e.g., flight shame), fear to the point of panic, rage or anger as well as love and hope (Neckel & Hasenfratz, 2021). However, apathy can also be a way to cope with dreadful events among contemporaries as well as tragic characters. Affects untypical of the tragic experience, such as pride, joy or enthusiasm, are equally unusual in the awareness of social-ecological transformations in the planetary emergency. Hope becomes highly ambivalent in the face of darkened prospects that need serious action rather than good faith. In tragic experience and the Anthropocene awareness alike, the affective relationship to the world of human agency is unsettled, ambivalent and alienated. Confidence gets lost.

(vi) Learning from suffering: The inability to learn practically within the tragic situation in order to correct the turn into misfortune is contrasted by a rather anthropological learning about the contingent existential risks of *vita activa*, often referred to with the formula *pathei mathos* from Aeschylus' *Agamemnon* (line 177). Any possible learning faces the challenge that "recognition" in Greek tragedy, *anagnorisis* (*Poetics*, 1452a), as well as the understanding of tragic dialectic in the Anthropocene comes often too late to avoid destructive processes that have already been ongoing for some time. Likewise, many consequences of the Great Acceleration were not hazarded as existential side effects of economic growth until the causal links between socio-economic trends and earth system trends were understood.

However, even though the Anthropocene is tragic, humans can intervene into its causation unless a final irreversible turn will have taken place which to assume or prospectively identify would be beyond our epistemic possibilities. What therefore urgently needs to be practiced is a planetary thinking and communication in knowledge economies that recognizes human beings in their complex dependence on and interconnectedness with their terrestrial contexts of agency: other human beings across nations, cultures, class, race, gender and other identities, the complex processes of the earth system including the lives of multiple species from the microbiome to livestock and wildlife. A radical and communal transformation of human self-understanding that is no longer defined by invalid oppositions such as culture vs. nature could help to identify and eschew tragic ironies that may be traced back to the beginning of the agricultural civilization (Morton, 2012). Moreover, dealing with suffering and eco-emotions can get orientation from tragedy and its rich poetic expressions of mourning, lamenting and compassion in the face of irreversible losses. On the other side, contemporary theatre needs to reflect its possibilities for negotiating and changing social reality in the Anthropocene (Raddatz, 2021) and unfold the critique of the tragic which tragedy itself comprises (Ette, 2011).

The Anthropocene has only begun; there are much more irreversible losses to come and tragic decisions will become harder and more far-reaching if urgently necessary transformations are addressed only tentatively. Furthermore, tragic turns will take on a new catastrophic quality when tipping points in the climate system are passed and their cascade will become unstoppable resulting in a much hotter global temperature. Global warming seems to be "a tragedy that is so much more tragic than all the earlier plays", but even if it seems plausible we likely are not for all of its changes "*too late* on the stage to have any remedial role" (Latour, 2014, p. 12–13). Resignation which refrains from volition and agency as favored by Schopenhauer in his account of tragedy's effect returns today in an acceptance of final failure and the plea to learn to die in the Anthropocene (Scranton, 2015). Tragic-apocalyptic thinking sees no realistic alternatives to aggravating the already ongoing disaster due to robust economic interests and notorious political shortcomings. In order to develop long-term perspectives for living in the damaged world of the Anthropocene (and possibly overcoming its catastrophic dynamics), it is necessary to re-think human beings as dependent agents in a more complex way. On the contrary to tragic characters, we are constantly gaining more knowledge about the various conditions and consequences of certain types of action in order to intervene in their repetition and prevent them from cannibalizing their enabling conditions (Fraser, 2022). Moreover, international collectives like Fridays for Future are holding governments accountable through public-performative claims, while the chorus in Greek tragedy barely has the courage to step up and intervene.

It is urgently important to contradict the modern idea of a post-tragic age and develop tragic narratives and sensitivities in order to resist the continuation of the tragic (Malkmus, 2020; Trautsch, 2020). We need to conceive of the human as a potential *homo tragicus* who has proven to be capable of collectively establishing life- and self-harming systems while failing to fully comprehend this dialectic, and simultaneously as a *homo discens* who is capable to learn from the varieties of tragic suffering, to work on epistemic limitations and comprehend the complex short- and long-term effects of practices, to self-correct and repair systems in order to maintain a "safe operating space for humanity" (Rockström et al., 2009). This cannot be done by individual heroes and heroines, but only as cooperative humanity on various local, regional and international levels (Ostrom, 2014). Interrupting a tragic Anthropocene requires a serious interdisciplinarity creating various forms of knowledge about humans and the earth from all sciences, practices, cultures and arts – not least tragedy.

References

Aristotle. (1995). *Poetics*. Ed. & Trans. S. Halliwell. Harvard University Press.

Buck, H. J. (2019). *After geoengineering. Climate tragedy, repair, and restoration*. Verso.

Ette, W. (2011). *Kritik der Tragödie. Über dramatische Entschleunigung*. Velbrück Wissenschaft.

Fraser, N. (2022). *Cannibal capitalism. How our system is devouring democracy, care, and the planet – and what we can do about it*. Verso.

Gardiner, S. M. (2011). *A perfect moral storm: The ethical tragedy of climate change*. Oxford University Press.

Horn, E. (2018). *The future as catastrophe: Imagining disaster in the modern age*. Columbia University Press.
Latour, B. (2014). Agency at the time of the Anthropocene. *New Literary History, 45*, 1–18. https://doi.org/10.1353/nlh.2014.0003
Malkmus, B. (2020). Safe conduct: The Anthropocene and the tragic. In G. Dürbeck & P. Hüpkes (Eds.), *The Anthropocenic turn the interplay between disciplinary and interdisciplinary responses to a new age* (pp. 93–112). Routledge.
McLachlan, R. (2019, June 4). Climate change is a fourfold tragedy. *Scientific American*. https://blogs.scientificamerican.com/observations/climate-change-is-a-fourfold-tragedy/
Morton, T. (2012). The oedipal logic of ecological awareness. *Environmental Humanities, 1*, 7–21. https://doi.org/10.1215/22011919-3609949
Neckel, S., & Hasenfratz, M. (2021). Climate emotions and emotional climates: The emotional map of ecological crises and the blind spots on our sociological landscapes. *Social Science Information, 60*(2), 253–271. https://doi.org/10.1177/0539018421996264
Ostrom, E. (2014). A polycentric approach for coping with climate change. *Annals of Economics and Finance, 15*(1), 97–134. https://doi.org/10.1596/1813-9450-5095
Panu, P. (2020). Anxiety and the ecological crisis: An analysis of eco-anxiety and climate anxiety. *Sustainability, 12*(19), 7836. https://doi.org/10.3390/su12197836
Raddatz, F.-M. (2021). *Das Drama des Anthropozäns*. Theater der Zeit.
Rockström, J., et al. (2009). A safe operating space for humanity. *Nature, 461*, 472–475. https://doi.org/10.1038/461472a
Scranton, R. (2015). *Learning to die in the Anthropocene. Reflections on the end of a civilization*. City Lights Books.
Szondi, P. (2002). *An essay on the tragic*. Stanford University Press. (Original edition 1961).
Toly, N. J. (2019). *The gardeners' dirty hands: Environmental politics and Christian ethics*. Oxford University Press.
Trautsch, A. (2020). *Der Umschlag von allem in nichts. Theorie tragischer Erfahrung*. de Gruyter.
Wallace, J. (2020). *Tragedy since 9/11: Reading a world out of joint*. Bloomsbury.

Asmus Trautsch is a writer and researcher in philosophy. He has a PhD in philosophy at Humboldt University Berlin and currently teaches at University of Greifswald. Before, he was guest lecturer and visiting scholar at Technische Universität Dresden, Leipzig University, Columbia University and Allegheny College. He is author of *A Theory of Tragic Experience* (Der Umschlag von allem in nichts. Theorie tragischer Erfahrung. Berlin/Boston: de Gruyter 2020).

Violence

David Porchon and Ludovic Aubin

Abstract Violence is a force that harms, hurts or kills. This force may go beyond the critical threshold and run out of control. The etymological root of the word violence comes from the Ancient Greek *bia* meaning "vital force". The word violence can be used to describe natural phenomena such as storms, waves and earthquakes. If there are some common charasterics of violence, force, power and authority, violence itself has its proper phenomenological aspect (Arendt L. Du mensonge à la violence: Essai de politique contemporaine. Calmann-Lévy, 1972). In that sense, violence appears as an unexpected event going against peace and order (Girard R. La violence et le sacré. Grasset, 1972). The concept of violence was defined by the Council of Europe in 1989 as follows: "Violence is characterized by any act or omission committed by a person, if it harms the life, the physical or psychological integrity or the freedom of another person or impairs the development of his person and/or harms his financial security" Thus, a typology can be made between physical violence, psychological violence, symbolic violence, sexual violence and metaphysical violence.

The difficulty in defining what kind of phenomenon is violent or not comes from the evolving norms of society. Because for any violent phenomenon, one must take into account the facts and their interpretations, i.e., practical and symbolic dimensions of the facts. For instance, in western societies, until modern thinkers (Rousseau, Locke) and contemporary legislation (United Nations, 1989), corporal punishments were considered a legitimate way for adults to punish children. Today it seems obvious that they are considered as forms of violence and thus punished by law.

D. Porchon (✉)
Catholic University of the West, Angers, France

L. Aubin
Federal University of Pernambuco, Recife, Brazil

© The Author(s), under exclusive license to Springer Nature Switzerland AG 2023

N. Wallenhorst, C. Wulf (eds.), *Handbook of the Anthropocene*,
https://doi.org/10.1007/978-3-031-25910-4_253

In this sense, a distinction can be made between a relativist approach contextualizing violence to specific cultural laws and norms, and a universalist approach considering that if an act of violence could be justified by traditions or laws it will never be legitimate. Thus, one ethical and complex question is: Is there any universal violent act that should systematically be denounced? Is there a risk of being violent (judgmental) when one criticizes some violent aspects of certain cultures?

Violence distinguishes itself from aggressiveness which is a self-conservation instinct. Violence could be a form of aggressiveness but characterized by deregulation. Contrary to animals, human relationships do not have any strong dominance patterns capable of preventing violence from escalating. In that sense, the violence of mankind is often characterized as Hybris: the loss of measure, the exceeding of limits.

Thus, the origin of human violence is not only in the satisfaction of needs nor does it stem from instincts (logic of interests), instead its roots can be found in desire. The mimetic dimension of desire is certainly one of the possible roots of violence (Girard, 1972). Indeed, humans often fight because they desire the same things. Not accidentally but almost necessarily. Then, violence occurs, when one is not able to peacefully partake in the object or make it circulate. In that sense, violence is an expression of powerlessness.

One of the most problematic aspects of violence is its contagious nature due to the highly mimetic potential of human beings. Once the violent partners imitate one another, violence assumes a self-feeding destructive dynamic. Each rival imitates the violence of the other. Moreover, when violence goes beyond a merely intersubjective relationship, it might contaminate a whole community through bonds of solidarity (kinship, loyalty, etc.). The mimetic violence then dives into the infinite circle of vengeance: each act of vengeance being vengeance for a previous one.

Then the question is, how did traditional societies survive their own endogenous violence? Mimetic theory responds that it was through prohibitions (preventing competitive desires to claim for same objects), obligations (to channel the desires to different objects), and originally and ultimately sacrifice, i.e., ritual repetition of a hypothetical founding murder, the effect of which is the expelling of tensions due to rivalries through the scapegoat mechanism.

Indeed, if violence were only a contagious principle of disorder, there would be no liveable nor sustainable world. The enigma of violence, and maybe the main reason why human societies have such difficulties in freeing themselves from it, is that traditionally it is also a principle of order. Through the scapegoat mechanism (in its spontaneous, ritual, religious, political, economic forms), unanimous violence towards a single victim is able to (abruptly and temporarily) cleanse the internal tensions of a social group, restoring peace, harmony and order and giving the persecutors (the mob or the community) the impression that the victim really deserved to be cast or killed (Girard, 1978). Here is the paradox of violence which can create order from chaos. This led societies to distinguish between good violence and bad violence. An important aspect of this scapegoat process is that violence is hidden behind myths, rituals, and institutions.

Among the deepest mysteries of modernity is undoubtedly that of the mass violence. The twentieth century in this regard caused the death of 191 million victims (OMS, 2002). Moreover, 164 million of them are due to the action of governments against their own citizens (Rummel, 1997; Sémelin, 2005). The states thus seem to have had a certain democidal tendency (Rummel, 1997). This demonstrates the historical and gradual loss of capacity of modern institutions to contain violence.

According to Girard, this phenomenon is due to the loss of legitimacy of violence itself. Because its arbitrary roots have been revealed by Christian revelation (Girard, 1978), violence in its religious, political and now economical form tends to lose its legitimacy and thus, its efficiency. We may apply this observation to any sector of modern societies. Minorities which were once oppressed and dominated through forms of structural and institutional violence now have rights. This is a major factor of change in contemporary societies.

A slow and deep process of desacralization has therefore been at work for two millennia. The relays and accelerators of which would be the Enlightenment, the European Revolutions, the Universal Declarations, the constitutions of modern democracies in general, etc. These texts would gradually prevent the reconciling and regenerating effectiveness of sacrifice and would work to reveal the scapegoat mechanism without knowing it or making it clear. It would thus produce paradoxical effects by guaranteeing everyone the right not to be taken as a scapegoat of his own group.

Consequently, systemic crisis happens when institutions are tempted to use violent means to reestablish order. Instead of being efficient, this worsens violence as a whole. Historically, deep crises have led to institutional system shift. How, then, can the modern world deal with violence if the scapegoat mechanism no longer works and the capacity for destruction increases? Maybe through economics (Dumouchel & Dupuy, 1979)? As always in this anthropology, economics is paradoxical: it instigates rivalries and simultaneously produces similar items in order to mitigate rivalries.

In the Anthropocene, new forms of violence seem to have emerged and continue to grow. Indeed, a number of authors (Gemenne, 2019; Klein et al., 2015; Perret, 2020; Serres, 2011; Welzer, 2012) show through numerous scientific data that the Anthropocene is an ecocide of which the first victims are millions of the most fragile human beings. The consequences of ecological disorders are the cause of emerging diseases (new viruses), massive climate migrations, wars, conflicts, famines, extreme meteorological events and in general a quarter of the global number of casualties (GIEC, 2019; OMS, 2007, 2015, 2018; Robin, 2021). Thus, if in the Anthropocene mankind appears to be a geological force, it is above all a geological force which harms, hurts or kills. The link between violence and the Anthropocene also appears in some way in the controversies over the beginning of this Era. Behind the colonizations linked to extractivism (Crutzen, 2002), the disappearance of millions of indigenous people during the colonization of the Americas (Lewis & Maslin, 2015) or the effects of radionuclides on populations (Zalasiewicz et al., 2017), human violence appears.

Thus, there would be a phenomenon of circular causality between destruction of the environment and destruction of human beings. In this circular logic, some human beings destroy the environment, and in turn, this environmental destruction kills human beings.

However, the violence of the Anthropocene does not seem to appear as an event for our societies, the human catastrophe disappearing behind the environmental catastrophe. Arendt noted such effects of metamorphosis of facts which have the consequence of concealing the violence and the function of evacuating responsibility for the subject emitting this violence. This metamorphosis then gives rise to a framework of rationality at the origin of a normality making violence banal. This process results in a rupture between intention and act, this disconnection providing the possibility of acts without consequences. Violence in the Anthropocene would therefore have its roots much more in an obliteration of violent consequences than in a series of original intentions.

Three attitudes towards violence in the Anthropocene thus seem to be possible: denial, justification, trivialization. But if the myth of our societies protects us from the reality of our violence, it also prevents us from facing it. Will humanity be able to survive its own violence?

References

Arendt, H. (1972). *Du mensonge à la violence: Essai de politique contemporaine*. Calmann-Lévy.

Crutzen, P. J. (2002). Geology of mankind: "The Anthropocene". *Nature, 415*, 23.

Dumouchel, P., & Dupuy, J.-P. (1979). *L'enfer des choses*. Seuil.

Gemenne, F. (2019). *Atlas de l'Anthropocène*. Sciences-Po Presses.

GIEC. (2019). *Réchauffement planétaire de 1,5 degré: Résumé à l'intention des décideurs*. GIEC.

Girard, R. (1972). *La violence et le sacré*. Grasset.

Girard, R. (1978). *Des choses cachées depuis la fondation du monde*. Grasset.

Klein et al. (2015). *Crime climatique stop : L'appel à la société civile*. Seuil.

Lewis, S., & Maslin, M. A. (2015). Defining the Anthropocene. *Nature, 519*, 171.

O.M.S. (2002). *Rapport mondial sur la violence et la santé.* W.H.O.

O.M.S. (2007). *Prévenir la maladie grâce à un environnement sain: une estimation de la charge mondiale de morbidité imputable à l'environnement.*

O.M.S. (2015). *Protéger la santé face au changement climatique: Evaluation de la vulnérabilité et de l'adaptation.*

O.M.S. (2018). *Un défenseur de la santé dans le monde face aux changements climatiques, à la pollution et à la résistance aux antimicrobiens, WHO*. Consulté le 10/05/20 sur https://www.who.int/publications/10-year-review/health-guardian/fr/index1.html

Perret, B. (2020). *Quand l'avenir nous échappe*. Desclée de Brouwer.

Robin, M.-M. (2021). *Les fabriques des pandémies: Préserver la biodiversité, un impératif pour la santé planétaire*. La découverte.

Rummel, R. J. (1997). *Death by government*. Transaction Publishers.

Sémelin, J. (2005). *Purifier et détruire: Usages politiques des massacres et génocides*. Seuil.

Serres, M. (2011). *La guerre mondiale*. Pommier.

United Nations. (1989). *Convention on the Rights of the Child.* Treaty Series, 1577, 3.

Welzer, H. (2012). *Les guerres du climat: Pourquoi on tue au XXIème siècle*. Folio.

Zalasiewicz, J., et al. (2017). The working group on the Anthropocene: Summary of evidence and interim recommendations. *Anthropocene, 19*, 55–56.

David Porchon is a psychologist and a Phd student in Educational science and political science. He is specialized in violence and conflict resolution. He is teaching at the Catholic University of the West since 2011. He is the author of some papers on the anthropology of violence in the Anthropocene. Papers (Selection): The author of the end: about some aspects of the anthropologival catastroph in the Anthropocene (Research & Education, 2021, *in French*). *Political Education in the Anthropocene* (Education & Socialisation, 2022, *in French*).

Ludovic Aubin is a sociologist, consultant and conflict mediator. He is Doctor in Sociology of Development (IEDES-Paris 1- Panthéon-Sorbonne, 2012). Specialized in the relationship between conflicts, contemporary crisis and sustainable development. He taught at the U.C.O. (Angers, France) from 2002 to 2010 and has worked with several professional audiences (teachers, caregivers, social workers, etc.) from 2003 to 2010. He has lived and worked in Brazil since 2010 where he worked with several training centers. From 2015 to 2020, he was a teacher-researcher under a post-doctoral contract at UFPE (Recife, Brazil). He published several papers on mimetic theory and sustainability and, recently, published a book about conflict resolution: “Conflict Prevention and Management in 12 steps”.

Vulnerability

Daniel Burghardt and Jörg Zirfas

Abstract The paper relates to concepts of vulnerability and the Anthropocene. Starting from anthropological vulnerability, different dimensions of inequality of ecological vulnerability and nature itself are considered. Then it addresses, competing concepts such as the Capitalocene and Chthulucene, which are closely related to concepts of vulnerability. The paper concludes with perspectives on education for sustainability.

The concepts of vulnerability and the Anthropocene share a similar fate in terms of scientific theory. They have both received a certain amount of public attention in recent decades – admittedly the concept of the Anthropocene more than vulnerability – and both promise to go beyond mere conceptual conjunctures to determine phenomena in an anthropological transformation of the relationship between nature and society. In this context, the concept of vulnerability has advanced across disciplines to become a central term, linking vulnerability to a series of quite different developments in modernity. Since then the vulnerability of objects, systems, groups, or individuals has been approached in different ways across many disciplines (cf. Burghardt et al., 2017).

As an anthropological thesis, it can be stated that people are vulnerable beings because they are bodily, socially, culturally and reflexively constituted living beings. People are physically and emotionally vulnerable, because they can be materially harmed or because they are denied recognition and participation. In this respect, the term vulnerability is understood to mean the violability of a person in the face of existing hazards, risks, crises or damaging events that have already occurred (Bürkner, 2010, p. 24). On the one hand, the reference to risk groups, i.e. particularly vulnerable groups such as children, sick people, people with disabilities, must

D. Burghardt (✉)
University of Innsbruck, Innsbruck, Germany
e-mail: Daniel.Burghardt@uibk.ac.at

J. Zirfas
University of Cologne, Cologne, Germany

© The Author(s), under exclusive license to Springer Nature Switzerland AG 2023

N. Wallenhorst, C. Wulf (eds.), *Handbook of the Anthropocene*,
https://doi.org/10.1007/978-3-031-25910-4_254

always be taken into account. And on the other hand, it should be pointed out that the term vulnerability speaks of a possibility or potential that refers to events or experiences that have not yet occurred. Therefore, questions about the concrete causes and effects of injuries can only be named after they have occurred (cf. Dederich, 2021, p. 240). The vulnerability discourse is, at its core, a risk discourse.

In relation to the Anthropocene, exploitation (extractivism, commodification, economization) and the littering of nature, the loss of biodiversity and habitats, and in particular global warming with its vulnerable consequences for humans and the environment are moving into the focus of the debate – the key words here range from storm surges to famine to climate refugees. Accordingly, if one speaks of the vulnerability of nature the following points are meant: the processes of the lack of regenerative capacity and sustainability of natural processes, the loss of uniqueness and diversity, the sensitivity and destructibility of ecosystems, the sacredness and beauty of nature, and also natural disasters, which are caused by humans.

From a historical perspective, the question of the extent to which humans themselves are responsible for changes in natural contexts has been discussed at least since the Lisbon earthquake disaster in 1755. Societies of the 16th to 18th centuries still use a classical "symbolic processing of natural disasters" (Walter, 2010, p. 22) and terms such as "plague" or "scourge" to inscribe natural calamities into an explanatory pattern "in which divine providence intervenes to admonish, punish, and mend people who have violated norms" (ibid.). But already in this age, there are indications that natural-catastrophic changes are classified as consequences of human transgressions, rather than by nature itself. "It is not God who is the punisher, but the hubris of human intervention in the world, which becomes counterproductive when it endangers the natural balance" (ibid., p. 23). Since then, questions of the vulnerability of nature are directly anthropological questions.

From then on, human intervention in nature, also in the course of industrialization, is further intensified – this is also where the concept of the Anthropocene has its historical starting point. In this sense, Michael Sonntag writes that "[v]irtually many of the 'natural disasters' [...] that have recently become more frequent are caused by human activity" (2003, p. 27), albeit indirectly: "Thus, as a result of the greenhouse effect and increasing global warming, there is a slowing of the Gulf Stream and an intensification of westerly winds from the Atlantic. The results include the phenomenon known as El Niño, increased avalanches, storm surges, hurricanes, river floods of the century, etc." (ibid.). Which also means: Humanity is becoming more and more the cause of natural vulnerabilities – of nature and of itself.

Despite the relative unambiguity of the consequences of global warming, a number of fuzzy relationship determinations arise from a vulnerability-theoretical perspective. First, it remains open so far who proves to be vulnerable and to what extent? Alternately, one speaks of the vulnerability of nature, of the earth, of systems/institutions and/or of humans. Anthropologically, it should be indisputable that humans are vulnerable beings who can be harmed and suffer by their circumstances, and who at some point will inevitably be confronted with their finitude and mortality. In the meantime, it can also be taken for granted that under climatic conditions the whole of humanity has come on the scene as a vulnerable risk group, which at the same time is working towards its own abolition (this topic has been virulent at the latest since the discussions about the atomic bomb, cf. Anders, 1980). However, blatant forms of inequality are present here, because it is well known that the poorest regions of the earth are most affected by the consequences of global warming. Here, the inhabitants of the respective region prove to be vulnerable to the same extent that nature and the earth are vulnerable. Conversely, the causative factors are mainly located in the rich industrialized countries, which suggests an opposing vulnerability ranking of those affected and those responsible.

Secondly, this inequality perspective leads to a critique of the conceptual apparatus. What are we specifically talking about when we speak of the Anthropocene and its vulnerabilities? And what are we precisely not talking about? While the term Anthropocene describes quite unspecifically the entry of humans into the geochronology, the term Capitalocene, for example, emphasizes that not all humans are equally driving this change, but that the impersonal logic of profit and growth of the capitalist system is the central influencing factor. Jason Moore therefore characterizes capitalism as "a way of organizing nature" (Moore, 2020, p. 9). Against this background, the notion of the Capitalocene offers the possibility of reconsidering this way as a powerful and unequal practice as well as the binarity that lies in the assumed opposition of nature and society. According to this, capitalism does not work against but through nature – and nature, as it were, through capitalism. With regard to the question of who is causal for the violation, the focus of interest is thus less on direct human action and the subsequent responsibility, but rather on the "Mute Compulsion" (Marx) of capitalist relations.

From an explicitly species-unspecific perspective, Donna Haraway's neologism of the "Chthulucene", inspired by the Cthulhu myth, criticizes the Anthropocene concept's focus on humans. Her vulnerable starting point here is the "vulnerable" and "wounded" earth (Haraway, 2018, p. 20). In a poststructuralist manner, the entities and genealogies and forms of kinship between humans, animals, and even bacteria and viruses are emphasized here. Haraway makes it clear that forms of resistance lie in "doing-with-becoming, composing-with the earth-bound" (ibid., p. 142)-and not humans alone. Her proposed counter-strategy consists in the appropriation of the entangling connection between humans and species of all kinds.

In this respect, reflections on an education for sustainability can focus on perspectives of a natural, social and anthropological minimization of vulnerability: 1. theoretically, in reflecting and criticizing existing (economic) structures and attitudes, and in developing new concepts – such as post-growth, interconnectedness with nature or eco-socialism; also in anticipating the dangers of social closure mechanisms in the face of the growing dangers of global warming (eco-fascism). 2. practical, in the exploration and experimentation with new forms of life – for example in product design, in consumption, or in the treatment of animals, 3. emotional, in the sensitization for problems and contradictions and for new qualities of life – in the social, political, or ecological (cf. Zirfas, 2011).

It remains to be said that the vulnerability of nature and human beings – that is, moments of their shakability, fragility, susceptibility to disruption, crisis and endangerment – ultimately remains ineluctable. Even in the case of more respectful attitudes toward nature, more sustainable economies, technological advances that make humans partially independent of natural resources and processes, diverse biological, medical, and technical immunization strategies, the vulnerability of ecosystems as well as humans remains (unless the dream of invulnerable humans dreamed by some transhumanists comes true). And more broadly, research available to date makes clear that each form of vulnerability reduction generates other and new forms of fragility, frailty, etc. (cf. Dederich & Zirfas, 2022).

References

Anders, G. (1980). *Die Antiquiertheit des Menschen*. Beck.

Burghardt, D., Dederich, M., Dziabel, N., Höhne, T., Lohwasser, D., Stöhr, R., & Zirfas, J. (2017). *Vulnerabilität. Pädagogische Herausforderungen*. Kohlhammer.

Bürkner, H.-J. (2010). *Vulnerabilität Forschungsstand und sozialwissenschaftliche Untersuchungsperspektiven*. Working Paper No. 43, Erkner, Leibnitz-Institut für Regionalentwicklung. Online verfügbar unter: https://leibniz-irs.de/fileadmin/user_upload/IRS_Working_Paper/wp_vr.pdf. (5.11.2019).

Dederich, M. (2021). Vulnerabilität. Eine Zwischenbilanz. In: *Weiterdenken – Perspektiven pädagogischer Anthropologie*, hg. Daniel Burghardt, Moritz Krebs, und Juliane Noack Napoles, 232–246. Beltz Juventa.

Dederich, M., & Zirfas, J. Hg. (2022). *Glossar der Vulnerabilität*. VS Verlag.

Haraway, D. J. (2018). *Unruhig bleiben. Die Verwandtschaft der Arten im Chthuluzän*. Campus. (Original edition 2016).

Moore, J. W. (2020). *Kapitalismus im Lebensnetz. Ökologie und die Akkumulation des Kapitals*. Matthes & Seitz. (Original edition 2015).

Sonntag, M. (2003). Katastrophen. Natur als "Feind" des Menschen. In *Natur. Pädagogisch-Anthropologische* Perspektiven, hg. Eckart Liebau, Helga Peskoller, und Christoph Wulf, Ch., 22–29. Deutscher Studien Verlag.

Walter, F. (2010). *Katastrophen. Eine Kulturgeschichte vom 16. bis ins 21. Jahrhundert*. Reclam. (Original edition 2008).

Zirfas, J. (2011). Der Geschmack an der Nachhaltigkeit. Ästhetische Bildung als Propädeutik und Regulativ einer Bildung für nachhaltige Entwicklung. In *Die unsichtbare Dimension. Bildung für nachhaltige Entwicklung im kulturellen Prozess*, hg. Gabriele Sorgo, 35–52. Forum Umweltbildung.

Daniel Burghardt is Professor of Education with a focus on inequality and social education at the University of Innsbruck. His work is concerned with educational theory, a materialist analysis of society, and ranges from anthropological questions about vulnerability and pedagogical tact to the history of aesthetic education. His publications include the Vierteljahrsschrift für Wissenschaftliche Pädagogik, Jahrbuch für Pädagogik, Paragrana. International Journal of Historical Anthropology, and in the Journal of Critical Theory.

Jörg Zirfas, Professor of Anthropology and Philosophy of Education at the University of Cologne. Head of the commission "Educational anthropology" (DGfE) and of the "Society of Historical Anthropology" at the Free University of Berlin. Historical and educational anthropology, philosophy and psychoanalysis of education, educational ethnography, aesthetics, culture and education.

War

Justin D. Cook

Abstract The ability of states to project power and wage war around the globe is unparalleled in history. Despite the absence of direct war among the world's superpowers, the international system has become an ever more tumultuous place as the over-reliance on deterrence postures, coercive policies, and the resort to armed conflict to *resolve* disputes have produced numerous protracted conflicts and wars instead of sustainable, positive peace and security. The objective of this contribution is to demonstrate that these coercive policies and the resort to armed force constitute a struggle for recognition. As such, the peaceful transformation of conflict relies upon policies of recognition in providing direct responses to the causes of conflict within the Anthropocene epoch.

Global peace has declined for a 13th consecutive year (Global Peace Index, 2021, p. 9). As of 2022, more than 84 million forcibly displaced persons worldwide have attempted to escape the worst of humanity: persecution, war, or ethnic cleansing. Extensive urban and environmental destruction prevent the return of the forcibly displaced and impose famines and maladies on those that remain. The world's military apparatus produces an estimated 5% of total Co2 emissions globally (Parkinson, 2020, p. 20). Massive deforestation often ensues with war, along with the degradation of soldiers' and local civilians' health due to burn pits and the use of depleted uranium munitions. The planet is facing its worst humanitarian crisis since World War II. War in the Anthropocene is congruent to man's responsibility for the large-scale devastation of the planet and condition of life (Wallenhorst, 2020, p. 83).

War – the instability generated by bellicose policies, the origins of war and the resort to armed force to achieve political objectives – is unquestionably detrimental to the stability of the international system. In a mere 20 years, the established European security architecture has been dismantled. Arms control treaties such as the CFE-A, the INF, Open Skies, and the Vienna Document of 1999 – pillars of

J. D. Cook (✉)
Catholic University of Paris, Paris, France
e-mail: j.cook@icp.fr

© The Author(s), under exclusive license to Springer Nature Switzerland AG 2023

N. Wallenhorst, C. Wulf (eds.), *Handbook of the Anthropocene*,
https://doi.org/10.1007/978-3-031-25910-4_255

military transparency, verification and peace – are no longer respected or in effect. Games of brinkmanship have renewed the great power competition between Russia and the United States over the future of NATO and Ukraine. The Russian-Georgian War of 2008, its subsequent annexation of Crimea in 2014 and full-scale military invasion of Ukraine in February 2022 have significantly jeopardized the sustainable, positive peace cultivated by Europe after centuries of war and increased the potential for the outbreak of armed conflict between the Moscow and Washington. Furthermore, escalatory policies and intense security competition between China and the US are driving the two nations toward a regional security dilemma. As nuclear disarmament has slowed globally, the end of the ABM Treaty of 1972 has paved the way for missile defense and proliferation, with the three superpowers intent on triggering a new international arms race. *Peaceful* protracted conflicts such as those between nuclear powers India and Pakistan, as well as North Korea have reinforced a pessimistic *status quo* of imposed peace. The age-old adage and key to ensuring security, *si vis pacem, para bellum*, is a historically erroneous logic by which countries defend their national security interests.

Furthermore, the recent trend of aggressive, interventionist policies in the post 9/11 era have been detrimental to international security, leaving a destructive path in their wake. Failed military interventions and stabilization missions to bring security and peace to Afghanistan, Libya, Iraq, Syria, Ukraine, or Yemen have worsened the security situations on the ground and created a humanitarian catastrophe for millions. The War on Terror – which has extended throughout 80 countries – has cost the US over $8 trillion and has led to over 900,000 deaths (Crawford, 2019, p. 1). Case in point: The US and its allies were unable to achieve their political objectives and fundamentally change the conditions on the ground in 20 years of war in Afghanistan and the 2003 War in Iraq dramatically increased terrorism.

Many of the wars taken place in the Anthropocene share a common social problematic: the struggle for recognition. Mis-recognition – to the extent that it is detrimental on a state's or people's agency and identity – can generate sentiments of insecurity. Perceived or real attacks on identity and status (Murray, 2019) i.e., discrimination (Lindemann 2010a, p. 41), stigmatization and exclusion, (Lindemann, 2012, p. 218), humiliations (Badie, 2017, p. 165), or the violation of socially accepted norms can serve as a justification for aggressive policies and armed violence as a means of restoring dignity or national honour. As a consequence, the struggle of recognition can be expressed through the resort to war.

Civil wars and wars of independence incited through cultural, ethnic, or linguistic discrimination and exclusion, such as those between Azerbaijan-Nagorno-Karabakh (1988), Moldova-Pridnestrovia (1990), Georgia-South Ossetia (1991), and Georgia-Abkhazia (1992) are conflicts that remain unresolved today (King, 2000; Kaufman, 2001; Souleimanov, 2013). The same forms of mis-recognition also led to genocides in Rwanda (1994) and in Sri Lanka (2006).

The causal role of humiliations as forms of mis-recognition and the outbreak of militarized disputes have also played a part in interstate conflicts. The 1975 Treaty of Algiers was considered a national humiliation for Iraq and a principal cause of the Iran-Iraq War of 1980 (Razoux, 2009, p. 6). The 9/11 attacks shattered the myth of American invincibility and were thus a humiliation and personal attack for President Bush (McCauley, 2017, p. 23). Argentina's seizing of the Falklands Islands in 1982 evoked a national humiliation for Britain, confirming that nation's second-rate superpower status behind the US and the USSR.

President Bush's 2003 ultimatum to Sadaam Hussein was the apex of a crisis turned militarized dispute, while the latter's public contempt of traumatic events of 9/11 triggered the symbolic escalation in the aftermath these attacks. Hezbollah's attack in 2006 marked a profoundly shocking and humiliating sentiment for Israel, provoking an ill-conceived, disproportionate militarized response by Israel in an effort to save face.

Prior to the 1967 Six-Day War, Egyptian leader Gamal Abdel Nasser's aggressive military posture towards Israel sparked a security dilemma by way of his desire to be recognized as the leader of the pan-Arab movement. Russia's militarized response to NATO expansion and Ukraine is better understood as an attempt to further national security *via* the recognition of its great power status (Cook, 2019, p. 203). Moscow's ultimatum prior to the February 2022 invasion symbolically elevated Russia's superpower status on par with the United States', though this illusion has since dissipated. China's superpower status ambitions have empowered it to challenge the United States' military preeminence and post-Cold War liberal order, explaining American hostility towards Beijing (Ward, 2017, p. 183).

Mis-recognition as a source of conflict and war notwithstanding, policies of recognition can prevail equally in the peaceful transformation of conflict and prevention of war (Cook, 2017). As compared to the traditional approaches based on deterrence, correction, or the resort to war alone, a policy of recognition can be extremely cost-effective in the sense that the positive transformation of perceptions and relations can be accomplished without the economic and ecological costs of mobilizing a military. Acts of recognition that result in the respect of national sovereignty and status, including those founded on inclusive measures or equality, and the acknowledgement of historical grievances, have been found to effective in the maintenance of peace and in the prevention of war (Clément et al., 2021, p. 7).

The face-saving approach utilized by US President John F. Kennedy with Nikita Khrushchev and recognition of the USSR's superpower status allowed for the de-escalation of the Cuban Missile Crisis (Lindemann, 2010b, p. 107). The Egyptian-Israeli Peace Treaty of 1979 codified recognition, and resulted in a significantly improved relationship between two countries that had fought five wars within 25 years, resulting in a Nobel Peace Prize for Egyptian and Israeli leaders Anouar El-Sadat and Menachem Begin.

The Moldovan-Pridnestrovian conflict achieved stability and transformed relations between them with policy of recognition within the 1990s based on inclusive measures on minority rights, status and *wrongs* committed during their civil war (Cook, 2017). Unlike separatist conflicts between Georgia, Abkhazia, and South Ossetia (2008), or that of Azerbaijan and Nagorno-Karabakh (2020), war has not returned to the region.

Policies recognizing the traumatic nature of the 9/11 attacks by Pakistani President Pervez Musharraf and Libyan leader, Colonel Muammar Gaddafi, help paved the way towards avoiding conflict and promoting the normalization of relations with the US (Quadri, 2016, p. 70). In fact, Gaddafi's decision to pay reparations and recognize Libya's role in the Lockerbie bombing was key to his country being *reinstated* into the international community (Jentleson & Whytock, 2005/06).

Further, US President Barack Obama's policies of rapprochement towards Russia allowed the two superpowers to reconcile after the Russian-Georgian War of 2008 and the subsequent recognition of Abkhazia and South Ossetia by Russia. Obama proposed to directly engage the United States' historical enemies, such as Cuba and Iran, at the table of negotiations. Though Iranian President Mahmoud Ahmadinejad was resistant to change, his successor President Hassan Rouhani was open to overtures and consequently agreed to the Joint Comprehensive Plan of Action in 2015, bringing transparency to Iran's previously opaque nuclear program. Obama's historical and identity-based recognition of Cuba helped dethaw 55 years of relations fraught with conflict and hostility; an unrealizable achievement with the traditional approach of containment and isolation alone (Leogrande, 2015).

Traditional approaches based on power and military domination – though still frequently utilized – have been challenged in both domestic and international opinion due to their human, financial and environmental costs. Military escalation in the name of ensuring national security inevitably jeopardizes others' national security interests, ultimately decreasing peace and security for those involved. *Endless* wars are a new norm that have demonstrated the ineffectiveness of waging armed conflict. The struggle for recognition, as a major cause of conflict, emphasizes the critical importance of pacific, transformative approaches to the maintenance of international peace and security. Recognition as a response to the root causes of conflict and security fears can only foster and cultivate the new-age adage to peace: *si vis pacem, para pacem.*

References

Badie, B. (2017). *Humiliation in international relations: A pathology of contemporary international system*. Hart Publishing.

Clément, M., Geis, A., & Pfeifer, H. (2021). Recognising armed non-state actors: Risks and opportunities for conflict transformation. In M. Clément, A. Geis, & H. Pfeifer (Eds.), *Armed non-state actors and the politics of recognition* (pp. 3–29). Manchester University Press.

Cook, J. D. (2017). *Faire la paix par la reconnaissance: la transformation des relations moldo-pridnestroviennes de 1989 à 1988*. Université de Lille II.

Cook, J. D. (2019). L'engagement russe aux Balkans: entre intérêts stratégiques et symboliques. In A. Pinot & C. Réveillard (Eds.), *Géopolitique de la Russie: approche pluridisciplinaire* (pp. 199–216). Editions SPM.

Crawford, N. C. (2019). *Pentagon fuel use, climate change, and the costs of war*. Costs of War Project. Boston University. https://watson.brown.edu/research/2019/pentagon-fuel-use-climate-change-and-costs-war. Accessed 20 Feb 2022.

Global Peace Index Overview & key findings. (2021). *The institute for economics and peace*. https://www.dmeforpeace.org/wp-content/uploads/2021/06/Global-Peace-Index-2021-DME.pdf. Accessed 20 Feb 2022.

Jentleson, B. W., & Whytock, C. (2005/06). Who won Libya? The force-diplomacy debate and its implications for theory and policy. *International Security, 30*, 47–87.
Kaufman, S. J. (2001). *Modern hatred: The symbolic politics of ethnic war*. Cornell University Press.
King, C. (2000). *The Moldovans: Romania, Russia and the politics of culture*. Hoover International Press.
Leogrande, W. M. (2015). Normalizing US-Cuban relations: Escaping the shackles of the past. *International Affairs, 91*, 473–488.
Lindemann, T. (2010a). *Causes of war: The struggle for recognition*. ECPR Press.
Lindemann, T. (2010b). *Sauver la face, sauver la Paix: Sociologie Constructiviste des crises Internationales*. L'Harmattan.
Lindemann, T. (2012). Concluding remarks on the empirical study of international recognition. In T. Lindemann & E. Rigmar (Eds.), *The international politics of recognition* (pp. 209–225). Paradigm Publishers.
McCauley, C. R. (2017). Toward a psychology of humiliation in asymmetric conflict. *American Psychologist, 72*, 255–265.
Murray, M. (2019). *The struggle for recognition in international relations. Status, revisionism, and rising powers*. Oxford University Press.
Parkinson, S. (2020). The carbon boot-print of the military. *Responsible Science, 2*, 18–20.
Quadri, M. A. (2016). U.S-Pakistan foreign policy during Musharraf's era. *Public Policy and Administration Research, 6*, 68–73.
Razoux, P. (2009). *La guerre Iran-Irak. Première guerre du Golfe 1980–1988*. Perrin.
Souleimanov, E. (2013). *Understanding Ethnopolitical conflict. Karabakh, South Ossetia, and Abkhazia wars reconsidered*. Palgrave Macmillan.
Wallenhorst, N. (2020). *La vérité sur l'Anthropocène*. Le Pommier.
Ward, S. (2017). *Status and the challenge of rising powers*. Cambridge University Press.

Justin D. Cook is Associate Professor at the Faculty of Social Sciences, Economics and Law (FASSED) within the Catholic University of Paris (ICP) and Affiliate researcher at The Lille Center for European Research on Administration, Politics and Society (CERAPS). He received his Ph.D. in Political Science from the University of Lille II (CERAPS) and his research focuses on the causes, management and the peaceful transformation of conflict.

Part XXI
The Challenge of International Institutions

Agenda

Claus Leggewie and Frederic Hanusch

Abstract Agenda are future oriented tools defining what to do and how to do it. In the Anthropocene, the future of agendas is ambiguous: while our room for manoeuvre shrinks, projections of futures scenarios are more nuanced than ever before in human history. To open up Anthropocene agendas and free them from presentism, novel ways of agenda making are needed that take the genuine temporalities the Anthropocene has disclosed, such as deep time, into account.

The Role of Agendas

Political decision makers routinely declare to be committed to the ever new, always adapting 'roadmaps' generated in the wide field of sustainability politics. While traditional road maps provide orientation to travellers, in particular drivers, by reducing the plasticity of a landscape to a one-dimensional representation, the 'roadmap' first emerged as a project management tool in the business world. The concept was soon adopted by those concerned with political planning processes who employ roadmaps, for instance, when defining milestones to be reached in resolving (international) conflicts—or, for that matter, stages of an intended decarbonization until the end of the twenty-first century.

The roadmap method is related to the broader approach of agenda setting which was originally developed for use by journalists and political elites. Both the media and career politicians are said to put topics "on the agenda," i.e., to bring certain issues to the public's attention and by reiterating them engage in a hierarchical selection among the endless matters people deal with in their everyday lives. Agenda

C. Leggewie (✉) · F. Hanusch
Justus-Liebig-University Giessen, Giessen, Germany
e-mail: claus.leggewie@zmi.uni-giessen.de; frederic.hanusch@zmi.uni-giessen.de

© The Author(s), under exclusive license to Springer Nature Switzerland AG 2023

N. Wallenhorst, C. Wulf (eds.), *Handbook of the Anthropocene*,
https://doi.org/10.1007/978-3-031-25910-4_256

setting means being capable of determining *what*—and, by means of framing, *how*—'the man on the street' thinks about the topic in question. Thus directing attention, for example via headlines or breaking news, was easier in the days of print and unidirectional electronic media than it is in our world of social media in which everybody can send messages as much as receive them and "influencers" utilize algorithms to have more impact on their "followers." As scholars of mass communication—from Walter Lippmann (1922) to Niklas Luhmann (1996) and McCombs and Reynolds (2002)—have argued all along, media are the source of everything we know about the world. What distinguishes current medial practice is that the influence of political elites and professional journalists has generally waned while that of laypeople has proliferated, with agenda setting power now primarily being measured by clicks.

Anthropocene Agendas

Nonetheless, political elites were very successful during the last three decades in setting one agenda in particular. Referencing increasingly robust scientific findings, they sensitized media and public to the necessity of protecting nature, climate, and species by reducing emissions within an exactly defined (though ever shorter) time period. This has resulted in very concrete adjustments to infrastructure planning in areas as diverse as transport and mobility, production and distribution, building and housing, agriculture and nutrition, right up to health, lifestyles, and private as well as public investments. Until the first years of the twenty-first century, climate protection had preoccupied only experts and was mentioned in everyday conversations with regard to the weather at best. Since then, it has become a major, and as of late the dominant, topic of public opinion. Also contributing to this phenomenon are "climate skeptic" positions voiced in social media, which either downright deny the existence of climate change or blame it on factors that are non-anthropogenic and cannot be affected by political or private intervention. This debate shows that, far from homogenizing a framing, agenda setting in fact helps polarize public opinion to a considerable degree.

Roadmaps are also regularly employed in futurology, which deals in identifying megatrends and breaking them down into a manageable timeframe. For the most part, futurology is a semi-scientific enterprise. Parts of it are highly speculative and presentist in orientation, exaggerating the significance of current vogues. The more reputable future researchers, however, map scenarios on a sounder empirical basis and point out a range of possible futures. Thought experiments, too, simulate such possible futures (say, for the year 2050 or 2100) and assess their (un)desirability for humans who will then be living—they thus represent a qualitative version, as it were, of the grammatical "Future Perfect" (What will something have been like?).

Agendas for Alternative Futures

The purpose of roadmaps, and of the political agenda setting that goes along with it, is to overcome presentism, put an end to humanity's careless handling of risks, and develop a preventive strategy for facing future perils, not least through technology assessment. From prognosis to prophylaxis: Agenda setting here serves as an instrument of writing a history of the future. All this is based on very specific conceptions of times: Presentism lives in and for the moment, relishing given possibilities, while millenarianism relies on salvation in the hereafter, disregarding the present with its pleasures and woes. Teleology pursues an 'end of history,' which is either utopian or dystopian in nature and, at least in the deterministic variant, cannot be averted by human actions. By contrast, thinking in terms of roadmaps means to extract a plurality of possibilities from the stock of humanity's experiences and expectations. The futures that thus emerge are not entirely open but contingent.

Agenda setting mirrors the expectations for the future of those who now live, even though its consequences mainly concern coming generations. This is where agenda setting touches on issues of intergenerational justice, then—which is something that Thomas Jefferson already understood (though he of course did not use these terms). Amidst the tumult of the French Revolution (and shortly before his return to the United States from Paris), Jefferson expressed his thoughts on "[t]he question whether one generation of men has a right to bind another" in a letter to his friend James Madison (Jefferson [1789] 1958). Back then the Western world was busy negotiating the nature of modern democracy against the backdrop of a new conception of time: The space of experience increasingly separated from the horizon of expectation so that the future seemed open, the present no longer God-given, and the past as just one of many versions (Koselleck [1979] 2004). This was taken as a *carte blanche*, along with a misconstrued idea of progress, but Jefferson insisted "that the earth belongs in usufruct to the living" ([1979] 1958). It cannot be utilized at will, that is, but must be handed over to the next generation in a state at least as good as we received it, and without debts. The future President derived a radical proposal from these deliberations: The constitution and laws were to expire with those who willed them into being—because else the dead would govern the living. Behind this idea is the assumption of contingent futures. But since the unborn cannot be represented politically, intergenerational justice consists not in an equality but in an independence of generations. Each of them must be free to begin anew; as Hannah Arendt once wrote, "that is the actualization of the human condition of natality" (Arendt [1958] 1998, 178). Jefferson's insight points to the temporality of democracy and open a perspective of "democratization of time": Each generation must obtain self-efficacy through an act of foundation and negotiate whether path dependencies (like the subsidization of fossil fuels) are compatible with the principle of usufruct. By letting go of the old and creating their own constitution, new generations would reinforce the democratic way of life. If inno- and exnovation thus merged, change would become normalized and time democratized.

To think about democratic agenda setting in planetary terms means to think (and act) in temporal dimensions: If time appears to be 'out of joint,' how are we to conceive of alternative futures to counter retrotopias? Which form of legislation could make political election cycles concur with the timescales of the Earth system? How to prevent generational conflicts and 'eternity costs' from arising, e.g., with respect to the long-term effects of mining or nuclear waste disposal? Could we use digital tools to preserve a pluralism of "selftimes" (*Eigenzeiten*; cf. Nowotny, 1996)? How does the desynchronization of socio-economic demand and natural regeneration impact on long-term prosperity? Once we thus expose the temporal dimension inherent to planetary conditions, new possibilities for action come into view as well. Humanity can free itself from its temporal dependency (or in Kantian terms: nonage). The democratization of time thus afforded amounts to an act of enlightenment; in fact, it is a fundamental prerequisite of planetary governance. But first we humans have to learn to interpret and change the world in temporal terms. To bring forth 'chronopolites' who are not just hospitable in the Kantian sense but mindful of the planet's enduring habitability, some normative guidance is needed. Equal chances of living a good life rely not only on spatial circumstances, like your place of birth, but just as much on temporal factors like generational membership and societal selftimes. If we bear this in mind, planetary governance will become more effective, e.g., by means of democratically negotiated arrangements of past, present, and future or a deliberate initiation of rupture, evolution, and preservation.

The concept of deep time is of particular importance for Anthropocene agendas. Originating in geology, the concept encompasses the entire time period of the universe's existence (Farrier, 2016; Glikson & Groves, 2015). As one of the chief distinguishing features of the planetary sphere, deep time allows for an embedding of human actions in the history of the planet; it thus demonstrates the (de-)coupling of natural and artificial rhythms. The concept of "timefulness" (Bjornerud, 2018) represents an attempt at capturing deep-time planet-human relations by determining what would bring about a more temporally aware society, and ultimately a polytemporal world view (Toulmin & Goodfield, 1982). There are good reasons for adopting an approach of "deep-time governance" (Hanusch & Biermann, 2020): For one thing, objects need to be stored securely, which concerns the preservation of nature (like wilderness areas) as much as artificial, human-made 'things' (like nuclear waste but also scientific insights) and hybrid objects (like infrastructures or farmland). Secondly, we must sustain populations—be it human ones, animals, or plants. In the process, the survival of one population must not rest upon the extinction of another; and we should retain as much as possible from doing it under conditions that restrict self-realization (as in zoos). And thirdly, novel kinds of relationships must be facilitated between humans and the Earth system, which make new orders in and after the Anthropocene possible.

References

Arendt, H. ([1958] 1998). The human condition. University of Chicago Press.
Bjornerud, M. (2018). *Timefulness: How thinking like a geologist can help save the world*. Princeton University Press.
Farrier, D. (2016). How the concept of deep time is changing. *The Atlantic* 31 Oct. 2016. https://www.theatlantic.com/science/archive/2016/10/aeon-deep-time/505922/
Glikson, A. Y., & Groves, C. (2015). *Climate, fire and human evolution: The deep time dimensions of the Anthropocene*. Springer.
Hanusch, F., & Biermann, F. (2020). Deep-time organizations: Learning institutional longevity from history. *The Anthropocene Review, 7*(1), 19–41. https://doi.org/10.1177/2053019619886670
Jefferson, T. ([1789] 1958). Thomas Jefferson to James Madison. Letter, 6 Sep. 1789. In Julian P. Boyd (Ed.), *The Papers of Thomas Jefferson* (Vol. 15). Princeton University Press. https://jeffersonpapers.princeton.edu/selected-documents/thomas-jefferson-james-madison.
Koselleck, R. ([1979] 2004). *Futures past: On the semantics of historical time* (Keith Tribe, Trans.). Columbia University Press.
Lippmann, W. (1922). *Public opinion*. Harcourt, Brace and Company.
Luhmann, N. (1996). *Die Realität der Massenmedien*. Springer Fachmedien.
McCombs, M., & Reynolds, A. (2002). News influence on our pictures of the world. In J. Bryant & D. Zillmann (Eds.), *Media effects: Advances in theory and research* (pp. 1–18). Lawrence Erlbaum Associates Publishers.
Nowotny, H. (1996). *Time: Modern and postmodern experience* (Neville Plaice, Trans.). Polity.
Toulmin, S., & Goodfield, J. (1982). *The discovery of time*. University of Chicago Press.

Claus Leggewie is holder of the Ludwig Börne professorship and director of the "Panel on Planetary Thinking" at Giessen University, Germany. He currently is co-editor of the book series "Climate & Cultures" (Brill) and the "Routledge Global Cooperation Series" (Routledge). Earlier affiliations include visiting professorships at the University of Paris-Nanterre and New York University (Max Weber Chair). From 2008 to 2016, he was a member of the German Advisory Council on Global Change (WBGU).

Frederic Hanusch is scientific manager of the "Panel on Planetary Thinking" at Giessen University, Germany. His research is focused on the intersections of democracy and planetary change. He is currently working on a book entitled "Deep-Time Governance" (Cambridge University Press) to explore how planetary temporalities can be institutionalized in governance systems. In his previous book "Democracy and Climate Change" (Routledge), he analyzed why more democracy leads to better climate performance.

Borders

Benjamin Boudou

Abstract This article examines the notion of borders in contemporary political theory and suggests innovative ways to tackle their lack of legitimacy for controlling migration.

Considered simply as devices of distinction, borders are everywhere. They can be physical or institutional, materialized or only apparent, defensive or aggressive, temporary or durable. "Border" is an all-encompassing concept. It cannot be understood in geopolitical terms alone as designating "elementary spatial structures, in linear form, functioning as a geopolitical discontinuity and a marking, a reference point" (Foucher, 1991, p. 38). Rather than a dividing line between different entities, a border is first and foremost a "site of difference" that assembles various borders to compose entities, categories or spaces (Abbott, 1995). It implies a constant need for maintenance (Barth, 1969, p. 17), i.e. the perpetuation of its function of distinction and identification with the norms and practices that organize separation, union and crossing (Busekist, 2016). A border connects and distinguishes different territories that are geographical and ecological, human and non-human (Espejo, 2020).

A better definition of a border could be based on the concept of apparatus [*dispositif*]. In its most generic sense, a border is an apparatus of circulation, interaction and separation. It organizes movements, codifies relationships and legitimizes distinctions. Like any apparatus, its existence is underpinned by "discourses, institutions, architectural forms, regulatory decisions, laws, administrative measures, scientific statements, philosophical, moral and philanthropic propositions" (Foucault, 1980, p. 194). This approach has led to the emergence of an entire epistemic community, the "border studies", which focuses on these interactions between norms and practices, border technologies and identity production, and the joint construction of external and internal borders. This has been called "borderization", i.e.,

B. Boudou (✉)
University of Rennes, Rennes, France
e-mail: benjamin.boudou@univ-rennes.fr

© The Author(s), under exclusive license to Springer Nature Switzerland AG 2023

N. Wallenhorst, C. Wulf (eds.), *Handbook of the Anthropocene*,
https://doi.org/10.1007/978-3-031-25910-4_257

the policies and justifications of the (social) construction of borders (Cuttitta, 2014; Wilson & Donnan, 2012). Achille Mbembe (2018) critically defines borderization as a "process by which the powerful agents of this world constantly transform certain spaces into impassable places for certain classes of populations." Because borders can be brutally enforced, borderization multiplies the "spaces of loss and mourning where the lives of so many people considered undesirable are shattered".

Borders can therefore be both tangible, like a barrier, and intangible, i.e., constituted by "discourses" that define them and make them operational. Indeed, the effectiveness of a border depends on its institutional capacity to produce rights, duties and authorizations to cross it. This is what John Searle (2005, 2010) calls "deontic power". While a barrier can prevent someone from crossing, the prohibition of unauthorized crossing makes it a border (Boudou, 2018, p. 57). Thus, the materiality of an object, such as a geophysical obstacle, may facilitate (rather than cause) its transformation into a border. For example, while the Pyrenees appear as a mountain range forming a natural border between France and Spain, the Pyrenees as a border are above all an opportunistic investment in an unstable, eminently political place where the French and Spanish national identities have been at stake (Sahlins, 1989). Moreover, if we stop seeing a border as a border, that is, when its deontic power loses its authority, the border disappears. This is what happened, for example, when the first Berliners started climbing on the Berlin Wall and the guards let them do it. They acknowledged the end of the legitimacy of the East German government and showed that the border no longer had any power. Its legality, the justifications for its existence and the population's consent were nothing more than empty shells. The border, which had become a wall in 1961, was nothing more than a ruin to be demolished in jubilation. In other words, the legitimacy of borders is equivalent to the legitimacy of the power that enforces it and therefore makes it exist (Boudou, 2018).

While it is therefore essential to understand how borderization takes place, the problem of legitimizing borders is all the more crucial when it concerns contemporary democracies. Democracies face the dual challenge of governing migration and justifying their policies. However, public policies relating to border control, the right of foreign residents to vote, the reception of refugees, the administration of residence permits and more generally the management of migration flows, may contradict democratic standards and values. Satisfying public opinion and bringing legal and administrative practices into line with respect for fundamental rights and more generally with standards of equality and freedom, is difficult and causes conflict. In other words, migration policies can certainly be politically legitimized, but they sometimes contradict domestic law, international norms and values considered fundamental in a liberal democracy (equality, freedom, justice, solidarity, etc.).

The main question is therefore the following: how, on the one hand, to believe in the freedom and equality of everyone, and, on the other, to limit in a more or less discretionary way the mobility or aspirations for a better life of those who move? This tension, long ignored in a world thought of as essentially made up of closed and autonomous nation-states (Sager, 2021), is now central for contemporary political theory. In a pathbreaking paper published in 1987, the political philosopher

Joseph Carens showed that none of the main theories of justice could coherently justify border control in liberal democracies and thus the exclusion of immigrants (Carens, 1987, p. 11–12). Various arguments have since been developed to justify the legitimacy of the conditions of access to territory, institutions and citizenship. This normative literature has proved particularly fruitful in the following areas: contrasting the motivations of individuals who migrate against the state's right to control its borders; assessing the relative weight of principles such as self-determination, the preservation of national culture or the continuity of democratic and liberal institutions; clarifying duties towards refugees; testing the coherence of a human right to migration; questioning the privilege of citizens to decide their own borders (Fine & Ypi, 2016). The aim of this literature is to adjust citizenship criteria and theories of democracy in the context of global migration. However, the results are limited to abstractly determining the conditions for an individual to exercise a moral right to migrate and for a state to control its borders.

A more innovative approach is to make the legitimacy of borders dependent on the inclusive logic of democracy: individuals whose interests are affected by a decision (especially regarding border control) must be involved in its elaboration, either by having their interests taken into account or by participating directly (Abizadeh, 2008; Bauböck, 2017). This principle not only justifies the right of foreign residents to vote, but also that borders must remain contestable and revisable.

At a time when borders are becoming mobile and increasingly dematerialised and mobile (Shachar, 2020), the problem of the legitimacy of their control does not therefore lie in the arbitrariness of how they were drawn or in the open/closed alternative. Rather, the issue is to determine when they cause injustice, unjustified exclusion and prevent consideration of those they affect. The threat to the integrity and dignity of individuals, the ignorance of the fundamental interests of migrants and the spread throughout society of the discrimination that borders induce, present reasons to reflect on their conditions of legitimacy and thus on criteria for judging the nature and extent of the power exercised there (Boudou, 2018).

References

Abbott, A. (1995). Things of boundaries. *Social Research, 62*(4), 857–882.

Abizadeh, A. (2008). Democratic theory and border coercion: No right to unilaterally control your own borders. *Political Theory, 36*(1), 37–65.

Barth, F. (1969). *Ethnic groups and boundaries: The social organization of culture difference*. Little, Brown and Company.

Bauböck, R. (2017). *Democratic inclusion*. Manchester University Press.

Boudou, B. (2018). *Le dilemme des frontières. Éthique et politique de l'immigration*. Éditions de l'EHESS.

Busekist, A.. (von) (2016). *Portes et murs. Des frontières en démocratie*. Albin Michel.

Carens, J. (1987). Aliens and citizens: The case for open borders. *The Review of Politics, 49*, 251–273.

Cuttitta, P. (2014). Borderizing the Island: Setting and narratives of the Lampedusa border play. *ACME: And International E-Journal for Critical Geographies, 13*(2), 196–219.

Espejo, P. O. (2020). *On borders: Territories, legitimacy, and the rights of place*. Oxford University Press.
Fine, S., & Ypi, L. (Eds.). (2016). *Migration in political theory*. Oxford University Press.
Foucault, M. (1980). *Power/knowledge: Selected interviews and other writings, 1972–1977*. Harvester Press.
Foucher, M. (1991). *Fronts et frontières. Un tour du monde géopolitique*. Fayard.
Mbembe, A. (2018). La démondialisation. *Esprit, 430*, 86–94.
Sager, A. (2021). Political philosophy beyond methodological nationalism. *Philosophy Compass, 16*(2), 1–11.
Sahlins, P. (1989). *Boundaries: The making of France and Spain in the Pyrenees*. University of California Press.
Searle, J. (2005). What is an institution? *Journal of Institutional Economics, 1*(1), 1–22.
Searle, J. (2010). *Making the social world. The structure of human civilization*. Oxford University Press.
Shachar, A. (2020). *The shifting border*. Manchester University Press.
Wilson, T., & Donnan, H. (Eds.). (2012). *A companion to border studies*. Blackwell.

Benjamin Boudou is professor of political science at the University of Rennes. He is the editor of the French journal of political theory *Raisons Politiques* and the author of *Politics of hospitality: A conceptual genealogy* (CNRS Éditions, 2017, in French) and *The dilemma of borders: Ethics and politics of immigration* (Éditions de l'EHESS, 2018, in French).

Boundaries

Claus Leggewie

Abstract When we talk about borders—or, more broadly, boundaries—in the planetary context, we should familiarize ourselves with their various, material as well as symbolical, meanings: the boundaries of the Earth system, the "limits of growth" as detected by the Club of Rome, borders as basic elements in international politics, boundaries between scientific disciplines or between age groups, and finally those lines we are accustomed to draw between nature and culture. But what are boundaries anyway, how are they drawn, where do they run? Wittgenstein once said that the limits of our language amount to the limits of our world. Accordingly, an etymological and semantic discussion of the concept may contribute to our understanding of *planetary boundaries* as well, a phrase that hails from Earth system science and is currently entering everyday language. Are they "real", are they "imagined"? The sociologist Georg Simmel once declared: "The boundary is not a spatial fact with sociological effects, but a sociological fact that is formed spatially." Is this also the case with physical boundaries and true for 400 parts per million in the atmosphere? Objectivist and constructivist views of the world mingle, clash or converge here.

Planetary boundaries demarcate areas where the resources and reproductive capacities of planet Earth are limited, i.e., not infinitely progressing. With all other pools of resources – water, energy, information – this is taken for granted; with respect to Earth: apparently not. Earth Overshoot Day is an attempt to locate this overexploitation at a temporal border. What is interesting about this visualization is how the planet's overshoot appears to leap into the infinite once it passes certain tipping points, even though the image of the donut does suggest a fixed frame. Kate Raworth's well-known diagram (6) consists of three areas. In the donut's hole, inhumane conditions prevail: shortages of food, education, and housing space, as well as substandard working conditions have to be confronted here. By contrast, in the donut's sweet spot, below the tasty glaze, there exists an ideal balance of ecology,

C. Leggewie (✉)
Justus-Liebig-University Giessen, Giessen, Germany
e-mail: claus.leggewie@zmi.uni-giessen.de

© The Author(s), under exclusive license to Springer Nature Switzerland AG 2023
N. Wallenhorst, C. Wulf (eds.), *Handbook of the Anthropocene*,
https://doi.org/10.1007/978-3-031-25910-4_258

politics, and economy. Of crucial importance is the pastry's outer boundary, or what Raworth calls its "ecological ceiling," which symbolizes the limited resource capacities of the Earth, as they manifest in climate change, polluted oceans, and a loss of biodiversity. Here, at the planetary boundaries, Earth system science and sustainability research put up a huge stop sign: This far and no further! While we are still facing that stop sign in terms of greenhouse gas emissions, in many other regards, we have ignored it and run it over.

The assumption of an infinite progression seems logical when we project the use of planetary resources into the Universe, which many consider to be itself infinite and thus elusive to intellectual inquiry, a sphere of spirituality or metaphysics. To engage in planetary thinking, however, means to acknowledge the infinity of the Universe and derive from that very infinity the finiteness of the planet. As much as the existence of *Homo sapiens* had a beginning, it will have an end, and according to Anthropocene theory, humanity is now massively accelerating this suicidal endeavour.

To return to the question at hand: What are boundaries? there is an early definition in Euclid's *Elements*: "A boundary is that which is the extremity of something". Most borders, by their very definition, create binary distinctions between a here and a there, and we could draw a line from ancient Greek mathematics to the late twentieth-century systems theory of Niklas Luhmann who was intent on showing that each inclusion entails an exclusion. Functional differentiation means separation: within or without, North or South Korea. In the real world of course, things are not as black and white, not even in Korea. Boundaries are open, moveable, malleable, hazy, fuzzy. The boundary of Earth is not that of its surface or of the anthroposphere alone. Instead, the boundedness of our planet's resources is fundamentally dependent on the state of its atmosphere, of an 'out there' with certain temporal as well as spatial conditions. Not just things have an extremity; events, too, come with a duration, with an end. Temporal and spatial boundaries may coincide, differ, work against each other.

Most commonly, boundaries, a term that goes back to processes of colonization and war, are defined (and thus literally 'limited') as territorial or national borders. *Granica*, the Old Slavic word for border, which was adopted as a loanword in German (*Grenze*) and Dutch (*grens*), designated the terrain to be marked by a line, a fence, or a wall. Also consider the medieval marches, borderlands whose land-*marked* borders were easily crossed; though in Swiss German, *übermarchen* still means to exaggerate. *Border*—from Old French *bordeure*, the band along the edge of a shield—was used in its current sense from the late fourteenth century onwards to denote the same thing, namely zones separating the own territory from what was foreign. This later developed into borders in a less physical sense: between civilization and wilderness (or civility and barbarism), between *Heimat* (home) and *Un-heim-lichkeit* (the unfamiliar and thus vaguely uncanny). Beyond the border, then, lies the unknown, the dangerous, the fascinating, the anarchic.

Borders may appear 'natural' but in general they are created (demarcated), subsequently institutionalized (managed), and eventually perpetuated, adjusted, or removed. They are social and political constructions, imposed by powerful elites.

And there are typical borderlands: zones of transition and hybridity where the territorially excluded may well coexist with the inner life of a region. When it comes to the constructedness of borders, we should recall that passport inspections and checkpoints controlled by the police or the military were only introduced in Europe in the mid-eighteenth century and that the border wall—*the* symbol of the 'German twentieth century'—was a rather unusual phenomenon. In the Middle Ages, walls mainly surrounded cities, defining who and what belonged. Spheres of influence outside the city walls were loosely staked out by border stones or other boundary markers, but travelers were free to pass these. Only when feudal overlords began expressing their mutual power and dependency relationships in spatial terms, did borders really take on a legal meaning. The often life-threatening flipside of inclusion now dramatically came to light in the exclusion of minorities, particularly of Jews and nomadic communities, from the territories of nation states. Equally problematic is the revisionist appeal to people who live as ethnic minorities in other 'cultural spheres', with the goal of incorporating such enclaves into the own territory. This corresponds to the geopolitical tradition, popular in postimperial contexts, to obliterate border lines stipulated in treaties (for example, that of Versailles) by establishing buffer zones.

The other English word for border, *frontier*, comes from the military realm where it specifically referred to the front lines of armies and the outposts built in such areas. Political borders often concur with natural boundaries: hard-to-surmount mountain crests, wide rivers, deserts, woodlands, etc., which also tend to roughly coincide with linguistic and cultural borders. By contrast, language regions, cultural areas, and trade routes get torn in two, potentially causing long-lasting territorial conflicts, when political borders are determined on the drawing board, with purely artificial factors, such as geographic coordinates, as criteria. Unlike officially protected and fortified borders, so-called green borders are permeable, either by design or due to negligence. The drawing up or opening of a border is always a highly symbolic act that represents and at the same time enforces the sovereignty and monopoly on violence of nation states. Signs that read *Caution! No trespassing!* or *Do not proceed beyond this point!* underline this executive power. Throughout history, territorial disputes have given rise to armed interstate conflicts, so-called border wars. And since geodetic point positioning is all but impossible in such 'terrain,' the demarcation of water bodies and airspace constitutes a particular challenge. Currently, there are still border disputes, for example, in the Caucasus, the Persian Gulf region, and Northern Ireland.

This brings us to the problem of borders' porosity and penetrability. Clearly, while borders separate two entities, they also connect them (to a degree not the case with more remote other entities). Are boundaries, in Leonardo da Vinci's words, "neither air nor water"? Or a twilight as in the paintings of the Impressionists? Is a boundary two-, one-, or even zero-dimensional? Is it at all 'real' or just a linguistic convention represented by symbolic markers? Which lines separate the North Sea from the rest of the Atlantic Ocean? Where runs the border between Asia and Europe within Eurasia? Where does the Sahara begin, a cumulus cloud, the Alpine foothills? When did the American and French Revolutions start, or the Copernican and Industrial Revolutions? These are but a few possible mereological questions.

In 1893, the historian Frederick Jackson Turner first published his seminal essay "The Significance of the Frontier in American History" in which he claimed that the United States' moving, open frontier had had the effect of promoting a unique American brand of democracy: it did not evolve through European-style military or missionary movements, based on a feudal social structure, Turner argued, but through the settlers' gradual westward advance along a frontier between civilization and wilderness. This topos has been transferred to expansions of all sorts—not least to the progression of the "electronic frontier" since the 1990s—and today, *space* is regarded as the 'last frontier,' with other planets potentially being terraformed in the future.

The topos of the open frontier could also be applied to a narrative of economic, political, and cultural globalization. Despite all inequalities between territorial blind spots and regional clusters, and although it is still grounded in nation states, globalization can be read as a success story of boundary dissolution. Meanwhile, the proliferating field of "Border Studies" within International Relations and Cultural Studies focuses less on borders as such than on their permanent transgression, their constructedness, and their effects on migration and cultural hybridization.

In conclusion, I would like to address the 1.5 or 2 degree 'threshold,' another version of a boundary that was defined as a result of the recognition and consideration of planetary boundaries. First introduced by the American economist William D. Nordhaus as part of a cost-utility analysis, the 2-degree threshold was taken up and turned into a normative goal for the mitigation of climate change by the IPCC in its assessment reports of the 1990s. It was also adopted by the German Advisory Council on Global Change (WBGU) in its political counseling activities, and in 2010, the target was incorporated into the United Nations Framework Convention on Climate Change. Since then, it has been further underpinned by empirical studies and was lowered to 1.5 degrees in view of the critical situation of low-lying coastal and small island states, organized in the AOSIS. The new goal of "pursu[ing] efforts to limit the temperature increase to 1.5 °C above pre-industrial levels" became part of the Paris Agreement in 2015.

Mike Hulme doubted the universality; ambiguity; doubtful achievability; and questionable legitimacy of the benchmark. A threshold is, first and foremost, a low obstacle that is easily crossed and overlooked. But as the German word "Schwellenangst" illustrates—denoting the fear one sometimes feels when entering a new place or situation—thresholds have symbolic meaning as well; one also thinks of glass ceilings and other discouraging barriers. And 0.5 degrees difference makes a real difference, as is shown in a recent projection of the IPCC. Arctic summers without ice would not happen once within a decade, but ten times; not fourteen but 37% of the world's population would be exposed to extreme heat waves every 5 years; the annual fishing catch would not decrease by 1.5, but by 3 million tons; and the proportion of plants and insects that would be reduced by half within 5 years would double or triple. The 0.5 degree difference is as real as a two degree increase in body temperature.

For this reason, the 1.5-degree target was quickly institutionalized by being incorporated into all kinds of conventions and laws. The merely symbolic stop sign was thus converted into a concrete politics of resolute decarbonization, with climate protection goals being radicalized along the way. In both its genesis and impact, then, this boundary/threshold/target is a hybrid of a symbolic marking on the one hand, and empirical evidence, obtained from physical data, on the other. It is based on various boundary or threshold values, combining historical measurements of temperature developments from 1880 to 2017 (relative to the mean value of 1951–1980) with cumulated carbon dioxide emissions and concentrations, so as to arrive at a remaining CO_2 budget, measured in billion tonnes. This in turn leads to an emissions curve which drops the more steeply the longer it takes for it to reach the tipping point.

When you cross *this* boundary, you do not get captured or shot by border guards on the spot – and life goes on. However, seemingly isolated extreme events, taking place all over the Earth and each concerning the lives of a relatively small group of people, add up to form the total picture of a mushrooming catastrophe. This dynamic has not been curbed, and yet the hybrid, empirical-symbolic construct that is the 1.5-degree target has resulted in specific measures on the national level as well as in comprehensive transnational plans, which do make a limitation of global warming seem realistic. Not just within democracies, the debates on climate protection goals are of course characterized by all sorts of compromises. It is a constant negotiation, as Hans-Joachim Schellnhuber put it, between what is "scientifically" and what is "economically desirable"—to which we must add what is politically feasible: in a world community made up of almost 200 states, an ideal circular and sustainable economy will hardly ever be realized.

To conclude: *Limits to growth, planetary boundaries, climat sans frontières*: since we recognized that the notion of unlimited progress was but an illusion, such boundary concepts appear to most of us like metaphors of self-denial, of prohibition, of restricted options. In a different light yet, self-limitation may bring forth new freedoms from old constraints, constraints imposed on us by the logics of growth and discounting. Climate change and species extinction have **no** boundaries—if we do not raise barriers to ensure that the future of coming generations is no longer looking so gloomy. The most critical boundary of the planet thus appears to be a temporal one: the often paper-thin line between age cohorts or generations whose life-defining biographical experiences tend to diverge like mountain valleys separated by just a steep ridge.

Claus Leggewie is holder of the Ludwig Börne professorship and director of the "Panel on Planetary Thinking" at Giessen University, Germany. He currently is co-editor of the book series "Climate & Cultures" (Brill) and the "Routledge Global Cooperation Series" (Routledge). Earlier affiliations include visiting professorships at the University of Paris-Nanterre and New York University (Max Weber Chair). From 2008 to 2016, he was a member of the German Advisory Council on Global Change (WBGU).

Brundtland Commission

Arnd-Michael Nohl

Abstract The establishment of the Brundtland Commission was preceded by a long period in which, against the background of the North-South conflict, environmental problems were discussed internationally and the concept of "sustainable development" was coined. This concept owes its popularity to the Commission's report, which it presented in 1987. The concept became the standard term of the relevant discussion, which can still be found in the discourse on the Anthropocene, but can also be criticised by the latter.

When UN Secretary-General Javier Pérez de Cuéllar set up a "World Commission on Environment and Development" in 1983 and appointed the Norwegian politician and physician Gro Harlem Brundtland as its chairperson, this did not happen out of the blue. As early as 1968, the Club of Rome, "a group of elitist businessmen, scientists, and philanthropists" (Egelston, 2013, 77), had come together and 4 years later published an influential report entitled "The limits of growth", which led to heated discussions, especially in the industrialised countries (see the respective chapter in this book).

Also in 1968, the Swedish government proposed to the United Nations that an international conference on environmental issues be convened (ibid., 61). However, organising such a conference proved to be difficult, as apart from the East-West conflict, it was primarily the countries of the Global South – at that time called developing countries – that had to be brought together with the industrialised countries. The former suspected the latter of wanting to use the new environmental discourse to put a stop to their striving for industrial development and to colonise them again (Borowy, 2017, 97). It was probably thanks to the negotiating skills of the conference leader, as well as the intellectual contribution of environment and development policy thinkers, that 113 states finally took part in the United Nations Conference on Environment and Development in 1972 (Egelston, 2013, 64). In a

A.-M. Nohl (✉)
Helmut Schmidt University, Hamburg, Germany
e-mail: nohl@hsu-hh.de

© The Author(s), under exclusive license to Springer Nature Switzerland AG 2023

N. Wallenhorst, C. Wulf (eds.), *Handbook of the Anthropocene*,
https://doi.org/10.1007/978-3-031-25910-4_259

preliminary report for the conference, the line of argumentation was to be found which, although it did not directly influence the results of the conference, did influence the further discussion. Joao Augusto de Araujo Castro summarised this line of argument as follows: "There is a pollution of affluence and a pollution of poverty" (Castro & de Araujo., 1972, 409). This meant that while the overproduction of goods in industrialised countries leads to environmental problems, in the global South it is caused by poverty (Egelston, 2013, 63).

One result of the Stockholm Conference was the founding of the United Nations Environment Program (UNEP), a UN sub-organisation whose functions were more advisory than executive, which was given its headquarters in Nairobi, Kenya. Environmental non-governmental organisations had already been very active at the 1972 Stockholm Conference. They, along with some states of the South, then also actively cooperated with UNEP (ibid.). Although it is impossible to identify a clear origin, the term "sustainable development" emerged in these discussion contexts. The "World Conservation Strategy", developed by the International Union for the Conservation of Nature and Natural Resources (IUCN) in cooperation with UNEP and the World Wildlife Fund in 1980, already had "sustainable development" in its subtitle. This NGO defined development as "the modification of the biosphere and the application of human, financial, living and non-living resources to satisfy human needs and improve the quality of human life." For development to be sustainable, "social and ecological factors, as well as economic ones; of the living and non-living resource base; and of the long term as well as the short term advantages and disadvantages of alternative actions" must be taken into account (IUCN, 1980). As will be shown, this concept, with a modified definition, would become highly significant for the Brundtland Commission and beyond.

When UNEP took stock of its environmental policy activities around the globe, a decade after its founding, the results were disappointing. In particular, little had been done in the member states (Egelston, 2013, 81). The program director then called for the establishment of a commission to develop the conceptual basis for intergovernmental and international agreements on environmental protection. The United Nations General Assembly adopted this call and in 1983 its Secretary-General appointed Ms. Brundtland to lead the World Commission on Environment and Development.

Gro Harlem Brundtland (* 1939), a public health expert, was Norway's minister of the environment from 1974 to 1979 and the country's youngest and first female prime minister in 1981. After the Commission work, the Labour Party politician would go on serve two more terms as prime minister and, from 1998 to 2003, as the secretary-general of the World Health Organization (cf. Palgrave Macmillan, 2019, 61). The commission assembled by Brundtland, which was to meet eight times over a period of 3 years, came from 21 countries, mostly from the Global South, and was able to draw on the cooperation of people from diverse professional and political backgrounds (Borowy, 2017, 100).

The report of the Brundtland Commission (UN-GA, 1987) is preceded by the foreword of the chairperson and a 21-page summary. The actual 300-page report is then divided into three parts. It begins with important conceptual foundations, including the concept of sustainable development, then covers the "Challenges" that are seen not only in the environmental field but also concerning development, and ends with the "Common Endeavours" that already exist or still need to be undertaken. In addition, the opinions of various stakeholders whom the Commission has consulted in different countries are printed in separate boxes.

The Commission sees the world as being at a turning point, where global crises are not only becoming more pressing but are closely interconnected. Against the background of these "interlocking crises", which include environmental degradation and the looming doubling of the world's population (ibid., 20), the authors define the concept of sustainable development as follows: "Sustainable development is development that meets the needs of the present without compromising the ability of future generations to meet their own needs" (ibid., 54). This development concept is oriented towards a balance between the interests of the present and those of future generations. Its basis is the concept of "needs", with which the Commission takes into account the concerns of the developing countries, insofar as it is particularly about "the essential needs of the world's poor", "to which overriding priority should be given" (ibid.). However, this is only possible – and this is the second argumentative basis of sustainable development – if the industrialised countries orient their growth towards the limits of ecological resources: "Living standards that go beyond the basic minimum are sustainable only if consumption standards everywhere have regard for long-term sustainability. Yet many of us live beyond the world's ecological means, for instance in our patterns of energy use" (ibid., 54–55). This concept is explained more concretely as follows: "At a minimum, sustainable development must not endanger the natural systems that support life on Earth: the atmosphere, the waters, the soils, and the living beings" (ibid., 55). The Commission then concludes its report with a series of proposals for changing and improving institutional and legal conditions, from the level of nation states to the United Nations (ibid., 303–338).

The report was widely discussed within the UN and also inspired the Conference on Environment and Development in Rio de Janeiro in 1992, where the concept of sustainable development was endorsed (Borowy, 2017, 102). This conference was obviously also a catalyst for the broad scientific reception of the Brundtland Report (Schubert & Láng, 2005), which contributed to sustainable development becoming an established term, both in English and in other languages. However, this broad reception also led to a blurring of the concept as defined by the Brundtland Commission. Borowy (2017, 103), for example, argued that in the UN's Millennium Development Goals (MDGs), environment and development have been divided and the latter has been regarded only as a matter for the global South. However, after a "World Summit on Sustainable Development" in Johannesburg in 2002 and the "United Nations Conference on Sustainable Development" in Rio de Janeiro, member states from the global South succeeded in transforming the MDGs into 17 Sustainable Development Goals, which were accepted by the UN in 2015 as part of its 2030 Agenda for Sustainable Development.[1]

[1] https://sdgs.un.org/goals [access: 11.8.21].

A discussion of the report of the Brundtland Commission and the concept of sustainable development from the perspective of the Anthropocene leads to the emergence of parallel argumentations. Just like "Our Common Future," the concept of the Anthropocene looks at the relationship between humans and the natural environment over long periods of time and across generations. While the Brundtland Report still considers the globe as a quasi-neutral geographical location where the relationship between humans and nature is to be brought into a permanently balanced relationship, Crutzen and Stoermer (2000) stated in their seminal paper that, since the invention of the steam engine in the late eighteenth century, humans have mutated into shapers of the globe itself (as is visible in climate change, for example), which is why they famously proposed referring to the "Anthropocene" (ibid., 17). However, the fact that they then call for "developing a world-wide accepted strategy leading to sustainability of ecosystems against human induced stresses" (ibid., 18) not only shows the closeness of the ways of thinking underlying both papers, but also marks the discursive success that the concept of sustainable development has had.

However, the subsequent discussion on the Anthropocene, which is no longer conducted only by natural scientists but also by social scientists, is also suitable for problematising a specific mode of argumentation of the Brundtland Report. Here, as in the Brundtland Commission, the discussion not only transcends the boundaries of natural and social sciences (Trischler, 2016, 318) but, without denying the human causation of environmental destruction, also questions the sharp dichotomisation between humans and nature (ibid., 320), which still characterises the concept of sustainable development. This concept made nature an object (worthy of protection) but considered humans alone as the acting subject. In contrast, the Anthropocene emphasises the involvement of human and non-human actors and thus revises sustainable development in the direction of a "more balanced and integrated vision of human-nonhuman relations" (Strang, 2017, 220).

References

Borowy, I. (2017). Sustainable development in Brundtland and beyond: How (not) to reconcile material wealth, environmental limits and just distribution. In E. Vaz, C. J. de Melo, & L. M. Costa Pinto (Eds.), *Environmental history in the making* (pp. 91–108). Springer.

Castro, J. A., & de Araujo. (1972). Environment and development: The case of the developing countries. *International Organization, 26*(2), 401–416.

Crutzen, P. J., & Stoermer, E. F. (2000). The Anthropocene. *Global Change Newsletter, 41*, 17–18.

Egelston, A. E. (2013). *Sustainable development. A history*. Springer.

IUCN=International Union for Conservation of Nature and Natural Resources. (1980). *World conservation strategy*. No place.

Macmillan, P. (2019). *The Statesman's yearbook companion. The leaders, events and cities of the world*. Palgrave Macmillan.

Schubert, A., & Láng, I. (2005). The literature aftermath of the Brundtland report 'our common future'. A scientometric study based on citations in science and social science journals. *Environment, Development and Sustainability, 7*, 1–8.

Strang, V. (2017). The Gaia complex: Ethical challenges to an anthropocentric 'common future'. In M. Brightman & J. Lewis (Eds.), *The anthropology of sustainability* (pp. 207–228). Palgrave.
Trischler, H. (2016). The Anthropocene – A challenge for the history of science, technology, and the environment. *NTM (Journal of the History of Science, Technology and Medicine), 24*(3), 309–335.
UN-GA=United Nations General Assembly. (1987). *Report of the world commission on environment and development: Our common future*. https://digitallibrary.un.org/record/139811#record-files-collapse-header. Accessed 10 Aug 2021.

Arnd-Michael Nohl is a professor at the Helmut Schmidt University, Hamburg, Germany. He received his PhD in 2001 and his second doctoral grade (habilitation) in 2006. He is the author of several books and numerous chapters and articles on education, transformative learning, socialization, migration, and qualitative methodology. His books include: *Education and Social Dynamics* (with Nazlı Somel, Routledge, 2019), *Interview and Documentary Method* (Springer VS 2017, 5th edition, in German), *Pädagogik der Dinge* (Klinkhardt, 2011, in German), and *Work in Transition* (co-authored, University of Toronto Press, 2014).

Closing Time

Bruno Villalba

Abstract Today, we face the looming threat of irreversible ecological meltdown (runaway climate change, biodiversity collapse, etc.), and technological disaster (in particular, the nuclear threat). To prevent these dangers from being realised, a change of direction is needed in how political actors think and behave. The political apparatus as we know it has been built on a 'continuist' vision, which is typical of the modern era: time is thought of as being utterly limitless. However, we may now be facing a very real limit: in what is referred to as the 'End Times', humanity's time on Earth may very well come to an end – we may go the way of countless species before us, into extinction and obscurity. The image of a closing window facilitates a political perspective of time that is adjusted to take account of these all-too-real threats. It serves as a counterweight in the political mechanism. Taking a realistic view in light of the upheavals currently in progress, the options available to us for action become clear. Representing a relatively short time, the concept of a closing window highlights the need to reassess the aims of political action, from a balanced perspective integrating today's people, future generations and non-humans alike. This article presents the main features of the concept and the political prospects which it opens up.

Philosophical and political thinking about time concerns how it is modulated, and, primarily, the alternating phases of continuity/progress and discontinuity/rupture (Hartog, 2016). Modern politics is founded on a perception of time which comprises the *entire duration* of human activities (Arendt, 2007). The temporal division

This article is an updated version of the entry '*Délai*' (Closing Window), published in Bourg D. and Papaux A., *Dictionnaire de la pensée écologique*, Paris, PUF, Quadrige collection, 2015, p. 255–258. We thank PUF and Dominique Bourg for their permission to edit and reprint the article.

B. Villalba (✉)
AgroParisTech, Printemps, Paris-Saclay, France
e-mail: bruno.villalba@agroparistech.fr

© The Author(s), under exclusive license to Springer Nature Switzerland AG 2023
N. Wallenhorst, C. Wulf (eds.), *Handbook of the Anthropocene*,
https://doi.org/10.1007/978-3-031-25910-4_260

of history helps imbue it with meaning. It is organised around the notions of *progress* (continuous improvement), *directionality* (the pursuit of happiness, notably in the form of material comforts), *linearity* (history runs in one direction only), *cumulativeness* (each passing step ushers in the next, right up to the final step), and *irreversibility* (once the wheels of development have begun to turn, there is no stopping them) (Rist, 2019). This widely held view is a universal way of looking at all of human history. Through instrumental rationality, technologies, and materialistic ideologies based on decades of growth-oriented politics, society has accepted this view of history as limitless. This results in time being seen as both *continuous and infinite*. In turn, this perception creates a certain political culture around time, which leads to the construct of *permanent* time (an unlimited window for political action to have an effect). The way in which we experience time feeds into the notion of *duration* as an everlasting period of time. In western democracies, the discourse and processes of political decision-making are traditionally rooted in the metanarrative of eternal human history. Politics facilitates the idea that humans are connected to the course of history, by giving *meaning* to the constant succession of events. Logical relations are essential in constructing our view of time, giving it direction and intelligibility (Arendt, 2007). The concept of duration represents the belief in political actors' unlimited ability to barter in relation to the goals and circumstances of their actions.

Finitude Becomes Apparent. Ecological Constraints and Technological Threats

On the threshold between the twentieth and twenty-first Centuries, a series of questions emerged about the breakdown of coordination between the natural and social timescales. This leads us (or should lead us) to think about constructing a 'political culture of the limits of human fate' (Chesneaux, 1996: 48). Now, as increasing attention is being paid to *nature* and *matter*, we are beginning to see that the political view of time described above, in which human society is so firmly grounded, is inaccurate. This reveals the fundamental *chasm* between the way in which we think of politics (over the long term) and the material reality, which demands to be addressed in relatively short order. The material reality includes ecological phenomena (such as runaway climate change, energy shortages, resource exhaustion, a sixth mass extinction event, poisoning of the environment, and so forth), and social ones (burgeoning social inequality, widespread surveillance, wars over resources such as Iraq's crude oil, etc.). Precisely because of this misconception of time as infinite, based on the idea of humanity's continuous and infinite progression towards an optimum achieved through voluntaristic policy (on technical and moral levels), modern societies have overseen such alienation in the interactions between humans and the natural world that, now, the very conditions which support human life on this planet are coming under threat.

This observation leads us to propose a slight revision of the political construct of time, based on the concept of a *fast-closing window*. The underlying idea is that, for societies which so voraciously consume energy and materials, there will soon *no longer be* any long term to speak of. The concept stems from the realisation that our world can no longer be built on a succession of different ways of organising which, while one apparently gives way to the next, all ultimately have enduring consequences. Today's societies are paying the price for yesterday's decisions (nuclear threats, environmental pollution, etc.), whose *inertial* effects are placing increasing pressure on the Earth system. Thus, we need to reconsider our relationship with time, thinking of it as limited rather than infinite. The concept of a dwindling timeframe is the logical extension, when we accept the idea of finitude, and acknowledge the irreversible thresholds our societies have crossed. The limited timeframe is a more appropriate means of measuring time, in view of the ecological and technological pressures we face. The emergency looming over us has become an unavoidable point of reference (Villalba, 2016).

The concept of a closing window is rooted in the work of German philosopher Günther Anders (Anders, 2006b, pp. 247–314). 1945 represents a definitive turning point in human history. Up until that point, time had the potential to be endless; thereafter, we have been living with the ever-present threat of annihilation (ever-present, but non-specific and unknowable). With the detonation of the nuclear bomb, the world moved into what is known as 'the *end times*': '*In "the end times" means in that age when, every day, we could bring about the end of the world. – "Definitively" means that, for all the time that is left to us, it will always be "the end times". No longer can another time come about: the end times will end only with our end*' (Anders, 2006a: 116). The notion of *duration* becomes obsolete: humanity is labouring under a *deadline*, and can never more get out from under it (Anders, 2006a: 247–314). Thinking about time in terms of a closing window helps reconcile the need to make major decisions in the short term with the need to manage the long-term consequences of those decisions. The window represents a period of time that is *finite*, beyond which there is nothing but oblivion. Either annihilation occurs, or we somehow manage to defer it, but the Sword of Damocles will forever be hanging over our collective heads.

This idea reveals one of the major temporal paradoxes afflicting modern societies. The emergency we now face has become something of a political leitmotif. Nevertheless, State authorities continue to think in relation to the long term, and act accordingly. Political authorities have recognised the skew between the timeframes on which nature and society operate; however, they still view it through a corrupting prism, failing to recognise the limitations of their own room for manoeuvre – their capacity to intervene. It is now imperative that we cease to view political timeframes through the lens of political projections (i.e. what we want). Rather, we must examine them realistically, in terms of the time remaining to us to enact policies which will ensure our continued existence. We also urgently need a paradigmatic shift in the way democracy works: going forward, it must be *founded upon* and *integrated into* this new context (both temporal and material) of impending resource exhaustion.

Key Features of the Closing Window

1. The window of which we speak cannot be considered a hypothetical construct; it is a material *reality* in today's world. Reports on the threat of climate meltdown all emphasise that we have a very short timeframe in which to address the problem (less than a decade, according to the IPCC). The increased destructive capacity of the atomic bomb means that, at any moment, the very idea of time could lose all meaning.
2. The window does *not* represent a specific and inescapable *fateful day*, whose date is set in stone. Rather, it is the movable slice of time remaining before we pass certain points of no return. However, the clock is already counting down, as we face the threat of irreversible damage: climate change is part of our day-to-day reality.
3. The idea can also help us understand the *spatial* dimension of reality. In spatial terms, the shrinking of the timeframe is related to the material constraints on our world. Not only do we need to re-examine our relationship with time; we need to think of the sequence of time as a physical reality, in relation to a physical territory. In so doing, we can differentiate our responses to the narrowing window of time to take action; we can tailor those responses to what is appropriate in a given geographical context. Such a spatial interpretation could help tear down the barriers between planning and actually taking action. At international climate events, there is a tendency to push back any concrete action (instead, governments make commitments to make commitments!). Typically, island nations bewail this can-kicking, because for them, there is a very real physical significance to what is happening: their islands will end up underwater. Thus, they think of the timeframe for action in terms of the fate of their own territory.
4. The closing window represents a *reduction in the available options*. According to traditional political thinking, there are an infinite number of possible combinations of factors. For example, social contractualism (Hobbes, Locke and Rousseau, among others) holds that the conditions in which actors associate with one another can constantly be renegotiated. In liberal democracy, all individuals have the ability to define themselves through their own unrestricted choices. However, faced with the threats of irreversible damage, we inevitably wonder whether that profusion of choice can possibly be maintained. We must implement realistic policies of choice, which take account of the material world's capacity to sustain those choices. The implication is that gradually, as the window closes, the range of choices available to us is falling away. In turn, our reduced capacity for choice will limit the possibility for other choices to emerge.
5. The concept implies that ultimately, we will find ourselves facing *inevitable* situations. Conversely, the hypothesis of *duration* implies that the emergency can always be deferred: we will always be able to renegotiate or develop a new solution. The idea of a closing window stems from a catastrophist – or at least alarmist – political standpoint. From such a perspective, it is crucial to reorder the priorities which guide political actions. What are the fundamental values which

must be safeguarded, not only to ensure that future generations *have* a future of which to speak, but also to serve the ideals of democracy?

6. The image of a closing window inevitably creates *anxiety*. Our inability to escape the ecological and technical constraints underlines how limited our ability to intervene actually is. Our days may be numbered, but we must make important choices. In order to do so, we need to leave behind the view of time which appeared to offer limitless new possibilities (for growth, innovation, etc.). Instead, we must begin continuously reassessing the time available to us (in full awareness of the dwindling scope for action). However, that anxiety may stimulate productivity, helping guide us to an appropriate response to the catastrophe we face (Dupuy, 2002).

Political Perspectives on the Closing Window

When we view the time-window in this way, it may seem as though politics and policy are toothless. How can we possibly take effective preventative action if, at this point, the mechanisms of irreversible change have already been triggered? However, by recognising the urgency of the threat, we give freedom to public decision-makers to take action. Such recognition can help reshape the way in which we relate to the future, and define the possibilities for effective action.

To begin with, *recognising the diminishing timeframe available is a political act in itself*, because it forces us to abandon a *conditional* view of the future, and adopt a *present-tense* approach to managing the climate emergency and the nuclear threat. Once we recognise the danger, we can no longer base our actions on an '*if*', but must act on the basis of a '*when*'. Decisions are no longer shaped by a context in which the threat could potentially be controlled indefinitely (the aforementioned 'if' scenario, which often leads to delay after delay after delay in reaching a decision). Instead, recognising the irreversible mechanisms already at work, we labour to prevent the consequences from becoming even worse (the 'when' scenario). Hence, political action is no longer a matter of our intention (what we want), but of what it is possible to do, within the time remaining, to defer the danger. Thus, the idea of a closing window reframes the time we have to take action, based on an understanding of what is *essential*. If we have little time left to make a decision, then we must focus on what is truly significant from a collective point of view.

Next, it should be noted that the window means we can still *continually regulate the threats*, and in so doing, keep the window open. When we acknowledge the reality of the irreversible threats, we see that by taking action within the time available, we can constantly wrangle the contributing factors and address the dangers, to push back the tipping points as far as possible. Thus, the concept of a closing window is a useful decision-making tool, because it forces us to integrate longer-term consequences into our short-term thinking. It therefore prevents us from putting off, for some distant future, the responsibility for dealing with the harmful effects of our present decisions (such as management of nuclear waste over thousands of years).

Consequently, the idea helps *transform democratic debate*. How can we come to decisions that are fair – i.e. which conform to an ideal of justice and solidarity that we wish to see perpetuated and extended across the globe? How can we pursue an emancipatory political agenda (driven by liberal individualism) whilst ensuring that we do not exceed the limits of what the planet can offer? Answers to these questions need to be found, in short order. They must be fair, for future generations as well as the present ones, and non-humans must be taken into account. Thus, we must think about the conditions for establishing liberal democracy: increasingly, our choices will be constrained by ecological factors (Villalba, 2017). Awareness of the closing window allows us to get ahead of this potential clash, by taking it into account in political debate and decisions made today.

Finally, on a theoretical level, the threat of running out of time to act *liberates the political imagination*. It opens the door to a school of political thought based on the prospect of drastic shifts – i.e. political action that considers the looming spectre of irreversible harm. Our thinking is freed from the *continuist hypotheses* (the idea of the long term, the 'if' scenario, etc.), allowing us to construct a relationship with the world as it actually is. Thus, we can envisage a different way of relating, politically, with non-humans (Pelluchon, 2021) and with the planet (Serres, 1995).

References

Anders, G. (2006a). *Le temps de la fin*. L'Herne.

Anders, G. (2006b). Le Délai (1960). In *La menace nucléaire*. Considérations radicales sur l'âge atomique, Éditions du Rocher/Le Serpent à plumes, (pp. 247–314).

Arendt, H. (2007). *The promise of politics*. Schocken.

Chesneaux, J. (1996). *Habiter le temps* (pp. 19–30). Bayard.

Dupuy, J.-P. (2002). *Pour un catastrophisme éclairé*. Seuil.

Hartog, F. (2016). *Regimes of historicity: Presentism and experiences of time*. Columbia University Press.

Pelluchon, V. (2021). *Les lumières à l'âge du vivant*. Seuil.

Rist, R. (2019). *The history of development: From Western origins to global faith* (5th ed.). zed books.

Serres, M. (1995). *The natural contract*. University of Michigan Press.

Villalba, B. (2016). Temporalités négociées, temporalités prescrites L'urgence, l'inertie, l'instant et le délai. In B. Hubert & N. Mathieu (Eds.), *Interdisciplinarités entre Natures et Sociétés* (pp. 89–109). Peter Lang.

Villalba, B. (2017). L'assèchement des choix. Pluralisme et écologie. *La Pensée écologique, 1*.

Bruno Villalba is Professor of Political Science, at AgroParisTech (Paris-Saclay) and Member of Printemps UVSQ (CNRS UMR 8085). His areas of research are Political Ecology, Sustainable Development and collapse studies. His research focuses on environmental political theory, notably through analysis of the capacity of the democratic system to reformulate its goals based on environmental constraints. He is the author or co-author of over ten books on politics and ecology. Books (selection): Collapsologists and their enemies (ed. Le Pommier, 2021, in French); Political ecology in France (Paris, La Découverte, 2022, in French).

Club of Rome

Roman Luckscheiter

Abstract Founded in 1968, the Club of Rome is a global network of experts committed to finding holistic solutions to key challenges of the Anthropocene. It caused a worldwide sensation with the report *The Limits to Growth*, which was prepared by a team from The Massachusetts Institute of Technology (MIT) led by Dennis Meadows. This controversial report modelled the threat of collapse if economic activity continued in an unsustainable manner and became a cautionary plea for a profound transformation to a more balanced global community. Fifty years later, in the context of the UN's 2030 Agenda for Sustainable Development, the new way of thinking promoted by the Club of Rome is more relevant than ever.

The founding of the Club of Rome took place in a year that has become a symbol of the protest movement worldwide: 1968 is regarded in the Western industrialized nations as the symbolic year of student unrest and the associated questioning of social values. The founding members of the Club, who met at the Accademia dei Lincei in Rome in 1968, did not have a revolt in mind, however, but rather—as one would say today—a 'transformation'. They pursued two goals: on the one hand, to make the worldwide and closely interrelated problems of humanity more transparent and more generally known ("die weltweiten und eng miteinander verknüpften Probleme der Menschheit besser durchschaubar und allgemeiner bekannt zu machen", Peccei, 1973:38) and, on the other hand, on the basis of such an awareness, to arrive at a social, political, and human development ("gesellschaftliche[n], politische[n] und menschliche[n] Entwicklung", ibid.) that would make a true peace rooted in the hearts and minds of men ("wahre[r] Frieden, der im Herzen und im Denken der Menschen verwurzelt ist", ibid.) possible. Unlike the '68 generation', the Club of Rome did not emerge as an extra-parliamentary opposition movement, but rather as an international circle from the centre of the academic and economic 'establishment'. Its foundation is to be understood in particular in the context of the

R. Luckscheiter (✉)
German Commission for UNESCO, Bonn, Germany
e-mail: Luckscheiter@unesco.de

© The Author(s), under exclusive license to Springer Nature Switzerland AG 2023
N. Wallenhorst, C. Wulf (eds.), *Handbook of the Anthropocene*,
https://doi.org/10.1007/978-3-031-25910-4_261

1960s, which were characterized by social modernization debates in the Western countries. The dissertation by Friedemann Hahn (2006) provides a good overview of this. The central figure in its founding was the 60-year-old Aurelio Peccei, a senior manager at an Italian automobile company, alongside his fellow campaigners Alexander King, a British chemist of almost the same age who worked for the OECD in Paris, and the 54-year-old Eduard Pestel, who had held a professorship in engineering at the Technical University of Hannover and was later to occupy high-ranking positions, for example at the German Research Foundation (Deutsche Forschungsgemeinschaft) and the Fraunhofer Society (Fraunhofer-Gesellschaft), and as Minister of Science in Lower Saxony.

In 1969, Peccei shook up his colleagues with the book *The Chasm Ahead.* It brought—as Eduard Pestel recalled in 1973—the realization that tinkering with the symptoms ("Herumdoktern an Symptomen", Pestel, 1973:45) would no longer help. These symptoms, which foreshadowed the 'chasm ahead', included uncontrolled population growth, widespread hunger, underemployment, and urban misery especially in the developing world, environmental pollution, foreseeable depletion of vital raw materials, growing social tensions, rapidly increasing urbanization, rising crime, inexorable polarization of military power, cultural uprooting, alienation between youth and the elderly, demand for co-determination, a widening gap between industrialized and developing countries, etc. ("unkontrolliertes Bevölkerungswachstum, weit verbreiteter Hunger, Unterbeschäftigung und Großstadtelend besonders in den Entwicklungsländern, Umweltverschmutzung, absehbare Erschöpfung lebenswichtiger Rohstoffe, wachsende soziale Spannungen, rapid zunehmende Verstädterung, ansteigender Kriminalität, unaufhaltsame Polarisierung militärischer Macht, kulturelle Entwurzelung, Entfremdung zwischen Jugend und Alter, Mitbestimmungsverlangen, wachsende Kluft zwischen Industrie- und Entwicklungsländern usw.", ibid.:45). The emerging Club of Rome distinguished itself not least of all by viewing this list of 'symptoms' as interrelated and tracing them back to a common origin: a global imbalance. On the threshold of the 1970s, a global mindset manifested itself that identified the growing discrepancy between industrialized and developing nations as the result of a misguided thinking with regard to progress. The named grievances, Pestel argued, were for the first time not problems deliberately caused 'by evil men and powers', but unintended consequences of a particular way of thinking. This laid down the Club of Rome's special line of approach: It was not a matter of fighting a particular evil, but rather of changing goals and acting anticipatorily ("antizipatorisch zu handeln", ibid.:49).

In order to effectively pursue its own goals, it was decided to commission comprehensive 'reports'. The very first study commissioned by the Club of Rome attracted maximum attention and is to this day directly associated with the Club as its definitive achievement: *The Limits to Growth* was published in 1972. The decision for this study was taken in 1969 at the first plenary meeting of the Club of Rome in Bern (where they met at the invitation of the Swiss Federal Council). The Massachusetts Institute of Technology (MIT) was commissioned to use its latest methods to show the consequences of population and economic growth, the dependencies that made up the 'world system', and the factors that could be used to influence them. The Volkswagen Foundation (Stiftung Volkswagenwerk) was enlisted to finance the study—which later caused Ernst Klett to show particular respect: The foundation had supported the project despite the fact that it was aware that the result would not exactly be the recommendation that the number of motor vehicles in the world should be multiplied ("obwohl ihr bekannt war, daß als Ergebnis nicht gerade die Empfehlung herauskommen werde, man solle die Zahl der Kraftwagen in der Welt vervielfachen", Klett, 1973:13).

The novelty of the study conducted by Dennis Meadows and his team lay in the so-called 'world model': This was based on "five basic quantities, or levels —population, capital, food, nonrenewable resources, and pollution—" (Meadows et al., 1972:89), their causal relationships to each other, and their interconnections through "interrelationships and feedback loops" (ibid.). Using computers, the researchers from various disciplines simulated developments for the period between 1900 and 2100 (ibid.:90)—not to make concrete predictions, but to identify and understand the causes and limits of growth (ibid.:91 f.). The central finding: Unchecked growth inevitably leads to the exhaustion of raw material supplies, then to the complete collapse of the entire system with a steeply rising mortality rate and shortened life expectancy (ibid.:125 ff.). Instead, Meadows and his team deliberately describe the 'state of global equilibrium' to be aimed for as a dynamic state that should not be confused with stagnation or "an end to progress" (ibid.:179). Rather, the desired stage would be characterized by constant population numbers and capital volumes and would create new free spaces, e.g., for technological progress to combat environmental pollution, etc. (ibid.:156 ff.). These were amply controversial basic assumptions; however, beyond the problematic models and calculations, the unchanged current message of the study is: "The equilibrium society will have to weigh the trade-offs […] not only with consideration of present human values but also with consideration of future generations" (ibid.:182).

The report was presented in advance for discussion at international conferences in Moscow and Rio de Janeiro. In its 'commentary' on the study, the Club of Rome summarizes the main objections, which relate above all to the model and the data situation (ibid.:186 ff.), but broadly endorses the conclusions of the MIT team. It emphasizes that "a basic change of values and goals at individual, national, and world levels" (ibid:195) is also crucial. At the same time, the board of the Club of Rome is at pains not to allow an overly pessimistic reading of the study (ibid.). "The question remains of course whether the world situation is in fact as serious as this book, and our comments, would indicate" (ibid.). In an 'update' by the Club on the thirtieth anniversary of the report, this latent optimism is interpreted as a symptom of the times: "In 1972 it seemed that humanity's population and economy were still comfortably below the planet's carrying capacity. We thought there was still room to grow safely while examining longer-term options. That may have been true in 1972; by 1992 it was true no longer" (Meadows et al., 2005:xii). Ernst Ulrich von Weizsäcker, Co-President of the Club of Rome from 2012 to 2018, recalls the 'shock' the report caused at the time: no one had thought of the long-term consequences of continuous growth ("Niemand hatte an die Langfristfolgen des Dauerwachstums gedacht.", Weizsäcker & Wijkman, 2019:11). In 2017, von

Weizsäcker wrote that the Meadows report remains relevant despite new analytical tools: Today's computer models are certainly much more sophisticated than the World3-model used at the time. Several ecologically favorable forms of growth in the last 50 years were not expected at that time. [...] And dangers that were hardly on the radar at the time have become burning issues, such as climate change, the scarcity of fertile soils, and the extinction of species ("Gewiss sind heutige Computermodelle viel raffinierter als das damals benutzte World3-Modell. Manche ökologisch günstige Wachstumsformen in den letzten 50 Jahren hatte man damals nicht erwartet. [...] Und Gefahren, die man damals kaum auf dem Schirm hatte, sind heute brennend aktuell, so die Klimaänderung, die Knappheit an fruchtbaren Böden und das Artensterben", ibid.).

There are detailed accounts of the extensive reception of the study and thus also of the perception of the Club of Rome around 1972. The focus of international criticism in general and of German criticism in particular was, on the one hand, that the model did not take social factors in account and, on the other hand, that the study was an expression of a naïve faith in computers ("naïve[n] Computergläubigkeit", Hahn, 2006:105 ff.). The accusations of a lack of scientific rigor, however, were also countered by appreciations that recognized the study's stirring character. For example, from the futurologist Robert Jungk (Hahn, 2006:107). When the study was also debated at the 1972 meeting of Nobel laureates, German Chancellor Willy Brandt argued in his speech, titled "Environmental Protection as an International Mission", that the study should be used as an opportunity for a new kind of planning—and that growth should not be abolished, but rather restructured, because anything else would sound like a mockery ("wie Hohn klingen", cited in Hahn, 2006:113), especially for developing countries (ibid.). In this context, it was tantamount to an avowal when the Association of the German Book Trade (Börsenverein des Deutschen Buchhandels) awarded its Peace Prize to the Club of Rome in 1973 as an explicit tribute to the courage and intellectual energy to design a future worth living ("Mut und die geistige Energie zum Entwurf einer lebenswerten Zukunft", Börsenverein, 1973:5), that the Club had demonstrated with its "report on the state of humanity" (ibid.). Support came from the world of politics—both from the laudator, Swiss Federal Councillor Nello Celio from Bern, and from the Lord Mayor of the City of Frankfurt am Main. The latter recalled the 1971 General Assembly of the Association of German Cities (Deutscher Städtetag) and its appeal to save cities in the face of haphazard economic growth ("planlose[s] wirtschaftliche[s] Wachstum[s]", Arndt, 1973:17 ff.).

The first report of the Club of Rome has become a milestone in the debate about the future of humanity in the face of the effects of its actions. Subsequent reports have not been able to follow up on this global attention, but they have continued to develop its objectives further and keep the spirit of the founding of the Club of Rome in 1968 alive. Over the years, the reports could be increasingly integrated into a global, institutionalized, and public engagement with questions regarding finite resources, such as *The New Limits to Growth*, which was published in 1992, the year of the Earth Summit in Rio de Janeiro. In 2015, the United Nations took an important step with the 17 goals of the 2030 Agenda for Sustainable Development. The

Club of Rome's objective thus finds a whole new resonance, in which the legitimacy of materialistic egoism ("Legitimität des materialistischen Egoismus", Weizsäcker & Wijkman, 2019:12) is now systematically called into question (ibid.). With the 2030 Agenda, the global community gives itself a framework that aims to overcome the 'silo' structure of economy, society, and environment (ibid:17) and also makes another idea of the Club of Rome from the founding year of 1968 a universal benchmark: Von Weizsäcker names balance ("Balance" ibid.:379) as a key feature of this new enlightenment ("neue[n] Aufklärung" ibid.), formulated in an up-to-date manner as harmony between the economic and the ecological SDGs ("Harmonie zwischen den ökonomischen und den ökologischen SDGs", ibid.). It is less the data and much more the spirit of the first report of the Club of Rome that is of considerable relevance today and is reflected, for example, in the UNESCO program Education for Sustainable Development. The Club of Rome itself ensures through various networks, from schools and think tanks to national, independent organizations, that the principle of balance has an international lobby that is simultaneously strengthened and challenged in these times of climate change.

References

Arndt, R. (1973). Oberbürgermeister der Stadt Frankfurt am Main. In *The Club of Rome. Ansprachen anlässlich der Verleihung des Friedenspreises*. Börsenverein des Deutschen Buchhandels.

Börsenverein des deutschen Buchhandels. Der Börsenverein verleiht seinen Friedenspreis 1973 an den Club of Rome. (1973). In *The club of Rome. Ansprachen anlässlich der Verleihung des Friedenspreises*. Börsenverein des Deutschen Buchhandels.

Hahn, F. (2006). Von Unsinn bis Untergang: Rezeption des Club of Rome und der Grenzen des Wachstums in der Bundesrepublik der frühen 1970er Jahre, PhD. University of Freiburg, 2006. Available at: https://freidok.uni-freiburg.de/data/2722. Accessed 19 Nov 2021.

Klett, E. (1973). Vorsteher des Börsenvereins. In *The club of Rome. Ansprachen anlässlich der Verleihung des Friedenspreises*. Börsenverein des Deutschen Buchhandels.

Meadows, D., et al. (1972). The limits of growth. In *A Report for the CLUB OF ROME'S Project on the Predicament of Mankind.* Potomac Associates.

Meadows, D., Randers, J., & Meadows, D. (2005). *Limits to growth. The 30- year update*. Earthscan.

Peccei, A. (1973). Dank für den Club of Rome. In *The Club of Rome. Ansprachen anlässlich der Verleihung des Friedenspreises*. Börsenverein des Deutschen Buchhandels.

Pestel, E. (1973). Arbeit und Ziele des club of Rome. In *The club of Rome. Ansprachen anlässlich der Verleihung des Friedenspreises*. Börsenverein des Deutschen Buchhandels.

Weizsäcker, V., Ulrich, E., & Wijkman, A. (2019). *Wir sind dran: was wir ändern müssen, wenn wir bleiben wollen*. Pantheon.

Dr. Roman Luckscheiter ist seit Januar 2020 Generalsekretär der Deutschen UNESCO-Kommission. Zuvor war der promovierte Germanist und Romanist für den Deutschen Akademischen Austauschdienst (DAAD) tätig, zu dem er 2008 von der Universität Heidelberg gewechselt war. Im DAAD war er in leitenden Funktionen unter anderem für die Förderung der deutschen Sprache weltweit, für Grundsatzfragen transnationaler Bildung und für zahlreiche Stipendienprogramme verantwortlich. Von 2014 bis 2018 leitete er die Außenstelle des DAAD in

Kairo. Roman Luckscheiter has been Secretary-General of the German Commission for UNESCO since January 2020. Previously he worked for the German Academic Exchange Service (DAAD) after finishing his Ph.D. studies at the University of Heidelberg. At the DAAD, Roman Luckscheiter was responsible for the promotion of the German language worldwide, for policy issues of transnational education, and for numerous scholarship programs. From 2014 to 2018, he headed the DAAD office in Cairo.

Critical Zone

Jeanne Etelain

Abstract This article examines the concept of Critical Zone – the skin of the planet Earth where rock meets life – that has recently emerged from the geoscience community. Outstanding research opportunities related to the ontological and epistemological problems of the Critical Zone also warrant expansions into artistic, anthropological, and political inquiry, while raising some significant philosophical questions about the space of the Earth and the human.

We have entered the Anthropocene, human activities transforming the Earth System itself. Global warming puts more and more pressure on the Earth's surface that supports nearly all terrestrial life, the human species, and our societies. Most of the life-sustaining processes and all living things exist in a narrow band close to the surface of the planet. Scientists have named this sensitive area the Critical Zone (CZ). In 2001, the US National Research Council (NRC) elected the study of the CZ as one of the most compelling research areas in Earth sciences in the twenty-first century. The NRC defines the CZ as the "heterogeneous, near surface environment in which complex interactions involving rock, soil, water, air, and living organisms regulate the natural habitat and determine the availability of life-sustaining resources" (2001, p. 2). Brantley et al. provide an alternative definition of the CZ as "the fragile skin of the planet defined from the outer extent of vegetation down to the lower limits of groundwater" (2007, p. 307).

The CZ, a term first introduced by the sedimentologist Gail Ashley (1998), is the thin outer veneer of the planet Earth's surface, lying between the sky and the rocks, that is the seat of life. It encompasses the lower atmosphere, vegetation canopy, water bodies (rivers, lakes, shallow seas), soil layers (pedosphere, vadose zone, the water table), and fresh groundwater. It is a porous medium resulting from interactions between biogeochemical-physical processes animated by solar energy, the

J. Etelain (✉)
New York Univerisity, New York City, NY, USA

Université Paris Nanterre, Nanterre, France
e-mail: jeanne.etelain@nyu.edu

© The Author(s), under exclusive license to Springer Nature Switzerland AG 2023
N. Wallenhorst, C. Wulf (eds.), *Handbook of the Anthropocene*,
https://doi.org/10.1007/978-3-031-25910-4_262

transformation of minerals in contact with gases giving birth to the land surface, the water cycle, and the living beings which populate Earth. Coterminous with the Gaia Hypothesis invented jointly by the geochemist James Lovelock and Lynn Margulis (1974), it is those same living beings who create and perpetuate the ideal conditions of their own existence and what is needed for life to thrive, namely the atmosphere, the oceans, and the crust—geology and biology being coexstensive to one another. Strongly affected by processes in the atmosphere, lithosphere, hydrosphere, and biosphere, the exchange of matter and energy that occurs within the CZ interacts with the other envelopes of the Earth System and determines the general habitability of the planet.

The CZ is perhaps the most heterogeneous portion of the Earth: it is where many physical, chemical, and biological components meet through complex linkages and feedbacks. It is a dynamic interface that makes the planet habitable: for example, by reacting with the rocks, the CO_2 from the air is neutralized on the continents to become limestone in the ocean. The CZ is all the more dynamic since it is the site of processes in which the involved elements are in constant transformation with each other through their interactions. The CZ is originally a very small portion of the Earth's surface where the activity of micro-organisms has developed and has gradually, over billions of years, generated the atmospheric composition, transformed the rocks, etc. The boundaries of the CZ are therefore not fixed but in motion. It results that the CZ plays a key role in the dynamics of the Earth System and its major biogeochemical cycles, and for its feedback on its global regulation. As a support for life, this sensitive layer is strongly reactive to humans (agriculture, industrialization, urbanization, etc.) who shape the landscape, extract its resources, and store their waste. It is critical in the physical sense of the term because it is one of the limit interfaces of the planet, a threshold that can cause abrupt changes. But it is also critical in the political sense of the term as it has become the "zone to defend."

Probably one of the most important applied aspects is the development of reliable data that will enable practices to protect the CZ. But, because of the coupled biogeochemical processes happening across spatial scales (from atomic to planetary) and temporal scales (from seconds to eons), the concept of CZ raises serious epistemological issues. As the geochemist Jérôme Gaillardet explains, there is a tension between lab experiments and field observations: when we try to reproduce how minerals in rocks are transformed into soil materials, we observe that the transformation rates obtained in the laboratory are much quicker (Latour & Weibel, 2020, p. 124). The reason for this difference is that all the factors that play a role *in situ*, such as the circulation of water or the composition of the soil, cannot be reproduced entirely in the laboratory. It is thus necessary to study each individual site in its entirety instead of studying one isolated reaction. This irreconcilable gap between the field and the laboratory challenges the scientific method—based on the assumption that it is the same everywhere and therefore can be reproduced experimentally—and paves the way for a science of the singular.

Drawing from Banwart et al. (2013, p. 20), the big science question raised by the CZ could be summarized as follows: how can complex variable processes be quantified by empirical observation, measured by sensitive instruments, aggregated into data sets, and predicted by mathematical modelling? Research communities argue that the study of the CZ depends on strategies to cross disciplinary boundaries (Brantley et al., 2007). For example, by combining the expertise of a geologist and a biologist, we can examine how one bacterium in the soil reacts to the presence of a metal. The concept of CZ thus challenges a conception of knowledge as object-centered and discipline-bounded, the study of a particular given object providing a particular discipline with its identity (matter for physics, life for biology, soil for geology, etc.). Process-oriented, the CZ involves entities which used to be studied by separate disciplines. The specialization of modern sciences is seen as a hindrance to an integrative—even holistic—scientific approach to the Earth that would facilitate the understanding of its behaviour in the face of great changes.

Processes in the CZ span wide spatiotemporal scales. If it takes a few seconds for a bacterium to reproduce in the soil, the degradation of minerals can take millions of years. If methane reflects atomic biochemical processes, it also involves global climatic factors such as precipitation. Interactions within the CZ thus require both long and frequent observations in selected ecosystems and at diverse locations. Numerous interdisciplinary programs called Critical Zone Observatories (CZOs) have been created, particularly in the USA, Europe, and Asia (Anderson et al., 2008; Giardino & Houser, 2015; Brantley et al., 2017; Gaillardet et al., 2018). Resulting in large amounts of data being collected across the entire planet by a wide range of disciplines, CZOs raise the question of the part-whole, both in terms of the relation of the parts to the whole and the relations between the parts themselves. How do the site-specific processes studied in the CZOs interact with the CZ wider system? Is the CZ more than its isolated parts and has properties of its own? Are the CZOs independent of each other and indifferent to their collection in the CZ? Are the global processes in the CZ just another local variable, making the CZ at once what encompasses all the parts and a part next to the parts? This problem is evident in the hesitation as to whether the term CZ should be employed in the singular, designating the entire Earth's surface, or in the plural, with reference to specific locations.

As a science for the Anthropocene, the CZ has gone beyond the strict framework of the natural sciences. For example, the exhibition organized by the artist Peter Weibel and the philosopher Bruno Latour addresses this concept from aesthetic, philosophical, and anthropological perspectives (2020). The CZ eludes traditional visualizing tools such as cartographies because of the complex interactions happening between various entities, processes, and scales. In collaboration with the historian of science and stage director Frédérique Aït-Touati, the Architect Alexandra Arènes has multiplied initiatives to offer new conceptual depictions that allow scientists to represent the Earth by including the dynamism of geochemical cycles and living things that shape the CZ while considering the human as fully immersed within it (Arènes, 2017; Arènes et al., 2018; Aït-Touati et al., 2019). She notably uses anamorphic projections of the planet as seen from the inside rather than from outer space.

Latour has put forward the CZ as a scientific concept that is highly geopolitical, in the literal sense of a politics of the Earth (2014). The CZ forces us to change at once our understandings of what is a land and what is a people, transforming our definitions of both territory and sovereignty. Within the coordinates of the CZ, a piece of land is composed by multiple forms of human behaviours (from EU legislation to agricultural practices to consumers habits) as well as by a diversity of non-human actors (from organisms to rocks to gases). The challenge is to reconcile the connection between land and people without falling into the reactionary trap of current rising nationalisms. Latour thus coins the notion of the terrestrial to designate the earthly politics of life forms to be undertaken in the CZ for the age of the Anthropocene (Latour, 2015, 2017; Latour & Weibel, 2020).

As a thin layer, the CZ is neither the Earth nor the Globe. As the seat of life, the CZ might seem close to the concepts of biosphere and ecosystem. But its focus on geochemical processes moves us away from both biocentric and ecocentric approaches.

The postcolonial historian Dipesh Chakrabarty has both extended and questioned the geocentrism of the CZ (2019). He advocates instead for the concept of the planet which must be distinguished from that of Gaïa (centered around that which makes life possible) and from that of world (centered around the ground for human dwelling). For him, these two concepts are not geocentric enough because they do not consider the long-term planetary processes that out-scale both the temporal horizons of life and the human. In addition to the Earth's surface, the concept of the planet integrates: (a) the deeper parts of the underneath Earth that include the rocky, hot, molten interior and function independently of the CZ; (b) the planetary pluralism that requires comparing the Earth with other planets like Mars and Venus in order to understand the conditions for the planet to become the seat of habitability processes in the first place. The planet thus requires a theory of politics that forces humans to face a dimension of their action that exceeds their own existence on Earth.

References

Aït-Touati, F., Arènes, A., & Grégoire, A. (2019). *Terra Forma*. Éditions B42.

Anderson, S., Bales, R. C., & Duffy, C. (2008). Critical zone observatories: Building a network to advance interdisciplinary study of earth surface processes. *Mineralogical Magazine, 72*(1), 7–10.

Arènes, A. (2017). Tracer le vivant. Enquête dans le territoire de Belval, Ardennes. *Billebaude, 10*, 88–93.

Arènes, A., Latour, B., & Gaillardet, J. (2018). Giving depth to the surface: An exercise in the Gaia-graphy of critical zones. *The Anthropocene Review, 5*(2), 120–135.

Ashley, G. (1998). Where are we headed? "Soft" rock research into the new millennium. *Geological Society of America Abstract/Program, 30*, A-148.

Banwart, S., Chorover, J., Gaillardet, J., et al. (2013). *Sustaining Earth's critical zone*. The University of Sheffield.

Brantley, S., Goldhaber, M., & Ragnarsdottir, V. (2007). Crossing disciplines and scales to understand the critical zone. *Elements, 3*(5), 307–314.

Brantley, S., McDowell, W., Dietrich, W., et al. (2017). Designing a network of critical zone observatories to explore the living skin of the terrestrial earth. *Earth Surface Dynamics, 5*(4), 841–860.

Chakrabarty, D. (2019). The planet: An emergent humanist category. *Critical Inquiry, 46*(1), 1–31.

Gaillardet, J., Braud, I., Gandois, L., et al. (2018). OZCAR: The French network of critical zone observatories. *Vadose Zone Journal, 17*(1), 1–24.

Giardino, J., & Houser, C. (Eds.). (2015). *Principles and dynamics of the critical zone* (Developments in Earth Surface Processes) (Vol. 19). Elsevier.

Latour, B. (2014). Some advantages of the notion of "critical zone" for geopolitics. *Procedia Earth and Planetary Science, 10*, 3–6.

Latour, B. (2015). *Face à Gaïa: huit conférences sur le nouveau régime climatique*. La Découverte.

Latour, B. (2017). *Où atterrir? Comment s'orienter en politique*. La Découverte.

Latour, B., & Weibel, P. (Eds.). (2020). *Critical zones: The science and politics of landing on earth*. ZKM/MIT Press.

Lovelock, J., & Margulis, L. (1974). Atmospheric homeostasis by and for the biosphere: The Gaia hypothesis. *Tellus, 26*(1–2), 2–10.

National Research Council. (2001). *Basic research opportunities in earth science*. The National Academies Press.

Jeanne Etelain is a Ph.D. candidate at New York University and Paris Nanterre University. Her dissertation is a philosophical investigation of the concept of zone in Western thought. She elects the zone as a basic category of contemporary thinking and as a key epistemological paradigm to the emerging field of the environmental humanities, for it sparks rich forms of comparative, transversal, and critical analysis. Her work has been published in *Les Temps modernes* and *La Deleuziana* (on whose editorial board she serves). She also contributed to Latour's edited catalog *Critical Zones* with an essay entitled "This Planet Which Is Not One."

IPBES

Alice B. M. Vadrot

Abstract The Intergovernmental Science-Policy Platform on Biodiversity and Ecosystem Services (IPBES) is a body that publishes global, regional, and thematic assessments about the state of biological diversity. Narratives and "epistemic selectivities"- as I call them- played an essential role in the making of this biodiversity knowledge body. Future (critical) research on IPBES is needed on knowledge integration practices at various sites and levels of policymaking.

The Intergovernmental Science-Policy Platform on Biodiversity and Ecosystem Services (IPBES) is a body that publishes global, regional, and thematic assessments about the state of biological diversity. The IPBES provides the Convention on Biological Diversity (CBD) and other Multilateral Environmental Agreements with timely and policy-relevant expertise. Like its counterpart for climate change, the Intergovernmental Panel on Climate Change (IPCC), the IPBES uniquely involves both experts and government representatives in its assessments.

Instead of applying a "linear model of expertise" (Beck, 2011) "speaking truth to power," IPBES structures ensure the participation of a broad spectrum of knowledge holders from diverse regions and scientific disciplines, including traditional knowledge (TK) of indigenous people and local communities. Thus, they go beyond the IPCC model, which has systematically favoured natural sciences as championed by Global North academia and limited the space for alternative representations of nature. This shift away from a Eurocentric, economistic, and anthropocentric understanding of nature has turned the IPBES into a popular research object for social scientists interested in global environmental knowledge production.

A. B. M. Vadrot (✉)
University of Vienna, Vienna, Austria
e-mail: alice.vadrot@univie.ac.at

© The Author(s), under exclusive license to Springer Nature Switzerland AG 2023

N. Wallenhorst, C. Wulf (eds.), *Handbook of the Anthropocene*,
https://doi.org/10.1007/978-3-031-25910-4_263

History

While the IPCC was formed in 1988 and thus supported the United Nations Framework Convention on Climate Change (UNFCCC, 1992), the IPBES came into being almost 20 years *after* the ratification of the Convention on Biological Diversity (CBD, 1992). Thus, for a long time, agreement-making on biodiversity could not rely on a globally agreed knowledge base, concepts, and indicators, unlike climate change.

Indeed, the success of the IPCC in popularizing climate change as both an environmental problem and a global policy issue spurred a debate on the lack of an "IPCC for biodiversity" and "Diversity without Representation" (Loreau et al., 2006). Not only did the IPCC provide a role model for interfacing science and policy in a complex, multilateral institutional landscape, its assessments coincided with parallel trends in the biodiversity field, most notably the Global Assessment Report (Heywood, 1995) and the Millennium Ecosystem Assessment (MA, 2005). While the former failed to become the CBD's knowledge base for decision-making, the latter was perceived as an important step towards providing policy-relevant knowledge. Involving a broad range of biodiversity scientists from all over the world in a UN-mandated writing process, it led to the formation of a community familiar with an assessment practice similar to the IPCC's.

The MA fostered the ecosystem approach, explicitly linking biodiversity, ecosystem services, and human well-being. This anthropocentric, quantitative method could potentially tackle the complexity of global biodiversity changes without dismissing regional or local specificities. Together with other initiatives, the MA formed the starting point for three intergovernmental stakeholder meetings, resulting in the creation of the IPBES in 2012.

A Conflictual Field

Narratives and "epistemic selectivities" played an essential role in bridging the partly competing interests of states, scientific communities, NGOs, international organizations (UNEP, FAO, UNDP, and UNESCO), and scientific associations (e.g., IUCN, DIVERSITAS, or ICSU) in relation to a future biodiversity knowledge body (Vadrot, 2014a; Vadrot, 2020). Knowledge fragmentation, the CBD's implementation deficit, and continuous loss of biodiversity provided common ground, enabling actors to reach a consensus after 7 years. Conflicts of interest between states, most notably from the Global North and Global South, significantly shaped the IPBES and continue to play out every time an assessment text is negotiated.

From the beginning, Global South governments were sceptical about the idea of an "IPCC for biodiversity," fearing that northern science would reinforce the dominance of the Global North in biodiversity negotiations under and beyond the CBD. Scientific capacities are unequally distributed, making it difficult for many developing countries to monitor and conserve biodiversity or nominate experts for IPBES assessments. Since the CBD's early days, international biodiversity politics have revolved around science and knowledge-related tensions and stubborn gaps between nations, all the more so after biotechnology emerged, accelerating the appropriation of genetic properties and traditional knowledge.

Beyond IPCC Practices

As a response, Global North governments agreed to include capacity building as a central function of IPBES, to recognize the contribution of TK, and to establish a Multidisciplinary Expert Panel (MEP) overseeing the selection of experts. The MEP was to ensure that assessments would be produced by a broad range of experts from different regions of the world and scientific disciplines (including social scientists, underrepresented at first, see Vadrot et al., 2018) and be gender-balanced. This created the conditions for an IPCC-like assessment practice while fixing several weaknesses that had promoted male natural scientists from the North.

The IPBES resembles the IPCC in many ways. Its members are states; they finance it through regular contributions, nominate 80% of experts and are central counterparts for scientific communities at the national level. In addition, the Secretariat and several Technical Support Units facilitate assessments and capacity building, often supported by national in-kind contributions. However, state actors are also directly involved at the scoping, writing, reviewing, summarizing, and adopting stages, both as regards final assessments and summaries for policy makers (SPMs). SPMs are negotiated line-by-line following UN rules and diplomatic rituals.

Partly owing to this iteration and interlinkage between science and policy, several social scientists (especially from International Relations, Environmental Sociology, and Social Studies of Science) developed an interest in the IPBES as a research object (Díaz-Reviriego et al., 2019). Hughes and Vadrot (2019) analyzed intense discussions of the notion of "biocultural diversity" during the adoption of the first IPBES thematic assessment on pollination, pollinators, and food production (2016) by more than one hundred governments. Ethnographic data collected during the plenary illustrate that specific terms give rise to conflicting ideas, are linked to state interests, and can shape the meaning of biodiversity both within and beyond the IPBES (Hughes & Vadrot, 2019). Wrangling over the meaning and representation of biodiversity is an essential component of IPBES assessment practice and the production of global biodiversity knowledge.

An Inclusive Framework

The concept of "ecosystem services" appealed to ecologists seeking to quantify biodiversity loss because it might increase political interest in tackling the impact of human activity on ecosystems and the economic costs thereof (Vadrot, 2020). While early on, governments generally supported the idea that the MA's coupling of biodiversity loss with ecosystem services and human well-being could serve as a framework, several critical voices emerged. In 2011, Bolivia's representative critically noted that "ecosystem services" emphasized the monetary value of biodiversity, potentially excluding other meanings and values (Vadrot, 2014a, p. 183). Nevertheless, the notions of "Mother Earth" and "ecosystem services" became —at least formally— equally accepted modes of understanding "nature's benefit to people" (Borie & Hulme, 2015, p. 495). Gustafsson and Lidskog (2018) describe this as the IPBES` endeavour to gain and sustain "epistemic authority."

However, "active opposition often based on the perceived risks of commodification of nature and associated social equity concerns" continued (Díaz et al., 2018, p. 271) and led to coining the term "nature's contributions to people," which is now used in IPBES assessments to underline the diversity of meanings and values associated with biodiversity. The first global assessment report even moved a step further by identifying both direct and indirect drivers of biodiversity loss; the latter include political, economic, demographic, and technological aspects of environmental degradation. The report estimated that more than one million species were under threat and called for "transformative change", anticipating a debate on societal transformation and how to achieve it.

Since the early history of the IPBES, competition between divergent conceptions of nature and its value has become an inherent part of global biodiversity knowledge production and assessment practice. As with the IPCC, IPBES concepts, terms, graphs, and numbers concerning biodiversity loss can influence public debate and increase awareness.

Future Research Directions

Even before its establishment, IPBES was already an object of research (Díaz-Reviriego et al., 2019). Early literature focused on the particularities of its history, revealing that tensions over the type of expertise deemed relevant shaped the body's institutional design, conceptual framework, and development (e.g., Granjou et al., 2013; Vadrot, 2014b; Borie & Hulme, 2015; Montana, 2017). In addition, some scholars contrasted the IPBES with the IPCC; they studied the incorporation of diverse knowledge forms and mechanisms designed to increase the involvement of diverse experts and stakeholders (e.g., Esguerra et al., 2017; Oubenal et al., 2017; Morin et al., 2017; Löfmarck & Lidskog, 2017; Gustafsson & Lidskog, 2018).

While early literature was mainly concerned with the centrality of "ecosystem services" and a strong emphasis on modelling and scenario-building to predict trends in biodiversity loss (e.g., Granjou et al., 2013; Brand & Vadrot, 2013; Turnhout et al., 2016), recent studies tend to focus on the integration of diverse knowledge systems, including traditional knowledge of local communities and indigenous people (e.g., Gustafsson & Lidskog, 2018; Tengö et al., 2017; Hill et al., 2013).

Future (critical) research is needed on knowledge integration practices at various sites and levels of policymaking, in particular on how to adjust assessment methods to avoid power asymmetries between knowledge forms and on ethical standards needed to avoid exploiting knowledge holders. The practices, discourses, processes, and power dynamics within the IPBES—and how these shape the production of global biodiversity knowledge both within and beyond it—are also worth exploring. Divergent interests and terminology preferences make the IPBES a valuable field site to empirically study the lively debate over ontologies, concepts, terms, and values attributed to nature and humans' relationship to it.

Finally, the IPBES itself has opened up new avenues for social scientists by extending biodiversity research to "values" and "transformative change", thereby also encouraging the active participation of experts from anthropology, sociology, political science, and ethics in IPBES assessments.

References

Beck, S. (2011). Moving beyond the linear model of expertise? IPCC and the test of adaptation. *Regional Environmental Change, 11*, 297–306.

Borie, M., & Hulme, M. (2015). Framing global biodiversity: IPBES between mother earth and ecosystem services. *Environmental Science and Policy, 54*, 487–496.

Brand, U., & Vadrot, A. B. M. (2013). Epistemic selectivities towards the valorization of nature in the Nagoya protocol and the making of the intergovernmental science-policy platform on biodiversity and ecosystem services (IPBES). *LEAD – Law, Environment and Development Journal, 9*(2), 202–222.

CBD. (1992). Convention on biological diversity. *United Nations, Treaty Series, 1760*, 79.

Díaz, S., Pascual, U., Stenseke, M., Martín-López, B., Watson, R. T., Molnár, Z., Hill, R., Chan, K. M. A., Baste, I. A., & Brauman, K. A. (2018). Assessing nature's contributions to people. *Science, 359*(6373), 270–272.

Díaz-Reviriego, I., Turnhout, E., & Beck, S. (2019). Participation and inclusiveness in the intergovernmental science–policy platform on biodiversity and ecosystem services. *Nature Sustainability, 2*(6), 457–464.

Esguerra, A., Beck, S., & Lidskog, R. (2017). Stakeholder engagement in the making: IPBES legitimization politics. *Global Environmental Politics, 17*(1), 59–76.

Granjou, C., Mauz, I., Louvel, S., & Tournay, V. (2013). Assessing nature? The genesis of the intergovernmental platform on biodiversity and ecosystem services (IPBES). *Science, Technology and Society, 18*(1), 9–27.

Gustafsson, K. M., & Lidskog, R. (2018). Organizing international experts: IPBES's efforts to gain epistemic authority. *Environmental Sociology, 4*(4), 445–456.

Heywood, V. H. (1995). *Global biodiversity assessment*. Cambridge University Press.

Hill, R., Halamish, E., Gordon, I. J., & Clark, M. (2013). The maturation of biodiversity as a global social-ecological issue and implications for future biodiversity science and policy. *Futures, 46*, 41–49.

Hughes, H., & Vadrot, A. B. M. (2019). Weighting the world: IPBES and the struggle over biocultural diversity. *Global Environmental Politics, 19*(2), 14–37.

Löfmarck, E., & Lidskog, R. (2017). Bumping against the boundary: IPBES and the knowledge divide. *Environmental Science and Policy, 69*, 22–28.

Loreau, M., Oteng-Yeboah, A., Babin, D., Barbault, R., Donoghue, M., Gadgil, M., Häuser, C., et al. (2006). Diversity without representation. *Nature, 442*, 245–246.

MA. (2005). *Millennium ecosystem assessment (program). Ecosystems and human well-being*. Island Press.

Montana, J. (2017). Accommodating consensus and diversity in environmental knowledge production: Achieving closure through typologies in IPBES. *Environmental Science & Policy, 68*, 20–27.

Morin, J., Louafi, S., Orsini, A., & Oubenal, M. (2017). Boundary organizations in regime complexes: A social network profile of IPBES. *Journal of International Relations and Development, 20*(3), 543–577.

Oubenal, M., Hrabanski, M., & Pesche, D. (2017). IPBES, an inclusive institution? Challenging the integration of stakeholders in a science-policy Interface. *Ecology and Society, 22*(1), 11.

Tengö, M., Hill, R., Malmer, P., Raymond, C. M., Spierenburg, M., Danielsen, F., Elmqvist, T., & Folke, C. (2017). Weaving knowledge systems in IPBES, CBD and beyond—Lessons learned for sustainability. *Current Opinion in Environmental Sustainability, 26-27*, 17–25.

Turnhout, E., Dewulf, A., & Hulme, M. (2016). What does policy-relevant global environmental knowledge do? The cases of climate and biodiversity. *Current Opinion in Environment Sustainability, 18*, 65–72.

UNFCCC. (1992). *United Nations Framework Convention on Climate Change. United Nations, FCCC/INFORMAL/84 GE.05-62220 (E) 200705*, Secretariat of the United Nations Framework Convention on Climate Change, Bonn, Germany.

Vadrot, A. B. M. (2014a). *The politics of knowledge and global biodiversity*. Routledge.

Vadrot, A. B. M. (2014b). The epistemic and strategic dimension of the establishment of the IPBES: Epistemic selectivities at work. *The European Jorunal of Social Science Research. Innovation, 27*(4), 361–378.

Vadrot, A. B. M. (2020). Building authority and relevance in the early history of IPBES. *Environmental Science & Policy, 113*(11), 14–20.

Vadrot, A. B. M., Rankovic, A., Lapeyre, R., Aubert, P.-M., & Laurans, Y. (2018). Why are social sciences and humanities needed in the works of IPBES? A systematic review of the literature. *Innovation. The European Journal of Social Science Research, 31*(sup1), 78–S100.

Alice Vadrot is Associate Professor for International Relations and the Environmental at the Department of Political Science of the University of Vienna, Visiting Research Fellow at the Centre for Science and Policy (CSaP) of the University of Cambridge, member of the Young Academy of the Austrian Academy of Sciences, and Senior Fellow of the Earth System Governance Platform. Her work addresses the role of knowledge and science in global environmental politics and agreement-making. She is an expert on the Intergovernmental Platform on Biodiversity and Ecosystem Services (IPBES). Since 2018, she leads the ERC project MARIPOLDATA, which develops a new methodological approach for grounding the analysis of science-policy interrelations in marine biodiversity politics in empirical research.

IPCC

Jonathan Lynn

Abstract The Intergovernmental Panel on Climate Change (IPCC) is the global body for assessing the science related to climate change. This article reviews the history of the IPCC and describes how it prepares its report, putting this in the context of global climate policy.

The Intergovernmental Panel on Climate Change (IPCC) is the global body for assessing the science related to climate change. It was set up in 1988 by two United Nations agencies, the United Nations Environment Programme (UNEP) and the World Meteorological Organization (WMO), to inform policymakers about the state of knowledge of the science of climate change.

While scientists have been discussing human-induced climate change for 200 years, a series of publications and conferences in the 1970s and 1980s led to growing awareness among governments. Policymakers needed an authoritative guide to the subject given the complexity of the science and at times differences within the scientific community. The move by UNEP and the WMO, endorsed that year by the U.N. General Assembly, established a body to tell policymakers what we know about climate change, what we don't know, and where further research is required.

In the three decades since then the IPCC has established itself as the gold standard for knowledge about climate change, its risks, and options for addressing it.

The mandate of the IPCC, as set out in the Principles Governing IPCC Work, makes clear that the IPCC covers all human activities affected by or contributing to climate change, and all scientific disciplines, including the life and social sciences. They also make clear that the IPCC is objective. It does not start from a particular position, and in discussing responses it is policy-relevant but never policy-prescriptive.

J. Lynn (✉)
Former Head of Communications, IPCC, Geneva, Switzerland
e-mail: jlynn@wmo.int

© The Author(s), under exclusive license to Springer Nature Switzerland AG 2023
N. Wallenhorst, C. Wulf (eds.), *Handbook of the Anthropocene*,
https://doi.org/10.1007/978-3-031-25910-4_264

It is important to understand that the role of the IPCC is to assess published scientific research and other relevant material to present the state of knowledge on the subject. The IPCC does not conduct its own original research, measurements of the climate or models of the climate.

Policy relevance means that IPCC reports also reflect the broader global policy framework. Response options are discussed in the context of the sustainable development goals, explicitly so in the reports of the current Sixth Assessment cycle.

IPCC reports have in turn had a large impact on global policymaking. The *First Assessment Report*, in 1990, led to the creation of the United Nations Framework Convention on Climate Change (UNFCCC), the global institution that convenes the annual climate negotiations, known as Conference of the Parties, or COP. The Fourth Assessment Report in 2007 led to the IPCC sharing that year's Nobel Peace Prize for its contribution to spreading understanding about climate change. The *Fifth Assessment Report* (IPCC, 2014), in 2013/14, was the scientific input into the 2015 climate conference COP 21, which led to the Paris Agreement. In setting out the Paris Agreement governments invited the IPCC to prepare a *Special Report on Global Warming of 1.5 °C* (IPCC, 2018). This report, appearing in 2018, galvanized and transformed public awareness of climate change and essentially set out the global policy framework within which climate change is now being discussed. The first part of the *Sixth Assessment Report* (IPCC, 2021), appearing in 2021, had an as yet unprecedented public and media impact, adding urgency to that year's climate conference COP 26.

IPCC reports run to many thousands of pages, including high-level Summaries for Policymakers and their Headline Statements, Technical Summaries, Frequently Asked Questions and Supplementary Material. But the current state of IPCC knowledge can be summed up in a few words developed by leading scientists on the *Special Report on Global Warming of 1.5 °C*:

- Every bit of warming matters.
- Every year matters.
- Every choice matters.

The IPCC draws its authority and credibility in part from a structure that ensures its findings are endorsed by both the scientific and policy communities. This structure was unique until recently, but the Intergovernmental Science-Policy Platform on Biodiversity and Ecosystem Services (IPBES), founded in 2012, draws strongly on the IPCC model.

It is important to note that the IPCC is an organization of governments, as its name – Intergovernmental Panel on Climate Change – implies. The IPCC's members are 195 governments, members of the United Nations or the WMO. Government representatives form the governing body, coming together in Sessions of the Panel to set the IPCC's rules, agree its budgets, decide which reports will be prepared, approve and accept IPCC reports and transact other business.

The governments also elect a Bureau of scientists, which mobilizes the scientific community to prepare the reports decided on by the Panel. The Bureau, currently numbering 34, includes the Chair, three IPCC Vice-Chairs, the two Co-Chairs and the Vice-Chairs of each of the three Working Groups, and the two Co-Chairs of the IPCC's Task Force on National Greenhouse Gas Inventories.

The three Working Groups assess different aspects of climate change. Working Group I – the Physical Science Basis – considers whether the climate is changing, what is causing it, how we know, and what will happen next. Working Group II – Impacts, Adaptation and Vulnerability – looks at how the climate change established by Working Group I is affecting humanity, farming and wildlife, and how these are adjusting to the changes already underway. Working Group III – Mitigation – considers options for stopping further climate change, in effect by reducing or halting emissions of greenhouse gases. The Task Force develops methodologies to enable governments to estimate their emissions of greenhouse gases and removals through sinks.

The IPCC completed the first part of the Sixth Assessment Report, the Working Group I contribution, in August 2021. The Working Group II and III report were released in early 2022. The Synthesis Report integrating the three working group contributions and the three special reports prepared earlier in the cycle was released in March 2023.

Once governments have decided that a report should be prepared – a special report on a particular subject or the main comprehensive assessment – the Bureau convenes a meeting of experts to scope out the report, in effect a table of contents with indicative bullets reflecting the area governments wish to examine and the relevant scientific literature available. Governments then discuss and approve this outline.

Governments then nominate experts to serve as authors of IPCC reports. The Bureau, or respective Working Group Bureau as a subset, selects the authors from these nominations looking for the necessary expertise while also respecting geographical balance, in particular between developed and developing countries, gender balance, and also balance between authors with past experience working on IPCC reports and those new to the process.

Here it is important to note that authors as well as the elected members of the Bureau serve in a voluntary capacity. The IPCC does not pay them for their work, although it does fund the participation of developing country authors and one representative per developing country in meetings when physical meetings are possible.

This voluntary principle also applies to the IPCC budget. Governments decide individually whether they want to contribute to the IPCC Trust Fund and if so how much and how frequently. There are no fixed or assessed contributions.

The preparation of IPCC reports is characterized by a process of repeated drafting and review. The authors prepare a first official draft, which is then open to expert comment. Experts nominated but not selected are invited to participate but the process is open to everyone that believes they have something to contribute, and it is managed through a self-declaration of expertise. The authors then prepare a second draft, including a first draft of the report's Summary for Policymakers, open to both experts and governments for comment. A final draft is then prepared for comments by governments on the Summary for Policymakers. Note that governments are explicitly invited to comment on the second and final drafts.

The authors are required to address every single comment received –though they may choose to ignore it. The drafts, review comments, and author responses are all published once the report is finalized to ensure transparency. Arguably this process makes the preparation of an IPCC report the largest peer review exercises undertaken. Dozens of experts – 200-250 on a working group contribution to a main assessment – serve as authors, with hundreds more as expert reviewers and contributing authors.

The process concludes with a plenary session in which representatives of the governments that have requested the report and will be working with it finalize the Summary for Policymakers with the scientists that have drafted it. The purpose of this discussion is to ensure that the Summary for Policymakers is an accurate and balanced reflection of the underlying report and formulated clearly for the use of policymakers. In that discussion governments proposing changes to the text must have the agreement of all other governments, as the IPCC works by consensus, and the sign-off of the scientists that the proposed change is scientifically accurate and does not introduce new material not already in the underlying report. Governments may invite the scientists to suggest appropriate wording. The scientists have the last word in this discussion.

The IPCC invests considerable effort and money in communicating its findings, from the timely release of reports, to an accessible website, active social media, and an ambitious programme of virtual and in-person outreach activities targeting diverse audiences all over the world. Many other organizations draw on IPCC products to conduct their own outreach. (For a discussion of recent development in IPCC communications, see Lynn & Peeva, 2021, *Communications in the IPCC's Sixth Assessment Report cycle*, in the Topical Collection on IPCC communications in *Climatic Change.*)

References

IPCC. (2014). *Climate change 2014: Synthesis report*. Contribution of working groups I, II and III to the fifth assessment report of the intergovernmental panel on climate change (Core writing team, R. K. Pachauri & L. A. Meyer (Eds.)). pp. 151. IPCC.

IPCC. (2018). Summary for policymakers. In: *Global warming of 1.5°C. An IPCC special report on the impacts of global warming of 1.5°C above pre-industrial levels and related global greenhouse gas emission pathways, in the context of strengthening the global response to the threat of climate change, sustainable development, and efforts to eradicate poverty* In V. Masson-Delmotte, P. Zhai, H.-O. Pörtner, D. Roberts, J. Skea, P.R. Shukla, A. Pirani, W. Moufouma-Okia, C. Péan, R. Pidcock, S. Connors, J.B.R. Matthews, Y. Chen, X. Zhou, M.I. Gomis, E. Lonnoy, T. Maycock, M. Tignor, & T. Waterfield (Eds.). World Meteorological Organization, 32 pp.

IPCC. (2021) *Climate change 2021: The physical science basis*. Contribution of working group I to the sixth assessment report of the intergovernmental panel on climate change. In V. Masson-Delmotte, P. Zhai, A. Pirani, S. L. Connors, C. Péan, S. Berger, N. Caud, Y. Chen, L. Goldfarb, M. I. Gomis, M. Huang, K. Leitzell, E. Lonnoy, J. B. R. Matthews, T. K. Maycock, T. Waterfield, O. Yelekçi, R. Yu, & B. Zhou (Eds.). Cambridge University Press. In Press.

Lynn, J., & Peeva, N. (2021). Communications in the IPCC's sixth assessment report cycle. *Climatic Change, 169*, 18. https://doi.org/10.1007/s10584-021-03233-7

Jonathan Lynn was Head of Communications of the IPCC from 2011 to 2021, during which he developed and implemented the IPCC communications strategy and managed the release of the Fifth Assessment Report and first part of the Sixth Assessment Report, as well as the associated Special Reports.

Precautionary Approach

Ortwin Renn and Pia-Johanna Schweizer

Abstract The debate on how to evaluate and manage risks focuses on three major strategies: (Renn O. EMBO Rep 8:303–305, 2007, Renn O. Risk governance. Coping with uncertainty in a complex world. Earthscan, London, 2008a; Stirling A. On 'Science' and 'Precaution' in the management of technological risk. Volume I: synthesis study, report to the eu forward studies unit by European Science and Technology Observatory (ESTO), EUR19056EN. IPTS, Sevilla. Available at: ftp://ftp.jrc.es/pub/EURdoc/eur19056IIen.pdf, 1999): (1) risk-based approaches, including, numerical thresholds (NOEL standards, performance standards, etc.); (2) reduction activities derived from the application of the precautionary principle (examples are ALARA, i.e., as low as reasonably achievable, BACT, i.e., best available control technology, containment in time and space, or constant monitoring of potential side-effects); and (3) standards derived from discursive processes such as roundtables, deliberative rulemaking, mediation or citizen panels. Experience demonstrates that there is no simple recipe for assessing, evaluating and managing risks. In view of complex cause-effect relationships, diverse attitudes and preferences as well as variations in interests and values, risks must be considered as physically as well as socially heterogeneous phenomena that preclude standardized evaluation and handling. Therefore, a coherent concept for evaluation and management is needed that ensures the integration of social diversity and multidisciplinary approaches into institutional routines and standardized practices. The main objective of this paper is to explore the potentials and the limitations of an approach to risk assessment and management that has been labelled the "precautionary principle". In its most simple version, precaution requires risk managers to err on the safe side. This includes rather accepting false negative (that risks are less severe than assumed) than false positive assessments (that risk are more severe than assumed). However, such a simple definition does not specify to what degree false negatives are socially tolerable, nor does it allow a discussion about future benefits that could potentially compensate for uncertain risks. The following sections will review main

O. Renn (✉) · P.-J. Schweizer
Research Institute for Sustainability - Helmholtz Center Potsdam (RIFS), Potsdam, Germany
e-mail: pia-johanna.schweizer@iass-potsdam.de

© The Author(s), under exclusive license to Springer Nature Switzerland AG 2023

N. Wallenhorst, C. Wulf (eds.), *Handbook of the Anthropocene*,
https://doi.org/10.1007/978-3-031-25910-4_265

positions on precaution and point out the present practice in the European Union, which has adopted the precautionary principle as a major legal guideline for its risk management practice. The paper concludes with suggestions for aligning the precautionary principle and the concept of responsible innovation.

The Different Implications of Precaution

The precautionary principle is one of the most contested strategies in both risk assessment and risk management. The Rio Declaration on Environment and Development offers one of the most often cited definitions of precaution:

> In order to protect the environment, the precautionary approach shall be widely applied by States according to their capabilities. Where there are threats of serious or irreversible damage, lack of full scientific certainty shall not be used as a reason for postponing cost-effective measures to prevent environmental degradation (https://www.unglobalcompact.org/what-is-gc/mission/principles/principle-7).

This approach is now applied in an increasing number of national jurisdictions, economic sectors and environmental areas, and constitutes a crucial principle for policy-making framed as precautionary principle in the European Union (van Asselt & Vos, 2006; Gilbert, 2020; Resnik, 2021).

Despite the intensity of the debate in the aftermath of the Rio Declaration, several fundamental ambiguities and queries remain concerning the nature and appropriate role of the precautionary principle in governance (Majone, 2002; Löfstedt, 2004; Marchant & Mossman, 2004; Wardman & Löfstedt, 2018). These are addressed – if not resolved – in a burgeoning academic literature (Sand, 2000; Fisher, 2001; Klinke & Renn, 2001; Stirling, 2003; van Zwanenberg & Stirling, 2004; Renn 2008b, 2009; Renn et al., 2009; Reber, 2018) as well as more policy-oriented publications (Stirling, 1999; Gee et al., 2001; O'Riordan et al., 2001; Renn & Elliott, 2011; Stirling & Coburn, 2018). For understanding the key issues of the debate, three positions towards risk analysis should be distinguished (see also Resnik, 2003, 2021; Renn, 2007):

> Within the frame of "scientific risk analysis", risk management relies on the best scientific estimates of probabilities and potential damages. It uses expected values as the main input to judge the tolerability of risk, as well as to design risk-reduction measures that are cost effective, proportional to the threat and fair to the affected population. In this frame, precaution may best be interpreted as being conservative in making risk judgements and choosing cautious assumptions when calculating exposure or determining safety factors (of 10 or multiples of 10) to cover inter-individual variability. In addition, as Martin Peterson (2007, p. 307) points out, 'the precautionary principle can be interpreted [in this frame] as an analogous epistemic principle which prescribes that it is always more desirable to avoid false negatives than false positives when it comes to assessing risks'.

Within the frame of 'precaution', the concept of risk is seen from the perspective of pervasive uncertainty as well as ignorance and non-knowledge. Precautionary risk management entails ensuring prudent handling of decision options in the contexts of high uncertainty about causes and effects and of highly vulnerable populations under conditions of risk. Instruments of precaution include minimization requirements, such as the principle of 'as low as reasonably achievable' (ALARA) or 'as low as reasonably practicable' (ALARP), diversification of sources of risk containment in time and space, and close monitoring (Pike et al., 2020). In the words of Andrew Stirling (2007, p. 311): 'Precaution does not automatically entail bans and phase-outs but calls instead for deliberate and comprehensive attention to contending policy or technology pathways.'

The frame of "deliberation" has been advocated both as an alternative as well as an addition to analytical procedures for assessing and managing risks (Berg & Lidskog, 2018). The task of risk management here is to organize, in a structured and effective manner, the involvement of stakeholders and the interested public in designing risk management strategies based on each stakeholder's knowledge and values. This strategy can be combined with both, risk analysis and the precautionary approach but has been advocated as an independent path to risk management as well as a policy-oriented implementation of the precautionary approach (van den Daele, 2000).

In the recent past, advocates of each side have launched a fierce debate over the legitimacy of their approaches. One side argues that precautionary strategies ignore scientific results and lead to arbitrary or inconsistent decisions (Peterson, 2007; Stefansson, 2019). The precautionary approach that it is better to be safe than sorry could, therefore, be interpreted as a mandate to ban everything that might result in negative side effects. Such a rule would logically apply to any substance or human activity and would lead to total arbitrariness (Majone, 2002). Accordingly, the principle has been labelled as ill-defined and absolutist, and is considered to lead to increased risk-taking, to be an ideology, to be unscientific, or to marginalize the role of science (Sandin et al., 2002). Some analysts claim that using the precautionary principle runs the risk that science might be held 'hostage to interest group politics' (Charnley & Elliott, 2002); others contest that policy-makers could abuse the precautionary principle to protect economic interests and to impede world trade (Majone, 2002).

Conversely, advocates of the precautionary approach argue that it does not automatically mean an immediate ban, but rather a step-by-step diffusion of risky activities or technologies until more knowledge and experience are available (Fisher, 2001; Stirling, 2003; Stirling & Coburn, 2018). They have accused their critics of ignoring the uncertainty and ambiguity of most hazardous situations, since they rely on data that often turn out to be insufficient for making robust judgements. They also claim that risk assessment techniques pretend to meet a high standard of scientific validity, rigour and reliability that cannot be sustained under scrutiny. Too many unpleasant surprises during the past few decades would therefore justify a more cautious and humble approach to claiming knowledge about possible risks (Gee et al., 2001).

The third approach has found wide acceptance among social scientists and risk analysts from academia (e.g, von Schomberg, 2006; Lepori, 2019); but so far had little impact on institutional risk management (Renn, 2007). There are some examples of community participation in risk decisions, such as in the US Superfund programme that cleans up contaminated waste sites, or US legislation negotiated among regulators, industry and non-governmental organizations (NGOS) (Coglianese, 1997). In recent years, policy-makers have acknowledged more and more that public participation provides many advantages because it transforms difficult issues of uncertainty into topics that can be negotiated. 'If society participates in the production of policy-relevant scientific knowledge, such "socially robust" knowledge is less likely to be contested than that which is merely reliable' (Funtowicz et al., 2000). Accordingly, the European Union as well as the OECD has highlighted the need for more stakeholder involvement and participation in risk management (OECD, 2020). However, concrete steps towards implementation into day-to-day risk management are still under dispute. Scholars also question the value of deliberative approaches in some settings, arguing that 'when there is trust in the regulator, a top-down form of risk communication (information transfer) may be better than dialogue'(Löfstedt, 2004; Wardman & Löfstedt, 2018).

Application of the Precautionary Principle in the EU

The precautionary principle has been adopted in a variety of forms at international, EU and national level (Renn & Elliott, 2011). It is applied across an increasing number of national jurisdictions, economic sectors and environmental areas. It has moved from the regulation of industry, technology and health risk to the wider governance of science, innovation and trade. As it has expanded in scope, so it has grown in profile and authority. In particular, as Article 174(2) in the European Commission Treaty of 2002, precaution now constitutes a key underlying principle in European Community policy-making. The 2000 Communication on Precaution of the European Commission provides evidence for the high significance that the precautionary principle has gained as a guiding policy of the EU in areas such as environmental, consumer and health protection. The document states in the first section: 'Applying the precautionary principle is a key tenet of its policy, and the choices it makes to this end will continue to affect the views it defends internationally on how this principle should be applied' (European Commission, 2000, p. 3). As Elisabeth Fisher (2002) points out, the Communication specifies some of the major conditions and requirements for applying the principle. There are two conditions mentioned: 'The measures, although provisional, shall be maintained as long as the scientific data remain incomplete, imprecise, or inconclusive, and as long as the risk is considered too high to be imposed on society' (European Commission, 2000, p. 21). In addition to the presence of remaining uncertainty, the Communication lists the condition that the risk must be too high to be imposed on society. This relates to the requirement of proportionality that is being mentioned as one of the major requirements of applying the principle.

The European Commission's communication highlights three important issues in developing the precautionary principle (Fisher, 2003). Each of these constitutes a means of ensuring that decision-making pursuant to the precautionary principle is not arbitrary (Articles 13, 16 and 22). E. Fisher (2003) has summarized these issues in form of three major messages. First, the precautionary principle 'should be compatible with the classic division of risk analysis: risk assessment, risk management, risk communication'. In the communication, the precautionary principle is largely understood as belonging to the stage of risk management; but it should also guide the risk assessment process. The 'implementation of an approach based on the precautionary principle should start with a scientific evaluation, as complete as possible, and, where possible, identifying the degree of scientific uncertainty at each stage of the risk governance process'. Second, the European Commission's communication highlights the need for 'proportionate, non-discriminatory and transparent actions'. It further stresses the benefit of using cost-benefit analysis, based on best scientific data. The communication also points out the need to 'involve all interested parties as early as possible and to the greatest possible extent'. Third, and finally, the communication allows for a wide range of risk management initiatives (Article 16). Such measures need not be legally binding but should serve as a yardstick that embraces adaptivity and social responsiveness (Fisher, 2003; van Asselt & Vos, 2006).

It is interesting to note that the assessment-based approach has been widely adopted by the official US regulatory bodies, while the precautionary approach has been widely advocated by the EU regulatory bodies (Wiener, 2018). There are, however, also numerous elements of precautionary approaches interspersed within the actual practices of US regulatory agencies, just as there are judgements about magnitudes of risk in the actual practices of regulators in the EU. A strict dichotomy between a 'precautionary' approach in Europe and an 'assessment based' approach in the US is therefore overly simplistic and fails to do justice to actual practice.

The Precautionary Principle Within the Paradigm of Responsible Innovation

In the past decade, the EU has fostered an innovation ecosystem, in which technologies (and other innovations) are not thought of as ends in themselves but are brought in line with fundamental values and principles upon which the European Union is built.[1] The current research and innovation strategy of the European Commission

[1] The following paragraph is adopted from the report "Guidance on the application of the precautionary principle in the EU" put forth by the project "REconciling sCience, Innovation and Precaution through the Engagement of Stakeholders" (RECIPES) funded by the Horizon 2020 Framework Programme of the European Union under Grant Agreement no. 824665: https://recipes-project.eu/sites/default/files/2022-01/RECIPES%20D3.2%20Report%20on%20Tools%20Guidelines%20and%20Recommendations.pdf. Credit goes to the authors of the report Johannes Andresen Oldervoll, Adriana Dimova, Marion Dreyer, Laura Elisabet Drivdal, Pia-Johanna Schweizer, Tijs Sikma, Jeroen van der Sluijs, Kristel de Smedt, Niels-Kristian Tjelle Holm, Dino Trescher and Ellen Vos.

(2021–2024) identifies research and innovation as a key driver in achieving European Commission goals that are geared towards a sustainable and prosperous future based on solidarity and respect for shared European values. The Commission's research and innovation strategy identifies the following tasks for research and innovation. Research and innovation shall help restore vulnerable ecosystems and the environment with the overall objective of Europe becoming the first climate-neutral continent. In addition, the aim is to improve public health, resilience against emerging threats and crisis preparedness. Longer term aims are building resilient democracies across the EU, fostering innovation, sustaining policies and institutions in support of democratic processes and enhancing trust in democratic institutions (https://ec.europa.eu/info/research-and-innovation/strategy/strategy-2020-2024_en).

Precaution in this context has been adopted as a crucial element for this value-based approach towards innovation that emphasizes non-economic as well as economic values. In that regard, the precautionary principle and precautionary thinking are considered tools that can guide technological development while providing regulatory guidelines for already established technologies. The overall aim is to sustain European values incorporated in normative anchor points such as a high level of protection of human health and environment, wellbeing and sustainable development. Furthermore, the precautionary principle and precautionary thinking are discussed as tools that support deliberative participatory democratic processes in EU innovation and technology governance. The notion of 'responsible research and innovation' (RRI) was launched by the European Commission and has been strongly promoted as an innovative governance concept in its former research and innovation program Horizon 2020 (2014–2020). RRI is committed to a responsible vision of innovation as laid down in the European Commission's current research and innovation strategy (Bowman et al., 2018; Gurzawska et al., 2017). RRI addresses the observation that innovation – as a goal in itself – does not always lead to results that are beneficial to society as a whole or else may be accompanied by negative side effects. Science and technology scholars, such as Stilgoe et al. (2013) have identified anticipation, reflexivity, inclusion and responsiveness as important characteristics of Responsible Innovation (RI). These four elements can be described as follows:

Anticipation: "Anticipation involves systematic thinking aimed at increasing resilience, while revealing new opportunities for innovation and the shaping of agendas for socially-robust risk research" (Stilgoe et al., 2013, p. 1570).

Reflexivity: "Reflexivity, at the level of institutional practice, means holding a mirror up to one's own activities, commitments and assumptions, being aware of the limits of knowledge and being mindful that a particular framing of an issue may not be universally held" (Stilgoe et al., 2013, p. 1571).

Inclusion: Inclusion could mean taking the time to involve different stakeholders as to lay bare the different impacts of a new technology on different communities (Reber, 2018).

Responsiveness: "Responsible innovation requires a capacity to change shape or direction in response to stakeholder and public values and changing circumstances" Stilgoe et al., 2013, p. 1572).

The use of the precautionary principle for policy and regulation can be seen as a safeguarding mechanism warding off uncertain, yet potentially negative risks. Seen in this light, the precautionary principle links especially with RI's dimensions of responsiveness and reflexivity. The use of the precautionary principle as a compass is an approach that helps innovation systems to improve anticipation of adverse effects. This interpretation of the precautionary principle speaks especially to RI's dimensions of anticipation and inclusion. Knowledge generated by using the precautionary principle as a compass (e.g., via technology assessment, foresight processes or risk research) can help promote a timely and more broadly informed application of the precautionary principle in EU risk policy and regulation. Application of the precautionary principle as a compass can stimulate and shape 'responsible innovation', e.g., by fostering sustainable production as well as safety-by-design and ethics-by-design approaches. The two ways use of exercising precaution can serve as an important mechanism for building capacity for anticipation, responsiveness, reflexivity and inclusion in risk governance.

A closer look at risk regulation in Europe reveals that the application of the precautionary principle does not conflict with the use of best available science and the need for regulatory stringency. On the contrary, EU risk management policy puts specific emphasis on scientific risk and hazard assessment.

Any regulatory regime based on precaution is faced with the question of how to make regulatory decisions in the contexts of uncertainty or even ignorance. In this regard, the differentiation proposed by Resnik (2003, p. 332) seems helpful:

decisions under certainty – the outcomes of different choices are known;
decisions under risk – probabilities can be assigned to the outcomes of different choices;
decisions under ignorance – as probabilities cannot be assigned to outcomes of unfathomable management options.

A similar distinction has been made by Stirling (2003) and by Renn (2009). One of the main conclusions has been that using precaution for the first two cases is neither necessary nor prudent, given that regulation needs to meet both objectives of protecting public health and the environment, and securing economic welfare. Therefore, precaution should be applied in instances of ignorance or other forms of prevailing uncertainties (such as "unknown unknowns" and "known unknowns").

To conclude, the main purpose of precaution is to avoid irreversible decisions (Renn, 2009; Nijsingh et al., 2020). Although being critical about the use of the precautionary principle itself, policy analyst G. Majone concedes that it does have its function in risk analysis – namely, where 'losses (or utilities) are unbounded' and where it is 'clearly impossible to calculate expected values' (e.g., when there is a threat of 'serious and irreversible damage'; Majone, 2002). In these cases, it is prudent to proceed with care so that decisions can be reversed promptly if outcomes or probabilities turn out to be higher than expected. In such cases, banning is only one regulatory option; other options include containing the risk, setting boundaries to ubiquitous spreading or bioaccumulation, developing alternatives, and/or minimizing exposure. Such a prudent approach to risk management would certainly be supported by representatives of all camps in this debate and would be compatible with the EU paradigm of responsible innovation.

References

Berg, M., & Lidskog, R. (2018). Deliberative democracy meets democratised science: A deliberative systems approach to global environmental governance. *Environmental Politics, 27*(1), 1–20.

Bowman, D. M., Garden, H., Stroud, C., & Winickoff, D. E. (2018). The neurotechnology and society interface: Responsible innovation in an international context. *Journal of Responsible Innovation, 5*(1), 1–12.

Charnley, G., & Elliott, E. D. (2002). Risk versus precaution: Environmental Law and public health protection. *Environmental Law Reporter, 32*(2), 10363–10366.

Coglianese, C. (1997). Assessing Consensus: The Promise and Performance of Negotiated Rulemaking. *Duke Law Journal, 46*, 1255.

European Commission. (2000). *Communication from the Commission on the Precautionary Principle* (COM(2000) 1). EU.

Fisher, E. (2001). Is the precautionary principle justiciable? *Journal of Environmental Law, 3*, 315–334.

Fisher, E. (2002). Precaution, precaution everywhere: Developing a 'common understanding' of the precautionary principle in the European Community. *Maastricht Journal of European and Comparative Law, 9*(1), 7–46.

Fisher, E. (2003). The legal dimension of developing a general model for precautionary risk regulation. In O. Renn et al. (Eds.), *The application of the precautionary principle in the European Union*. University of Stuttgart. Available at: http://www.sussex.ac.uk/spru/environment/precaupripdfs.html

Funtowicz, S., Shepherd, I., Wilkinson, D., & Ravetz, J. (2000). Science and Governance in the European Union: A contribution to the debate. *Science and Public Policy, 5*(27), 327–336.

Gee, D., Harremoes, P., Keys, J., MacGarvin, M., Stirling, A., Guedes Vaz, S., & Wynne, B. (2001). *Late Lesson from Early Warnings: The Precautionary Principle 1898–2000*. European Environment Agency.

Gilbert, S. G. (2020). Precautionary principle. In P. Wexler (Ed.), *Information Resources in Toxicology* (pp. 489–494). Academic. https://doi.org/10.1016/B978-0-12-813724-6.00045-1

Gurzawska, A., Mäkinen, M., & Brey, P. (2017). Implementation of Responsible Research and Innovation (RRI) practices in industry: Providing the right incentives. *Sustainability, 9*(10), 1759.

Klinke, A., & Renn, O. (2001). Precautionary principle and discursive strategies: Classifying and managing risks. *Journal of Risk Research, 4*(2), 159–173.

Lepori, M. (2019). Towards a new ecological democracy: A critical evaluation of the deliberation paradigm within green political theory. *Environmental Values, 28*(1), 75–99.

Löfstedt, R. E. (2004). *The swing of the Pendulum in Europe: From precautionary principle to (Regulatory) impact assessment* (AEI-Brookings Joint Center for Regulatory Studies Working Paper 04-07). Kings College.

Majone, G. (2002). What price safety? The Precautionary Principle and its Policy Implications. *Journal of Common Market Studies, 40*(1), 89–109.

Marchant, G. E., & Mossman, K. L. (2004). *Arbitrary and Capricious: The precautionary principle in the European Union Courts*. The AEI Press.

Nijsingh, N., van Bergen, A., & Wild, V. (2020). Applying a precautionary approach to mobile contact tracing for Covid-19: The value of reversibility. *Journal of Bioethical Inquiry, 17*(4), 823–827.

O'Riordan, T., Cameron, J. C., & Jordan, A. (Eds.). (2001). *Reinterpreting the precautionary principle*. Cameron May.

OECD. (2020). *Innovative citizen participation and new democratic institutions: Catching the Deliberative Wave*. OECD Publishing. https://doi.org/10.1787/339306da-en

Peterson, M. (2007). The precautionary principle should not be used as a basis for decision-making: Talking Point on the precautionary principle. *EMBO Reports, 8*(4), 305–308.

Pike, H., Khan, F., & Amyotte, P. (2020). Precautionary principle (PP) versus as low as reasonably practicable (ALARP): Which one to use and when. *Process Safety and Environmental Protection, 137*, 158–168.

Reber, B. (2018). RRI as the inheritor of deliberative democracy and the precautionary principle. *Journal of Responsible Innovation, 5*(1), 38–64. https://doi.org/10.1080/23299460.2017.1331097

Renn, O. (2007). Precaution and analysis: Two sides of the same coin? *EMBO Reports, 8*, 303–305.

Renn, O. (2008a). *Risk governance. Coping with uncertainty in a complex world*. Earthscan.

Renn, O. (2008b). Precaution and ecological risk. In S. E. Jörgensen & B. D. Fath (Eds.), *Human ecology. Volume 4 of the encyclopedia of ecology* (5. Volume) (pp. 2909–2916). Elsevier.

Renn, O. (2009). Precaution and the governance of risk. In N. Adger & A. Jordan (Eds.), *Governing sustainability* (pp. 226–258). Cambridge University Press.

Renn, O., & Elliott, E. D. (2011). Chemicals. In J. B. Wiener, M. D. Rogers, J. K. Hammitt, & P. H. Sand (Eds.), *The reality of precaution. Comparing risk regulation in the United States and Europe* (pp. 223–256). Earthscan.

Renn, O., Dreyer, M., Klinke, A., Losert, C., Stirling, A., van Zwanenberg, P., Müller-Herold, U., Morosini, M., & Fisher, E. (2009). The application of the precautionary principle in the European Union. In O. Renn, P.-J. Schweizer, U. Müller-Herold, & A. Stirling (Eds.), *Precautionary risk appraisal and management. An orientation for meeting the precautionary principle in the European Union* (pp. 2–117). Europäischer Hochschulverlag.

Resnik, D. B. (2003). Is the precautionary principle unscientific? *Studies in History and Philosophy of Biological and Biomedical Sciences, 34*, 329–344.

Resnik, D. B. (2021). Precautionary reasoning and the precautionary principle. In *Precautionary reasoning in environmental and public health policy* (The International Library of Bioethics) (Vol. 86, pp. 11–128). Springer. https://doi.org/10.1007/978-3-030-70791-0_5

Sand, P. (2000). The precautionary principle: A European perspective. *Human and Ecological Risk Assessment, 6*(3), 445–458.

Sandin, P., Martin Peterson, M., Ove Hansson, S., Rudén, C., & Juthé, A. (2002). Five charges against the precautionary principle. *Journal of Risk Research, 5*(4), 287–299.

Stefánsson, H. O. (2019). On the limits of the precautionary principle. *Risk Analysis, 39*(6), 1204–1222.

Stilgoe, J., Owen, R., & Macnaghten, P. (2013). Developing a framework for responsible innovation. *Research Policy, 42*(9), 1568–1580.

Stirling, A. (1999). *On 'Science' and 'Precaution' in the Management of Technological Risk*. Volume I: Synthesis Study, Report to the EU Forward Studies Unit by European Science and Technology Observatory (ESTO), EUR19056EN. IPTS: Sevilla. Available at: ftp://ftp.jrc.es/pub/EURdoc/eur19056IIen.pdf

Stirling, A. (2003). Risk, uncertainty and precaution: Some instrumental implications from the social sciences. In F. Berkhout, M. Leach, & I. Scoones (Eds.), *Negotiating change* (pp. 134–167). Edward Elgar.

Stirling, A. (2007, April). Precaution and risk assessment: towards more measured and constructive policy debate. In: *European molecular biology organisation reports* (Vol. 8, pp. 309–315).

Stirling, A., & Coburn, J. (2018). From CBA to precautionary appraisal: Practical responses to intractable problems. *Hastings Center Report, 48*, S78–S87.

Van Asselt, M. B. A., & Vos, E. (2006). The precautionary principle and the uncertainty paradox. *Journal of Risk Research, 9*(4), 313–336.

Van den Daele, W. (2000). Interpreting the precautionary principle – Political versus legal perspectives. In M. P. Cottam, D. W. Harvey, R. P. Paper, & J. Tait (Eds.), *Foresight and precaution* (Vol. 1, pp. 213–221). A.A. Balkema.

Van Zwanenberg, P., & Stirling, A. (2004). Risk and precaution in the US and Europe. *Yearbook of European Environmental Law, 3*, 43–57.

Von Schomberg, R. (2006). The precautionary principle and its normative challenges. In E. Fisher, J. Jones, & R. von Schomberg (Eds.), *Implementing the precautionary principle: Perspectives and prospects* (pp. 19–42). Edward Elgar.

Wardman, J. K., & Löfstedt, R. (2018). Anticipating or accommodating to public concern? Risk amplification and the politics of precaution reexamined. *Risk Analysis, 38*(9), 1802–1819.

Wiener, J. B. (2018). Precautionary principle. In *Elgar encyclopedia of environmental law* (pp. 174–185). Edward Elgar.

Ortwin Renn has been the former scientific director of the Research Institute for Sustainability - Helmholtz Center Potsdam (RIFS, formerly IASS), Germany and a retired professor for environmental sociology and technology assessment at the University of Stuttgart. He also directs the non-*profit company DIALOGIK,* a research institute for the investigation of communication and participation processes. Renn is *Adjunct Professor for "Integrated Risk Analysis"* at Stavanger University (Norway), *Honorary Professor* at the Technical University Munich and Affiliate Professor for "*Risk Governance*" at Beijing Normal University. His research interests include risk governance (analysis perception, communication), stakeholder and public involvement in environmental decision making, transformation processes in economics, politics and society and sustainable development.

Pia-Johanna Schweizer leads the research group "Systemic Risks" at the Research Institute for Sustainability - Helmholtz Center Potsdam (RIFS, formerly IASS), Germany Dr. Schweizer is also speaker for the research area "Global Implications of Socio-technical Change" at the IASS. Dr. Schweizer has gained a PhD in sociology from Stuttgart University (summa cum laude). She has authored and co-authored more than 30 peer-reviewed publications and serves on the editorial and scientific advisory board of peer-reviewed journals. She was awarded the Chauncey Starr Distinguished Young Risk Analyst Award by the Society of Risk Analysis in 2015. Dr. Schweizer's main fields of research are systemic risk, risk analysis, risk communication, risk perception, risk governance and deliberative decision-making.

Security

Hajo Eickhoff

Abstract The principle of connecting and stabilizing seems to be universal. The universe shows this in its incessant physical and chemical processes. Connections of elements are more stable than their single elements. What is connected seems to be automatically secured. This is also true for biological processes. Humans continue these processes by means of culture. They treat things and processes under the terms of optimization and security. Optimizing is the search for solutions for a set task and its realization, as well as its securing against changes. Modern cultures have not been able to solve the problem of security because they focus primarily on optimizing and securing limited technical problems, while, in terms of the whole, they have confronted humanity with abysses that they are now trying to circumvent.

Safety Culture and Optimization

There is no safety or security. Nowhere. For anything or anybody. The word does not even appear in Romance languages or in German. Only as a negation. The basic Latin word for security, *cura*, means worry, but only its negation *securus* gives the meaning 'secure' – without (*sed*) worry (*cura*).

Security or safety designates a state with regard to its potential for change. It can refer to things, people, and processes such as working, mountain climbing or travelling. States that do not change are 'secure'. Negatively formulated, security denotes states which are marked by the absence of dangers and risks.

Western cultures are safety cultures. They want to avoid risks and dangers. In them, everything has to be monitored and insured so that what is, remains. Or becomes better. Property is preserved by safes, works of art are protected by copyrights, websites are encrypted by software, identities are checked by security staff. Everything can become an object to be kept safe: things, art, possessions, and people (cf. Eickhoff, 2015 p. 13).

H. Eickhoff (✉)
Interdisziplinäres Zentrum für Historische Anthropologie, Freie Universität, Berlin, Germany

© The Author(s), under exclusive license to Springer Nature Switzerland AG 2023

N. Wallenhorst, C. Wulf (eds.), *Handbook of the Anthropocene*,
https://doi.org/10.1007/978-3-031-25910-4_266

After the physical needs of food and water, security is the strongest need of people in Maslow's scheme of needs. This includes stability, security, freedom from fear, order and protection; also, myths and religion or the preference of the known over the unknown; as well as world views, since people want to be able to explain all kinds of phenomena to themselves; also tradition and habit are forms of security. Living safely creates trust, and a familiar environment is a basis for protecting oneself against uncertainty.

Biological, social, psychological, economic and cultural security create different forms of immunity. Biological security is our immune system against internal and external invaders, psychological security means having confidence and social security means belonging to a group. Economic security is the long-term guarantee of basic needs. Cultural security is peace.

Military parades with their huge arsenal of weapons and soldiers, the existence of the atomic bomb, terrorist activities and dictatorial states show how insecure people's lives are. New dangers are always emerging, for which new security procedures are being developed, such as IT security. Anyone online is exposed to cyberattacks that can cause enormous damage, to the point of incapacitating companies and governments. Operations in war can be carried out drone-assisted from outside the theatre of war without the presence of soldiers.

But life on Earth was always uncertain even without humans. Life could be extinguished at any time by astronomical events such as meteorite impacts or changes in the chemical balance of the atmosphere, by volcanic eruptions, changes in climate, or by viruses and bacteria.

Western cultures are optimization cultures. Optimization is a measure to improve a given condition with a particular goal in mind. It is considered to be the best possible solution to a task taking into account important conditions. Consumer goods, tools and processes, playing the piano and skiing can all be optimized. People are visionaries and have ideas about things, processes and themselves that they want to improve.

Cosmic Being – Circling, Colliding, Collapsing, Drifting

What is, should stay. Secure its existence. Or improve it. Both phenomena of physics, chemistry, biology and culture. Optimization and security are related not just to the intentions of humans. Comparable processes were already to be seen in the universe before the beginning of terrestrial life, because everything moves toward energetically stable, i.e. secure conditions.

The universe is potentiality. An incessant becoming and changing. An eternal drifting. Nothing can stay in one place. Physical and chemical phenomena appear constructive as well as destructive. Destructive, because meteorites hit the earth and wiped out life, suns explode or black holes devour everything, constructive, because the earth and life came into being and because systems like atomic nuclei, atoms or molecules are stable and held together by a binding energy. This energy makes them more secure against change. Physically and chemically, compounds are more stable, more secure than their individual parts.

Creation in the Christian account is also potentiality. At the end of each day, God judges his creation with "It is good." But looking at the result of all the days, he says, "It is very good." The whole appears better than its parts, more optimal, more perfect. But when God realized that the creatures were not perfect after all, he punished them with the Flood. Noah, the perfect one, saved creation, which got a new chance through the Flood (cf. Erbele-Küster, 2021 p. 156f).

Earthly Being – Optimizing Life and Making It Safe

Life is moving and exchanging. It must assert and secure itself. Anorganic molecules combined to become organic molecules, from which the building blocks of life like amino acids, nucleic acid and DNA emerged. Further chemical evolution produced molecules capable of metabolism, stimulation and reproduction.

Unicellular organisms are the first life forms. Unprotected, the DNA lay in the cytoplasm until they brought other cells or cell elements into themselves and made them into cell organs, such as the nucleus, which now protected the DNA. Unicellular organisms joined together to form multicellular organisms and thus gained evolutionary advantages because they lived more safely in the association. This makes it clear that life already knows competition at the cellular level. Emanuele Coccia speaks of war as a social phenomenon in the context of non-human contexts. "War is the social bond that allows non-human society to improve and spread" (Coccia, 2018 p. 152). In fact, multicellular organisms evolved and differentiated themselves in competition, learned, stored what they learned, and developed an infinite variety of life forms. The fact that the forms that are well adapted to the environment survive is a principle of optimization and protection.

With the growing complexity of living beings the central organ, the brain has developed. Living beings with small brains adapt only conditionally to the environment and as a species are in danger of dying out, with complex brains they adapt better to new conditions and secure their species. Humans have a brain that is so optimally equipped that their existence has a high degree of security.

But not all species push for improvement. Living things like bacteria work according to patterns that are billions of years old. Cyano-bacteria invented photosynthesis and introduced the resulting oxygen into the atmosphere. Even today, they generate the largest proportion of oxygen in the atmosphere. In their kind many bacteria are perfect. In contrast, vertebrates evolved from fish to amphibians, reptiles and birds to mammals: an evolution from the horizontal of fish to the vertical of man, and to the endowment of a multiform organ of control and cognition. An optimization, which makes it possible for humans to control all living beings and to biologically secure their own species, in the end even to transcend it.

Securing Culture

In humans, stimulus and reaction are decoupled. They can pause and harbour intentions, they do not have to respond to stimuli, and can thus give their own form to their lives. Their dexterous hands, intellect, memory and consciousness ensured constant expansion of their knowledge and skills from the very beginning, which they passed on to subsequent generations. They discovered fire, made simple tools, and fed on what they could find.

A break in the process of improvement came when they settled down. About twelve thousand years ago, the first humans stopped and gave up their existence as hunter-gatherers. They built houses and practised agriculture and animal husbandry, which was a biological, social, spiritual and economic process followed by a revolution of the way of life: In a narrower sense, culture emerged – as soil care (*culturare*). Food production began, as well as an immense production of things such as consumer goods, tools and materials for enclosures, paths and camps.

Production surpluses were used for trade, which gave rise to trade routes and markets. As the population grew and prosperity increased due to a regular harvest, secured meat supply and trade, settling was an economic optimization.

The fixed location of the house, the property and the wealth made it possible to explore the region around the house and expand the sphere of influence. With sedentarization began the colonization of the earth and with it a constant decimation of the humus layer.

Expanding and Safeguarding Knowledge

Knowledge begins with single-celled organisms. Amoebae recognize each other as prey and hunters. An amoeba recognition. If unicellular organisms join together to form multicellular organisms, their knowledge expands and is refined by the fact that all cells are in exchange with each other and store their individual and collective experiences. Despite the development of the brain and the formation of a consciousness, part of knowledge remains unconscious, because the inner structures work autonomously. Nevertheless, the formation of consciousness is an optimization process, which is promoted and secured by human language.

Science is a continuation of the optimization of personal knowledge. It begins with general considerations of what appears as world – that is philosophy. From this gradually the individual sciences have emerged, for example, when an area of knowledge had deepened and widened to such an extent that it went beyond the scope of philosophy. Although the progress of knowledge takes place in both intellectual and material phenomena, the progress of knowledge and societies is primarily measured by the state of technology and the natural sciences. It is more about consumption than about quality of life. The major subsidies are awarded to projects that are considered to have a practical, i.e. technical, relevance, and humanities scholars are rarely consulted on explosive questions of the time and the future. Knowledge about the scarcity of resources or climate change would have long been sufficient to make preparations to protect humanity from the dangers that lie ahead.

Human Beings Lose Their Security Through Their Actions

Settlement turned hunter-gatherers into farmers. From the peasantry developed craftsmen and traders, who soon left the land and founded cities. The first great empires arose as an expression of growing prosperity. The climax was the Roman Empire, whose entire territory was based on a rectangular street grid and optimized the transport of goods, weapons and people. When Christian nuns and monks withdrew from society and produced the goods of their lives very efficiently, the monasteries prospered. The inhabitants worked ceaselessly and consumed little, so they established markets around the monasteries with merchants and farmers, turning monasteries into great economic centres. By the thirteenth century, trade and crafts in European cities had also developed to the point where banks had to facilitate large-scale projects such as the shipping of goods. Most importantly, insurance policies were established to limit the risks of commercial ventures.

In the Renaissance era, craft transformed into manufacturing, and in the nineteenth century, into industry. During that time, people were considered to be beings who first had to develop into what they could be. The individual was considered free, but was called to self-improvement through education and practice. They were free also for the labour market. It was this freedom that interested industrial capitalism. It strove for permanent improvement of production and product. The history of labour is a history of increasing division of labour. Professions became specialized, and within the professions, work steps became more and more small-scale until machines could be used. It was precisely the division of labour that optimized work. The machines and apparatuses with their speed, endurance and precision and the output of immense quantities of goods became idols that changed everyday life, which at the same time was influenced by the monotonous, alienated and empty work that was always the same day after day. Life itself became strange and empty, and there remained an ambivalence of admiration and emptiness that gave neither security nor safety.

Growth, acceleration and optimization progressed in all areas of life at the expense of natural resources. The earth became a mere object from which material could be extracted at will and indissoluble materials returned as waste. The products of digitalization have continued the trend. Alienation at work has become an existential feeling of strangeness, so that people are alienated from their environment, but also indifferent to it. It is this astonishment that in a crisis this immense technosphere suddenly becomes perceptible. This mass of all man-made products, which is greater than the mass of all living beings. "While the biomass continues to decrease due to deforestation and destruction of soils and oceans and species extinction, the man-made mass is growing faster and faster" (Welzer, 2021 p. 12). The threat to living beings shows that incessant optimization does not lead to security, because constant growth with limited resources must be avoided for logical as well as factual reasons.

Culture of Optimization and Safety

Optimization Potential Enhancement

Enhancement serves to optimize human characteristics that are worthy of improvement. For better performance and more beauty, humans exhaust themselves in sports or take doping substances, are stimulated by psychotropic drugs, take Ritalin for exams, or are surgically brought into shape. Behind the self-improvement and the improvement of the body lies the optimization of the personality as well as the effort to be able to withstand the systematic optimization of everyday life (cf. Rosa, 2013 p. 550).

In neuro-enhancement, it is healthy people who are optimized by pharmacological and technical interventions in the brain that make them posthuman. Nick Bostrom contends that a mentally and physically completely healthy, productive person who draws conclusions quickly and precisely, who has a good memory and enjoys life and responds appropriately to situations, is already a posthuman person, since these characteristics can only be achieved through probes or micro-chips, i.e. through neuro-enhancement (cf. Bostrom, 2018, pp. 146–147). Neither the individual nor the human species is safe from modern technology.

Dark Sides of Human Optimization

The techno-sphere is an image of the growth myth of modern societies and presents the dark sides of the permanent optimization processes: optimization and security also fail to achieve their goals due to unintended side effects: soil clearing destroys humus, deforestation reduces oxygen in the atmosphere, plastic waste in the oceans causes fish to ingest microplastics, radioactive substances remain a danger for hundreds of thousands of years, melting glaciers and climate change cause seas to rise and weather disasters to increase. And space debris is proliferating – thousands of satellites and over a hundred million pieces of debris are orbiting the Earth, threatening life.

Safety and Perfecting Personality

For 200 years, machines have been gradually replacing people. Precise, fast and persistent, they do not get sick and do not demand wages. Nevertheless, societies focused on optimization and safety have not yet succeeded in avoiding or mitigating catastrophes. Weather disasters, floods, glacier melting and pandemics and, in 2022, the start of a war in Europe are causing worry and fear, dampening optimism and fuelling hatred in commercial social media and hatred and violence in the streets. Societies are also becoming fragile and unstable and losing their secure cohesion. In the attempts to constantly adapt to mechanical and digital requirements, people lose sight of their immediate needs and the meaning of their lives. In the overload, they become irritable and ill: on edge, dissatisfied, rebellious or depressed. It is the lack of attention to themselves that makes it possible for them to exploit themselves by trying to adapt and to contribute thoughtlessly to the exploitation of the earth and the animal and plant world. Cognitively and emotionally overwhelmed people do not find rest and the uncertain, uncontrollable and gloomy future does not give a feeling of security and well-being.

Let Lie and Postpone – Really, Really Act

The optimization of life has brought a lot of good into the world. In detailed areas of life, many things have become better and safer: fewer traffic fatalities and more freedom of movement, less illiteracy, a decrease in alcohol consumption and infant mortality, an expansion of women's rights and lower crime. A smartphone is a medium that combines dozens of tools and functions in a single device. It has come a long way in terms of optimization, incorporating devices such as a camera, computer and telephone, facilitating and speeding up many functions, and saving an enormous amount of material. On the other hand, it has produced enormous waste dumps and consumed resources on its way through many intermediate products. In terms of the whole – the earth, the animal and plant world and humanity – the optimization projects have not made life safe.

Much of the current misery is human-made, so humans could eliminate or mitigate it. For this, however, they must quickly take clever political and ecological measures. They would have to get off the wrong path of growth, efficiency and compulsion to constantly improve, must stop geopolitically exploiting the unequal distribution of resources and would have to turn their backs on the relentless competition of individuals, companies and states. Since the human species was created by gaining power by separating stimulus and response and gaining the ability to pause, they could now do so consciously by examining stimuli before they are perceived. You could postpone reactions and ask what benefit a reaction brings to the endangered world.

A real action would be one that comes from pause and critical reflection. Such a pause can be a real action. It can break with convention or open up necessary perspectives.

The misery of the present state of the world forces new questions: people are plundering the planet and, realizing that they are in the process of destroying it, they come up against the question of how to save it. Women have fought for rights without being equal to men, raising critical questions about patriarchy. The failure of the security of life despite technical optimizations has led to critical questions about security.

Current disasters have arisen from attitudes of power – the power of economics, gender, education and property. These must be challenged. If development processes are to incorporate our security, the economy must be brought and kept in cycles. The circular economy is the only form of production that is safe. To achieve this, products must be manufactured in such a way that they can be and remain recycled. Trust proves to be the safest form of communication of persons and societies, the hierarchy-free relationship of the sexes the safest form of power distribution, and participation is the safest cycle of respect and recognition.

We need an awareness of what is simple: on what (soil), in what (atmosphere) and with whom (plants and animals) we live. The obvious, that is obscured by competition, optimization, perfection and security constraints. To make this 'simple' perceptible and to recognize its meaning and beauty and to fight for its preservation is a global, social and personal task, which is a way to a safer future.

References

Bostrom, N. (2018). *Die Zukunft der Menschheit*. Suhrkamp.

Coccia, E. (2018). *Metamorphosen. Das Leben hat viele Formen. Eine Philosophie der Verwandlung*. Carl Hanser Verlag.

Eickhoff, H. (2015). Unsicherheitsabsorption und Resilienz. Strategien zur Bewältigung von Unsicherheit. In: *Paragrana. Internationale Zeitschrift für Historische Anthropologie*, 24/1, S. 13–25.

Erbele-Küster, D. (2021). Und siehe – es ist sehr gut (Genesis 1,31). Eine ästhetisch-ethische Vision und ihre Aussagekraft angesichts des Imperfekten und der Korrumpiertheit. In: *Paragrana. Internationale Zeitschrift für Historische Anthropologie*, 30/1, S. 150–159.

Rosa, H. (2013). *Beschleunigung und Entfremdung. Entwurf einer Kritischen Theorie spätmoderner Zeitlichkeit*. Suhrkamp.

Welzer, H. (2021). *Nachruf auf mich selbst. Die Kultur des Aufhörens*. S. Fischer.

Hajo Eickhoff, Dr. Phil, works as a cultural historian, exhibition organizer and cultural and business consultant in Berlin. Studied philosophy, history and art history in Freiburg, Berlin and Aachen. His research focuses on postures, the inner design of the human being, the intelligence of cells, the visual arts and the structure of occidental cultural development.

Societal Boundaries

Ulrich Brand, Barbara Muraca, Éric Pineault, Marlyne Sahakian, Anke Schaffartzik, Andreas Novy, Christoph Streissler, Helmut Haberl, Viviana Asara, Kristina Dietz, Miriam Lang, Ashish Kothari, Tone Smith, Clive Spash, Alina Brad, Melanie Pichler, Christina Plank, Giorgos Velegrakis, Thomas Jahn, Angela Carter, Qingzhi Huan, Giorgos Kallis, Joan Martínez Alier, Gabriel Riva, Vishwas Satgar, Emiliano Teran Mantovani, Michelle Williams, Markus Wissen, and Christoph Görg

Abstract The notion of *societal boundaries* aims to enhance the debate on planetary boundaries. The focus is on capitalist societies as a heuristic for discussing the expansionary dynamics, power relations, and lock-ins of modern societies that impel highly unsustainable societal relations with nature. While formulating societal boundaries implies a controversial process – based on normative judgments, ethical concerns, and socio-political struggles – it has the potential to offer guidelines for a just, social-ecological transformation.

The planetary boundaries concept has profoundly changed the vocabulary and representation of global environmental issues. We bring a critical social science perspective to this framework through the notion of societal boundaries and aim to provide a more nuanced understanding of the social nature of thresholds, as complementary to the entry on planetary boundaries (in this textbook). Given the ever more destructive escalatory dynamics of capitalism, including the expansion of commodity frontiers, the commodification of nature and space, and neoliberal/austerity governance (Muradian et al., 2012; Temper et al., 2018; Calvário et al., 2017), a crucial question of our time is: How can the social metabolism of contemporary societies be de-escalated, in a way that is socially just? Critical social science has contributed

This book chapter is an excerpt of the full article Brand et al. (2021), available in open access. Contact: U Brand, University of Vienna, ulrich.brand@univie.ac.at

U. Brand (✉) · B. Muraca · É. Pineault · M. Sahakian · A. Schaffartzik · A. Novy · C. Streissler · H. Haberl · V. Asara · K. Dietz · M. Lang · A. Kothari · T. Smith · C. Spash · A. Brad · M. Pichler · C. Plank · G. Velegrakis · T. Jahn · A. Carter · Q. Huan · G. Kallis · J. M. Alier · G. Riva · V. Satgar · E. T. Mantovani · M. Williams · M. Wissen · C. Görg
University of Vienna, Vienna, Austria
e-mail: ulrich.brand@univie.ac.at

© The Author(s), under exclusive license to Springer Nature Switzerland AG 2023

N. Wallenhorst, C. Wulf (eds.), *Handbook of the Anthropocene*,
https://doi.org/10.1007/978-3-031-25910-4_267

over the years not only to a better understanding of the drivers of unsustainability, but also to identifying possible entry points for more sustainability and for what far-reaching social-ecological transformations might look like. In doing so, it amplifies otherwise neglected voices and emphasizes already existing alternatives.

In asking what kind of worlds "we" want to live in, critical social science discusses how the "we" participating in the deliberation is or can be constituted, and how the conditions for well-being or a good life are to be defined and framed, in concrete historical moments and contexts beyond a solely Western understanding of prosperity. As such, the societal boundaries concept refers to collectively defined thresholds, that societies establish as self-limitations and conditions for a "good life *for all*." Societal boundaries imply a contested and controversial process and are based on normative judgements, ethical concerns, and sociopolitical struggles. They have the potential to act as guidelines for a just, social-ecological transformation, as further discussed in Brand et al. (2021) and as summarized below.

Societal boundaries are necessary towards coping with the deepening ecological crisis and its devastating socioeconomic impacts – especially for those who already live under precarious conditions. Instead of being objectively given by biophysical processes, societal boundaries, as we understand them, emerge from contested societal processes that lead to collectively defined thresholds that societies commit not to trespass. These limits pertain to poverty, inequality, ecological destruction, injustices, subordination, exploitation, consumption, defense of the commons, and so forth. Societal boundaries are structural boundaries, particularly set by political rules within societies, that secure the material and energy prerequisites that enable substantial conditions for a good life for all. There is no guarantee that societies would democratically decide a path toward self-limitation nor that this can be achieved via consensus formation. This is where progressive social movements and other political actors, political education, alternative projects come in: they reinforce and support sociocultural values and norms rooted in social justice considerations, which in turn must be embedded in social relations and institutions. To become socially relevant, the value of such boundaries is more or less accepted throughout societies, and it informs policy-making processes.

Critical social science work can help to formulate *politics of self-limitation* and demarcate *societal boundaries,* social conditions, and sociopolitical measures to respect these constraints – for example by keeping fossil fuels in the ground and organizing social life around alternatives with lower emissions and less devastation of livelihoods. It also keeps open space for critical questioning.

The central idea of *societal boundaries* is a change of analytical and political perspectives: rather than thinking of the planet as bounded, we insist to think of the planet as potentially abundant—as long as we limit ourselves collectively, and make space for others to share the resources it has to offer in a responsible way among current living and future generations. This is also a perspective of respectful cohabitation with non-human others (Kallis, 2019; Akbulut et al., 2019; Hickel, 2019). Boundaries, planetary or societal, are not given; rather they are always relational, a function of human intentions, actions, practices, and interactions – and it is these factors that should be bounded to make space for all. Shifting the focus from boundaries and limits to *self-limitation* emphasizes that this is a social challenge and a process rooted in forms of participation, collective self-determination, and democratic deliberation. For centuries, the democratic governance of the natural commons, as common wealth in the global North and South, gave us practice in self-limitation. Self-limitation questions the idea of considering environmental problems in terms of a technocratic challenge to address, which can implicitly include shifting the boundaries or continuing with expansion all the way up to the "no-trespass" point.

The term self-limitation echoes the literal meaning of *autonomy*, or giving to oneself one's own laws or rules as an act of self-government. In the traditional liberal understanding, autonomy is rooted in the idea of an independent, individual self that is not determined by external norms and therefore free. Yet in the radical tradition of autonomism, it is intended as a social relation and a collective process of self-determination via local, horizontal, anti-authoritarian practices (Alcoff & Alcoff, 2015). What is common to both is the idea that freedom implies giving oneself rules of conduct and therefore limits, instead of following arbitrarily or externally imposed ones. It constitutes the very foundation of democracy as self-rule. Rather than *ending* where someone else's freedom begins, freedom as autonomy *begins* with the self-imposition of limits to make space for others to simply be (Alcoff & Alcoff, 2015). When considered in its societal dimension, autonomy resists its opposite, heteronomy, or the functional regulation of conduct according to given principles, such as the so-called law of the market, or the mantra of austerity and growth. As such, autonomy as collective self-limitation and self-determination requires taking responsibility for one's own destiny and giving to oneself, as a community, self-imposed norms instead of following external impositions (Gorz, 1980; Castoriadis, 2010; Fuchs et al., 2021; Muraca, 2013). In this way, autonomy implies *liberation* from the structural and mental constraints of the capitalist imperatives.

The idea of autonomy as self-limitation is present in different variations in many traditions, societies, and communities across the world. For example, Gandhi's notion of *swaraj* implies autonomy and freedom of the individual and the community as bound by responsibilities and duties toward other individuals and communities, and thereby necessarily encompassing spiritual or ethical living within limits and non-violence, including toward nature (Shrivastava, 2019). Embracing autonomy as guiding principle also implies making space for other world-making practices in a pluriverse of socio-natural configurations (Escobar, 1995), instead of forcing them into the so-called "one-world world" of the Western dominant model of development *as* growth. The pluriverse is "a world where many worlds fit" (as the Mexican Zapatista movement prominently coined it (Holloway & Peláez, 1998)), that enables alliances across different social and environmental movements and resisting communities coming together. It involves relational, spiritual, and affective dimensions of well-being rooted in the principles of equity, solidarity, cooperation, participation, ability to redistribute, and co-habitation of diverse modes of living.

In recent years, a growing body of literature suggests that some form of social "boundary" is required for sustainability transformations. One visual framework which combines planetary boundaries with social foundations is Kate Raworth's (2017; see also Spash, 2020) "doughnut economics". An interdisciplinary team of social and environmental scientists have applied national data to the doughnut framework to determine where and in what way "a good life for all within planetary boundaries" might be attained at a national level (O'Neill et al., 2018). No country is currently able to respect planetary boundaries and guarantee the right to a "good life for all" as defined in the study, although an analysis based on cities or regions might provide different results. Two strategies are suggested by the authors to reduce resource use: (1) to follow degrowth strategies and a steady-state economy in rich nations and (2) to restructure and improve physical and social provisioning systems.

Beyond the doughnut model and the calculations provided by O'Neill et al. (2018), the question of how to relate human needs to resources is one that requires social debate and participatory approaches. Such an approach is proposed in the notion of "sustainable well-being" (Gough, 2017) and is the main thrust of the "Living Well Within Limits" project which emphasizes how citizens might engage in deliberation around identifying how their needs can be met or satisfied in relation to energy sources and systems of distribution (Brand-Correa & Steinberger, 2017). To address this situation, the notion of consumption minima *and maxima* has been developed recently around the notion of "consumption corridors" that join notions of environmental justice and well-being (Fuchs et al., 2021; Sahakian et al., 2021; Wiedmann et al., 2020). Consumption corridors is a societal boundary proposal based on the assumption of generalizable needs common to all.

Alternatives to growth-driven and consumerist capitalist modernization must pursue diverse strategies by strengthening the pluriverse of radical or systemic alternatives that exist across the world and/or by aiming at transforming the state, be it from outside or from within, wherever possible. Many alternatives are reassertions of ancient and traditional approaches, emerging from marginalized peoples and movements of resistance to the dominant system. Others arise from within modern or industrialized societies, often from sections of the middle class or elite urban population that are disillusioned with their own lifestyles and sensitive to the inequities and unsustainability they perpetuate (see for more details Kothari et al., 2019, Brand et al., 2021). Individually and collectively these conceptions embody *alternatives* in worldviews and practices that challenge the structures of inequality, oppression, and unsustainability, and replace them with those that promote justice, equality, and sustainability. They share a rejection of neoliberal globalization, and embrace forms of selective economic deglobalization (Bello, 2008; Novy, 2020), which involves dismantling the "one big market" (Polanyi, 2001) coordinated by global financial markets and sustained by fossil-fuel logistics of airports, motorways, and cargo shipping.

The role of the state in social-ecological transformations is ambiguous. Due to the strategic selectivity of the state in capitalism, it tends to be part of the problem. Its dependence on growth and taxes pushes state agency toward securing unsustainable structures, processes, and power relations even with respect to policies that, at first sight, intend to deal with the ecological crisis. In many countries, the repressive side of the state to defend the interests of elites is much stronger that its distributive side, often acting openly and in a one-sided manner in the interests of capital and oligarchies.

But the state can also be part of the solution, as a terrain for contestation. This depends, however, on changing the concrete form of the state through strengthening decentralized units (municipalities) and democratizing both public institutions of basic provision (education, health, care) as well as economic policy making.

And finally, radical social-ecological transformations would require putting in place socially sustainable degrowth strategies at multiple levels of governance in the global North, and various radical well-being strategies in place of the development model in the global South. Degrowth has been depicted as an equitable and democratically-led selective downshifting of production and consumption levels that sustains human well-being, social justice, and ecological conditions, while reducing commodification and marketization of social life (Schneider et al., 2010; Sekulova et al., 2013; Jackson, 2017; Chertkovskaya et al., 2019). Degrowth does not only challenge the material and ideological foundations of growth economies, but also questions the cultural infrastructure that justifies it (Muraca, 2013; on important differences about degrowth, see Eversberg & Schmelzer, 2018; Spash, 2020).

Societal boundaries, as further discussed in Brand et al. (2021), are necessary for coping with the deepening ecological crisis and its devastating socioeconomic impacts – especially for those who suffer the most from these impacts and who remain underrepresented and absent from decision-making processes. With the notion of societal boundaries, we bring together procedural questions, with an explicit recognition of the need for self-limitation at the collective level or, in other words, of freedom *as* autonomy – autonomy not defined as independency but as ability for self-determination, leading to social and political rules that guarantee the substantial conditions for a good life for all.

References

Akbulut, B., Demaria, F., Gerber, J., & Martínez-Alier, J. (2019). Who promotes sustainability? Five theses on the relationships between the degrowth and the environmental justice movements. *Ecological Economics, 165*, 106418.

Alcoff, L., & Alcoff, J. (2015). Autonomism in theory and practice. *Science & Society, 79*(2), 221–242.

Bello, W. (2008). *Deglobalization: Ideas for a New World economy*. Zed Books.

Brand, U., Muraca, B., Pineault, E., Sahakian, M., Schaffartzik, A., Novy, A., Streissler, C., Haberl, H., Asara, V., Dietz, K., Lang, M., Kothari, A., Brad, A., Pichler, M., Plank, C., Velegrakis, G., Jahn, T., Carter, A., Qingzhi, H., Kallis, G., Alier, J. M., Riva, G., Satgar, V., Spash, C., Mantovani, E. T., Williams, M., Wissen, M., & Görg, C. (2021). From planetary to societal boundaries: An argument for collectively defined self-limitation. *Sustainability. Science, Practice and Policy, 17*(1), S.265–S.292.

Brand-Correa, L., & Steinberger, J. (2017). A framework for decoupling human need satisfaction from energy use. *Ecological Economics, 141*, 43–52.

Calvário, R., Velegrakis, G., & Kaika, M. (2017). The political ecology of austerity: An analysis of socio-environmental conflict under crisis in Greece. *Capitalism Nature Socialism, 28*(3), 69–87.
Castoriadis, C. (2010). *A society adrift: Interviews and debates, 1974–1997*. Fordham University Press.
Chertkovskaya, E., Barca, S., & Paulsson, A. (2019). *Toward a political economy of degrowth*. Rowman and Littlefield.
Escobar, A. (1995). *Encountering development: The making and unmaking of the third world*. Princeton University Press.
Eversberg, D., & Schmelzer, M. (2018). The degrowth Spectrum: Convergence and divergence within a diverse and conflictual Alliance. *Environmental Values, 27*(3), 245–267.
Fuchs, D., Sahakian, M., Gumbert, T., Di Giulio, A., Maniates, M., Lorek, S., & Graf, A. (2021). *Consumption corridors: Living well within sustainable limits*. Routledge.
Gorz, A. (1980). *Ecology as politics*. South End Press.
Gough, I. (2017). *Heat, greed and human need: Climate change, capitalism and sustainable Well-being*. Edward Elgar.
Hickel, J. (2019). Degrowth: A theory of radical abundance. *Real-World Economics Review, 87*, 54–68.
Holloway, J., & Peláez, E. (1998). *Zapatista! Reinventing revolution in Mexico*. Pluto Press.
Jackson, T. (2017). *Prosperity without growth: Foundations for the economy of tomorrow* (2nd ed.). Routledge.
Kallis, G. (2019). *Limits: Why Malthus was wrong and why environmentalists should care*. Stanford University Press.
Kothari, A., Salleh, A., Escobar, A., Demaria, F., & Acosta, A. (Eds.). (2019). *Pluriverse: A post-development dictionary*. Tulika and Authors Upfront. https://radicalecologicaldemocracy.org/pluriverse
Muraca, B. (2013). Décroissance: A project for a radical transformation of society. *Environmental Values, 22*, 147–169.
Muradian, R., Walter, M., & Martinez-Alier, J. (2012). Hegemonic transitions and global shifts in social metabolism. *Global Environmental Change, 22*(3), 559–567.
Novy, A. (2020). The political trilemma of contemporary social-ecological transformation – Lessons from Karl Polanyi's *the great transformation*. *Globalizations*, published online December 4.
O'Neill, D., Fanning, A., Lamb, W., & Steinberger, J. (2018). A good life for all within planetary boundaries. *Nature Sustainability, 1*, 88–95.
Polanyi, K. (2001 [1944]). The great transformation: The political and economic origins of our time. : Beacon Press.
Raworth, K. (2017). *Doughnut economics: Seven ways to think like a 21st-century economist*. Random House.
Sahakian, M., Lorek, S., & Di Giulio, A. (2021). Advancing the concept of consumption corridors and exploring its implications. *Sustainability: Science, Practice and Policy, 17*(1), 305–315.
Schneider, F., Kallis, G., & Martinez-Alier, J. (2010). Crisis or opportunity? Economic degrowth for social equity and ecological sustainability. *Journal of Cleaner Production, 18*(6), 511–518.
Sekulova, F., Kallis, G., Rodríguez-Labajos, B., & Schneider, F. (2013). Degrowth: From theory to practice. *Journal of Cleaner Production, 38*, 1–6.
Shrivastava, A. (2019). Prakritik swaraj (Natural Self-Rule). In A. Kothari, A. Salleh, A. Escobar, F. Demaria, & A. Acosta (Eds.), *Pluriverse: A post-development dictionary* (pp. 283–285). Tulika and Authors Upfront.
Spash, C. (2020). Apologists for growth: Passive revolutionaries in a passive revolution. *Globalizations, 17*(8), 1–26.
Temper, L., Walter, M., Rodriguez, I., Kothari, A., & Turhan, E. (2018). A perspective on radical transformations to sustainability: Resistances. *Movements and Alternatives. Sustainability Science, 13*(10), 385–395.
Wiedmann, T., Lenzen, M., Keyßer, L., & Steinberger, J. (2020). Scientists' warning on affluence. *Nature. Communications, 11*, 3107.

Ulrich Brand works as a Professor of International Politics at the University of Vienna. He obtained his doctoral degree at Goethe University Frankfurt and wrote his post-doctoral thesis (second monograph) on the internationalisation of the state at the University of Kassel. He is the author of books and articles on critical international politics, ecological crisis, environmental politics, social-ecological transformations, the imperial mode of living and Latin America. His recent books: *The Imperial Mode of Living. Everyday Life and the Ecological Crisis of Capitalism* (with Markus Wissen, Verso 2021, translated into nine languages) and *Capitalism in Transformation: Movements and Countermovements in the twenty-first Century* (as co-editor, Edward Elgar 2019).

UN Institutions

Pierre-Yves Cadalen

Abstract This article discusses the position of UN institutions regarding the political and social challenges the Anthropocene raises. After a presentation of the main international institutions or agreements linked to environmental issues, it questions the possible evolutions of the international order under the Anthropocene. The article's conclusion mainly pays homage to Joel Wainwright and Geoff Mann's book, *Climate Leviathan*¸ published in 2018.

International politics of the environment have grown bigger. What was coined as marginal or defined "low politics" has been reconsidered over the last years: international society is ruled by interdependent forces, and interdependence is an unavoidable characteristic of environmental matters (Keohane & Nye, 1997; Morton, 2019). Since 1992's Rio Summit, environment has been regarded as an important international issue, and this tendency has deepened as the problem was manifestly becoming perceivable. Indeed, the Anthropocene is not only an idea or a norm that would enjoy favourable conditions to extend and produce an international consensus amongst the states (Finnemore & Sikkink, 1998); it is also and above all a concrete phenomenon which threatens the very conditions of life on Earth (Wallenhorst, 2020), and actually produces a negative unity of humankind (Chakrabarty, 2009). This latter dimension is far from neutral as international relations are concerned: it means that the Anthropocene gives humanity a negative definition that objectively equalizes the relations between human beings and invites to a positive definition which would implicate a concrete equalization of human conditions in order to avoid disaster (cf. Amazon).

UN Institutions under the Anthropocene cannot be defined outside this double contextualization: 1) the consciousness linked to the Anthropocene has produced an international ecological regime of power and 2) the Anthropocene is addressing all power levels and their potential capacity to mitigate the current destructive

P.-Y. Cadalen (✉)
CERI Sciences Po, CRBC UBO and AMURE UBO, Brest, France
e-mail: pierreyves.cadalen@sciencespo.fr

© The Author(s), under exclusive license to Springer Nature Switzerland AG 2023

N. Wallenhorst, C. Wulf (eds.), *Handbook of the Anthropocene*,
https://doi.org/10.1007/978-3-031-25910-4_268

dynamics. In other words, UN institutions matter, but so do States, NGO's, transnational companies, workers unions or political forces (Badie, 2013, 2018). UN institutions are not a homogenous force regarding the Anthropocene. Let's briefly examine the main agreements linked to environmental matters and then discuss the role UN institutions play in this context. This will lead us to question the international regime and the perspectives of the international system at the light of the Anthropocene and facing its destructive dynamics.

First and foremost, the fleet's flagship of environmental agreements is by far the United Nations Framework Convention on Climate Change (UNFCCC) created in 1992 by the Rio Summit and entered into force in 1994 (Aykut & Dahan, 2015). Indeed, climate change was immediately considered as the top-priority of all ecological issues, and the negotiations of the Kyoto protocol in 1997 and for the Paris Agreement in 2015 have been widely mediatized (Gemenne, 2015). This strong institutionalization of a climate international regime gave birth to some market-based mechanisms – as the carbon market, mostly implemented in Europe – which was criticized for the principles it relied on, as well as for its inefficiency (Lohmann, 2010, 2012; Tordjman, 2021; Vogel, 2009). Another decisive mechanism of this process is the Reducing Emissions from Deforestation and Forest Degradation (REDD) and REDD+ mechanisms, which aimed at protecting global forests for the "ecosystemic service" they are delivering as carbon sinks. This conception has been criticized for similar reasons (Friis Lund et al., 2017).

A second important international agreement is the Convention on biological diversity (CBD). The 15th conference of the parties (COP) of this agreement takes place in 2021, while the UNFCCC knows its 26th COP during the same year. The convention on biological diversity aims at protecting biodiversity by creating incentives for the adoption of biodiversity's national protection plans. It is worth underlining that the CBD has also been criticized for the commodification of nature its principles imply – such as, notably, the economic valuation of the genetic resources linked to biodiversity.

Indeed, the issue of biodiversity leads us to the key question of intellectual property. Indeed, both agreements were mentioning the necessary technological transfers from the North to the South. This move was directly prevented by the World Trade Organization (WTO) and the World Intellectual Property Organization (WIPO), as one of the top-priorities these organizations have come to defend is the international harmonization of intellectual property regimes on capitalist and liberal economic and legal grounds. Despite the demands coming from developing countries, WIPO's collaboration with the United Nations Development Program (UNDP) and the United Nations Conference on Trade and Development (UNCTAD) did not give birth to massive technological transfers from Northern industrialized countries to Southern developing or underdeveloped countries (May, 2006). Instead, the WTO and the WIPO work more as politicized institutions which aim at expanding the liberal model of property rights, highly orientated towards the protection of private property rights. The current dynamics of technical and capitalist accumulation of the Nanotechnology, Biotechnology, Information technology and Cognitive science (NBIC) sector are linked to the commodification of the living and are not presently regulated by international institutions, if not directly supported by the general defence of intellectual property rights (Tordjman, 2021).

Another key issue regarding UN institutions is scientific information. Both the UNFCC and the CBD created intergovernmental panels in charge of synthesizing and producing international research on climate change for the former, mass extinction and biodiversity loss for the latter: the Intergovernmental Panel on Climate Change (IPCC) and the Intergovernmental Science-Policy Platform on Biodiversity and Ecosystem Services (IPBES). In 2019, the IPBES released its first report on biodiversity, pointing the extremely worrying species extinction rates. Five main factors are distinguished, from the most important to the least: "land and sea use, direct exploitation of organisms, climate change, pollution, and invasive alien species". Each of these dynamics is negatively affecting biodiversity, which had led some biologists to the assessment we have entered the sixth mass extinction. The fifth one produced the annihilation of dinosaurs (Broswimmer, 2010). In 2021, the IPCC released a pre-version of its sixth report, to be published in 2022. One of the mains conclusions issued from this pre-version is the acceleration of climate change which would lead in the next 20 years to a global average warming of 1,5oC, compared to the pre-industrial area (1850–1900). The reduction of carbon dioxide, as well as methane and nitrous oxide emissions, is considered a necessity by this international panel of scientists, given that the anthropogenic nature of climate change is by far undeniable. The links between politics, sciences and biodiversity are also key to the Biodiversity Beyond National Jurisdiction (BBNJ) Treaty, which should come soon into force, without "undermining" the United Nations Convention on the Law of the Seas (UNCLOS) though. In 2017, the General Assembly of the UN required a new Treaty that would protect "marine biological diversity of areas beyond national jurisdiction", that "should not undermine existing relevant legal instruments" (Resolution 72/249). It is the most recent UN input to environmental international law.

As the UN institutions tackle international society's matters, all its institutions can be linked to the Anthropocene. The structural dynamics of the Anthropocene are economic, which means the World Bank (WB), the International Monetary Fund (IMF) and the World Trade Organization (WTO) are first-rank actors of Anthropocene international politics. For instance, the IMF annual budget approaches 1 billion dollars, while the United Nations Environment Program (UNEP) annual budget is around 450 million dollars. The difference between both organisms is even more significant as the IMF manages almost 700 billion dollars and is lender of many States. This situation gives the IMF a concrete power over national public policies. Indeed, it is important to consider the diversity of international institutions and the asymmetry of power between them. Anthropocene international politics have to be linked to the driving forces of economic globalization, which is also politically produced (Strange, 1996). Other UN institutions must be mentioned, such as UN-Women and the International Labour Organization (ILO). The women being, particularly in the peasantry, strongly affected by climate change, they are considered by the former as a key actor of mitigation of and adaptation to climate change. As for the latter, the 169 Convention of ILO is a key element that allowed the autochthonous people to become an international actor, notably about environmental issues (cf. Amazon).

Meanwhile these agreements entered into force, the sixth mass extinction has gone faster, and so have the global emissions of greenhouse gas emissions. Therefore, it is important not to forget the Anthropocene is a dynamic and objective process whose orientation cannot only determined by ideas or international agreements, but rather by concrete changes or inertias in economic, social, and political structures. A discussion is then necessary about the international regime relative to environmental matters to conclude this article.

Regarding this question, the Climate Leviathan, a book published in 2018 by Joel Wainwright and Geoff Mann, is particularly stimulating. They basically consider the current dynamics of international institutions as a process which would reproduce structural international inequalities while creating an international regime of climate management, which they precisely name "Climate Leviathan". It is dominated by the reconduction of liberal capitalism and the faith in technical progress to solve the huge problems linked to the Anthropocene – a faith that is basically grounded in Jacques Ellul's assumption that the technician system engineers reality as if political and social problems could not find other solutions than technical ones (Ellul, 1977).

As one of the main stakes Anthropocene raises is to analyze contemporary international dynamics to wonder if there is an alternative possibility, this book is a precious contribution to the Anthropocene debate. It identifies four distinct possibilities, the first being the "Climate Leviathan". A second one is the Behemot, based on denial and capitalist defense: Trump and Bolsonaro are two variations of this "reactionary capitalism". A third one is the "Climate Mao", an authoritarian and anticapitalist regime, which would be likely to emerge from China: the 2021 crisis over energy prices, partly linked to the priority given by the Chinese State to carbon dioxide emissions' reduction, might be a sign of this phenomenon emergence. Lastly, the authors long for a "Climate X" democratic and anti-capitalist solution to the current problems raised by the Anthropocene, which is normatively desirable. One of the elements unaddressed by the authors is the possible articulation between State politics, UN institutions, and this "Climate X" scenario. In other words, we don't think international institutions are doomed to be part of a "Climate Leviathan" scenario, nor that a sovereign anti-capitalist and democratic policy is impossible. This book, however, has the merit to inscribe the Anthropocene debate in political and economic realities international institutions, as States, must directly and firmly address.

References

Aykut, S. C., & Dahan, A. (2015). *Gouverner le climat? : vingt ans de négociations internationales*. Presses de Sciences Po.

Badie, B. (2013). *La fin des territoires. Essai sur le désordre international et l'utilité sociale du respect*. CNRS éditions.

Badie, B. (2018). *Quand le Sud réinvente le monde: essai sur la puissance de la faiblesse*. La Découverte.
Broswimmer, F. J. (2010). *Une brève histoire de l'extinction en masse des espèces*. Agone.
Chakrabarty, D. (2009). The climate of history: Four theses. *Critical Inquiry, 35*, 197–222.
Ellul, J. (1977). *Le système technicien*. Calmann-Lévy.
Finnemore, M., & Sikkink, K. (1998). International norm dynamics and political change. *International Organization, 52*, 887–917.
Friis Lund, J., Sungusia, E., Bukhi Mabele, M., & Scheba, A. (2017). Promising change, delivering continuity: REDD+ as conservation fad. *World Development, 89*, 124–139.
Gemenne, F. (2015). *L'enjeu mondial : l'environnement*. Presses de Sciences Po.
Keohane, R., & Nye, J. (1997). *Power and interdependence*. Pearson.
Lohmann, L. (2010). Commerce du carbone : justice et ignorance. *Écologie & politique, 39*, 135–145.
Lohmann, L. (2012). *Mercados de carbono*. La neoliberalización del clima.
May, C. (2006). The world intellectual property organization. *New Political Economy, 11*, 435–445. https://doi.org/10.1080/13563460600841140
Morton, T. (2019). *La pensée écologique*. Zulma.
Strange, S. (1996). *The retreat of the state: The diffusion of power in the world economy*. Cambridge University Press.
Tordjman, H. (2021). *La croissance verte contre la nature: critique de l'écologie marchande*. La Découverte.
Vogel, J.-H. (2009). *The economics of the Yasuní initiative*. Anthem Press, London.
Wallenhorst, N. (2020). *La vérité sur l'anthropocène*. Le Pommier.

Pierre-Yves Cadalen is currently post-doc researcher at the CRBC – Université Bretagne Occidentale and associate to the CERI-Sciences Po. He recently obtained his PhD in Political Science and International Relations. His works are mainly related to the relations of power around the Environmental Commons. His former studies, which also included a Bachelor in Philosophy, led him to theorize the new forms of power related to the Anthropocene era. Recent publications: "L'Amazonie et le vivant à l'épreuve de l'écopouvoir", *Raisons politiques*, n° 80, 2020, pp. 77–90; "Le populisme écologique comme stratégie internationale", *Critique internationale*, n° 89, 2020, pp. 165–183; "Republican populism and Marxist Populism: Perspectives from Ecuador and Bolivia", in *Discursive Approaches to Populism Across Disciplines – The Return of Populists and the People*, Palgrave MacMilan, 2020.

UNESCO

Maria Böhmer

Abstract E for Education, S for Science, and C for Culture – UNESCO has not only one of the broadest mandates of all UN institutions in terms of content, the organization also has a huge impact on large sections of the population through programmes and initiatives implemented by and with civil society. This article describes UNESCO's course of action regarding the anthropogenically caused and interconnected crises that are threatening Earth's habitats and the future of humankind. The discussion draws on best practice examples to show how UNESCO addresses these challenges through education, science, and culture and on the basis of the 2030 Agenda for Sustainable Development. The case of World Heritage sites being in danger underscores in particular the negative impact of the Anthropocene, whilst also demonstrating how, through the protection and preservation of these sites, UNESCO and its networks seize opportunities for positive change and lay down a path to a peaceful, more just, and more sustainable world community.

The Effects of the Anthropocene on UNESCO World Heritage Sites

400 species of coral, 1500 species of fish, and 4000 species of molluscs—off the coast of Australia lies one of the world's most fascinating marine ecosystems, with the world's largest concentration of coral reefs. But this extraordinary ecosystem, the UNESCO World Heritage Site Great Barrier Reef, is under threat—and this threat is human-made: Ongoing water pollution, the expansion of ports and maritime traffic, but above all the serious effects of climate change, especially global warming, pose an existential threat to this unique World Heritage site. With the Reef 2050 Plan, the Australian government has initiated a variety of conservation measures since 2015. But the results are sobering: The long-term prospects for the reef

M. Böhmer (✉)
German Commission for UNESCO, Bonn, Germany
e-mail: Boehmer@unesco.de

© The Author(s), under exclusive license to Springer Nature Switzerland AG 2023

N. Wallenhorst, C. Wulf (eds.), *Handbook of the Anthropocene*,
https://doi.org/10.1007/978-3-031-25910-4_269

have deteriorated further—and this is not an isolated case. Increasingly, UNESCO World Heritage sites are at risk from climate change and insufficient sustainable action.

The case of the Great Barrier Reef exemplifies the challenges of the Anthropocene and possible solutions. First, it shows how the direct influence of humans on nature and their intervention in the Earth's habitat threaten to cause irreversible damage to our planet. Secondly, it shows how an economic development characterized by strong growth, which has led to progress and improvement in the living conditions of many people since industrialization, is now threatening to reverse itself and call our future into question. Third, it illustrates that the challenges of the Anthropocene can only be met together: through multilateral cooperation, through binding global agreements such as the United Nations' Agenda 2030, and finally through the willingness of each individual state and civil society to implement them.

It is a fundamental task of politics and economics to enable lifestyles that are sustainable and globally applicable. For every human being is called upon as an individual to contribute to this change through his or her actions. The big question is: Will humanity succeed in actively and consciously assuming the role of "Stewards of the Earth System"? As Will Steffen, Paul Crutzen, and John McNeill had already demanded in 2007? Time is running out!

What contribution can UNESCO make towards this goal? Is it an effective forum for sustainable transformation in education, science, and culture? Is it a driving force vis-à-vis governments and civil society?

Peace and Sustainability—UNESCO's Mission and Image of Humanity

"Since wars begin in the minds of men, it is in the minds of men that the defences of peace must be constructed"—this is UNESCO's guiding principle. Initially, the focus was on education as the key issue. But it quickly became clear that culture is inseparably linked to this, and that science must be integrated as the third pillar in order to fulfill the founding mission (see UNESCO, 1946).

UNESCO places its faith in the possibility of a peaceful future for humankind, in the ability of humans to develop and transform themselves, and thus in a new era in the history of humankind (for more on this, see Vermeren, 2003). To this day, this hopeful, future-oriented humanistic view of humanity and the world continues to shape the work of UNESCO for a peaceful, more just, and more sustainable world community, making it a central corrective to the negative consequences of the developments of the Anthropocene.

With the adoption of the 2030 Agenda by the United Nations General Assembly in October 2015, UNESCO expanded its guiding principle to include sustainability. This was not new, since the goal of sustainable development has been an integral part of its work from the very beginning. With the increase in anthropogenically caused, intertwined crises, it became increasingly clear how serious the influence of humans on atmospheric and biological processes and the entire living environment of the Earth is. Consequently, the 2030 Agenda states that holistic sustainable action is indispensable.

It is a core political task to resolve the associated tension between economic, ecological, and social goals. With the 2030 Agenda, a global action plan was presented for this purpose. But implementation is slow. We need to act faster, more comprehensively, more consistently, and thus more effectively. To this end, the UN has proclaimed the Decade of Action. The preamble of the 2030 Agenda emphasizes that sustainable development is the key prerequisite for peace: "We are determined," it states, "to foster peaceful, just, and inclusive societies which are free from fear and violence. There can be no sustainable development without peace and no peace without sustainable development." (United Nations, 2015).

Peace and sustainability are interdependent. This is of great importance for UNESCO. Through its broad mandate in the thematic areas of education, science, and culture, it is directly involved in the implementation of nine of the altogether seventeen goals of the 2030 Agenda.

UNESCO's Effectiveness and Fields of Action in the Anthropocene

UNESCO not only has the broadest mandate of all UN organizations in terms of content, it is also broadly positioned in terms of impact, modes of operation, and fields of action. It works at the political level through conventions[1] that are binding under international law and sets global standards for monitoring, for example through the obligation to inform all member states about the respect, protection, and implementation of the human right to education (cf. UNESCO, 2008).

UNESCO networks experts and has an impact on large sections of the population through programs and initiatives implemented by and with civil society. Examples include the networks of the 1157 UNESCO World Heritage Sites, the 738 UNESCO Biosphere Reserves, the 177 UNESCO Geoparks, and the more than 12,000 UNESCO Project Schools. UNESCO is represented in all member states by national committees, which serve as interfaces with civil society. This is unique for a UN organization and gives UNESCO a special potential for the imperative, challenging transformation.

[1] For example, the 2005 Convention on the Protection and Promotion of the Diversity of Cultural Expressions (UNESCO, 2005).

Education: The Key to Transformation

According to the founding idea of UNESCO, education is the key to a peaceful future, and it continues to occupy a central position in its work to this day. It was therefore logical to give UNESCO responsibility for implementing Sustainable Development Goal 4 (SDG 4) of the 2030 Agenda: "To ensure inclusive and equitable quality education and promote lifelong learning opportunities for all."

Learning to change the world: Since more than 15 years, UNESCO has been contributing to this goal with Education for Sustainable Development (ESD). In 2020, it launched a new decade with the "ESD 2030" program. At the UNESCO World Conference on ESD 2030 in May 2021, the 193 UNESCO signatory states pledged in the Berlin Declaration to anchor ESD in all educational contexts in their countries—as an essential key to achieving the goals of the 2030 Agenda. ESD is implemented in an exemplary way within the UNESCO networks. These include not only the UNESCO project schools but also World Heritage sites, biosphere reserves, and geoparks, where the educational work is shaped by ESD approaches. UNESCO is thus further expanding its global pioneering role for ESD.

Science: Global Networking in Science and the Free Availability of Knowledge

Water is the basis of human life on the planet Earth. And water knows no borders. As early as 1963, scientists from the field of hydrology met under the umbrella of UNESCO to establish international structures and networks for worldwide cooperation. The conference report states: "Human activities in recent decades have led to exploitation of resources in regions that previously had been largely underdeveloped […] where water is a critical resource whose availability would be a key to development. Even highly developed areas in well-watered zones are not immune from water problems, partly because development and management have not been rational" (UNESCO, 1963:3).

The subsequent hydrological decade contributed to the establishment of the International Hydrological Program in 1975 (for more on the IHP, see Makarigakis & Jimenez-Cisneros, 2019), through which UNESCO has since acted as a key player in the implementation of sustainable water management and resilience at local and regional levels—also with a view to SDG 6 of the 2030 Agenda, "Clean Water and Sanitation". International hydrological research addresses a plethora of highly complex issues, such as water quality and pollution, water scarcity, the occurrence of flooding, sea-level rise, and ocean acidification.

UNESCO has been actively conducting ocean research for more than 60 years: In 1960, it established the Intergovernmental Oceanographic Commission (IOC), the global forum for coordinating and promoting ocean research and observation, and it was the key driver behind the UN General Assembly's designation of

2021–2030 as the UN Decade of Ocean Exploration. Sustainable development relies on international research. Through its initiatives, UNESCO supports SDG 14 of the 2030 Agenda: "Conserve and sustainably use the oceans, seas, and marine resources for sustainable development."

In 1970, UNESCO founded the "Man and the Biosphere" (MAB) program as the first global program to address human-environment relations. The goal of creating model regions for a sustainable coexistence of humans and nature, and thus preserving biological diversity, was convincingly achieved and radiates. UNESCO biosphere reserves now cover almost ten million square kilometres, or roughly 2% of the Earth's surface. Economic, ecological, and social needs must be balanced, and local solutions to global challenges must be developed—a permanent field of tension of negotiation and trial and error. Insights from sustainability science and ESD are incorporated (see UNESCO, 2017). Through their global network, biosphere reserves contribute to exploiting innovative potential, mutually benefit from their knowledge and experience, and share this with others. This is how transformation succeeds.

Culture: UNESCO's Concept of Culture—Opportunities for Sustainability

UNESCO advocates a broad concept of culture: Culture is the "set of distinctive spiritual, material, intellectual, and emotional features," which also includes "modes of life, the fundamental rights of the human being, value systems, traditions, and beliefs." It is through culture, it continues, that humans "question their achievements" (UNESCO, 1982). It thus places cultural identity and heritage at the centre of what determines action and values. In UNESCO's view, cultural resources have the potential to be engines of change and transformation and represent a central force in anchoring sustainability as a core principle in society.

Culture and nature are inextricably linked—this is one of the essential foundations of the UNESCO World Heritage Convention of 1972, which already drew attention five decades ago to the fact that "natural and cultural heritage is increasingly threatened with destruction not only by the traditional causes of decay, but also by changing social and economic conditions which aggravate the situation with even more formidable phenomena of damage or destruction" (UNESCO, 1972). The World Heritage Convention obliges the signatory states to protect and preserve the World Heritage sites.[2] In this context, the negative consequences of climate change, detrimental interventions such as uncoordinated large-scale projects, mass

[2] It is thus the task of signatory states to "adopt a general policy which aims to give the cultural and natural heritage a function in the life of the community and to integrate the protection of that heritage into comprehensive planning program" (UNESCO, 1972).

tourism, and non-sustainable management plans are increasingly coming into focus. They often lead to irreversible damage.

UNESCO had already pointed out these dangers in 2006, but the response in most countries was low. In the meantime, the alarm signals are increasing, and the awareness of the dangers and the willingness to change course are growing. There is no more time to lose.

The General Conference of UNESCO is expected to adopt a policy paper on the impact of climate change on World Heritage sites. Based on this, the Operational Guidelines of the World Heritage Convention will be further developed. The aim is, first, to be better able to counter climate risks and increase the resilience of World Heritage sites worldwide. And second, to make World Heritage sites best practice examples of sustainable development. To achieve this, the management plans of each World Heritage site must be made sustainable.

An encouraging message is connected with a new study (UNESCO, 2021). It provides evidence that marine UNESCO World Heritage sites such as the Wadden Sea help mitigate climate change by absorbing significant amounts of carbon from the atmosphere and storing it in the sea. Roughly half of the natural World Heritage sites help prevent natural disasters such as floods or landslides.[3]

UNESCO's impact opportunities and initiatives underscore: We are responsible for the future of planet Earth. It is in our hands whether the necessary fundamental changes will occur, whether there will be a positive Anthropocene that reconciles humanity and nature and gives our entire planet a real perspective.

References

Makarigakis, A. K., & Jimenez-Cisneros, B. E. (2019). UNESCO's contribution to face global water challenges. *Water, 11*, 388. https://doi.org/10.3390/w11020388

Steffen, W., Crutzen, P. J., & McNeill, J. R. (2007). The Anthropocene: Are humans now overwhelming the great forces of nature? *Ambio: A Journal of the Human Environment, 36*, 614–621. https://www.jstor.org/stable/25547826. Accessed 5 Oct 2021.

UNESCO. (1946). *Report of Dr. Julian Huxley, executive secretary of the preparatory Commission of the United Nations Educational, scientific, and cultural organisation, to the general conference held in Paris, November 1946.* https://digital.archives.unesco.org/en/collection/governing-documents/detail/7df00726-962f-11e8-8718-d89d6717b464/media/a0d34886-ccfa-dc93-acf0-de8e26e9bb5e?mode=detail. Accessed 5 Oct 2021.

UNESCO. (1963). *Preparatory meeting of experts in the field of scientific hydrology*. https://unesdoc.unesco.org/ark:/48223/pf0000138698_eng. Accessed 5 Oct 2021.

UNESCO. (1972). *Convention concerning the protection of the world cultural and natural heritage*. https://whc.unesco.org/archive/convention-en.pdf. Accessed 5 Oct 2021.

UNESCO. (1982). *Mexico City declaration on cultural policies, world conference on cultural policies*. https://unesdoc.unesco.org/ark:/48223/pf0000052505. Accessed 5 Oct 2021.

---

[3] For example, the 2200-kilometer-long Sundarbarns mangrove coastline in India and Bangladesh provides effective flood protection that would otherwise require a $300 million investment in human-made infrastructure.

UNESCO. (2005). *The 2005 convention on the protection and promotion of the diversity of cultural expressions*. https://en.unesco.org/creativity/sites/creativity/files/passeport-convention2005-web2.pdf. Accessed 5 Oct 2021.

UNESCO. (2008). *The right to education. Monitoring standard-setting instruments of UNESCO*. https://www.right-to-education.org/sites/right-to-education.org/files/resource-attachments/UNESCO_The_Right_To_Education_Monitoring_Standard-Setting_Instruments_of_UNESCO_2008_En.pdf. Accessed 5 Oct 2021.

UNESCO. (2017). *A New Roadmap for the Man and the Biosphere (MAB) Programme and its World Network of Biosphere Reserves*. https://en.unesco.org/mab/strategy. Accessed 5 Oct 2021.

UNESCO. (2021). *Custodians of the globes's blue carbon assets*. https://unesdoc.unesco.org/ark:/48223/pf0000375565. Accessed 5 Oct 2021.

United Nations. (2015). *Resolution adopted by the General Assembly on 25 September 2015*. https://www.un.org/en/development/desa/population/migration/generalassembly/docs/globalcompact/A_RES_70_1_E.pdf. Accessed 5 Oct 2021.

Vermeren, P. (2003). *La Philosophie saisie par l'UNESCO*. https://unesdoc.unesco.org/ark:/48223/pf0000132733. Accessed 5 Oct 2021.

Prof. Dr. Maria Böhmer has been President of the German Commission for UNESCO since 2018. Previously, she was Minister of State at the Federal Foreign Office and Special Representative for UNESCO as well as the Chair of the UNESCO World Heritage Committee. She negotiated the Agenda 2030 Sustainable Development Goals at the United Nations. From 2014 to 2018, she was Chairwoman of the Advisory Board of the Tarabya Cultural Academy and a Senator of the Leibniz Association. Previously, Böhmer served as Minister of State to the German Chancellor and Federal Government Commissioner for Migration, Refugees, and Integration and as a member of the German Bundestag. She has been Professor of Education in Heidelberg since 2001.

Part XXII
The Challenge of Peace

Activist

Peta J. White

Abstract In these Anthropocene times, there is much to (re)consider about how our societies evolve. Being an activist implies that you challenge the current or normalised ways of being and acting in society. Taking action to re-direct, intervene, and change social practices and/or address issues can be undertaken either individually or as part of a collective. Activist practice may take many forms and aspire to a variety of outcomes from raising consciousness through to encouraging others to change practices or to influence the redesign or enactment of policy.

Activists are people who act in public ways, motivate others to take action, or socially organise for change. We often associate the term with visible, vocal, highly motivated, articulate and transformative leaders of movements. Most of us carry images of political strike action, or people using their bodies as physical barriers, or placard wielding gatherings of individuals. Social movement theory defines an activist as a person holding to a collective identity involving collective action (Hunt & Benford, 2004). However, Baumgardner and Richards (2000) broadened ideas about activist work suggesting that it include "everyday acts of defiance" (p. 283). In the Anthropocene we are learning to dissociate identity and activism (Bobel, 2007), therefore 'you don't have to be female to be a feminist' and you can do activism without being labelled as an activist.

In 2000, Sachs described the Activist Professional as teachers focussing their classroom encounters and learning towards activist outcomes. There are many narrative accounts of classrooms taking up local (through to global) issues by learning and acting to generate change in the school community, or more broadly. The activist focus may include student or staff wellbeing, school materials, or relevant environmental, social justice, economic or political issues. Many learning programs encourage the enactment of activism (White, 2010) supporting the teacher (or group leader) to enact skills and protocols that scaffold and support the exploration of a

P. J. White (✉)
Deakin University, Burwood, VIC, Australia
e-mail: peta.white@deakin.edu.au

© The Author(s), under exclusive license to Springer Nature Switzerland AG 2023

N. Wallenhorst, C. Wulf (eds.), *Handbook of the Anthropocene*,
https://doi.org/10.1007/978-3-031-25910-4_270

challenge/issue and the engagement with deciding how to act and who to involve. In many education organisations this practice tends to not be named as activism and students and teachers not identified as activists. For example, the New South Wales government maintains a policy about 'Controversial issues in schools' (NSW Government, 2021) stating that schools are neutral places for rational discourse and objective study where any investigation of an issue must follow protocols. Highlighting the need for activists to engage in discourse and to work with community to be given opportunities to be heard.

The notion of an 'active and informed citizen' forms part of the Australian Curriculum, with one aim being, "the capacities and dispositions to participate in the civic life of their nation at a local, regional and global level and as individuals in a globalised world" (Australian Curriculum, n.d.). Global citizenship and sustainable lifestyles forms part of UN Sustainable Development Goal 4.7 as a 2030 target. The indicator for this is, "Extent to which (i) global citizenship education and (ii) education for sustainable development, including gender equality and human rights, are mainstreamed at all levels in: (a) national education policies, (b) curricula, (c) teacher education and (d) student assessment" (United Nations, n.d.). Both documents highlight the imperative for active engagement in citizenry practices, which often requires activist practice.

Educators recognise that knowledge exploration or generation alone will not change attitudes, behaviors, or values (Kollmuss & Agyeman, 2002). Dewey (1938) described schools as representative of the surrounding social environment acknowledging the individuality of learners and the importance of place-based, action-orientated, socially engaging learning opportunities. In Anthropocene times, there are many locally expressed, yet global issues or 'wicked problems' where the complexity of these challenges may include various interacting systemic (social, political, environmental, economic) elements. It is challenging to represent the complexity in a learning environment; however, encouraging individuals to explore personal responsibility and personal actions may be a solution (even though the choice of action usually involves low-impact actions (Tolppanen et al., 2020)). Additionally, the immediate consequences of action and in-action are often the same (Gardner, 2006), reinforcing the lack of impact from considerable effort. Therefore, aspiration towards longer term, systemic change agendas need to be embedded in our activist/education practices.

The international School Strike for Climate (SS4C, 2021) was inspired by one student, Greta Thunberg (2021) whose first success was persuading her parents to reduce their carbon impact and to allow her to take action by striking from school outside the Swedish parliament, igniting a global *kolstrejk för klimatet* (School Strike for Climate). The "Greta effect" (Nevett, 2019) resulted in the SS4C which involves

> … school students of all ages, races, genders, backgrounds and sexualities from every part of Australia. United by our concern for the future of the planet that we live in, we have bridged the thousands of kilometres that separates us to create one of the biggest movements in Australian history. We are striking from school to demand that our politicians take our future seriously and treat climate change as what it is: a crisis.

Demands of the October 2021 strike include:

1. No new coal, oil and gas, including Adani's mega coal mine and net zero emissions by 2030
2. 100% renewable energy and exports by 2030.
3. Funding for a just transition and jobs for fossil fuel workers and their communities. (SS4C-About, 2021)

Students with knowledge or the lived experience of the impacts of the climate crisis comprehend climate science and have become experienced communicators implementing activists' strategies to bring change to government policy and practice to ensure a safe and just future for all. These activist efforts of the SS4C, led by school students, have changed the face of education (White et al., 2021), even if this is not yet acknowledged in the school education systems.

The Extinction Rebellion (XR, 2021) suggests that we share a moral duty to take action as life on earth is in crisis. Our emergency is the changing climate (faster than scientists predicted) which results in biodiversity loss, crop failure, social and ecological collapse, leading to mass extinction. Their demands include governments telling the truth to communicate the urgency, governments acting now to halt biodiversity loss and reduce greenhouse gas emissions to net zero by 2050, and for governments to create and be led by citizen assemblies on climate and ecological justice.

Activist work shapes public policy, approaches to education, employment, and welfare, as well as everyday interactions throughout communities. We are witnessing an emergence of new rights including ecological, sexual, and Indigenous rights (Isin, 2009). Activist foci relevant to the Anthropocene include civil rights, race, gender, animal ethics, disability rights and many more focussed agendas highlighting the negative and socially damaging practices, prejudices, discriminations, and oppressions applied broadly across societies. Key activist campaigns of the Anthropocene can be found on the regularly updated Wikipedia list https://en.wikipedia.org/wiki/List_of_social_movements

Activist practices may include voicing and demonstrating, through to acts of civil disobedience (Thoreau, 1849). Few activists act illegally, yet many are tarnished by the actions of a few or the exaggerated claims of reactionary opponents of progressive change, who blame violence on groups such as XR, Black Lives Matter, and anti-fascist group ANTIFA. Replacing the tradition of a 'manifesto', we now see novels detailing ways of taking up activist practices, such as 'The Activist' by Ojaide (2006). The rise of social media has changed activist communication, reducing the power of media corporations who have enjoyed a long history of message control. Content can now be disseminated through personal networks and community associations (Poell & van Dijck, 2015). Activists' posts can be shared and may 'go viral' meaning shared across many social media platforms and networks generating momentum with many witnesses. This can lead to acceleration and amplification of the message (Holmes, 2008).

References

Australian Curriculum. (n.d.). *Civics and citizenship*. Retrieved from https://www.australiancurriculum.edu.au/f-10-curriculum/humanities-and-social-sciences/civics-and-citizenship/aims/

Baumgardner, J., & Richards, A. (2000). *A manifest: Young women, feminism, and the future*. Farrar, Straus and Giroux.

Bobel, C. (2007). 'I'm not an activist, though I've done a lot of it': Doing activism, being activist and the 'perfect standard' in a contemporary movement. *Social Movement Studies, 6*(2), 147–159. https://doi.org/10.1080/14742830701497277

Dewey, J. (1938). *Experience and education*. Macmillan.

Gardner, S. M. (2006). A perfect moral storm: Climate change, intergenerational ethics and the problem of moral corruption. *Environmental Values, 15*(3), 397–413. https://doi.org/10.3197/096327106778226293

Holmes, B. (2008). Swarmachine. *Third Text, 22*(5), 525–534. https://doi.org/10.1080/09528820802440003

Hunt, S. A., & Benford, R. D. (2004). Collective identity, solidarity and commitment. In S. Snow, S. Soule, & H. Kriesi (Eds.), *The Blackwell companion to social movements* (pp. 433–457). Blackwell.

Isin, E. (2009). Citizenship in flux: The figure of the activist citizen. *Subjectivity, 29*, 367–388. https://doi.org/10.1057/sub.2009.25

Kollmuss, A., & Agyeman, J. (2002). Mind the gap: Why do people act environmentally and what are the barriers to pro-environmental behavior? *Environmental Education Research, 8*(3), 239–260. https://doi.org/10.1080/13504620220145401

Nevett, J. (2019). The Greta effect? Meet the schoolgirl climate warriors. *BBC News*. Archived from the original on 22 July 2019.

NSW Government. (2021). *Controversial issues in schools*. Retrieved from https://education.nsw.gov.au/policy-library/policies/pd-2002-0045

Ojaide, T. (2006). *The activist*. Farafina.

Poell, T., & van Dijck, J. (2015). Social media and activist communication. In C. Atton (Ed.), *The Routledge companion to alternative and community media* (pp. 527–537). Routledge.

Sachs, J. (2000). The activist professional. *Journal of Educational Change, 1*, 77–94. https://doi.org/10.1023/A:1010092014264

SS4C. (2021). *School strike for climate – About*. Retrieved from https://www.schoolstrike4climate.com/about

Thoreau, H. D. (1849). Civil disobedience. In J. Bessant, A. M. Mesinas, & S. Pickard (Eds.), *Citizenship* (When students protest: Secondary and high schools) (pp. 125–142). Rowman & Littlefield Publishers.

Thunberg, G. (2021). *Greta Thunberg – Wikipedia*. Retrieved from https://en.wikipedia.org/wiki/Greta_Thunberg

Tolppanen, S., Claudelin, A., & Kang, J. (2020). Pre-service teachers' knowledge and perceptions of the impact of mitigative climate actions and their willingness to act. *Research in Science Education*, 1–21.

United Nations. (n.d.). *Sustainable development: The 17 goals*. Retrieved from https://sdgs.un.org/goals, https://sdgs.un.org/goals/goal4

White, P. J. (2010). *Action learning group project*. Retrieved from https://sites.google.com/site/actionlearninggroupproject/

White, P. J., Ferguson, J. P., O'Connor, S. N., & O'Shea Carré, H. (2021). School strikers enacting politics for climate justice: Daring to think differently about education. *Australian Journal of Environmental Education, 38*, 26–39.

XR. (2021). *The extinction rebellion*. Retrieved from https://rebellion.global/

Peta White is an Associate Professor in science and environmental education at Deakin University. She educated in classrooms, coordinated programs, supported curriculum reform, and prepared teachers in several jurisdictions across Canada and Australia. Her PhD explored learning to live sustainably as a platform to educate future teachers. Peta continues her commitment to initial teacher education leading courses, units, and programs and in-service teacher education through research-informed professional learning programs. Peta's current research follows three narratives: science and biology education; sustainability, environmental, and climate change education; and collaborative/activist methodology and research.

Buen Vivir

Jörg Zirfas

Abstract The concepts of the art of living and Buen Vivir are historically-culturally and content-intentional very different models. The term "art of living" can be understood as a philosophical school of thought that raises questions about the design and shaping of a happy life. In its more recent variants, it draws on models from antiquity and centers them around the question of the extent to which human beings can realize a good, successful, and beautiful life on the basis of their own wise choices. In contrast the model of "Buen Vivir" focuses on care for the other and care for nature: it is concerned with the search for good alternatives of life or the utopia of a right way of living together. This concept comes from theindigenous peoples of the Andean region.

By self-care, of course, more is meant than the trivial fact that one is always taking care of oneself in some way or another. The art of living is a demanding program. In addition to a very far-reaching hermeneutics of oneself, which includes interpretations of oneself in addition to understanding, and an action-based practice oriented to specific ethical categories, which is definitely associated with work, practice, and discipline, it is also accompanied by an aesthetic style, which in turn should meet original aesthetic-artistic criteria (cf. Brenner & Zirfas, 2002). Today, the art of living means forming one's own laws, inventing the guidelines of one's life oneself, letting aesthetic self-invention converge with playful-ethical self-control (cf. Hadot, 1991; Foucault, 2007). The idea of a beautiful life performatively links production with work, but also with reception. It is concerned with a fulfilled life, with the unity of consciousness and experience, enjoyment and use of an affirmatively beautiful and original existence (cf. Gödde & Zirfas, 2014).

J. Zirfas (✉)
University of Cologne, Cologne, Germany
e-mail: joerg.zirfas@uni-koeln.de

© The Author(s), under exclusive license to Springer Nature Switzerland AG 2023
N. Wallenhorst, C. Wulf (eds.), *Handbook of the Anthropocene*,
https://doi.org/10.1007/978-3-031-25910-4_271

The art of living is always called for when the validity of traditional customs, traditional norms, and significant claims of ought begin to waver (cf. Fellmann, 2009). This is not only the case with regard to the Anthropocene. But the Anthropocene radically questions the focus on self-care, which in some respects appears to be the effect of a neoliberal era associated with optimization, deregulation, competitiveness, globalization, and privatization of the market and the social, and therefore loses sight of the idea of sustainability, for example. In this respect, an ecological art of living is also still too strongly centred around an (existential) self-care, even if it understands nature not only as a means for its own ends, but also as an end in itself (Schmid, 1998, p. 424–428).

In contrast, a sustainable art of living would have to be developed that encompasses the relationships to other people, generations and to nature as a whole and refrains from centering around self-care. The term "sustainability" is understood to refer to the development possibilities of the future, in which the interrelationships of economy, ecology, education, sociality, politics and culture (should) be given greater consideration. The aim is also to improve the quality of life of future people, to preserve their scope for freedom, to enable positive social development, and to ensure future economic growth and ecological compatibility (cf. German UNESCO Commission, 2008). Education for sustainable development (ESD) therefore aims to "enable people to actively shape an ecologically compatible, economically efficient and socially just environment, taking global aspects into account" (BMBF cited in ibid., p. 7).

The aforementioned concept of sustainability links questions of an almost unpredictable future with questions of generational justice, social inequality, health and quality of life, political, economic and ecological reason. In addition, these fields of reflection and action are furthermore to be gathered as an integrating cross-sectional task in an intercultural, global scale under aspects of a lifelong learning process by all people.

Here, the art of living takes on an almost utopian character. And because all utopias have the tendency to limit the diversity of human life in favour of a social order that is considered good. The development aimed at today for sustainability is more diverse than all historically known utopias. But if the goals for sustainable development were to be realized, this would certainly require the restriction of fundamental rights (cf. Wulf, 2018, p. 180). In this respect, the "grand narrative" (Lyotard) of sustainability can also be subjected to critical reflection, which, in relation to a better life in the future, assumes not only global, but also differential forms of a beautiful and successful life. Insofar as the idea of sustainability is also committed to a concept of quality of life and well-being of future generations as well as their possibilities of freedom, it must also place the different possibilities of life arts under its protection. That is, the idea of sustainability is committed to the art of living insofar as it should also include in its normative ideas the possibilities of a life that succeeds from different perspectives (cf. Zirfas, 2011).

In summary, three ways in which the art of living relates to sustainability can be identified (cf. Gödde & Zirfas, 2018): (1) theoretically, in reflecting on and critiquing existing structures and in developing new concepts (such as post-growth or interconnectedness with nature). (2) practical, in exploring and experimenting with new ways of living (for example, in consumption, product design, or the treatment of animals); (3) emotional, in raising awareness of problems and contradictions and of new qualities of life (in the social, political, or ecological).

While the art of living emphasizes self-care, the model of "Buen Vivir" focuses on care for the other and care for nature: it is concerned with the search for good alternatives of life or the utopia of a right way of living together (Acosta, 2017). This concept comes from the indigenous peoples of the Andean region; it has found its way into the constitutions of Ecuador (2008) and Bolivia (2009). It is a plural model that can be interpreted in many ways. In Article 275 of the Constitution of Ecuador, we find the following substantive formulations: respect for diversity, by which is meant not only biodiversity, but also cultural, plurinational diversity; the human right not only for individuals, but also for communities of persons (for example, communities); and coexistence with nature. Life is related not only to humans and animals, but to the cosmos; nature and cosmos acquire a complementary subject status with independent legal claims. In this context, the rebirth of Mother Earth (Pachamama) and her reproductive possibilities play a central role (cf. Poma, 2011).

Ramírez (2010) describes the shift in thinking and acting from "anthropocentrism" to "biocentrism" inherent in the concept of Buen Vivir. In his perspective, the good life is based on the rejection of (Western) individualism with its personal need satisfaction and neoliberal capitalism with its colonialism and extractivism towards ethical values of nature and sufficiency. If we follow the considerations of Eduardo Gudynas (2012, p. 30), then there are the following similarities between the present models of Buen Vivir: (1) Buen Vivir does not operate with a linear or even completed model of development and progress, but with an alternative to these models. (2) Buen vivir advocates a harmonious relationship with nature and grants it rights. In contrast to an anthropocentric view of opposing or dominating nature, Buen Vivir emphasizes the connectedness between humans and nature. In order to express this bond, there is also the much-cited giving back of part of the food to Pachamama, to the great Mother Earth – who as goddess gives life (cf. Rösing, 1997, p. 14). (3) Quality of life is not understood in terms of income, material wealth and material goods, but in terms of a fulfilled life. (4) Buen vivir does not focus on materialism but on spirituality and emotionality. (5) Social relations and things are not reduced to economic facts; a demercantilization of nature and society is demanded.

In addition to concern for nature and connectedness with it, the concept of Buen Vivir focuses above all on concern for others and connectedness with them – also with regard to a good life for future generations. Influenced by Andean ways of thinking and living of pre-Hispanic traditions, especially of the Kichwa population dominating in Ecuador, this concept simultaneously enters into new mixtures with globalized, critical social movements and thus aims at a new shaping of social relations. Buen Vivir as a form of social action and organization aims at a practical knowledge of living together, in which the idea of mutual aid plays a central role (Giebeler, 2018). This model breaks with an art of living strongly centred around self-care by focusing on questions of integration of others, co-care, and mutual relatedness. Against the backdrop of a dismantling of capitalist, paternalistic, and homogenizing structures, it calls for comprehensive democratization and collective citizenship, recognition and practice of multiculturalism and interculturality, sufficiency and responsibility, and the establishment of solidarity, integrality, and inclusion (cf. Acosta, 2017, p. 145). It is about a shared art of living in which the mutual and dialogical interdependence of humans, community, and nature is central.

References

Acosta, A. (2017). *Buen Vivir. Vom Recht auf ein gutes Leben*. 5. Aufl. Oekom.

Brenner, A., & Zirfas, J. (2002). *Lexikon der Lebenskunst*. Reclam.

Wulf, C. (2018). Die Bildung von Kindern in der Weltgesellschaft. In *Pädagogische Anthropologie der Kinder*, hg. Gerald Blaschke-Nacak, Ursula Stenger, und Jörg Zirfas, S. 178–186. Beltz Juventa.

Deutsche Unesco-Kommission e.V. (2008). *UN-Dekade Bildung für nachhaltige Entwicklung 2005–2014*. Eigenverlag.

Fellmann, F. (2009). *Philosophie der Lebenskunst zur Einführung*. Junius.

Foucault, M. (2007). *Ästhetik der Existenz. Schriften zur Lebenskunst*, hg. Daniel Defert, und François Ewald. Suhrkamp.

Giebeler, C. (2018). Die Kinderzentren des Buen Vivir in Ecuador. In *Pädagogische Anthropologie der Kinder*, hg. Gerald Blaschke-Nacak, Ursula Stenger, und Jörg Zirfas, 138–162. Beltz Juventa.

Gödde, G., & Zirfas, J. Hg. (2014). *Lebenskunst im 20. Jahrhundert. Stimmen von Philosophen, Künstlern und Therapeuten*. Wilhelm Fink.

Gödde, G., & Zirfas, J. Hg. (2018). *Kritische Lebenskunst. Analysen – Orientierungen – Strategien*. Metzler.

Gudynas, E. (2012). buen vivir. *Das gute Leben jenseits von Entwicklung und Wachstum*. https://www.rosalux.de/fileadmin/rls_uploads/pdfs/Analysen/Analyse_buenvivir.pdf. S. 5–33 (Abruf: 16.4.21).

Hadot, P. (1991). *Philosophie als Lebensform. Geistige Übungen in der Antike*. Gatza.

Poma, M. (2011). Vivir Bien (Gut leben). *Zur Entstehung und Inhalt des Guten Lebens*. https://amerika21.de/analyse/42318/vivir-bien. (Abruf: 16.4.21).

Ramírez, R. (2010). La transición ecuatoriana hacia el Buen Vivir. En *Sumak Kawsay/BuenVivir y cambios civilizatorios*, 2da Ed., Eds. Irene León, 125–141. FEDAEPS.

Rösing, I. (1997). *Jeder Ort – ein heiliger Ort. Religion und Ritual in den Anden*. Benziger.

Schmid, W. (1998). *Philosophie der Lebenskunst: eine Grundlegung*. Frankfurt/M.

Zirfas, J. (2011). Der Geschmack an der Nachhaltigkeit. Ästhetische Bildung als Propädeutik und Regulativ einer Bildung für nachhaltige Entwicklung. In *Die unsichtbare Dimension. Bildung für nachhaltige Entwicklung im kulturellen Prozess*, hg. Gabriele Sorgo, 32–52. Forum Umweltbildung.

Dr. Jörg Zirfas, Professor of Anthropology and Philosophy of Education at the University of Cologne. Head of the commission "Educational anthropology" (DGfE) and of the "Society of Historical Anthropology" at the Free University of Berlin. Historical and educational anthropology, philosophy and psychoanalysis of education, educational ethnography, aesthetics, culture and education.

Care

Renaud Hétier

Abstract The philosophy of care involves thinking of our moral relationship to everything that requires care (vulnerable people, objects, the world, etc.) in a different way to the conventional categories of abstraction and generalization. The relationship of care (giving and receiving) is marked by interdependence, and in that context, autonomy takes a back seat. We humans are beings in relation, and beings of relation. We communicate, and we expect answers. The ethics of care invites us to engage, express ourselves adequately, heed any form of expression that helps understand what is important (for the other), and be open-minded towards the specificity of every given context, and the uniqueness of any recipient of care.

The word *care* is enormously versatile, covering a vast range of activities and sentiments. In fact, in philosophical discussions in other languages, including French, the English word is used, as there is no translation which adequately conveys all that *care* entails. The theory of care is rooted in Carol Gilligan's, 1982 book, *In a Different Voice* (1982). Gilligan attaches a great deal of importance to the female voice in resolving a predicament posited by Kohlberg (should Heinz steal medicine to save his wife if he cannot afford it?). Whereas Jake, the boy interviewed for the study, responds abstractly and presents a logical hierarchy (it is better to steal – and risk going to jail – than to allow his wife to die), the girl (Amy) believes it would be better to try and convince the druggist to save Heinz's wife's life, and avoid prison. Kohlberg argues that this sensitivity to context and to relationships is confused. Gilligan analyses Amy's argument anew, demonstrating that it has merit, and holding that a different (female) voice should also be heard: 'a mode of thinking that is contextual and narrative rather than formal and abstract' (Gilligan, 1982, 19). This distinction can finally be generalised: 'While women thus try to change the rules in order to preserve relationships, men, in abiding by these rules, depict relationships as easily replaced' (Gilligan, 1982, 77).

R. Hétier (✉)
UCO, Angers, France
e-mail: Renaud.hetier@uco.fr

© The Author(s), under exclusive license to Springer Nature Switzerland AG 2023

N. Wallenhorst, C. Wulf (eds.), *Handbook of the Anthropocene*,
https://doi.org/10.1007/978-3-031-25910-4_272

As one might imagine from the above quote, Gilligan's critique is from a feminist point of view: the 'different voice' to which she refers is a female voice. Following in Gilligan's footsteps, Joan Tronto extended the range of ethics of care, giving it a political dimension, beyond feminism. Certainly, it is true that the majority of 'caregivers' are women, but the lack of recognition shown to them, more generally, needs to be addressed. Any caregiver (women and immigrants making up a disproportionate majority of this group) is worthy of greater recognition. By shining a spotlight on them, we also shine a light on humanity's need for care. The problem is one of distribution (of roles) and recognition (of the care dispensed). Thus, it is the social and political aspects of care which are highlighted. Tronto's definition, which is in keeping with Berenice Fischer's thinking, is as follows: 'we propose that care be considered *a generic activity that includes everything we do to maintain, perpetuate and repair our "world" so that we can live in it as well as possible*' (Tronto, 2009, 244). This description extends the definition of care beyond the confines imposed by certain critics, whose criteria restrict it to caring for loved ones, in the way that a parent cares for a child. Now, caring for our environment can also be considered an activity of *care*.

Tronto rationalises the work of care by describing it in three phases (Tronto, 2009, 248–250): 1. *caring about*, which is being aware of a problem and feeling concerned. 2. *taking care of*, which is acting to provide means (for example, giving a person money). 3. *care giving*, which is 'making contact with the (Garrau, 2014) bject of the *care*' (Tronto, 2009, 249), getting directly involved. 4. *care receiving* means being concerned about the effects of the *care* on the object in question. When that object is a subject (a person), this means lending them an ear. This description plainly highlights the fact that care is a type of work – an activity (rather than simply a personal, private emotion), though as Molinier (Paperman & Laugier, 2011, 343) points out, 'in order to be effective, the work of care should not be perceived as a task'.

Having briefly outlined what care means in the world of philosophy, we can distinguish three main avenues for discussion of it: the first relates to expression, the second to the logical structure, and the third to the concrete position that we adopt in dealing with the world around us. To begin with, there is a difficulty of expression. It may not always be easy to articulate one's position. Yet this difficulty in expressing it does not, in any way, reflect on the validity of that position. Gilligan writes of the female interviewee, Amy: 'as the repetition of the questions during the interview suggests to her that she is giving wrong answers or being misunderstood, she begins to lose confidence in herself, and becomes increasingly shy and uncertain' (Gilligan, 1982, 54). Later on, we shall revisit the question of expression of views, which is ripe for discussion in an educative work. Then – and this is the heart of Gilligan's argument – the logical structure of thought may not be one of abstraction and generalisation. The girl speaks of her 'vision of the world [which] is formed of interwoven human relationships, together forming a cohesive whole, rather than of isolated, independent individuals whose relations are governed by systems of rules' (Gilligan, 1982, 54). On this point, it is worth noting that Wittgenstein, in his 'blue book', points to a 'general impulse'. Finally, it is important to take account of every individual's position in life in relation to others. What does it mean 'to be independent'? Who is 'independent'? When we look at the matter of care, we begin to see the vulnerability and interdependence of all. These may be kept hidden, precisely by playing down, and keeping a distance from, everything on which the claim of independence is based.

In particular, it was Sandra Laugier who reawakened the question of expression. For Laugier, it is a question of 'spotting what is important' (Molinier et al., 2009, 178). This requires an education – for example, through literature – that provides 'learning of appropriate expression, and education in sensitivity'. In reference to Wittgenstein and Foucault, the crucial thing for Laugier is to 'see what can be seen', to see what is right in front of our eyes, but which we fail to see because we are not paying attention. Therefore, 'paying attention to private individuals can alter the way in which we see reality' (Laugier, 2012, 156). This culture of attention can also be cultivated through cinema, as Stanley Cavell showed, or with television series, as Sabine Chalvon-Demarsay demonstrated. It is expressed by Marie Garrau: 'indeed, the appropriate action will depend on detailed perception of context' (Garrau, 2014, 47). In addition to attention, expression is up for discussion: it is a matter of 'finding contact again with experience, and finding a voice with which to express it' (Laugier, in Pascale et al., 2009, 185). Put differently, it could be argued that culture trains us to find 'differences', and thus, protects us from indifference.

Ultimately, one might justifiably wonder whether the ideal of independence, which has been prevalent in education since the Age of Enlightenment, is actually contributing to a lack of attention paid to care receivers, who have become an absolute priority in today's Anthropocene world.

In his book *Resonance*, Hartmut Rosa argues in favour of the prevalence of relationships over the existence of either the subject or the world: 'the subject and the world do not pre-exist the relationship as isolated entities, but are themselves, so to speak, "products of a relationship" ' (Rosa, 2018, 172). Bernard Aspe echoes this view, pointing to a 'transindividual relationship [which] goes beyond individuals, incorporating them as part of a reality that is broader than themselves: a resonance system' (Aspe, 2013, 78). François Flahaut, meanwhile, stresses that relationships are indispensable: 'we can do better than to simply deny our humiliating dependence on something that does not depend, in turn, on us (corporeal life, what we have inherited from the previous generation, relationships with others, the social environment, material resources, collective and natural resources [etc.]'. If we take relationships to be so crucially important, it undeniably flies in the face of the ideal of independence, espoused since the Age of Enlightenment. This ideal not only serves as the goal of education, but as a particularly potent driving force in western societies.

Taking account of the priority of relationships, but simultaneously a relationship of resonance with the world (as opposed to alienation) requires a rather broader definition of *care* – notably, a political concept, as advocated by Tronto (to recap, she defines care as '*everything we do to maintain, perpetuate and repair our "world" so that we can live in it as well as possible*'. However, the conditions in which we find ourselves may lead to a slight reformulation of the orientation of the activity of care: '*everything we do to maintain, perpetuate and repair our "world" so that we can* continue to *live in it as well as possible,* and furthermore, so our children can continue to live in it as well as possible'. Consideration for future generations, as pointed out by Hans Jonas (1986), is now an integral part of our moral horizon. In addition, our moral duty in today's world is to foster relationships (rather than the idea of individualistic independence) and resonance (rather than alienation, particularly in the acceleration of life and in consumption), even from childhood.

Therefore, there is a connection to be drawn between care, relationship and resonance. This does not pose major theoretical problems. Authors writing about care stress our interdependence, and the importance of relationships in dealing with the moral problems we encounter. In his work on resonance, Rosa makes no mention of the ethics of care. However, we can establish a certain link between the fourth phase of care and resonance: in both cases, it is a matter of looking for a response (Rosa speaks of a 'responsive' world), and being attentive. In that sense, a relationship is not limited to being a relationship between two subjects: it can just as well be a relationship with any object, or a relationship with the world. Undeniably, having been plunged into the Anthropocene, we are in a position where we must take care of the world, if we wish to continue living in it, and maintain lively relationships. In the wake of Rosa's work on resonance, we must ask ourselves whether the relationship of care, which is so necessary today on a large scale, would actually encourage resonance. Indeed, care can be understood as an activity which produces a response, in that it is directed at someone, or something, outside of the self.

When we consider the place of childhood and the purposes of education, we find ourselves grappling with philosophical and anthropological issues. Is it fair to make our children responsible for taking care of a world which has been inadequately cared for by earlier generations? Is it right to place children in a situation where they must accept responsibility for care, although by their very nature, they should still be care receivers? The first of these questions leads to a certain degree of vigilance: what we need to do is make children aware of the issues – not place on their shoulders a responsibility which they are not equipped to bear. In answer to the second question, it should be remembered that, as Martine Menès points out, 'children who are of interest to no-one have interest in nothing – not even in themselves' (Menès, 2012, 50). Children should never cease to be the object of care. In addition, there is an obvious connection between caring for children and caring for the world of the future. However, it may be that there is no better way, in preparing for a threatening future, than placing children in a position where they can feel capable of taking care of something (see the entry 'Bringing Life' in this compendium).

References

Aspe, Bernard. 2013. 'Simondon et l'invention du transindividuel'. *La Revue des livres*, n° 12 74–79.

Garrau, M. (2014). *Care et attention*. PUF.

Gilligan, C. (1982). *In a different voice. Psychological theory and women's development*. Harvard University Press.
Jonas, H. (1986, [1979]). *Le Principe responsabilité : une éthique pour la civilisation technologique*. Cerf.
Laugier, S. (2012, [2006]). *Éthique, littérature et vie humaine*. PUF.
Menès, M. (2012). *L'enfant et le savoir*. Seuil.
Molinier, P., Laugier, S., & Paperman, P., (2009). *Qu'est-ce que le care?* Payot.
Paperman, P., & Laugier, S. (2011). *Le souci des autres. Éthique et politique du care*. EHESS.
Rosa, H. (2018, [2016]). *Résonance. Une sociologie de la relation au monde*. La Découverte.
Tronto, J. (2009). Un monde vulnérable. Pour une politique du care. La Découverte.

Renaud Hétier is Professor at the Catholic University of the West (UCO). He is Doctor of Educational Sciences. He is the author of books on education, and anthropology in the Anthropocene. Books (selection): *Humanity versus Anthropocene* (PUF, 2021, *in French*); *Cultivating attention and care in education* (PURH, 2020, in French); *Create an educational space with the fairy tales* (Chronique sociale, 2017, *in French*); Education between presence and mediation (L'Harmattan, 2017, *in French*).

Freedom

Roland Bernecker

Abstract The planetary impact of human agency is the signature of the Anthropocene. This not only gives the concept of freedom a central place in thinking about the Anthropocene, but presents us with this challenging question: to what extent have conceptualizations of freedom and liberty contributed to the dramatic acceleration in the depletion of natural resources, and more generally to a mindset alienating humans from ways of sustainably inhabiting the biosphere, that sustains them and of which they are an integral part? Can freedom and liberty be upheld as those prime and fundamental values, extensively enshrined in constitutional foundations of our societies and institutions, e.g. in the United Nations Charter (preamble and art. 1, 3)? Human agency being the main driver of the ecological crises we are facing, do we not have reason to reconsider freedom as an incongruous ideologeme to be discarded, or at least its conceptual core thoroughly reassessed? Interestingly, this question lies outside the scope of recent geoscientific research dealing with the Anthropocene, which "reframes the notion of the human species in a particularly naturalist fashion. […] instead of individuals with free will, it views us as a united species with a simple history" (see *Human Condition*, this vol.).

While this naturalist framing might seem plausible in theories primarily concerned with geology and the physical system of the biosphere, its reductionism misses a central point in the problem it is dealing with: the impact of contingent choices in human governance. It is only with the full acknowledgement of human freedom, and in consequence of the voluntarist perception of our responsibility, that we access the political dimension of the questions for which the Anthropocene only serves as a conceptual catalyst: the impact of human freedom and human governance on a global scale. In the theory of international relations, the recent shift towards the concept of global governance is precisely intended to "help us understand where we came from and why we have got to where we are, as well as a way to develop strategies for where we should be going" (Weiss & Wilkinson, 2018, 12).

R. Bernecker (✉)
Brandenburg University of Technology, Cottbus-Senftenberg, Germany

© The Author(s), under exclusive license to Springer Nature Switzerland AG 2023

N. Wallenhorst, C. Wulf (eds.), *Handbook of the Anthropocene*,
https://doi.org/10.1007/978-3-031-25910-4_273

The naturalist reduction tends to suppress the dimension of human freedom, to shift the focus away from how humans have used the spaces available to them and what new lessons we have to consider in view of fundamental choices we will have to make.

A Shift in Values

Thinking the Anthropocene leads to a critical perception of freedom. While freedom depends on the availability of spaces and resources that sustain options for agency and development, the emblematic notion of the Anthropocene emphasizes the accelerating corruption of these resources, and the shrinking, and possibly vanishing, of available spaces. We encounter geological boundaries that create categorical limitations to human freedom. Planet earth is the space in which human agency unfolds. Attempts to extend this sphere in astronautic explorations have only confirmed our dependency on what we learned to visually recognise as a tiny "pale blue dot" (Carl Sagan) in an enigmatically inhospitable cosmos. The inverted perception of earth from a cosmic perspective has revealed the immensity of space as a probably final limit to the spatial expansion of human agency. It has led to a "new geotropism" (Bajohr, 2020, 12), a renewed focus on our confined terrestrial domain. Our sense of freedom may be infinite, the life-sustaining resources for its realizations are not. In light of the scale of *homo faber's* imprint on the geological space, it is becoming ever more evident that "[…] the Anthropocene reduces society's options, curtailing human agency and freedom to make meaningful […] choices." (Orr & Brown, 2020, 257).

In this conflicting constellation, not only space, but time is becoming an increasingly relevant factor. A basic principle for the reasonable limits of individual freedoms is the point at which my actions curtail the freedom of another person. In his contemporary debate on *enclosures*, Locke, a foundational figure of liberal thought, argued that "he that leaves as much as another can make use of, does as good as take nothing at all. Nobody could think himself injured by the drinking of another man, though he took a good draught, who had a whole river of the same water left him to quench his thirst: and the case of land and water, where there is enough of both, is perfectly the same." (Locke, [1690] 2012, 18). In the Anthropocene, there isn't enough anymore of both. And, given the deep impact of human agency on life supporting systems, we have to consider the distributional justice on a diachronic axis as well: *Sustainability* includes the impact of our actions and choices on future generations whose "[ability] to meet their own needs" must not be compromised by the choices we make (United Nations, 1987, 15). Möllers points out that the theory of liberalism, "since it operates with constructed subjects, can project them into the future and reason with future generations" (2021, 266). The concrete modalities of how to equitably and effectively consider the impact of contemporary actions on more distant points on the time-axis, is an intricate economic, political and ethical challenge. It lies at the core of the concept of sustainability. Some consider we behave like in a Ponzi scheme, paying the prize for the present with the future: "We borrow environmental capital from future generations with no intention or prospect of repaying" (United Nations, 1987, 14). Behind these reflections on the accelerating deterioration of the life-supporting planetary biosphere, however, looms another, more fundamental threat. If not of immediate relevance, it cannot be entirely disregarded as an element of the human condition in the broader picture: the calculable definitive limits for life on earth in an unstoppably ageing solar system.

In defining the geological space of earth as one united and interdependent system, the Anthropocene vigorously challenges the concept of freedom. Like no previous generation, we have to learn to acknowledge the impact of our agency on people and societies in distant regions, as well as in still unbroken futures. We have to confront the basic ethical question in which ways our choices limit or condition their freedom. Which of our choices are still innocent? "There are few, if any, neutral acts" (Orr & Brown, 2020, 257).

In light of these considerations, the value of freedom and the political order of liberal democracy are exposed to increasingly critical assessments. The current crisis of liberal thought is fed by different sources: from the political right as conservative republicanism, focusing on the decline of what it perceives as traditional values; from the left for the blatant deficits in equity and social justice liberal capitalism has engendered; and geopolitically by strengthening and newly emerging authoritative "political systems with a fundamentally different understanding of the relationship between state authority and civil freedoms [that] now confidently and assertively challenge the primacy of a West" (Bernecker & Grätz, 2021). The ecological crisis may be seen as another substantive argument against the case of freedom and liberty as fundamental human values, seen that the societies enjoying the greatest freedoms in terms of governance, options for lifestyles and self-realization, are those with the deepest ecological footprint, while the devastating effects of a deteriorating biosphere are heaviest in less privileged regions. From here, one may conclude that the option to attach more equitable calculations of costs and prizes to the consumption of resources will not be sufficient, and that a much deeper transformation is needed: "a shift in values" (Ziegler, 2020, 147).

A Critique of Freedom

Are our concepts of freedom and liberty the drivers that have led to the current scenario of dramatic ecological threats? For some, unsustainable concepts of development and progress were the consequence of a biased emphasis on human freedom and liberty, which, therefore, would need to be unfalteringly reconsidered: "A strong national or global consensus to pursue any sustainable pathway with real vigour (and hence any real chance of success) must entail strong enforcement of policies and regulations, and consequent restrictions on the freedom of individual or corporate actors to undermine the selected policy path. There is no shying away from the fact that avoiding involuntary degrowth and collapse necessarily entails restrictions on economic freedom, and is probably not compatible with current forms of capitalism." (Hensher, 2020, 214–215). Some see freedom and liberty less as the cause, than as a symptom of the period of fossil-driven over-consumption: "Can the concepts of liberty characteristic of modernity (however inadequately realized) survive into the deep future? Or was modernity a fossil-fuelled blip in human history to be followed by some kind of relapse?" (Quilley, 2020, 234). In this line of thought, freedom and liberty appear as historically contingent values which, by stimulating lifestyles of excessively unsustainable individualism, have failed to prove their worth and deserve to be ditched, or will naturally disappear again in the course of history.

This confrontation of ecological responsibility in the Anthropocene with the concept of freedom and liberty is paradoxical. Ecology and sustainability are concepts expressing the ambition to preserve "options for future generations" (United Nations, 1987, 15), to keep spaces available for human agency, to sustain for future generations the same freedom we claim for us, the "freedom to decide what we have reason to value" (World Commission on Culture and Development, 1995, 26). Sustainability is concerned with the preservation of freedom. "Freedom is central to the process of development", as Amartya Sen has emphasized, "for two distinct reasons. (1) The evaluative reason: assessment of progress has to be done primarily in terms of whether the freedoms that people have are enhanced; (2) The effectiveness reason: achievement of development is thoroughly dependent on the free agency of people." (Sen, 2000, 4). Freedom is a pivotal notion for the ethical construct of human agency, and has been a vital source of action and reflection that "like no other keeps moving human thought" (Brunner & Conze, 1972, 456).

The critique of freedom motivated by ecological concerns builds its argument on an intentionally limited conceptualization of freedom. In this vein, McCready defines freedom as "the absence of impediments to the actions of self-interested individuals conceived in isolation from one another" (2020, 67). Equating freedom with the selfish actions of isolated individuals leads to a reductionist and misleading perspective in which freedom cannot be recognized anymore as a fundamental and inescapable quality of the human condition, which is translated into the contingencies of cultural choices on different levels of our individual, social and political existence. Freedom is an unavoidable cultural factor. Limiting freedoms for individuals by enforcement or regulation necessarily goes along with the endowment of selected other individuals to take decisions on their behalf. This is the core of Kant's notion of enlightenment as the emergence of people from their self-imposed *Unmündigkeit*, which can be translated as *nonage*, *minority* or *immaturity*. In illiberal governance models, the freedom to take decisions does not disappear. All policies are based on decisions taken by individuals. Virtually, every decision can be traced back to a place where choices are made and competing options are discarded. In terms of governance, it is highly relevant who the people are that inhabit these places, what their interests, motivations and worldviews are, what their legitimacy is to impose regulations and restrictions upon the community they govern, and what processes have led to their selection as well as to their policy decisions. As rightly pointed out by Möllers, in a governance perspective the emergency of the ecological question must not lead to "dropping all political processes of mediation, even if one was in the position to do so" (2021, 267).

It is not an easy question if limiting individual freedoms and abandoning the liberal order in favour of a more authoritative, possibly expert-driven governance, will bring about greater efficiency in solving the ecological crisis. Solutions that are needed will probably contain technological innovations more likely to occur in the competitive spin of liberal governance. The current geopolitical situation seems to suggest that authoritative governance tends to go along with abusive forms of concentration of power for political elites focused on preserving their power and privileged access to resources. The highly influential concept of *negative and positive liberty*, coined by Isaiah Berlin in 1958, precisely addressed this systemic danger in light of the experiences of the twentieth century (Berlin, 2002).

Dimensions of Freedom

In a perspective relevant for the Anthropocene and considering previous reflexions (cf. Arendt, 2018; Berlin, 2002; Constant, 1819; Skinner, 2012; Taylor, 1985, and recently Möllers, 2021), the concept of freedom can be explained more comprehensively in five distinguishable, though interconnected dimensions. On the individual level, freedom signifies spaces and options for self-realization and the exploration of ways to experience oneself in the world (Berlin, 2002). With our knowledge about the Anthropocene, we have reason to develop new educational modalities to conceive experiences of self-exploration as a process leading to greater consciousness of our deep *entanglement* with all other forms of life (Sheldrake, 2020). The second layer of freedom is the attribution of resources and property as preconditions to the enjoyment of freedoms. A community needs to produce and to effectively distribute resources that can nourish and sustain its members and their ambition for freedom (Constant, 1819). A third layer of freedom is the actual involvement of citizens in the political decision-making for their own affairs, the transparency and legitimacy in the distribution of power, and the enjoyment of fundamental civic freedoms (Arendt, 2018). The fourth freedom is the freedom of a formally constituted community, today mostly in form of a nation or state, to decide on their own affairs and not to be governed out of the sphere of interest of another community, e.g. under colonial rule. This level is concerned with the freedom inherent to sovereignty. It is dealt with in foreign policy, economic interactions and military resilience. In antiquity, this was a main denotation of freedom (Skinner, 2012). Foreign policies are highly relevant for the ecological question. Not only are the solutions we have to find international, they have to be negotiated, adopted, implemented and monitored on a global scale, in dealing with a whole range of antagonistic national interests; the defences of these national interests may as well easily trigger wars with disastrous consequences not least in an ecological perspective.

Interestingly, these four dimensions of freedom relate to Franklin D. Roosevelts points of the famous "Four Freedoms speech" in his 1941 *State of the Union address*:

"Freedom from fear" corresponds to the dimension of foreign policy – the absence of war; "Freedom of speech" corresponds to internal governance, the right to speak up and freely participate in the public debate; "Freedom from want" corresponds to the dimension of effective distribution of resources; "Freedom of worship" is one aspect of the freedom to individual self-realization.

However, a fundamental fifth dimension, emphasized for example by Höffe (2015), is of particular relevance in the Anthropocene: the acquired sovereignty in the human species' struggle for survival in a harshly competitive biosphere. It is important to consider that, generally, the dominant position of humans in the Anthropocene has created exceptional conditions for hugely enhanced enjoyment of freedoms, through highly effective civilizational securitizations, e.g. the supply of food, healthcare, social organization, education, research. However greatly they differ in regions and social strata, the Anthropocene is a direct effect of the dominant position of our species, allowing humans to effectively overcome many of the immediate pressures linked to the daily fight for survival.

The anthropologization of the biosphere has been an important element in the history and the development of human freedom. In their struggle for persistence, humans had to compete with a broad range of other creatures for limited resources, under a permanent threat of annihilation by predators, diseases or climatic conditions. The particular human disposition for fear and the considerable role of fear in our overall psychological condition (Fuchs, 2020, 102), are likely to be the late gravitational ripples of ancestral experiences of permanent extreme existential pressure, more than being caused by a modern *vertige of freedom* (Kierkegaard). This view is supported by the impression that the frantic civilizational drive of humanity, climaxing in the planetary domination of the Anthropocene, is in part fuelled by fear, an anguish deeply imprinted in our species during a long period of prehistoric fragility. The Anthropocene is the direct effect of a powerful immunisation against the most imminent risks and threats inherent to life in the competitive physical system of the biosphere. The ecological crisis and the Covid-19 pandemic remind us that we are still integral part of this highly dynamic and basically unpredictable system, and subjected to its vagaries. Notwithstanding the anthropogenic imprint we have stamped on the surface of our planet, there is no dualism in the physical sphere. We cannot escape nature with freedom. But the inverse is also true: we cannot escape freedom with nature.

More than ever, we have choices to make, in all of the dimensions of freedom explicated above: the freedom from war and foreign domination; the freedom to take part in transparent political decision-making in our communities; the freedom resulting from being allotted an equitable part of the resources produced by these communities; and the freedom to follow our own intuitions in the direction of self-realization. Lastly, the freedom to experience in greatest possible sovereignty our dependency and profound entanglement with the mysterious physical system of terrestrial life. The commitment to sustainability signifies, in its essence, the mission to uphold these freedoms to the greatest possible extent, for humans of our generation as well as for generations to come.

What these five layers of freedom reveal, however, is the effect of an eventual collapse of this architecture of freedoms, starting from the top: the biosphere is the first and fundamental resource we depend on for lives with choices in our communitarian sovereignty, for inclusive and mediated governance, for the production and effective distribution of resources as well as for personal self-realization. Foremost in the perspective of freedom, the preservation of our life-sustaining biosphere is the inescapable and overriding priority.

The Anthropocene is a powerful concept revealing to what extent the effects of human choices increasingly weigh on the biosphere with which our human lives are entangled. Our cumulated actions have a scale of consequences that transcends what we are accustomed to. As an exponentially grown quantitative factor, the choices humans make individually become systemically more relevant. The ecological question is a question of freedom, and the fundamental human aspiration for freedom has become inseparably linked to the ecological question. As Möllers points out, the solution of the ecological crisis of the Anthropocene lies in a positive political project "that provides enough motivation for collective actors to do more than a perhaps equitable rule would require" (2021, 269).

References

Arendt, H. (2018). The freedom to be free: The conditions and meaning of revolution. In J. Kohn (Ed.), *Thinking without a banister: Essays in understanding, 1953–1975*. Random House.

Bajohr, H. (Ed.). (2020). *Der Anthropos im Anthropozän. Die Wiederkehr des Menschen im Moment seiner vermeintlich endgültigen Verabschiedung*. De Gruyter.

Berlin, I. (1958/2002). Two concepts of liberty. In H. Hardy (Ed.), *Liberty*. Oxford University Press.

Bernecker, R., & Grätz, R. (2021). Culture and liberty. In *Global perspectives*. University of California Press. https://doi.org/10.1525/gp.2021.25631

Brunner, O., & Conze, W. (Eds.). (1972). *Geschichtliche Grundbegriffe: historisches Lexikon zur politisch-sozialen Sprache in Deutschland.* (Vol. 2). Ernst Klett Verlag.

Constant, B. (1819). *The liberty of the ancients compared with that of the moderns*. https://www.earlymoderntexts.com/assets/pdfs/constant1819.pdf

Fuchs, T. (2020). Warum gibt es psychische Krankheit? Grundlagen der psychiatrischen Anthropologie. In *Paragrana. Internationale Zeitschrift für Anthropologie* (Band 20, Heft 2). De Gruyter.

Hensher, M. (2020). A beginner's guide to avoiding policy mistakes in the Anthropocene. In C. J. Orr, K. Kish, & B. Jennings (Eds.), *Liberty and the ecological crisis. Freedom on a finite planet*. Routledge.

Höffe, O. (2015). *Kritik der Freiheit. Das Grundproblem der Moderne*. Beck.

Locke, J. (1690/2012). *Second treatise of government*. Digitized by Dave Gowan. https://resources.saylor.org/wwwresources/archived/site/wp-content/uploads/2012/09/POLSC2012.3.pdf

McCready, A. R. (2020). Are freedom and interdependency compatible? Lessons from classical liberal and contemporary feminist theory. In C. J. Orr, K. Kish, & B. Jennings (Eds.), *Liberty and the ecological crisis. Freedom on a finite planet*. Routledge.

Möllers, C. (2021). *Freiheitsgrade. Elemente einer liberalen politischen Mechanik* (3rd ed.). Suhrkamp.

Orr, C. J., & Brown, P. G. (2020). From the ecological crisis of the Anthropocene to harmony in the Ecozoic. In C. J. Orr, K. Kish, & B. Jennings (Eds.), *Liberty and the ecological crisis. Freedom on a finite planet*. Routledge.
Quilley, S. (2020). Liberty, energy, and complexitiy in the longue durée. In C. J. Orr, K. Kish, & B. Jennings (Eds.), *Liberty and the ecological crisis. Freedom on a finite planet*. Routledge.
Sen, A. (2000). *Development as freedom*. Anchor Books.
Sheldrake, M. (2020). *Entangled life. How fungi make our worlds, change our minds and shape our futures*. Random House.
Skinner, Q. (2012). *Liberty before liberalism*. Cambridge University Press.
Taylor, C. (1985). What's wrong with negative liberty. In *Philosophy and the Human Science* (Philosophical Papers 2). Cambridge University Press.
United Nations. (1987). *Report of the World Commission on Environment and Development*. Our Common Future [Brundtland Report]. https://www.are.admin.ch/are/en/home/media/publications/sustainable-development/brundtland-report.html
Weiss, T. G., & Wilkinson, R. (Eds.). (2018). *International Organization and Global Governance* (2nd ed.). Routledge.
World Commission on Culture and Development. (1995). *Our creative diversity* (Report of the World Commission on Culture and Development, 1995). https://www.gcedclearinghouse.org/sites/default/files/resources/%5BENG%5D%20Notre%20diversit%C3%A9%20cr%C3%A9atrice.pdf
Ziegler, R. (2020). Liberation from excess: A post-growth economy case for freedom in the Anthropocene. In C. J. Orr, K. Kish, & B. Jennings (Eds.), *Liberty and the ecological crisis. Freedom on a finite planet*. Routledge.

Roland Bernecker is Visiting Professor, Chair of Cultural Management at Brandenburg University of Technology. He was Secretary-General of the German Commission for UNESCO, Director of the French-German Cultural Institute and Lecturer at the University of Nantes, France. He was German delegate in the negotiations of UNESCO's international conventions for the Intangible Cultural Heritage in 2003, and for the Diversity of Cultural Expressions in 2005. His research interests include the history and future of Liberalism, cultural leadership, global governance and narrative theory.

Hope

Florina Guadalupe Arredondo-Trapero

Abstract The purpose of this article is to analyze the concept of hope in the Anthropocene era, viewed from the perspective of the global climate crisis caused by human activity. The paper explains how hope is related to ideas of utopia, since it is the mobilizing function of utopia. Hope is an open invitation to undertake a long march towards the future and to be ready for the new.

Although the concept of hope has been criticized in literature for being a source of unrealizable dreams, it has been academically related to attitudes such as motivation and desire. Hope has also been the subject of in-depth philosophical study, and is a concept which has been treated differently by diverse schools of thought (Bloeser & Titus, 2017). The concept of hope today is one which is particularly relevant in the face of the current climate crisis, a crisis which has prompted a re-evaluation of what hope brings to human life. One of the contributions of this re-evaluation is the search for a sustainable world.

The relevance of hope is radical, particularly in response to the current anthropocene crisis. The discussion of climate change and global warming motivates us to hope for a better future, a utopian future. The Socratic question "How shall we live?" has been transformed into: "How shall we live in a more sustainable way?" In this sense, every human being should ask themselves this philosophical question, and also question their own existence, their way of relating to the planet, and what each individual can do in order to not compromise the viability of the human species and all other forms of life. In order to seek new answers to the climate crisis, he have to have a belief in utopia. Since hope is the function of utopia, the latter is closely attached to hope (Bloch, 1986). Without hope, utopia remains an image of a desired future.

F. G. Arredondo-Trapero (✉)
Tecnológico de Monterrey, Monterrey, Mexico
e-mail: farredon@tec.mx

© The Author(s), under exclusive license to Springer Nature Switzerland AG 2023

N. Wallenhorst, C. Wulf (eds.), *Handbook of the Anthropocene*,
https://doi.org/10.1007/978-3-031-25910-4_274

The etymological root of utopia is topos, meaning place: its two closest prefixes are eu, "the best" and ou, negation. Utopia is the place that "is not yet" real, but "could be". The source of utopia is in daydreams, and it manifests itself in expressions of human culture such as art and literature (Bloch, 1986). Art and literature express what may be the reflection of the future of the climate crisis, but also propose in their expressions the possibility of new realities, of a new world. Examples of this are the works of David Hockney (MFAH, 2021) who, through his creations, conveys magical visions of forests, and those of Yayoi Kusama, who captures our imagination about how humans relate to reality, a reality open to endless possibilities (The Guardian, 2017).

In other words, utopia is the best place that does not exist. Or rather, it is the best place that does not yet exist, but which could in the future (Gálvez-Mora, 2008). In this sense, utopia, together with hope, acquires a dynamic dimension. Both utopian thinking and hope are present in various spheres in a broader way than is commonly thought, and one of these spheres is in the search for a sustainable world.

Utopia is born out of dissatisfaction with current living conditions, and one of these conditions is the unsustainability of life on the planet as we currently know it. Utopia constitutes an open proposal against the status quo, i.e., from a reality that reflects climate crisis. In relation to the issue of climate change, it is a matter of analyzing the current unsatisfactory conditions in order to propose viable alternatives. Thus, utopia linked to hope has a shattering function and requires militant optimism. Utopia and hope are not just about wishful thinking, but also about concrete and determined effort. Utopia and hope generate an optimism within themselves that mobilizes humans to strive to create something tangible from a daydream.

According to Gálvez-Mora (2008), Ernst Bloch's utopia has three functions in reality: (a) it allows a critique of the current reality; (b) it gives a direction to follow and (c) it analyzes the possibility of its content. Such functions are supremely relevant in the context of the climate crisis; for example, allowing a critical review of the current situation, outlining strategies for change and evaluating whether, in the light of the steps taken, what was dreamed of is possible. Hope makes utopia functional, i.e. it leads human beings to take steps towards it.

Utopia however is necessary as a prerequisite for adaptation and mitigation strategies in the face of climate change. Without utopian thinking, we lose sight of the horizon. As Bloch (1986) states, every castle was first imagined in the air, in the human imagination, for without the ability to imagine nothing would have been created in the world (Gálvez-Mora, 2008). In the same way, without the ability to imagine a sustainable world, it will not be possible to achieve the changes that this challenge demands. It is necessary to imagine that humanity is capable of leading a different and more sustainable way of life.

It is a matter of generating utopia as an open possibility of reality. To think of a sustainable world is to think of the best place that does not yet exist, but which could be created given the right conditions. Utopia is the objective basis of what is to come. Bloch states that utopia leads to the novum, that is, to the new, to what is yet to be. Both utopia and hope imply knowing the conditions of reality, and in this case, knowing the details of the environmental crisis. Science allows humans the possibility of understanding how the world is today, but also to imagine how it could be in the future. It is this possibility that generates hope for a better, sustainable world.

According to Levitas (2016), the concept of "Utopia" was proposed by Thomas More (2002), who stated that it implies the process of imagining the world in a different way; it is a process that does not replicate the previous state but is rather something which has not yet been invented. This process is necessary to address the consequences of climate change, conflicts, and the growing inequalities caused by a predatory and devastating economic system. Reinforcing this idea, Fassbinder (2016) theorizes that when it comes to linking utopia with climate change, there are forces that may limit or even prevent this link from being made. Current attempts to mitigate climate change have not been as successful as they could be because humanity is trapped in an unbridled market economy.

Philosophers who criticize modernity have failed to be encouraged by the steps currently being taken towards creating a sustainable world. For Sloterdijk (2019) modernity, and the climate crisis that has been generated primarily in the last stage of this period, is akin to an aircraft heading inexorably towards disaster. The aircraft is trying to repair itself mid-flight, but there is no landing strip. Modernity has been set in motion, but without brakes; there is no device that can slow the trajectory and lead to a successful outcome. We are condemned to acceleration and with it the devastation of the environment.

In a direct criticism of capitalism, Klein (2014) expresses his belief that the failed vision of the modern era is the conviction that humans could dominate nature and place their power above the natural world. The utopia of industrialization that promised to take humans wherever they wanted, regardless of context, has broken down. A dangerous utopian narrative has led humans to see themselves as outside the environment, when in fact they are the environment. After excessive fossil fuel-based industrialization, humans today realize that they are not free, not in charge, and not in a one-way conversation with the environment.

In addition to this critique of modernity, the anti-humanist theories of Gray (2002) question the idea of progress, since environmental destruction is believed to result from the defective nature of humanity. Gray also considers that morality and human will are only illusions, and that humanity is a voracious species focused on eliminating other forms of life. Finally, he postulates that humans will ultimately be unable to survive, that the human age will come to an end; he opines that the earth will continue, devoid of humanity. Hope is meaningless in the face of anti-humanist discourse.

Conversely however there are humanistic voices which encourage the belief that changes towards a sustainable world are possible. This belief opens up the possibility of hope. In opposition to anti-humanist discourse, Valcárcel (2015) uses poetic expression to challenge us to take the next step and believe in the continued existence of humans. When confronted with the idea of humanity in decline and a collapsed planet, the human race is able to debate and find consensus, launch projects, undertake challenges, fight for a sustainable world and seek to reconcile through democracy, equity, and human rights. As a result, hope for the future continues to exist.

Valcárcel (2015) encourages us to believe in the potential of humanity, arguing that humans are in fact capable of looking far beyond what they are actually able to do and harness power, modesty, humility and awareness in the desire to create a better world. Humans should take care not to blur the line between being mere sons of gods and wishing to be gods themselves, as this is an unrealizable and unsustainable desire. The risk of this taking place is real, since the West is Faustian and forgets its vulnerability. This assertion is based upon the myth of the magician who sells his soul to the devil in order to obtain wisdom (Nicol, 2022). It refers to the fact that men and women are capable of abandoning the search for sustainability in order to obtain recognition, wealth and power. For this reason, it is important not to lose sight of the dream of a sustainable world and the hope that mobilizes humanity to work towards to this challenge.

In this quest for change, a break with reality is required in order to make space for utopia. If we fail to break with our current reality, we are merely perpetuating the cycle of self-destruction. The problem of climate change implies not only a break with reality but demands that we think of other ways of life, and other forms of production and consumption. We must not only rethink them but construct an entirely different relationship between human beings and nature.

Thus, by breaking with reality, human beings are forced to make radical changes in the way they have previously related to the world. For centuries, humans have cultivated a vision of anthropocentricity which places themselves at the center of all existence. In a certain way, the relationship between human beings and the planet has resulted in a split, one in which human beings have come to see themselves as separate from the rest of nature, when in reality we are only one small part of the natural world (Gándara-Tovar & Arredondo-Trapero, 2021).

Hope is not only a motivation or desire but something that implies a projection into a future that is achievable. While utopia is the projection of the imagined future, hope implies movement towards something which has not yet come into being. In this sense, hope is the antithesis of passivity. Hope implies striving towards an imagined future, making changes in reality and at the same time evaluating such changes to confirm whether we are or not moving in the direction of an imagined future.

Hope requires moral impulse. This moral impulse is in the entrails of the human being, ignoring it counteracts the dynamism of one's own life to seek something beyond. The moral impulse is in the search for the new world, and it originated with the first humans, those who crossed rivers, plains, seas and mountains. Those first humans sought in their footsteps not only new lands, but also the future. Within each human being there is an original spark that impels them to the new, to the realization of the possible. But the moral impulse for the new world, an impulse which is necessary to confront the climate change crisis, does not remain within the individual or among individuals. The moral impulse must penetrate the web of life, seek to remove its structures and rethink what we have been up until now and the way we have tried to move in and about the world.

Transformative thinking is concerned with the dynamics of reality and how reality in motion generates new realities. It also presents the diverse situation of humanity as we face different lived realities in the context of climate change, so transformative thinking must be flexible and adaptive. Hope, in the face of the planetary crisis, must be a fire whose spark burns everywhere, and manifests as transformative thinking, the search for a new reality that permeates each human being in the quest for planetary sustainability. It is an invitation to think of a new promised land. It is a new exodus, but paradoxically located in the same place. The Earth is so far the only place available for humanity to inhabit.

It is not a question of emigrating to a different land, but of emigrating from the way we are in this world, of finding a new way of thinking, one that accommodates a viable world for human beings and the diversity of life. It is about imagining that in the future the planet will continue to spin, and that the human race will continue to inhabit it. It is about imagining, in spite of everything, that humanity will not fade into oblivion and that the possibility of continued human life is infinite. Another awakening of consciousness is necessary, not only a consciousness of oneself as a human being, but a planetary consciousness (Arredondo-Trapero & Gándara-Tovar, 2021).

Hope invites us to seek a different future relationship with the earth and to be open to welcoming the new. As a new planet is not possible, at least not with the same conditions as the one we currently have, it is about imagining a completely new world. Although the dream is not real, it is achievable. It is not an overflowing and unattainable dream, but a waking dream.

Humanity must envision an alternative world by moving on from the current consciousness and creating a new reality, one very different from the present. Our current dissatisfaction with reality belies human achievements and leads us to seek not only a better tomorrow, but a better today. Transformational thinking traces a new path, one based on rationality but at the same time passionately linked to hope, a hope that does not give way to the disregard of the imagined (Arredondo-Trapero & Gándara-Tovar, 2021).

Actions which seek to respond to climate change are realities in process. Hope anticipates the future reality of a sustainable world. Reality is a process by which utopia is possible, by allowing humans to become open to the future. It is not about partial achievements that lead us to an unrealized destiny, but about the final achievement of a sustainable life.

For Ernst Bloch, hope is the constitutive element of human life; it is not something alien or secondary to it. Hope is the condition in which we ourselves are and exist: human beings are arguably utopian by nature, and by working towards utopia we differentiate ourselves from animals because anticipation is our power and our destiny.

Thus, hope is understood as not waiting passively for reality to change. Climate change adaptation and mitigation strategies represent exactly this: being proactive, not waiting passively for global warming to resolve itself. Nor should we believe that adaptation to climate change will happen automatically: we need to take concrete steps to change our behaviours. Hope forces human beings to step outside of themselves and helps them to assume responsibility for creating a brighter future.

Human beings are focused on the future, one which they are powerless to escape, but which they can make better through design. We can argue that not only is hope a purely human construct, but it is the only one that offers humanity a lucid horizon and the chance of future survival. Hope is constitutive of reality, it is not outside of it, distant from it. Hope is at the ultimate root of what is possible, what is not yet, what is unfinished but capable of being finished (Aguirre, 2008). By hoping, human beings can broaden their horizons, they do not restrict them. In this way, hope is oriented towards a reality that is imminent, but which has not yet been realized.

We can argue that the future direction of humanity depends largely upon the path we take today, and that the human race cannot afford the luxury of trying to apportion blame for the current crisis but should instead focus on taking concrete steps towards a future which we can build together. Although hope cannot guarantee success, its value lies in having made us walk. It makes us walk because it shows us a vision of a new world, of a sustainable world, and foments the enthusiasm necessary for us to work towards a better future.

References

Aguirre, J. M. (2008). Razón y Esperanza. Pensar con Ernst Bloch. *Cuadernos de Ideas*, 20. UCSH.

Arredondo-Trapero, F. G., & Gándara-Tovar, A. J. (2021). Climate change, cultural manifestation and the principle of Hope, based on Ernst Bloch's theoretical proposal. In F. W. Leal, J. Luetz, & D. Ayal (Eds.), *Handbook of climate change management*. Springer. https://doi.org/10.1007/978-3-030-22759-3_226-1

Bloch, E. (1986). *The principle of hope*. Basil Blackwell.

Bloeser, C., & Titus, S. (2017). Hope. In E. N. Zalta (Ed.), *The Stanford encyclopedia of philosophy* (Spring 2017 Edition). Recovery from: https://plato.stanford.edu/archives/spr2017/entries/hope/

Fassbinder, S. D. (2016). Climate change mitigation in fantasy and reality. *Knowledge Cultures, 4*, 250–271.

Gálvez-Mora, I. M. J. (2008). La función Utópica en Ernst Bloch. Recovery from: http://www.posgrado.unam.mx/filosofia/wpcontent/uploads/2018/09/04galv.pdf

Gándara-Tovar, A. J., & Arredondo-Trapero, F. G. (2021). Climate change and awareness of the I-world relationship from Erwin Schrödinger's non-dualist conception. In F. W. Leal, J. Luetz, & D. Ayal (Eds.), *Handbook of climate change management*. Springer. https://doi.org/10.1007/978-3-030-22759-3_224-1

Gray, J. (2002). *Straw dogs: Thoughts on human and other animals*. Granta. ISBN 1-86207-512-3.

Klein, N. (2014). *This changes everything: Capitalism vs. the climate*. Simon & Schuster.

Levitas, R. (2016). Less of more. *Utopian Studies, 27*, 395–401.

MFAH. (2021). Hockney – Van Gogh the joy of nature. Recovery from: https://static.mfah.com/documents/hockney-van-gogh-exhibition-labels.7438312592610155850.pdf

More, T. (2002). *Utopia*. Cambridge University Press.

Nicol, E. (2022). El mito fáustico del hombre. *Revista Filosofía y Letras. UNAM*. Recovery from: http://revistafyl.filos.unam.mx/faustico/

Sloterdijk, P. (2019). La irreverencia del pensar. Centro de Estudios Públicos (CEP), *Biblioteca Nacional de Chile*. Recovery from: https://www.youtube.com/watch?v=d9n5nFB5dsQ

The Guardian. (2017). Yayoi Kusama's infinity mirrors reignites art's selfie debate. *The Guardian*. Recovery from https://www.theguardian.com/artanddesign/2017/nov/09/yayoi-kusama-infinity-mirrors-art-the-broad-los-angeles

Valcárcel, A. (2015). Construyendo el Humanismo para el Siglo XXI. *Cátedra Alfonso Reyes*. Recovery from: http://www.catedraalfonsoreyes.org/videoteca/seminario-construyendo-el-humanismo-del-siglo-xxi-3a-sesion/

Florina Arredondo earned her Doctorate degree in Economics and Business Management from the University of Deusto, Spain (2007) and a certification in Ethics from the Doctorate in Humanistic Studies (2008) of the Tecnológico de Monterrey. She is currently a research professor in the Department of Humanistic Studies at Tecnológico de Monterrey, where she teaches Ethics, Sustainability and Social Responsibility. She is the author of several articles and reviewer of national and international journals. She coordinates the Sustainability and Anthropocene group of the School of Humanities and Education of the Tecnológico de Monterrey. Finally, she is part of the research team in Ethics & Human Flourishment of the Tecnológico de Monterrey and she belongs to the National System of Researchers Level I.

Human Rights

Tracey Skillington

Abstract Ideally, human rights exist independently of any state regime but, in truth, have little effect if not allocated on the basis of membership of at least some community, state or otherwise. Without such claims to membership and the assignment of duties to specific agents to fulfil rights to a safe environment, to health, life and security of person, there is no binding obligation on authorities to provide all climate vulnerable peoples with the means to survive ecological adversity. Although of immense humanitarian importance in a context of declining resources worldwide, universal rights to health and a dignified existence still await further social realization. This chapter assesses how tensions arising between the ideal and the real social life of rights today, set against a background of accelerating climate crisis, are being addressed via new democratic struggles.

Actualizing Human Rights

Human rights are rights enjoyed by individuals by virtue of being human and by extension being vulnerable in their humanness to suffering. Vulnerability to suffering thus defines the common basis of human rights, including the right to life, health, freedom from persecution and inhumane treatment (Turner, 2006, p. 2). However, it is difficult to enforce human rights without the support of a legal institution, if not a sovereign state, and an accompanying set of citizenship rights. In *The Origins of Totalitarianism*, Hannah Arendt (1951) explores how rights that, ideally, exist independently of any state regime, in truth, have little effect if not allocated on the basis of membership of a particular community (e.g., citizenship rights). Without such claims to membership, the individual is vulnerable to human rights abuse. Arendt draws critical attention to the importance of distinguishing grounded legal rights from the abstract rights of individuals, noting how enjoyment of the latter is

T. Skillington (✉)
Department of Sociology & Criminology, University College Cork, Cork, Ireland
e-mail: t.skillington@ucc.ie

© The Author(s), under exclusive license to Springer Nature Switzerland AG 2023

N. Wallenhorst, C. Wulf (eds.), *Handbook of the Anthropocene*,
https://doi.org/10.1007/978-3-031-25910-4_275

utterly dependent on the individual's abilities to exercise a 'right to have rights'. Without the support of an institutional order of legal and political justice, rights to a safe and clean environment, for instance, or sufficient food and water reserves remain unspecified. As long as specific agents are not identified in the assignment of duties to fulfil such rights, there is no pressing or binding obligation to provide all members of climate vulnerable communities with the means to survive ecological adversities. Although of immense importance in a context of declining natural resource availability worldwide, universal rights to health, food, or a dignified life still await further social realization, confined as they are at present to the category of 'manifesto rights' (Feinberg, 1980, p.7; Skillington, 2012, p. 1197).

Much of the content of current climate change disputes are focused on enabling a greater actualization of these and related rights concerns or addressing what are by now classic tensions arising between the boundaries of state sovereign responsibility, citizens' concerns and human rights ideals. However, it is perhaps important to bear in mind that early variants of most of the institutionally relevant rights that are hotly contested today emerged initially from the historical struggles of carbon democracies, where criteria for inclusion and integration into the democratic project included, crucially, support for the legitimacy of ongoing projects of mass resource extraction and consumption (i.e., support for largescale industrialization, urban development, transport, agriculture, the advancement of science, technology, global trade and communication). Over the course of nineteenth century and most of the twentieth century, employees of capitalist worlds achieved political and social inclusion thanks to their crucial support for and role in enabling (both directly and indirectly) carbon fuelled industries to thrive. Only because of the centrality of their labour to the growth of the global infrastructure of carbon capitalism were voting rights, freedom of speech, equality before the law, etc., granted to certain workers and gradually extended to other social groups (e.g., women, minorities). In many ways, what we are still dealing with today are the legacies of carbon democracies (in terms of the continuity of the imperatives of unlimited growth and extraction) whose uncritical foundations are now subject to challenge by citizens who assert these same rights against the basic enabling premises of carbon democracies (support for the further expansion of extractive economies). However, a carbon neutral model of democratic justice is unlikely to emerge fully at this historical point in the development of the Anthropocene without a more fundamental transformation of current ecological, social, political, economic, and legal arrangements. That said, growing numbers are taking it upon themselves to challenge the ongoing legitimacy of governments' commitments to carbon economies and assert their rights to self-determination in the shaping of their futures. In doing so, these actors demonstrate how it is still possible for democracies to act upon themselves with a view to changing the models by which societies represent themselves and act (i.e., actualize historicity, as French sociologist, Touraine (1988), refers to it). The desired model reinforces commitments to human rights when addressing climate change but in ways that also protect wider nature (i.e., adopting a consciously less anthropocentric human rights approach).

Climate Change as a Human Rights Concern

The shift towards a more reflexive examination of the human rights dimensions of climate change initially met with considerable opposition, particularly amongst more powerful players. For example, the US, fearing a stronger official recognition of linkages between climate change and human rights violations would bolster the case for further unwanted 'extra-territorial' legal regulations, made its reservations clear in a submission to the UN Office of the High Commissioner for Human Rights in 2009 (prior to the publication of the OHCHR's report on the relationship between climate change and human rights). Here the US laid out a series of objections to a human rights approach to climate change (Skillington, 2012, p. 1205). Climate change, it noted, is 'one of many natural and social phenomena that may affect the enjoyment of human rights' and, therefore, cannot be singled out as 'the cause' of human rights violations, particularly those arising internationally. Restricting resource rights eligibility to 'legitimate' claimants, particularly those with a legal contractual right to precious reserves of minerals, oil, gas, seeds, forests, arable lands, etc., and striking 'a balance' between environmental harm and the benefits of the activities causing it were asserted instead as primary concerns.

Causing particular angst for the US were the efforts being made legally to identify a 'collective or self-standing right to a safe and secure environment'. It noted in its report to the Office of the High Commissioner for Human Rights in 2009 how this right would most certainly be used as a 'political or legal weapon' against the US. For instance, the petition submitted a few years earlier by the Inuit, under the auspices of the Inuit Circumpolar Conference (the Inuit Circumpolar Council since 2006) to the Inter-American Commission of Human Rights (IACHR) claiming that the United States had violated the rights of the Inuit people to life, food, and culture, by failing to refrain from actions that would decrease US CO_2 emissions. A similar case was taken by the Arctic Athabascan Council against Canada for violating Athabascan rights through its air pollution practices, especially its contribution to high levels of black carbon, widely considered an important driver of Arctic climate change. Although both cases were unsuccessful due to a lack of sufficient evidence to prove categorically the traceability of these harms, they did, nonetheless, help to raise the profile of a human rights approach to climate justice considerably and strengthen resolve amongst those most adversely affected by climate change to pursue the matter further through domestic and international court settings. Considering that by May 2020, over a thousand climate-related legal cases had been initiated against various federal state agencies in the United States (see Setzer & Byrnes, 2020), it would seem that the concerns of the US were, indeed, well founded, even if its desire to restrict the institutional relevance of a human rights approach to climate change proved not to be fully justified. Equally, the progress achieved more generally with human rights based legal challenges against states around the world feed into wider communities of discourse on the human rights implications of climate change (Skillington, 2017).

Another important development in this regard has been the launch of a number of cases recently against various alliances of states. For instance, the complaint lodged by six Portuguese children and young adults with the European Court of Rights in September 2020 against 33 Council of Europe Member states (the EU27 plus the UK, Norway, Russia, Ukraine, Switzerland and Turkey) on the grounds that these states are in breach of Article 2 (right to life), Article 8 (right to respect for private and family life) and Article 14 (discrimination on the grounds of age) of the European Convention on Human Rights. Acting on the belief that there is 'reasonable and convincing evidence' that several of their fundamental rights and freedoms have been violated and in ways that cannot be justified as 'necessary in a democratic society' (p. 9 of the Complaint), the youths in this case argue that this group of states share responsibility for 'indivisible injuries' in their ongoing exportation of fossil fuels, importation of embodied carbon and financing of fossil fuel extraction elsewhere, injuries that are exacerbated further by the narrowing margin of appreciation applicable in the area of climate change mitigation. Such juries are said to also constitute a serious violation of universal rights to health, security, and self-determination. In promoting a 'trans-territorial' argument on the question of legal responsibility for climate destruction, the applicants in this case hope to pre-empt efforts on the part of states to shift responsibility for climate change impacts onto vague or loosely defined categories of guilt (i.e., climate harms generated by 'the global community'). Usually, the rule of exclusive attribution precludes situations of multiple or shared responsibility. However, the hope in this case is to challenge this tendency by arguing that this group of states, in fact, shares responsibility under the terms of the ECHR to 'extricate' themselves from a position of wrongdoing.

In filing legal complaints, youth help to reinvigorate the relevance of human rights norms and further enhance the 'living spirit' of the democratic community. States are reminded of the fact that the democratic community is triply drawn, not only in terms of historical legacies (e.g., revolutionary pasts, rights struggles, carbon expenditure), geographical scope (increasingly transnational) but, also, in terms of its temporal reach as a partnership that extends across generations. Arguably, one of the most important contributions made by youth's legal campaigns to date has been in terms of the challenge they pose to 'the illusion of separation' of peoples, states and generations (Shue, 2020) and that of the separation of state powers when determining the scope of responsibility for climate harm. Instead what are emphasized are the shared dimensions of our lives and with that, shared duties to protect planetary or transboundary resources, as much as rights to exploit the same (see principle 21 of the 1972 Stockholm Declaration).

References

Arendt, H. (1951). *The origins of totalitarianism*. Harcourt Brace.

European Court of Human Rights in September. (2020). *Request No. 39371/20, Claudia DUARTE AGOSTINHO and others against Portugal and 32 other States*. http://hudoc.echr.coe.int/eng?i=001-206535. Accessed 29 June 2021.

Feinberg, J. (1980). *Rights, justice and the bounds of liberty: Essays in social philosophy*. Princeton University Press.
Setzer, J., & Byrnes, R. (2020). *Global trends in climate change litigation 2020 snapshot*. https://www.lse.ac.uk/granthaminstitute/wp-content/uploads/2020/07/Global-trends-in-climate-changelitigation_2020-snapshot.pdf. Accessed 17 Feb 2023.
Shue, H. (2020). Distant strangers and the illusion of separation: Climate, development and disaster. In T. Brooks (Ed.), *The Oxford handbook of global justice* (pp. 259–276). Oxford University Press.
Skillington, T. (2012). Climate change and the human rights challenge: Extending justice beyond the borders of the nation state. *The International Journal of Human Rights, 16*(8), 1196–1212.
Skillington, T. (2017). *Climate Justice & Human Rights*. Palgrave.
Touraine, A. (1988). *Return of the actor: Social theory in postindustrial society*. University of Minnesota Press.
Turner, B. S. (2006). *Vulnerability and human rights*. Penn State University Press.

Tracey Skillington is Director of the BA (Sociology), Department of Sociology & Criminology, University College Cork, Ireland. Recent monographs include *Climate Justice & Human Rights* (Palgrave), *Climate Change & Intergenerational Justice* (Routledge) and forthcoming, *A Critical Theory of Climate Trauma* (Routledge). Her publications have appeared in many journals over the years, including the *European Journal of Social Theory*, the *British Journal of Sociology*, the *International Journal of Human Rights, Distinktion: Journal of Social Theory*, *Sociology*, the *Irish Journal of Sociology* and *Sustainable Development*.

Peace

Simon Dalby

Abstract The Anthropocene has been shaped by violence, conquest and ecological disruptions related to warfare over the last few centuries. Some of the key dates being discussed as the starting point of the Anthropocene are related to warfare and conflict. To sustain civilization in future, the causes of war have to be transcended, and fears of environmental disruptions triggering conflict must be dealt with by international cooperation and the extension of current efforts at environmental peacebuilding.

In many ways the Anthropocene demands a recognition that things in the future have to be different, and as such the notion of peace, or should that be a peaceful Anthropocene, has to be different from the history that has given rise to the notion of the Anthropocene in the first place. Moving beyond the patterns of violence that Bonneuil and Fressoz (2016) term the Thanatocene, where fossil fuelled industrial states threatened and used technological violence on the large scale, especially in the twentieth century, is now an essential part of what needs to be done to secure a habitable earth for future humans and many other species too.

Security understood in terms of the preparation to use violence is no longer functional in a rapidly changing world where rival states threaten one another, both directly in terms of military capabilities and indirectly by the disruptions caused by fossil fueled economic systems (Dalby, 2020). Or, to be more precise, if humanity is to survive in the long-term then ending the violence that has been key to the transformation of the earth system in recent centuries is essential (Oswald, 2020). In a nuclear armed world facing rapid geological transformation peace is a precious condition that needs to be actively cultivated.

If the future stages of the Anthropocene, and the aspirational statements of earth scientists concerning a future sustainable earth, are to be fulfilled (Steffen et al., 2018), then politics will have to be done differently. The novel situation we face is

---

S. Dalby (✉)
Wilfrid Laurier University, Waterloo, ON, Canada

© The Author(s), under exclusive license to Springer Nature Switzerland AG 2023

N. Wallenhorst, C. Wulf (eds.), *Handbook of the Anthropocene*,
https://doi.org/10.1007/978-3-031-25910-4_276

one where the rich and powerful among humanity are operating on a geological scale as change agents, a novel telluric force, in the earth system. Failure to move towards a stable configuration of the earth system will likely lead to a "hothouse earth" with great disruptions to human societies and likely violent consequences. Avoiding this trajectory is essential to a peaceful future for much of humanity.

In the discussion of the Anthropocene it is important to note that the processes that have led to the new circumstances frequently start long before they appear widely in the sediments around the planet. This discrepancy between when processes started, and when they occur on a large enough scale to mark a new geological period, marks the disjuncture between historians of humanity and the historians of sediments. Stratigraphy and history don't meet easily. But this difficulty is actually useful in terms of thinking about peace, war and the transformation of the earth system.

If one follows Lewis and Maslin (2018) and dates the Anthropocene from what they call the Orbis spike, the dip in global carbon dioxide levels that reached its lowest point in 1610, then this start date is as a result of the European conquest of the Americas. The direct use of military violence subdued the conquered peoples. The related death of millions of indigenous peoples because of disease brought by Europeans, led to large scale reforestation of abandoned lands and a drop in global carbon dioxide levels. There was little peaceful about the European colonization of the Americas. If in fact the wars of the seventeenth century in Europe, which shaped the beginning of the modern states system, were caused by the related climate disruptions then this part of the Anthropocene was anything but peaceful (Parker, 2013).

Paul Crutzen's (2002) initial suggestion of the starting point for the Anthropocene was the industrial revolution in England. This involved the expansion of the use of coal as an industrial fuel in steam engines and soon after their use as a mode of propulsion in trains and ships. But the industrial revolution was intimately connected to the rise of European empires and the transformation of landscapes and their peoples into parts of the growing integration of the world economy. This was frequently a violent process of dislocation and conquest too as resources were extracted and plantations imposed on landscapes (Moore, 2015). It is these violent processes, and in places the wholescale destruction of indigenous societies as European colonization expanded, that makes Jarius Grove (2019) suggest that the last half millennium might better be called the Eurocene rather than the Anthropocene.

If one prefers to view the start date of the Anthropocene as 1945 and the first detonations of nuclear weapons, then here too large scale organized violence is key to this dating. There are compelling reasons to use this date, not least the matter of geological traces of radioactive fallout elements from the bombs simultaneously showing up in sediments around the world, a clear indication of a sedimentary golden spike (Zalasiewicz et al., 2019). A couple of decades later, and the year 1964 in particular, when the amounts of radiological material in the atmosphere reached their apogee, prior to the partial nuclear test ban treaty stopping most atmospheric nuclear testing, and hence reducing fallout quantities, might also mark an obvious starting point.

Regardless, it's clear that violence and war are key parts of the processes that both caused the changes now called the Anthropocene and provide the geological markers that denote its arrival. Failure to grapple with the forces unleashed may yet lead to further major wars, and if that is the case then further radiological depositions in the sedimentary record of the planet will occur. Andrew Glikson (2017) suggested that perhaps the term Plutocene might be more appropriate given the long term residence of plutonium and its radioactive derivatives in future sediments. Preventing further nuclear warfare or weapons testing is key to a future peaceful Anthropocene.

The challenge now is to transition to a much more ecologically sustainable world and manage the transition so that the inevitable disruptions in our rapidly changing world are handled without the resort to war. The more climate change happens and the less biodiversity left in the world, the tougher the future is going to be for the majority of humanity, whatever about the very affluent elites who may have the option to use their wealth to avoid the consequences of their actions. Living in a rapidly changing and increasingly artificial world, this new context of the Anthropocene suggests both that the causes of wars need further investigation and the necessity of avoiding them, given both the fragility of human systems and the scale of the destructive capabilities of novel technologies, has to become a much higher priority in international politics to ensure a sustainable transition to a peaceful future (Brauch et al., 2016).

The frequent invocations of climate change as a threat multiplier and as a catalyst of conflict in Washington in particular, suggest that peace is increasingly challenged by environmental disruptions, and hence the likelihood of warfare and political violence is actually increasing in the current phase of the Anthropocene (Werrell & Femia, 2017). This view from the metropolitan North, and its strategic thinktanks, suggests that climate change disruptions will both make people more vulnerable directly due to extreme weather, agricultural disruptions and migration, and also provide fertile ground for insurgencies and terrorist recruitment in places where state governments are weak or widely seen as illegitimate.

This view of the Global South as inevitably insecure and the source of terrorism and other dangers, both misconstrues the causal relationships, because climate change is mostly a matter of metropolitan consumption, rather than peripheral activities, and perpetuates security policies in "fortress" mode where the rich parts of the world try to cordon off the poorer sections to prevent migration and other problems understood in terms of security (White, 2014). If Northern "security" is based on this geopolitical vision of a perpetually divided world, then effectively the rich world is consigning much of the rest to a violent future of ongoing conflict (Welzer, 2012).

Avoiding this nightmarish view of the future is the impetus behind the rapidly growing scholarly and policy community work on environmental peacebuilding. But fortunately the alarms about likely small scale wars as a result of climate change are, it seems, frequently overstated, and policy makers are increasingly thinking about efforts at dealing with environmental change as opportunities to simultaneously make peace (Busby, 2020). The necessity of working together to deal with common dangers caused by ecological disruptions, suggests a reinvention of the old arguments about functionalism in international matters, where habits of working together operate to reduce the likelihood of conflict in situations where interests are shared.

But these ideas of environmental peacemaking do need to engage numerous groups in particular places to deal with inequities in resource access, gender and ongoing ethnic disputes. In doing so they will also need to avoid replicating many of the problems of previous development projects, which simply extend earlier colonizing practices that disrupt rural social systems (Ide, 2020). Cooperative environmental peacebuilding initiatives fit well with Wallensteen's (2015) suggestion that "quality peace" requires the removal of likely causes of violence, especially in post-conflict situations so as to prevent the resumption of violence in future crises.

While the precise mechanisms that cause conflict, and need to be overcome to ensure peace, are an empirical matter for investigation in particular cases, the movement to link peacemaking and environmental innovation suggests the possibilities of a much more peaceful next stage of the Anthropocene. At the biggest scale of the earth system and international politics such policies are now key to making the next phase of the Anthropocene one of a sustainable peace.

References

Bonneuil, C., & Fressoz, J.-B. (2016). *The shock of the Anthropocene: The earth, history and us*. Verso.

Brauch, H. G., Oswald Spring, Ú., Grin, J., & Scheffran, J. (Eds.). (2016). *Sustainability transition and sustainable peace handbook*. Springer.

Busby, J. W. (2020). Beyond internal conflict: The emergent practice of climate security. *Journal of peace research early view*, 10.

Crutzen, P. J. (2002). Geology of mankind–the Anthropocene. *Nature, 415*(6867), 23.

Dalby, S. (2020). *Anthropocene geopolitics: Globalization, security, sustainability*. University of Ottawa Press.

Glikson, A. (2017). *Plutocene: Blueprints for a post-anthropocene greenhouse earth*. Springer.

Grove, J. (2019). *Savage ecology: War and geopolitics at the end of the world*. Duke University Press.

Ide, T. (2020). The dark side of environmental peacebuilding. *World Development, 127*, 104777.

Lewis, S., & Maslin, M. (2018). *The human planet: How we created the anthropocene*. Pelican Books.

Moore, J. (2015). *Capitalism in the web of life: Ecology and the accumulation of capital*. Verso.

Oswald, U. (2020). *Earth at risk in the 21st century: Rethinking peace, environment, gender, and human, water, health, food, energy security, and migration*. Springer.

Parker, G. (2013). *Global crisis: War, climate change and catastrophe in the seventeenth century*. Yale University Press.

Steffen, W. R. J., & Richardson, K. et.al (2018). Trajectories of the earth system in the anthropocene. *Proceedings of the National Academy of Sciences, 115*(33), 8252–8259.

Wallensteen, P. (2015). *Quality peace: Peace building, victory and world order*. Oxford University Press.

Welzer, H. (2012). *Climate wars: Why people will be killed in the twenty first century*. Polity.

Werrell, C. E., & Femia, F. (Eds.). (2017). *The epicenters of climate and security: The new geostrategic landscape of the Anthropocene*. Center for Climate and Security.

White, R. (2014). Environmental insecurity and fortress mentality. *International Affairs, 90*(4), 835–851.

Zalasiewicz, J., Waters, C. N., Williams, M., & Summerhayes, C. P. (Eds.). (2019). *The Anthropocene as a geological time unit a guide to the scientific evidence and current debate*. Cambridge University Press.

Simon Dalby is a Professor at Wilfrid Laurier University, Waterloo, Ontario, where he teaches in the Balsillie School of International Affairs, and a Senior Fellow at the Centre for International Governance Innovation. His research is focused on geopolitics, environmental security and the Anthropocene. He is coeditor of *Reframing Climate Change* (Routledge, 2016) and *Achieving the Sustainable Development Goals* (Routledge, 2019) and author of and most recently, *Anthropocene Geopolitics: Globalization, Security, Sustainability* (University of Ottawa Press, 2020).

Values

Alice B. M. Vadrot

Abstract In the Anthropocene debates about "values" are connected to the idea that the importance of nature for society can be expressed quantitatively. However, as this chapter shows, the increased uptake of political instruments drawing on the monetary value of nature has led to contestation and increased political debate on how to renegotiate the complex values of nature. Negotiations over the values of nature have the potential to shape social and political order and are an important future object of research for social science scholars.

In simple terms, values determine the relationship between human beings and their natural environment in two important ways: Firstly, people hold subjective beliefs about nature its importance for life on Earth, and right and wrong behaviour toward nature. The sustainability debate thus heavily draws on the idea of behavioral change, which is closely connected to a call to shift values and beliefs about our relationship with the environment. Societal transformation may involve modifications to the political, legal, and economic structures of a country, but in order to induce institutional change, the values expressed by people's relation to nature must change, too. The second way in which "values" matters in relation to the Anthropocene is connected to the idea that nature's importance for society can be expressed hypothetically, most notably, but not exclusively, in economic and monetary terms.

This chapter focuses on the latter notion of "values" and the ongoing debate on the financialisation of nature. It argues that the struggle over the value of nature is closely connected to the way in which we know and govern global environmental change. Estimates of the economic value of nature aspire to protect the environment by revealing the hidden costs associated with the destruction of ecosystems and

A. B. M. Vadrot (✉)
University of Vienna, Vienna, Austria
e-mail: alice.vadrot@univie.ac.at

© The Author(s), under exclusive license to Springer Nature Switzerland AG 2023

N. Wallenhorst, C. Wulf (eds.), *Handbook of the Anthropocene*,
https://doi.org/10.1007/978-3-031-25910-4_277

provide policymakers with a solid tool to resolve trade-offs between different uses of land and resources. However, as this chapter will show, using the case of biological diversity, the increased uptake of political instruments drawing on the monetary value of nature in national and global environmental policy-making has led to contestation by some actors. They reject this anthropocentric, economistic approach and support the idea that nature should be protected for its own sake. Indeed, the struggle over values, for instance within the framework of the Intergovernmental Science-Policy Platform on Biodiversity and Ecosystem Services (IPBES), emerged as a response to an overly narrow understanding of nature's value in economic terms.

The Value of Nature

Humans rely on healthy terrestrial and marine ecosystems that provide food, water, clean air, and recreational activities contributing to human health, well-being, and happiness. We attribute different values to nature depending on the regional, cultural, and political contexts within which our relationship with nature unfolds. The recognition of nature's value for humanity has a long tradition in the science of ecology; while many would agree that nature has an inherent value, views on what this value is—and how to represent or justify it— diverge: can a species or ecosystem have a value that is independent from the humans who define it? In other words, if humans are always those who ascribe some value to the natural environment, how can the environment itself have a value? Nature's values are doomed to be anthropogenic in some way, even if a society decided to recognize the rights of nature and protect the environment for its own sake.

The economic value of biodiversity largely relates to three levels: ecosystems, species, and genes. It has changed with the advent of new technologies and attempts by scientists to develop methods revealing the hidden values of plant and animal species beyond their current market value. Since the 1980s, there have been several attempts to differentiate between economic and non-economic values, and to assess the monetary importance of specific species and ecosystems. The rise of biotechnology and life-science industries altered the perception of biological diversity by extending scientific interest to the potentially valuable and patentable genetic properties of species (e.g., wild maize and wild tomatoes). Natural scientists became increasingly aware of the economic impacts of their research: Once discovered, a species might gain a value as a commodity, but only through research would this value be revealed, which was a source of concern among scientists: "The benefits of even the most unimportant research are often quite unexpected. Who would have predicted that these tiny, slimy seeds of a useless, ugly weed, stuck to an old newspaper costing no more than a few dollars and 30 minutes of our time, might enrich the U.S. economy by tens of millions of dollars [...]" (Iltis 1988, 103).

Calculating the Value of Nature

One early, comprehensive approach distinguished between commodity, amenity, and moral value (Norton, 1988, 201). "Commodity value" is attributed to a species that is turned into a tradable product. While direct commodity value is generated when a species, or member of a species, are bought or sold on the market, indirect commodity value arises when copies of the original (e.g., imitation of alligator skin for bags or shoes) are traded. "Amenity value" implies that an ecosystem or a species improves human well-being in a non-material way, such as through recreation or aesthetics (e.g., hiking, hunting or fishing), and accounts for vast market opportunities. In contrast, "moral value" implies that nature has an intrinsic value. According to Norton, "species have a moral value even if that moral value depends on us." However, is there a way to translate this moral value into numbers, i.e. measure the economic value of a species or ecosystem independently from its monetary value as a commodity and amenity? And if so, wouldn't this be a strong argument for its protection?

In the 1980s, scholars started developing methods to estimate the economic value of biological diversity and ecological economics gained prominence. Pearce, for instance, assessed the value of the African elephant by summating the value of ivory exports and the additional amount that tourists would be willing to pay to ensure that they could watch an elephant in its natural habitat (Pearce, 1993, 82). Jacquemot and Filion assessed the value of birds by focusing on direct benefits derived from bird watching-related activities. They discovered that some 100,000 persons participated in such activities in Canada; related expenditure added up to $2.4 billion of total GDP in 1986.

In the case of plants used for pharmaceutical purposes, Pearce differentiated between three types of value that could be added up to assess the actual economic value of a specific plant: the actual market value of the plants when traded; the market value of the drugs for which they are the source material; and the drugs' value in terms of their life-saving properties, using the monetary value of a "statistical life" (Pearce, 1993, 83). In 1997, for the first time, scientists tried to estimate the total value of nature by focusing on the world's ecosystem services; these were valued at around $33 trillion (Costanza et al., 1997).

The "Pay to Conserve Logic"

One important driver of scholarly attempts to measure the value of biodiversity is the hope that transparency about the actual market value of a species or ecosystem could prevent their overexploitation or rapid destruction. The idea is simple: If it were possible to calculate the real costs associated with those services that nature provides to society for free (e.g., climate regulation, food, pollination, or soil formation), this could provide strong arguments for the internalization of environmental costs and conserving biodiversity—rather than converting it into raw material inputs into the economy. Hence the concept of "ecosystem services" has played an important role in recent attempts to apply economic thinking for the sake of nature conservation and sustainable use. Developed to assess the benefits provided by healthy ecosystems to humans (Daily, 1997; Costanza et al., 1997), it gained popularity after the publication of the Millennium Ecosystem Assessment (MA) in 2005. One central aim of the MA was to analyze and quantify the importance of ecosystems to human well-being as far as possible in order to improve decision-making and environmental management.

Since the MA's publication, the strengthened link between biodiversity and ecosystem services has contributed to a wide-ranging debate among policymakers and scientists from different backgrounds (such as economics, ecology, political science, human geography, and political ecology) about the scientific credibility and operability of the notion, and about the impact of both an economistic framing of nature and the implementation of related policy instruments (e.g. Gómez-Baggethun et al., 2010; Dempsey & Robertson, 2012; Büscher et al., 2012; Brand & Vadrot, 2013). The performative effects of this epistemic choice and communication strategy have been subjected to critical scrutiny because it was also used instrumentally by many environmentalists to convince policymakers to act (Vadrot, 2020). Dempsey and Suarez (2016) describe related trends as "international for profit biodiversity conservation," whereby conservationists frame and communicate their research *inter alia* by relying on a market–environmentalist rhetoric and discursive successors of the "pay to conserve logic" of the late 1980s—what McAfee (1999) called "selling nature to save it."

Struggle Over the Value of Nature

Contention over the value of nature shaped the establishment of the IPBES and continues to affect negotiations between states. The strong emphasis on the ecosystem-service approach in the early days of the IPBES gave rise to opposition by some governments, most notably Bolivia and some Latin American states that had written the rights of nature into their constitutions (Vadrot, 2014). In response, the IPBES started expanding its conceptual framework and acknowledging the varied values and framings associated with biodiversity, including *pachamama* and *buen vivir*. The Pollination Assessment (2016), for instance, includes both an overview of the cultural significance of bees, butterflies and other pollinators, and an economic estimation of their annual market value: some USD235–577 billion (in 2015).

Furthermore, governments agreed on the production of a "methodological assessment regarding the diverse conceptualization of multiple values of nature and its benefits, including biodiversity and ecosystem functions and services." The assessment will be adopted in 2022 and may well significantly shape the political debate and scientific discourse on the value of nature.

Future Research Directions

A central point of contention over "values" in the Anthropocene is how to mediate conflicts between actors who ascribe different or competing values to nature. Negotiations over the values of nature – such as in the IPBES case – have the potential to shape social and political order. Critical analysis is called for to unpack the construction and use of value claims by both science and policy. The IPBES has significantly contributed to opening a discursive space within which the many, complex values of nature can be renegotiated, but there is a need for more research on competing interests associated with different values and for critical assessments of the impact of economic value claims on environmental science and policy. The ways in which ideas that developed and were promoted at the international level then spread to the local level and now shape society-nature relationships is equally important, as are the rise and translation of notions such as the "rights of nature" into implementable policy.

References

Brand, U., & Vadrot, A. B. M. (2013). Epistemic selectivities towards the valorization of nature in the Nagoya protocol and the making of the intergovernmental science-policy platform on biodiversity and ecosystem services (IPBES). *LEAD – Law, Environment and Development Journal, 9*(2), 202–222.

Büscher, B., Sullivan, S., Neves, K., Igoe, J., & Brockington, D. (2012). Towards a synthesized critique of neoliberal biodiversity conservation. *Capitalism Nature Socialism, 23*(2), 4–30.

Costanza, R., d'Arge, R., de Groot, R., Farber, S., Grasso, M., Hannon, B., Limburg, K., et al. (1997). The value of the world's ecosystem services and natural capital. *Nature, 387*(6630), 253–260.

Daily, G. C. (1997). *Nature's services: Societal dependence on natural ecosystems*. Island Press.

Dempsey, J., & Robertson, M. M. (2012). Ecosystem services: Tensions, impurities, and points of engagement within neoliberalism. *Progress in Human Geography, 36*(6), 758–779. https://doi.org/10.1177/0309132512437076

Dempsey, J., & Suarez, D. C. (2016). Arrested development? The promises and paradoxes of 'Selling nature to save it'. *Annals of the American Association of Geographers, 106*(3), 653–671. https://doi.org/10.1080/24694452.2016.1140018

Gómez-Baggethun, E., de Groot, R., Lomas, P. L., & Montes, C. (2010). The history of ecosystem services in economic theory and practice: From early notions to markets and payment schemes. *Ecological Economics, 69*(6), 1209–1218.

Iltis, H. H. (1988). Seredipity in the exploration of biodiversity. In E. O. Wilson (Ed.), *Biodiversity* (pp. 98–105). National Academy.

McAfee, K. (1999). Selling nature to save it? Biodiversity and green Developmentalism. *Environment and Planning D: Society and Space, 17*(2), 133–154.

Norton, B. (1988). Commodity, amenity, and morality. The limits of quantification in valuing biodiversity. In E. O. Wilson & F. M. Peters (Eds.), *Biodiversity* (pp. 200–205). National Academy Press.

Pearce, D. W. (1993). *Economic values and natural world*. Earthscan.

Vadrot, A. B. M. (2014). *The politics of knowledge and global biodiversity*. Routledge.

Vadrot, A. B. M. (2020). Building authority and relevance in the early history of IPBES. *Environmental Science & Policy, 113*(11), 14–20.

Alice B. M. Vadrot is Associate Professor for International Relations and the Environmental at the Department of Political Science of the University of Vienna, Visiting Research Fellow at the Centre for Science and Policy (CSaP) of the University of Cambridge, member of the Young Academy of the Austrian Academy of Sciences, and Senior Fellow of the Earth System Governance Platform. Her work addresses the role of knowledge and science in global environmental politics. She has conducted extensive research on the Intergovernmental Platform on Biodiversity and Ecosystem Services (IPBES) and has developed the concept of "epistemic selectivities". Since 2018, she leads the ERC Starting grant project MARIPOLDATA, which develops and applies a new methodological approach for grounding the analysis of science-policy interrelations in ocean politics in empirical research.

Wisdoms

Aurélie Choné

Abstract Given the failings of the technological, economic and political solutions that have been suggested to fight against climate change and biodiversity loss, what can we learn from the world's wisdoms, be they from the East or the West, the North or the South, from indigenous people or philosophers, linked to a religious system or not? What type of anthropological theory and concrete practices do they offer, that could keep humankind from self-destruction?

With global warming, biodiversity loss and the Earth's sixth mass extinction, the prospect of an end to the human species has become less and less theoretical. The human species is guilty not only of crime against life (through intensive livestock farming for example) and crime against the planet (through its lack of respect for nature, as shown in water pollution and deforestation) but also of crime against the future: what kind of planet are we bequeathing to our environment and to our children? At the same time, the gap is becoming ever wider between the haves (who have too much, in fact, in the sense that they cannot make reasoned use of their goods) and the have-nots (a child dies of hunger every 5 seconds). This is all common knowledge, and hardly anyone denies the figures anymore, except for a few climate sceptics.

Although individuals and the States are aware of this serious situation, they have been unable to alter their lifestyles and production patterns. Everyone is still living "life as normal", even though the number of vegetarians has been slightly increasing in rich countries (meat production is surpassing the transportation industry as the most polluting industry) and even though sustainability has become a media buzzword. If it does not rapidly change directions, humankind will reach a dead end, which may well result in implosion and war, with large-scale environmental migration by 2050. Several solutions have been put forward to meet this severe crisis

A. Choné (✉)
University of Strasbourg, Strasbourg, France
e-mail: achone@unistra.fr

© The Author(s), under exclusive license to Springer Nature Switzerland AG 2023
N. Wallenhorst, C. Wulf (eds.), *Handbook of the Anthropocene*,
https://doi.org/10.1007/978-3-031-25910-4_278

(which has been lasting so long that the name cannot fit anymore) in this, the Anthropocene epoch.

First, a technological solution: resorting to even more technology, for example by creating an ecosystem on Mars or burying nuclear waste there. This solution is based on the belief that human beings will always manage to find a way out thanks to their knowledge and skills, but it is illusory as it boils down to reenacting on another planet the colonization of the earth, thus only displacing, obscuring and aggravating the issue. It is typical of the Promethean worldview that has prevailed in the West since the Industrial Revolution – an attitude of domination over nature leading up to Western "naturalism" (Descola, 2005) and ultimately to the current ecocide.

Ancient Greek philosophy clearly identified one of human beings' worst impulses, hubris, i.e. their claim to equal divine abilities or attributes (for instance through feats of technology). Heraclitus wrote that "insolence (*hybris*) is more to be extinguished than a conflagration" (Kirk et al., 1983, p. 211.) When man tries to compete with cosmic powers, he is led to entertain the dream of playing God, of changing into a "Homo deus", endowed with almighty powers through the manipulation of big data, the use of nanotechnology… Communication with the beyond-which mystics in all the world's cultures have pursued – then gives way to a desire for transcendence based on the belief that science can help solve all challenges, including death (as transhumanism asserts). Yet this desire rests on an illusion, as are aware those who know that the more knowledge they have mastered, the more remains to be understood. A wise man knows that he knows nothing, as expressed by the Greek philosopher Socrates: *scio me nihil scire.*

Faced with the limits of their knowledge, human beings have turned to belief. The desire for scientific transcendence has found an echo in the desire for religious transcendence. Established creeds, beliefs, religious institutions or superstitions can give one the feeling of having absolute mastery over the invisible or the supernatural. The embrace of what others consider to be improbable is the other face of hyper-knowledge, the quest for moral perfection the other face of the quest for cognitive perfection. The dangers posed by such an attitude are well known: sectarianism, fanaticism, fundamentalism, dogmatism.

Another more political solution has been to try to institute collective norms applying to individuals (for example, banning diesel cars from cities, on a national or local scale) and to States (world climate conferences …) in order to reduce pollution. Faced with the failure of such measures, Greenpeace, Oxfam, the Nicolas Hulot Foundation for Nature and Man and the association "Notre Affaire à tous" have filed a lawsuit against France for climate inaction. Such initiatives are certainly useful, but very difficult to implement: what politician would dare impose such unpopular measures? What sanctions should be meted out to States who refuse to comply with international agreements or to commit to them?

Economic solutions have also been considered. The advocates of degrowth are convinced that man must turn his back to the capitalist culture of excessive consumption and production and to an advertising system designed to continually arouse new desires. The anthropology of the modern Promethean *homo oeconomicus* is no longer relevant according to them. Distinguishing themselves from such "fundamentalists", "reformist" movements are calling for the reformation of the current capitalist system towards more sustainability and favouring "green growth". Yet both stances raise problems: it is difficult to impose degrowth and voluntarily give up the comforts of modern life. On the other hand, it is impossible not to be aware of the contradictions inherent in so-called "green growth" when one observes the ideological appropriations of environmental discourses (the development of eco-tourism or eco-districts also has less visible negative impacts).

Given the failings of all the aforementioned solutions, what can we learn from the world's wisdoms, be they from the East or the West, the North or the South, from indigenous people or philosophers, linked to a religious system or not? What type of anthropological theory do they offer, that could keep humankind from self-destruction? First, they can help us realize that to change our lives, we first need to cultivate consciousness, that the problem is not located outside of us but inside, that human beings need to redefine what "human" means lest hubris (excess) and bestiality (i.e. the need to satisfy desires that are in essence insatiable) should lead humankind to its doom through a boomerang effect.

The different bodies of wisdom can also help us recreate links with the Earth and the universe, recover that spirit of harmony and cosmic consciousness which lie at the core of Native American wisdom or indigenous wisdom (which has given rise to permaculture). Indigenous beliefs have also strongly inspired eco-psychology, which seeks to heal the psyche and restore the Earth conjointly. First of all, then, one needs to understand that everything is connected together and that the pollution of the outside world is a reflection of inner pollution (Roszak et al., 1995). There are today numerous psycho-physical therapies based on the world's systems of wisdom that can help cleanse one's desires and passions. Yet these healing and self-development techniques, necessary though they may be, cannot be enough, and may even prove harmful as they tend in fact to bolster our arrogant and selfish ego.

The solution that is needed is above all of an ethical nature (Hess, 2013) and involves a decentering of the self as advocated by founder of deep ecology Arne Naess. The Norwegian philosopher relies on Spinoza and Gandhi in his deconstruction of anthropocentrism in order to lay the foundations for an "ecosophy", "a philosophy of ecological harmony or equilibrium" (Alan & Yuichi, 1995). It is also necessary to revive ancient philosophers' virtue ethics since it emphasizes individuals' concrete motives instead of focusing on norms and only decreeing prohibitions and obligations. It can thus help bridge the particularly dramatic gap between theory and practice (Pelluchon, 2017) Virtues such as humility, consideration and attention are profoundly ecosophical.

Above all, the different systems of wisdom call on us to accomplish an act of inner transformation, transition or revolution, to strip away the self and reach towards self-realization. Wisdom is not the same as religion. It does not involve faith or any established religious system. It has more to do with a spiritual process, an inner experience, leading to life change and a higher level of consciousness. Wisdom generally originates in a specific spiritual tradition: one should not trust self-claimed gurus teaching syncretic blends of different traditions, since they are probably promoting self-development or wellbeing strategies aiming to boost the ego rather than acting as guides onto a spiritual path involving suffering, trials and loss of ego.

There are many different bodies of wisdom – ancient, pagan, Chinese, Egyptian, Christian, gnostic, Sufi, Hindu, Orthodox, and so on. Beyond their distinctions, one can define a wise person as someone who knows how to master his desires, embodies and teaches perfect balance, is connected to his inner nature and to outside nature, does not live according to ego but to another divine, transcendent dimension, in the respect of self and others, the respect of life, with an awareness of the sacred and of the fact that everything is connected together. How is such a state to be reached? All the different traditions point to the need to work on oneself, to go through an initiatory journey involving confrontation with one's own dark, opaque nature.

Lao-Tzu writes at the beginning of the *Tao Te Ching*: "The Tao that can be told is not the eternal Tao./ The name that can be named is not the eternal name./ The nameless is the beginning of heaven and Earth./ The named is the mother of the ten thousand things./ Ever desireless, one can see the mystery./ Ever desiring, one sees the manifestations./ These two spring from the same source but differ in name; this appears as darkness./ Darkness within darkness./ The gate to all mystery." (Lao Tzu, 1989). The darkness mentioned in *Tao Te Ching* echoes St John's "dark night of the soul", Meister Eckhart's "nothingness" or Nicolaus Cusanus's "learned ignorance": no evidence can hold when viewed in relation to God, Being or Tao.

In the Indian tradition, in the wisdom of yoga, *artha* (material prosperity) and *kama* (desire, including erotic desire) must be checked so as not to violate *dharma* (the socio-cosmic law) and ensure the preservation of biological and cultural diversity. If, on the other hand, man claims to be the master of the world, his hubris will inevitably have a serious impact on the environment. The selfish desire to "own the world", as promised by travel agencies, reflects a Promethean, noxious and disrespectful attitude typical of spoiled Westerners. The main difficulty lies in actually changing one's lifestyle by voluntarily embracing frugality, simplicity and "the return to the obvious" (Del Vasto, [1945]2014). What is the definition of a right action, of *karma yoga?* Indian traditions teach the path of unselfish action, of "non-action": "The poets rightly teach that Sannyas is the foregoing of all acts which spring out of desire, and their wisest say Tyaga is renouncing fruit of acts" (Bhagavad Gita, 2018 p. 75).

In Europe, poet Lanza del Vasto, a disciple of Gandhi, taught the path of non-violence on his return from India. The communities of the Ark he founded after World War II promote values such as sharing, mutual aid and the return to a simple life close to nature. More recently, farmer Pierre Rabhi has celebrated "happy sobriety" (Rabhi, 2010); the movement "les Colibris" he initiated in 2006 advocates eating locally-grown organic food, using clean transportation means or composting food waste. Writing in *Le Monde diplomatique,* J.-B. Malet explains that "Rabhi derives his iconic popularity from a mythical figure: that of the old peasant, the wise grandfather rooted in a village community shattered by capitalism but whose ancestral knowledge proves irreplaceable in the face of impeding dangers. In a context of environmental disasters and celebration of consumption, his call to frugality and his attacks against productivist agriculture chime in with the collective sense that modernity has spun out of control. In response, the leader of the 'colibris' calls for 'an insurrection of consciences', spiritual regeneration, harmony with nature and the cosmos, an alternative, local model of non -mechanized, organic agriculture. These ideas flow through the media, charmed by his attractive personality (…)" (Malet, 2018) Malet is here denouncing the media treatment of Rabhi as an "old wise man" and pointing to the possible appropriation of pre-modern, "ancestral" wisdoms.

To be able to discriminate between true and false wisdom and avoid deceptions (Choné, 2017), it is necessary to follow a path of personal transformation. Swiss psychologist Carl Gustav Jung described it as "an individuation process", involving the integration of the opposite (conscious-unconscious, male-female, good-evil, human-non-human) poles of the human psyche (Jung [1963], 1995). According to Jung, "the future of humanity will depend to a large extent on the recognition of the shadow. Evil is – psychologically speaking – terribly real." (Jung, 1949). Jung considers that self-knowledge is the only way "if humanity is not to destroy itself through the might of its own technology and science" (Jung, 2002, p. 78).

Thus the prospective anthropology emerging from the encounter with the world's wisdoms will need to be based on introspection: "Know thyself", the motto which was inscribed on the pediment of the temple of Delphi, is an encouragement to introspection, meditation and contemplation. Such an inward turn, and reconsideration of one's lifestyle and daily habits, if it is truly engaged in, will dispel any trace of cognitive dissonance. This introspective anthropology implies coming to terms with one's "shadow", limitations, one's death (including the death of the human species): being a philosopher means learning to die, Cicero wrote. Philosophy can then be understood once more as a form of wisdom, as a way of life, as a spiritual practice as defined by philosopher Pierre Hadot (Hadot, 2002). Then will the "Orphic" attitude (Hadot, 2004) to nature prevail, laying the foundations for a relational anthropology based on immersion in nature and on a double process of self-realization and self-renunciation.

References

Alan, D., & Yuichi, I. (Eds.). (1995). *The deep ecology movement: An introductory anthology*. North Atlantic Publishers.

Bhagavad Gîtâ. (2018). Translated by Edwin Arnold. Global Grey (ebook). (Original edition 1885).

Choné, A. (2016). Ecospirituality. In A. Choné, I. Hajek, & P. Hamman (dir.) (2017). *Rethinking nature. Challenging disciplinary boundaries* (pp. 38–48). Routledge.

Del Vasto, L. (2014). *Principes et préceptes du retour à l'évidence*. Desclée de Brouwer. (Original edition 1945).

Descola, P. (2013). *Beyond nature and culture*. University of Chicago Press. (Original edition 2005).

Hadot, P. (2002). *Exercices spirituels et philosophie antique*. Albin Michel.

Hadot, P. (2004). *Le voile d'Isis*. Gallimard.

Hess, G. (2013). *Éthiques de la nature*. Presses Universitaires de France.

Jung, C. G. (1949). *Letter to father Victor White*, 31. december 1949. In Jung, Carl Gustav. 1993 (Original edition 1949). Correspondance 1941–1949. Albin Michel.

Jung, C. G. (1995). *Memories, dreams, reflections* (A. Jaffe Ed.). Fontana Press/Harper Collins. (Original edition 1963).

Jung, C. G. (2002). *The undiscovered self*. Routledge.

Kirk, G. S., Raven, J. E., & Schofield, M. (1983). *The pre-Socratic philosophers* (2nd ed.). University Press.

Malet, J.-B. (août 2018). Le système Pierre Rabhi. *Monde diplomatique* n°773.

Pelluchon, C. (2017). L'éthique des vertus: une condition pour opérer la transition environnementale. *La Pensée écologique, 1*(1), e. https://doi.org/10.3917/lpe.001.0101

Rabhi, P. (2010). *Vers la sobriété heureuse*. Actes Sud.

Roszak, T., Gomes Mary, E., & Kanner, A. D. (Eds.). (1995). *Ecopsychology: Restoring the earth, healing the mind.* Sierra Club Books.

Tzu, L. (1989). *The complete Tao Te Ching* (Gia-Fu Feng and Jane English, Trans.). Vintage Books. https://terebess.hu/english/tao/gia.html#Kap01

Aurélie Choné is a Professor in German Studies at the Institute of German Studies within the Faculty of Languages, University of Strasbourg, France, Editor of the international academic peer-reviewed journal *Recherches germaniques*. Her main research fields center on German-speaking literature, the history of ideas and cultural history, Cultural transfers between East and West, Nature-Culture relations, Ecocriticism, Literary and Cultural Animal Studies, and the Environmental Humanities. Her recent publications include: *Destination Inde. Pour une géocritique des récits de voyageurs germanophones (1880–1930).* Paris: Honoré Champion, 2015. *Rethinking Nature. Challenging Disciplinary Boundaries*. London: Routledge 2017. "Les Humanités environnementales: circulations et renouvellement des savoirs en France et en Allemagne/Environmental Humanities: Wissenstransfer und Wissenserneuerung in Frankreich und Deutschland". *Revue d'Allemagne et des Pays de langue allemande* 51(2), 2019 and *Le végétal au défi des humanités environnementales/Die Pflanzenwelt im Fokus der Environmental Humanities*. Berlin: Peter Lang, 2021 and Les esprits scientifiques. Les agricultures alternatives entre savoirs et croyances, Presses Universitaires Grenoble Alpes, collection « Ecotopiques », 2022.

Correction to: Handbook of the Anthropocene

Nathanaël Wallenhorst and Christoph Wulf

Correction to: N. Wallenhorst, C. Wulf (eds.), *Handbook of the Anthropocene*, https://doi.org/10.1007/978-3-031-25910-4

The book was inadvertently published with an incorrect affiliation of Prof. Michel Bourban in Chapter 87 and Chapter 168 as University of Warwick, Coventry, UK, whereas it should be University of Twente, Enschede, Netherlands, and the e-mail address m.bourban@utwente.nl.

The updated original version for these chapters can be found at
https://doi.org/10.1007/978-3-031-25910-4_87
https://doi.org/10.1007/978-3-031-25910-4_168

© The Author(s), under exclusive license to Springer Nature Switzerland AG 2023
N. Wallenhorst, C. Wulf (eds.), *Handbook of the Anthropocene*,
https://doi.org/10.1007/978-3-031-25910-4_279

Correction to: Bioeconomics

Sylvie Ferrari

Correction to:
Chapter 178 in: N. Wallenhorst, C. Wulf (eds.), *Handbook of the Anthropocene*, https://doi.org/10.1007/978-3-031-25910-4_178

The book was inadvertently published with an error on page 1096 as follows: "According to this law, also called the second principle of thermodynamics, the entropy of a closed system (which exchanges energy but not matter with its environment) is continuously increasing."

This has been corrected to "According to this law, also called the second principle of thermodynamics, the entropy of an isolated system is continuously increasing."

The updated version of this chapter can be found at https://doi.org/10.1007/978-3-031-25910-4_178

© The Author(s), under exclusive license to Springer Nature Switzerland AG 2024
N. Wallenhorst, C. Wulf (eds.), *Handbook of the Anthropocene*,
https://doi.org/10.1007/978-3-031-25910-4_280

Correction to: Anxiety

Konrad Oexle and Thomas Reuster

Correction to:
Chapter 153 in: N. Wallenhorst, C. Wulf (eds.), *Handbook of the Anthropocene*, https://doi.org/10.1007/978-3-031-25910-4_153

The book was inadvertently published with errors on page 935. In the first paragraph, in the sentence "Armstrong's heart rate rose to 150 bpm", the number should be "156 bpm". This has been corrected.

Also, in the sentence "Thus, although he had only 20 s of fuel left to land,...", the number should be "18 s". This has been corrected.

The updated version of this chapter can be found at
https://doi.org/10.1007/978-3-031-25910-4_153

© The Author(s), under exclusive license to Springer Nature Switzerland AG 2024
N. Wallenhorst, C. Wulf (eds.), *Handbook of the Anthropocene*,
https://doi.org/10.1007/978-3-031-25910-4_281

Printed by Printforce, the Netherlands